GRAMMATICAL INFERENCE

The problem of inducing, learning or inferring grammars has been studied for decades, but only in recent years has grammatical inference emerged as an independent field with connections to many scientific disciplines, including bio-informatics, computational linguistics and pattern recognition. This book meets the need for a comprehensive and unified summary of the basic techniques and results, suitable for researchers working in these various areas.

In Part I, the objects of use for grammatical inference are studied in detail: strings and their topology, automata and grammars, whether probabilistic or not. Part II carefully explores the main questions in the field: what does learning mean? How can we associate complexity theory with learning? In Part III the author describes a number of techniques and algorithms that allow us to learn from text, from an informant, or through interaction with the environment. These concern automata, grammars, rewriting systems, pattern languages and transducers.

COLIN DE LA HIGUERA is Professor of Computer Science at the University of Nantes.

GRAMMATICAL INFERENCE

Learning Automata and Grammars

Colin de la Higuera
Université de Nantes

CAMBRIDGE UNIVERSITY PRESS
Cambridge, New York, Melbourne, Madrid, Cape Town, Singapore,
São Paulo, Delhi, Dubai, Tokyo

Cambridge University Press
The Edinburgh Building, Cambridge CB2 8RU, UK

Published in the United States of America by Cambridge University Press, New York

www.cambridge.org
Information on this title: www.cambridge.org/9780521763165

First published 2010

Printed in the United Kingdom at the University Press, Cambridge

A catalogue record for this publication is available from the British Library

ISBN 978-0-521-76316-5 Hardback

Contents

Preface

Young men should prove theorems, old men should write books.

Godfrey H. Hardy

There is nothing to writing. All you do is sit down at a typewriter and bleed.

Ernest Hemingway

A zillion grammatical inference years ago, some researchers in grammatical inference thought of writing a book about their favourite topic. If there was no agreement about the notations, the important algorithms, the central theorems or the fact that the chapter about learning from text had to come before or after the one dealing with learning from an informant, there were no protests when the intended title was proposed: *the art of inferring grammars*. The choice of the word *art* is meaningful: like in other areas of machine learning, what counted were the ideas, the fact that one was able to do something complicated like actually building an automaton from strings, and that it somehow fitted the intuition that biology and images (some typical examples) could be explained through language. This 'artistic' book was never written, and since then the field has matured.

When writing this book, I hoped to contribute to the idea that the field of grammatical inference has now established itself as a scientific area of research. But I also felt I would be happy if the reader could grasp those appealing *artistic* aspects of the field.

The artistic essence of grammatical inference was not the only problem needing to be tackled; other questions also required answers...

Why not call this book 'grammar induction'?

When one wants to search for the different published material about the topic of this book, one finds it associated with many different fields, and more surprisingly under a number of names, such as 'grammar learning', 'automata inference', 'grammar identification', but principally 'grammar induction' and 'grammatical inference'. Even if this is not formalised anywhere, I believe that 'grammar induction' is about finding a grammar that can explain the data, whereas grammatical inference relies on the fact that there is a (true or only possible) target grammar, and that the quality of the process has to be measured relatively to this target.

Practically this may seem to make little difference, as in both cases what probably will happen is that a set of strings will be given to an algorithm, and a grammar will have to be produced. But whereas in grammar induction this is the actual task, in grammatical inference this is still a goal but more a way of measuring the quality of the learning method, the algorithm or the learning setting.

In other words, in the case of grammar induction what really matters is the data and the relationship between the data and the induced grammar, whereas in grammatical inference the actual learning process is what is central and is being examined and measured, not just the result of the process.

Is this about learning languages or learning grammars?

Even if historically the task has been associated with that of learning languages (and typically with that of children acquiring their first language), we will concentrate on learning by a machine, and therefore (as in any computational task) a representation of the languages will be essential. Given this first point, a simple study of formal language theory shows us that not all representations of languages are equivalent. This will justify the choice of presenting the results with a specific representation (or grammar) system in mind.

Why use terms from formal language theory, like finite automata, when you could use more generic terms like finite state machines or alternative terms like hidden Markov models?

Actually, there is a long list of terms referring to the same sort of object: one also finds Mealy and Moore machines and (weighted) finite state transducers. In some cases these terms have a flavour close to the applications they are intended for; in others the names are inheritances of theoretical research fields. Our choice is justified by the fact that there are always some computable transformations allowing us to transform these objects into deterministic or non-deterministic finite automata, with or without probabilities. In the special case of the transducers, separate theory also exists.

By defining everything in terms of automata theory, we aim to use a formalism that is solidly established and sufficiently general for researchers to be able to adapt the results and techniques presented here to the alternative theories.

Why not introduce the definitions just before using them?

A generally good idea is to only introduce the definitions one needs at the moment one needs them. But I decided against this in certain cases; indeed, the theories underlying the algorithms and techniques of grammatical inference deserve, at least in certain cases, to be presented independently:

- Chapter 3 introduces concepts related to strings, and also to distances or kernels over strings. If the former are well known by formal language theory specialists, this is not always the case with researchers from other fields interested in grammatical inference. It seemed interesting to have these definitions in one separate chapter, in order for these definitions to be compared and depend on each other.
- Most of the material presented in Chapter 4 about grammars and automata can be found in well-known textbooks, even if providing uniform notations is not customary. When considering probabilistic finite state automata, this is even worse as the definitions are far from trivial and the theory is not straightforward. We have also chosen to give here the main algorithms needed to deal with these objects: parsing, estimating, distances, etc.
- The reader just looking for an algorithm might go directly to the chapter corresponding to the class of grammar or the type of presentation he is dealing with. But to understand the convergence issues associated with the algorithm he wants to use, he might want to read Chapter 7 about learning models, and more specifically be interested in the inclusion of complexity questions in these models.
- The panorama concerning probabilities and grammatical inference would have been very difficult to understand if we had not devoted Chapter 10 to this issue: whether we use the distributions to measure classification errors or to actually learn the grammars that generate them, we try in this chapter to give a uniform view of the field, of its definitions and of its difficulties.
- For very different reasons, Chapter 6, about combinatorics, groups a number of more or less well-known results concerning automata, deterministic or non-deterministic, probabilistic or not. These are usually hardness results and rely on special constructions that we give in this separate chapter.

Did the author invent all these things?

No, of course not. Many scientists have contributed the key ideas and algorithms in this book. Some are clearly identifiable but others are not, the idea having 'been around for a while' or being the synthesis of thoughts of several people. Moreover, in such a field as this one, where papers have been written in a great diversity of areas and where up to now no text book had been produced, it is difficult, if not impossible, to know who has the paternity of what idea. I have tried to *render unto Caesar the things which are Caesar's*, and have consulted widely in order to find the correct authorships, but there are probably going to be researchers who will feel that I have wrongfully attributed their result to someone else. This I hope to have avoided as much as possible and I wish to express my apologies for not having found the correct sources. A specific choice to increase readability has been to leave the citations outside the main text. These are discussed at the end of each chapter.

Is the reader right to say that he or she has seen these proofs before?

Yes, of course he or she is right. In some rare cases, our knowledge today of the objects and the problems allows us to propose alternative proofs that somehow fitted in better. In other cases (as duly acknowledged as possible), the original proof seemed unbettered. And for many reasons, most important of which is usually just the fact that the proof does add that useful information, I chose to include them, with just the necessary changes of notation.

These do not seem to be the standard notations. Why didn't the author use the standard notations?

In such a new field as grammatical inference there is no such thing as standard notations. Grammatical inference has borrowed notations from many fields: machine learning, formal language theory, pattern recognition and computational linguistics. These are even, in many cases, conflicting. Moreover the authors of the many papers all have different backgrounds and this diversity is reflected in the variety of notations and terms that are used. Choices had to be made for the book to be readable. I have used notations that could adapt smoothly to the different grammatical inference settings reported in this book. A chief goal has been to make algorithmic ideas from one case reusable in another. It is nevertheless fair to say that this has a price (for example where introducing automata with two types of final states) and specialists, used to working with their own notations, may be inclined to disagree with these proposals.

Is this the right moment to write a book on grammatical inference?

That is what I believe. It certainly is not too early, since most of the key questions were asked by Noam Chomsky or Ray Solomonoff 50 years ago! Mark Gold's results date back another 40 years, and Dana Angluin's first contributions to the field correspond to work done 30 years ago. The series of conferences known as ICGI have now been running for more than 15 years. The theme today is present in a number of conferences and the main algorithms are used in many fields. A noticeable thing is that the field is broadening: more complex classes of grammars are being learnt, new learning paradigms are being explored and there are many new applications each year. But no basic books with the main definitions, theorems and algorithms exist. This is an attempt to help register some of the main results (up to now) to avoid them being rediscovered over and over.

Why exercises?

Even if this is not a text book, since grammatical inference is not being taught in a curriculum, there are lectures in machine learning, text mining, and other courses where formal languages and generalisation are taught; there have also been attempts to hold lectures on the topic in summer schools and doctoral schools; moreover, I have received sufficient encouragement to initiate a list of exercises in order to be able to teach in the near future. The exercises are intended for this purpose. They are also here to give a flavour of a few of the interesting questions in the field.

What about trees and graphs?

The original goal was to cover extensively the field of grammatical inference. This of course meant discussing in detail tree automata and grammars, giving the main adaptation of classical string algorithms to the case of trees, and even dealing with those works specific to trees. As work progressed it became clear that learning tree automata and grammars was going to involve at least as much material as with strings. The conclusion was reached to only sketch the specificities here, leaving the matter largely untouched, with everything to be written. This of course is not justified by the importance of the question, but only by the editorial difficulty and the necessity to stop somewhere. Of course, after trees will come the question of graphs...

Is this the final book?

No, yet even more preposterously, I hope it to be an initial book. One that allows fellow researchers to want to write their positions in new books expressing the variety of points of view of a community made up of colleagues with such different interests, whether in machine learning, computational biology, statistics, linguistics, speech recognition, web applications, algorithmics, formal language theory, pattern analysis...

Acknowledgements

It all started in July 2005, in a meeting with David Tranah who easily convinced me that I really wanted to write a book on grammatical inference. During the next few years he was to encourage me in many ways and somehow remind me that this was something I really wanted to do.

During the years it took to write it, I had an unexpected (by me) number of problems I would not have been able to solve without the expertise of Thierry Murgue, who helped me solve a number of technical questions, but also provided his knowledge about probabilistic finite state machines.

A lot of what I know of the topic of grammatical inference is due to the patient efforts of Laurent Miclet and Jose Oncina, with whom, over the years, I have interacted and learnt.

On various occasions I have had to teach these things or organise workshops related to these topics, and it has always been a pleasure to prepare these with colleagues like Tim Oates, Pieter Adriaans, Henning Fernau, Menno van Zaanen and the aforementioned.

Most of the material I am presenting here comes either from work done by others, or, when I have been actually involved, from work done with students and a number of collaborators. Let me thank Leo Becerra Bonache, Rafael Carrasco, Francisco Casacuberta, Pierre Dupont, Rémi Eyraud, Jean-Christophe Janodet, Luisa Micó, Frédéric Tantini, Franck Thollard, Enrique Vidal and a number of students including Cristina Bibire, Anuchit Jittpattanakul and Émilie Samuel.

I am grateful also to several of the people thanked above for reading, with a lot of care, the different chapters in many of the preliminary versions I prepared. But I also received help in this from Hasan Akram, Stefan Gulan, François Jacquenet, Anna Kasprzik, Satoshi Kobayachi, Etsuji Tomita and Takashi Yokomori. Of course, I take entire blame for all the remaining errors.

Most of the work was done while I held a position at Saint-Étienne University, first in the EURISE team and later in the Laboratoire Hubert Curien. The conditions were great and I am pleased to use this occasion to thank all my colleagues there.

When one has not written a book, the customary acknowledgement to the author's spouse and children seems meaningless. When one has spent endless evenings and weekends being between entirely absent and half-absent from the family activities and spirit, it reaches its full meaning. Lindsey, Boris and Vikki, all my thanks.

1
Introduction

> The grand aim of all science is to cover the greatest number of empirical facts by logical deduction from the smallest number of hypotheses or axioms.
>
> ***Albert Einstein***

> Todos os problemas são insolúveis. A essência de haver um problema é não haver uma solução. Procurar um facto significa não haver um facto. Pensar é não saber existir.
>
> ***Fernando Pessoa**, Livro do Desassossego, **107***

The problem of inducing, learning or inferring grammars has been studied now for some time, either as a purely mathematical problem, where the goal is to find some hidden function, or as a more practical problem of attempting to represent some knowledge about strings or trees through a typical representation of sets of trees and strings, such as an automaton or a grammar.

1.1 The field

Historically, one can trace back the problem to three main different sources: the first is computational linguistics, the second is inductive inference and the third is pattern recognition. The historical lineage, to be fair, requires us to discuss the important contribution of at least two other fields, not entirely independent of the three historical ones: machine learning and bio-informatics.

1.1.1 (Computational) linguistics

The key question of language acquisition (meaning the acquisition by a child of its first language), when considered as a formal language learning task, takes its root in Noam Chomsky's pioneering work. The fact that the human being should be able to discover the syntactic representations of language was key to many of the formalisms introduced to define formal languages, but also to study grammatical inference. Since that time there has been continuous research activity linked with all aspects of language learning and acquisition in the many topics related with computational linguistics.

A special idea of interest that emerged from the early work is identification in the limit, as the first paradigm for learning languages, possibly because of a strong belief that probabilities were not a plausible answer. The particular situation is that of *identifying* a *target* language from *text*, i.e. only strings from the language, also called positive examples.

There is an important thing we should insist on here, which is that one should make a distinction between *language learning* and *language acquisition*: it is generally admitted that language acquisition refers to the first language a child learns, whereas *language learning* covers the learning of another language, given a first language to support the thought and the learning process.

A second point of interest is that there is a general agreement that negative examples should not be part of the learning process even if some type of interaction between the learner and the teacher (through corrections for example) can be added.

A number of problems related with grammatical inference have been addressed in this field: parsing with formalisms that may be much more complex than grammars from formal language theory, tagging, solving many grammatical tasks... An interesting point concerns the fact that probabilistic settings have not been very popular in computational linguistics. A notable exception concerns the question of language modelling: language models intervene when different sequences of individual words or utterances have been found and one wants to decide which is the one which most probably is an actual sentence of the chosen language. The normal way to solve this is through some local decisions ('given the last two words, the new word is most likely to be...') but in that case the long-term dependencies are not taken into account. One of the major challenges to grammatical inference scientists is to come up with methods that learn these long-term dependencies.

1.1.2 Inductive inference

Inductive inference is about finding some unknown rule when given some elements of an infinite presentation of elements. Take for instance the problem of guessing the next element of the following increasing sequence: 2, 3, 5,... In this 'game' one does not actually need to find the actual function (no one is explicitly asking for that) but it is easy to see that without getting the function right we will continue to be in a situation of just guessing as the game goes on: if we don't find the actual function (or at least a function that *is like* the generating device), the task of *consistently* predicting the next element is impossible. Notice also that being right just once doesn't *prove* anything.

One of the key issues in this example is that we are not only interested in finding some mechanism – or algorithm – to do the learning, but we will also want to study the parameters of the problem: how do we get hold of our data? In this case it just arrives in some specific order, but that may not be systematically the case. How fast can we learn? How do we know that we have learnt?

Inductive inference is the research area interested in these issues. Obviously, the rules of the game may be very complex as one can play with the different variables of the problem:

- the class of functions,
- the way the data is presented,
- the rules used to measure success.

Grammatical inference is only concerned with a small sub-case of this problem: the class of functions corresponds to grammars generating or recognising strings, trees, graphs or other structured objects. On the other hand, all types of presentations may be of interest. In certain cases we will be given examples and counter-examples; in others, only some strings that belong to the language. In certain cases the learning algorithm will have to deal with examples arriving one by one, in some predefined order (or not); in another case we will be asked for a batch treatment with all the data given at once; and in an interactive setting the data will be sampled or 'bought' by the learning algorithm itself. If anything, motivation will probably reside in the fact that these presentations may correspond to practical issues. And more importantly, success will depend on some provable convergence with only a limited quantity of resources, whether computation time, quantity of examples or number of accesses to an oracle.

The inductive inference field has continued to introduce and analyse key ideas and has led to the development of a field known as algorithmic learning theory. The name of the field is possibly slightly misleading, because in most cases the algorithms are not developed with an interest in breaking polynomial boundaries but more with the mind set on showing what is decidable and what is not.

1.1.3 Pattern analysis, pattern recognition

In the problem of recognising handwritten characters (for example the postal codes on an envelope), we may be given images of each of the possible letters and digits, construct some sort of a model for each symbol, and then, when a new character is presented, we can compare the new character to the models and find the model that fits best. This idea requires robust models, a training phase to set the parameters of the models and some way to measure how adequate a model is for a given character.

One may also consider that a character, once pixellated, can be described by a string (if we use either four or eight directions). To take into account the variety of lengths and the variations of figures, one possibility is to associate with the character 'A' a language of all strings corresponding to this character. Then by parsing a new string, one should be able to decide if the current written character is an 'A' or a 'B'. There are obviously many variations of this question, where, for example, the language may include probabilities so that eventual intersections can be dealt with. There will also be a need to study distances between strings or between strings and languages to take better decisions.

The above is one of the many attempts to use grammar induction in pattern recognition. We use here the term *grammar induction* and not *grammatical inference*: there is actually no reason to believe that there is a hidden grammar to be discovered. In this case we are only believing that somehow a grammar can represent in a reasonable way a set of strings corresponding to the letter ‘A’.

A second point one can make is that if the task we describe is a pattern recognition task, since the induced grammar has a meaning *per se*, it is tempting to analyse this grammar, and therefore to consider that grammatical inference techniques enable us to discover *intelligible* patterns, and thus that they permit us to tackle *pattern analysis* tasks.

1.1.4 Machine learning

From the inductive inference perspective, classes and paradigms were analysed with a distinct mathematical flavour.

On the other hand, more pragmatically oriented researchers were working on practical problems where techniques were tested enabling them to ‘induce’ an automaton or grammar from some data.

The meeting point was going to be provided by the work done in another field, that of machine learning.

Machine learning has developed as an important branch of artificial intelligence. The great idea of machine learning is to enable a system to get better through some form of interaction with its environment in which new knowledge is extracted from experience and integrated in the system. Some examples are learning from observation, from examples, from mistakes, and from interaction with a teacher.

Work consisted of developing ideas, algorithms and even theories of what was learnable and what was not, whilst at same time considering important applications where *very nice heuristics* were not going to provide a suitable answer: provable algorithms were needed, in which the quantity of necessary information requested in order to be able to learn some satisfactory function had to be minimal. All the notions associated with the key words in the previous sentence need to be defined. This led to the introduction of models of learning of which the probably approximately correct (PAC) model is the best known, and unarguably the most inspiring.

These efforts to re-understand the learning algorithms through the eyes of complexity theory (polynomial time and polynomial space required), statistics (given some known or unknown distribution, can we bound the error the classifying function will make over unseen data, or can we bound the error over the possibility that our error is very small?) have also influenced the special problem of learning grammars and automata.

At first, the influence of researchers in the growing field of computational learning theory (COLT) led to some essential results and analysis around the central problem of learning deterministic finite state automata (DFA) in the important and specific setting of PAC learning. The task was proved to be impossible using results on learning from equivalence queries. The hardness of the minimal consistency problem (for DFA) was proved in 1978; but the mathematical links between complexity theory and computational learning theory led to elaboration on that result: the closely related problem of finding a small consistent

DFA (not more than polynomially larger than the smallest) was proved to be intractable. This on one hand forbade us to have hopes in Occam approaches but also opened the path to finding other negative results. These somehow seemed to *kill the field* for some time by showing mathematically that nothing was feasible even in the case of what most formal language theoreticians will consider as the easiest of cases, that of DFA.

The introduction of alternative learning models (like active learning), the capacity of probabilistic automata and context-free grammars to model complex situations and the necessity to learn functions with structured outputs allowed a renewal of interest in grammatical inference techniques and are some factors indicating that future cross-fertilisation between grammatical inference and machine learning scientists could be a key to success.

1.1.5 Computational biology

The basic elements in computational biology are strings or sequences describing DNA or proteins. From these a number of problems require the analysis of sets of strings and the extraction of rules. Among the problems that have suggested that grammatical inference could be useful, we can mention the one of secondary structure prediction. Context-free grammars can describe partly some well-known structures, which has motivated several attempts in the direction of learning context-free grammars. But, because of the size of the data, other researchers argued that grammars were going to prove to be too complex for computational biology. Pattern languages have been favoured by these researchers, and works in the direction of extracting patterns (or learning pattern languages) have been proposed.

There are a number of good reasons for wanting to use grammatical inference in this setting, but also a number of reasons why the results have so far been on the whole disappointing.

It is clear that we are dealing with strings, trees or graphs. When dealing with strings, we will typically be manipulating a 4-letter alphabet $\{A, T, G, C\}$ (standing for the four basic nucleotides) or a 20-letter alphabet if we intend to assemble the nucleotides 3 by 3 into amino-acids in order to define proteins. There are trees involved in the patterns or in the secondary structure, and even graphs in the tertiary structure.

Then, on the other hand, the strings are very long, and the combination 'very long strings' + 'recursivity' + 'no noise' makes the task of learning quite hard. One particularity of formal languages is that, if they do allow recursive definitions, they are not very robust to noise, especially when the noise is defined through edit operations. There are nevertheless a number of attempts to learn automata, grammars and transducers for biological applications.

1.1.6 Grammatical inference as an independent field

In 1994 the first *International Colloquium in Grammatical Inference* took place, arranged as a follow-up to a workshop that had been organised the year before. From then on, grammatical inference has emerged as an independent field. Little by little the conference acquired its rules and the community organised itself. This did not lead to the researchers working in the field isolating themselves: terms like *grammatical inference* and *grammar induction* have progressively been accepted in many communities and grammatical inference is today a meeting point between researchers from many areas, with very different backgrounds, most of whom have a primary research interest in one of many other fields. Nevertheless, whether for children's language acquisition, automatic music composition or secondary structure prediction (to name just three very different tasks), the objects remain common and the algorithms can be similar.

1.2 An introductory example

Let us introduce the problems and techniques of grammatical inference through a simple problem from a multi-agent world, where the goal is to discover the strategy of an adversary agent.

1.2.1 The problem

Let us consider the following situation: in an interactive world an agent has to negotiate with (or against) other agents. He or she may be wanting to acquire or sell resources. The agent's decisions (to buy or not to buy) will depend on what he or she expects to win but also on the other agent's attitude: should he or she keep on negotiating in order to reach a better price? Is the opponent going to accept the lower price?

Let us suppose that the adversary agent is driven by a finite state machine which describes its *rational strategy* (see Figure 1.1 for a trivial first example). The automaton functions as follows: at any one time the agent is in a given state and will act according to the label of that state. At the same time the agent discovers the action his or her opponent makes and moves on to the state reached through reading the action as a label of an

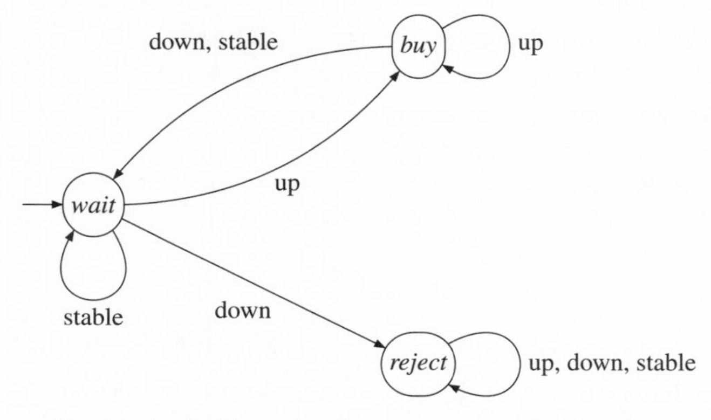

Fig. 1.1. A primitive rational strategy for a buying situation.

Table 1.1. *The game matrix for the buying game.*

	buy	*wait*	*reject*
up	5	1	-80
stable	5	-1	-80
down	10	3	-80

edge. The agent is then ready to proceed. Obviously the model is rather primitive and some sort of non-determinism/probabilistic outcome would be better suited. If we consider the automaton depicted in Figure 1.1 this agent will start by being doubtful (action *wait*). If our first offer is a raise (action up) then the agent will start buying (action *buy*), but we will need to raise each turn (action up) for him or her to stick to the buying policy, which is presumably what we want. If not, the agent will get back into the neutral initial state from which, if we lower our offer, he or she will actually get into a refusal state (action *reject*) in which no more negotiation can take place.

1.2.2 The gain matrix and the rules of the game

But there is more to it than just trying to see what the opponent is doing. Finding the right action for an optimised immediate gain is one thing; ensuring long-term gain is another! This is the question we are interested in now. Notice also that the reasons for which the adversary adopts a particular policy and his own gain are not of direct importance here, since only our own long-term gain matters.

So the sort of game we are considering is one where each player has to choose between a finite number of *moves* or *actions*. Both players play their move simultaneously. To simplify, if we have players A and B, each player is allowed to play moves from Σ_A and Σ_B respectively, where Σ denotes the alphabet, or set of allowed moves.

Each time an event takes place (an event being a simultaneous action by each agent) there may be a gain or a loss. This is usually represented through a gain table, such as in Table 1.1 where $\Sigma_A = \{\text{up, down, stable}\}$ and $\Sigma_B = \{\textit{buy, wait, reject}\}$: we have represented what we expect to gain out of an event. For example if simultaneously we (as player A) raise and our opponent (player B) is *neutral* (*wait*), then our gain is of +1; in all cases if the game is terminated by the opponent (*reject* state) our loss will be of -80.

1.2.3 The prisoner's dilemma

The best known and studied variant of the above game is called the prisoner's dilemma. The story can be told as follows:

Table 1.2. *The game matrix.*

	s	a
s	1	5
a	3	0

There has been a crime for which two persons have been arrested. In the morning, the judge addresses each of the suspects separately: *"We believe you are guilty but have no proof of this. So, we propose the following deal:*

If you admit guilt but your colleague does not, then you will be considered as an informer, get out free and your colleague will be jailed for five years.

If you decide to remain silent, on one hand your colleague might decide to be less stupid and accept the offer (remember that in this case you will suffer the entire burden alone, and go to jail for five years!).

If you both plead guilty we won't need an informer, but you both get a three-year sentence.

In the very unlikely case where you should both choose to remain silent, we have against you the fact that you were found doing grammatical inference in the dark, so you will both suffer a one-year sentence..."

This is a symmetrical game as both the opponents have the same gain matrix represented in Table 1.2: we will denote being **s**ilent by 's', whereas 'a' stands for **a**dmitting and thus defecting. By analysing the game it will appear that the Nash equilibrium[†] is reached when both players decide to defect (admit), reaching a far from optimal situation in which they both get imprisoned for three years! The game has been studied extensively in game theory; a simplified analysis goes as follows:

- Suppose the other person remains silent. Then if I am also silent I get one year in prison, whereas if I admit I get out free: better to admit.
- Suppose on the contrary he admits. If I am silent I will spend the next five years in prison: better admitting since in that case it will only be three.

The natural conclusion is therefore to admit, in which case both players reach the *bad* situation where they both 'pay' three years; if they had cooperated they would get away with a one-year sentence. The game has been much more thoroughly analysed elsewhere. We may now ask ourselves what happens when the game is iterated: the goal is only to score points and to score as many[‡] as possible over time. The game is played over and over against the same opponent.

Suppose now that the opponent is a machine. More specifically interesting to us is the case where this machine follows a *rational strategy*. In this case the object is no longer to win an individual game but to win in the long term. This is defined by means of a gain

[†] A Nash equilibrium is met when no player can better his situation for the opponent's current move.
[‡] In this case 'as many' requires spending as few years as possible in prison!

function computing the *limit of means*: the best strategy, against a fixed adversary, will be the one which has the highest mean gain, for an arbitrarily long game. With this, a finite number of initial moves which may seem costly will, in the long run, not count. Formally, the gain of strategy one (S_1) playing against strategy two (S_2) is defined by first noting that if both strategies are deterministic, a unique game is described. This unique game is an infinite string over an alphabet $\Sigma_A \times \Sigma_B$: $\langle m_0^A, m_0^B \rangle, \ldots \langle m_i^A, m_i^B \rangle, \ldots$ where $\langle m_i^A, m_i^B \rangle$ is the combination of the ith moves made by players A and B. In such a sequence, let us denote by $G[m_i^A, m_i^B]$ the gain for player A at step i. This in turn leads to a unique outcome described by the game matrix: $g(S_1, S_2, i) = \sum_{j \leq i} G[m_i^A, m_i^B]$.

$$g(S_1, S_2) = \lim_{n \to \infty} \frac{\sum_{j \leq n} g(S_1, S_2, n)}{n}$$

Remember that in this setting we are not directly interested in how much our opponent may make out of the game, only in what we are going to make.

1.2.4 What is a rational strategy?

We will say that a strategy is rational if it is dictated by a Moore machine. Informally (formal definitions will be given in Chapter 4) a Moore machine is a finite state machine whose edges are labelled by the actions made by an opponent and whose states have an output corresponding to our action when in the given situation. Consider for instance the rational strategy depicted in Figure 1.2. This corresponds to the iterated prisoner's dilemma as described above. Then it follows that an agent using this strategy will start by playing a, independently of anything else (remember that both players play simultaneously, so this is quite normal).

Now the agent's next call will depend on our first move. If we are silent, then the agent follows the transition labelled by `s` into the middle state: his or her move is deterministically a result of our move. His or her second action will be to be silent also, since that is what is indicated in that state. If on the contrary we had defected (action `a`), then the agent would have remained in the left-hand state and would repeat a.

Again, from Figure 1.2 it follows that if the first three calls in some game are $\langle \texttt{a}, a \rangle$, $\langle \texttt{a}, a \rangle$, $\langle \texttt{s}, a \rangle$ then the agent using this strategy will necessarily play s as his or her fourth move since having parsed his or her adversary's first three moves (the string `aas`) the agent is in the central state corresponding to output s.

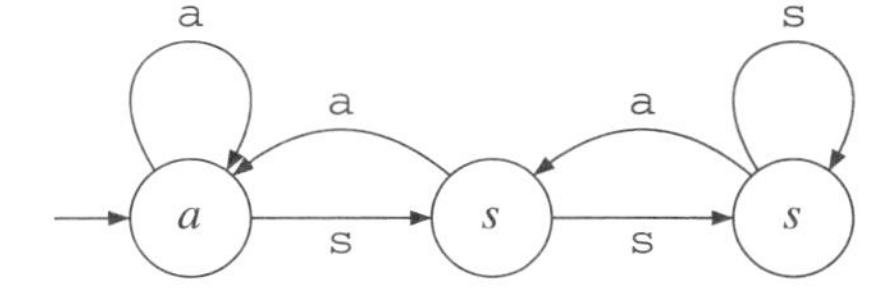

Fig. 1.2. A rational strategy.

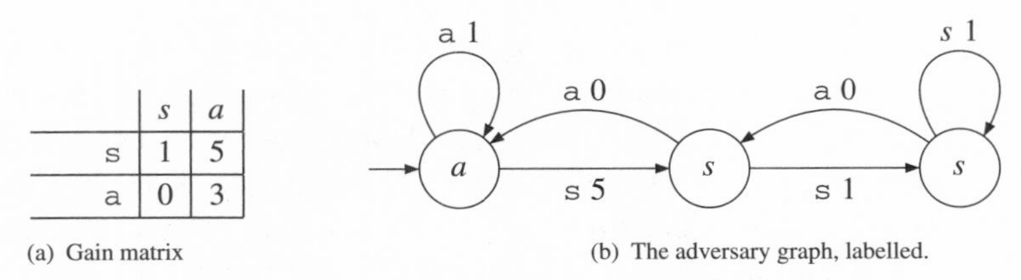

	s	a
s	1	5
a	0	3

(a) Gain matrix (b) The adversary graph, labelled.

Fig. 1.3. Using the graph to find the best path.

1.2.5 Beating a rational strategy that we know

We suppose that the strategy of the opponent is rational, i.e. it is given by a deterministic finite automaton as above (Figure 1.2). The first question is then: can we imagine an optimal strategy against it?

Suppose first that we know the opponent's strategy: then we can use game theory to compute the winning strategy. Let us consider for that the opponent's graph in which we value the edges by our own gain. Take for example the situation from Figure 1.3 and the strategy represented in Figure 1.2. The edges are weighted according to our values from the game matrix recalled in Figure 1.3(a). The result is shown in Figure 1.3(b).

What is a winning strategy in this case? As the adversary has imposed moves, it consists of finding a path that has minimal average weight. As the graph is finite this path is obtained by finding a cycle of minimum mean weight and finding the path that is best to reach this cycle. Both problems can be solved algorithmically.

Algorithm 1.1 computes the best strategy (in the sense of our criterion) in polynomial time. In this case the best strategy consists of being silent even when knowing that the opponent is going to defect, then being silent once more, and then alternating defection and silence in order to beat the more natural collaborative approach.

Algorithm 1.1: Best strategy against a rational opponent.

Data: a graph
Result: a strategy
Find the cycle of minimum mean weight;
Find the path of minimum mean weight leading to the cycle;
Follow the path and stay in the cycle

Summarising, if we know the adversary's strategy, we can run a simple algorithm that returns our own best strategy against that opponent. All that is now needed is to find the opponent's strategy, at least when it is described by an automaton!

So the question we now want to answer becomes: 'Having seen a game by this opponent can we reconstruct his strategy?'

Table 1.3. *The data to learn from: a game.*

$\lambda \rightarrow a$
a $\rightarrow a$
as $\rightarrow s$
asa $\rightarrow a$
asaa $\rightarrow a$
asaas $\rightarrow s$
asaass $\rightarrow s$
asaasss $\rightarrow s$
asaasssa $\rightarrow s$

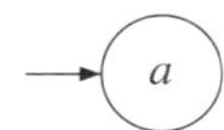

Fig. 1.4. Automaton obtained after using the information $\lambda \rightarrow a$.

1.2.6 Reconstructing the strategy

Let us suppose that we have seen a game played by the rational agent (following some unknown rational strategy). We will leave open for the moment the interesting point of how we got hold of the data from which we hope to reconstruct the adversary's strategy, and what would happen if we had to pay for this data (by playing against him or her, typically). To simplify the discussion, we suppose the data corresponds to a game between him or her (the rational agent, player B) and us (player A).

Suppose the game we are looking at went as follows: $\langle$a, $a\rangle$, $\langle$s, $a\rangle$, $\langle$a, $s\rangle$, $\langle$a, $a\rangle$, $\langle$s, $a\rangle$, $\langle$s, $s\rangle$, $\langle$s, $s\rangle$, $\langle$a, $s\rangle$, $\langle$s, $s\rangle$, with the adversary's moves coming second: first he or she played a and we played a, next he or she again played a while we played s,...

An alternative presentation of the game is given in Table 1.3. Indeed the fact that the rational agent played a as a fourth move means that this move will be triggered in any game where our first three moves are a, s and a in that order. Better said, we can associate to the input string asa the output a. The empty string λ corresponds to the first move: no previous move has been made. Notice that our last move (we played s) is of no use for learning: we can't infer anything from it, since player B did not react to it.

The algorithm we are going to follow is a loose version of a classical grammatical inference algorithm as described in Chapter 12. We intend only to introduce the key ideas here.

Step 1 At first, analysing Table 1.3, as our adversary's first call is a (information $\lambda \rightarrow a$), the initial state of his or her automaton is necessarily labelled by a, resulting in the partial strategy depicted in Figure 1.4. This strategy is partial because it can only cover part of the situation: in this case it only allows our adversary to play the first move.

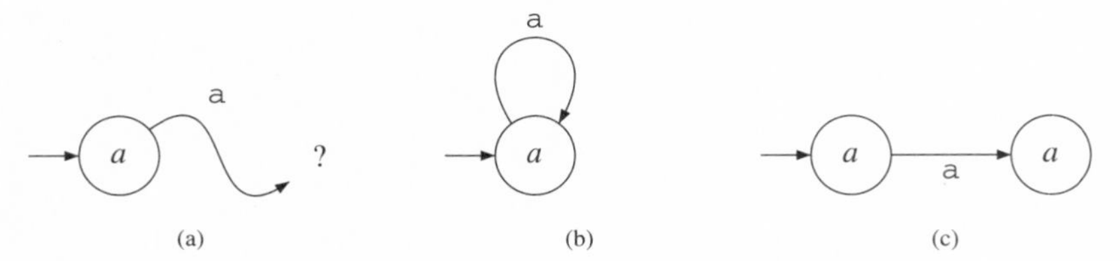

Fig. 1.5. How to parse a $\rightarrow a$.

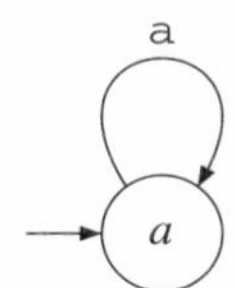

Fig. 1.6. After Step 2.

Step 2 The next string we want to parse (i.e. to read in the automaton) is a, corresponding to the information a $\rightarrow a$ in Table 1.3. This means that we are trying to find an end to the loose edge depicted in Figure 1.5(a).

There are two possibilities. The first is to create a new state, under the possibly sensible argument that there is no reason that the state the agent is in after reading a is the same as before (Figure 1.5(c)). The second is the more risky option of using the same state, taking the optimistic point of view that ***if there is nothing to forbid us doing so, we might as well do it*** (Figure 1.5(b)). A first point in choosing the second strategy (and thus considering the strategy depicted in Figure 1.6) is that if we don't do this now there will be every reason to create a new state each time the problem is raised, and we will end up with an enormous automaton that will only be able to parse the learning data without being able to generalise, i.e. to parse unseen or unknown moves. The principle behind this idea is sometimes referred to as the Ockham or parsimony principle.

There is a second point here, which concerns the fact that we will want to prove things about the sort of algorithms we develop. And here we can argue (we will) that if the second idea (of using just one state) was wrong, then surely we would have noticed this immediately (with for instance (λ, a) and (a, s) in the data, a loop with a would have created a conflicting situation: what output should we have in the state?). And if no such conflict arises, and still it would have been a bad idea to create the loop, then we should consider ***blaming the data***: there was no information in the data prohibiting the decision we were certain to make.

We therefore continue reading the data from Table 1.3 with the partial solution depicted in Figure 1.6, which is consistent with the data seen so far.

Step 3 We now need to parse the next bit of data: as $\rightarrow s$. Following the idea from the previous step, we want to consider the automaton from Figure 1.7.

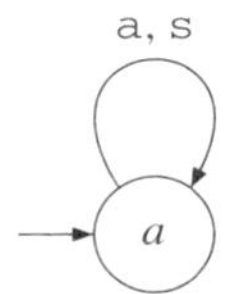

Fig. 1.7. A (bad) candidate to try to parse as → s.

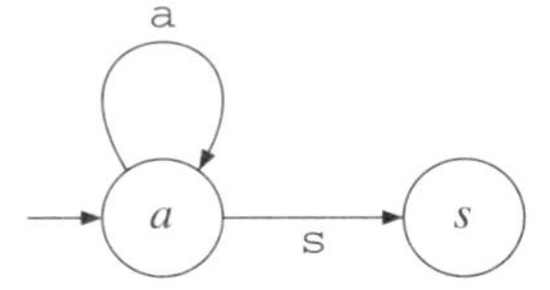

Fig. 1.8. Accepting to create a new state to parse as → s.

But this time things don't work out right: The candidate strategy indicates that the agent, after seeing a, will answer a, but the data tells us otherwise. So what can we do?

(i) Keep the automaton and see more strings. In this actual case, in fact, the algorithm would halt and return the strategy corresponding to Figure 1.7 as answer, since this automaton can parse all the strings. But this is not a convincing answer since the automaton will be playing a all the time, so it is clearly a wrong model for the strategy of our opponent. Actually, the determinism of the structure makes it impossible to recover from an early error.

(ii) Consider that we have made an error somewhere and backtrack. We could consider two places to backtrack to: after Step 1 or after Step 2. The algorithm we are going to follow here is a greedy algorithm: it will not backtrack, just delete the last choice and consider the next possibility. There remain plenty of possible lines here that could (or could not) better the algorithm.

At this point we may want to discuss the difference between an algorithm built to mathematically identify a target and an artificial intelligence algorithm: in a mathematical identification algorithm the decision to do something will be taken whenever there is no proof that it is a bad idea to do so. In other words, if we were not *supposed* to take the decision, we could have hoped to find a refutation in the data. This is of course an optimistic assumption. On the other hand an argument closer to artificial intelligence (but also perhaps to common sense) will tell us to decide ***if there is evidence for it to be a good idea***.

We create a new state and thus are left with the strategy depicted in Figure 1.8.

Step 4 The next string that cannot be parsed is string asa (whose output is a, because of asa → a). Once again there are three possibilities: automata from Figures 1.9(a), 1.9(b) and 1.9(c). The first is clearly inconsistent with the data since as → s. But in fact the previous argument is good enough: ***Do not create a new state if you don't have to***. Furthermore, checking the possibilities in some predetermined order is essential: if needed, the information allowing us to discard a choice can then be forced into the learning data. Obviously an exhaustive beam search could be performed, even if there will be a cost that

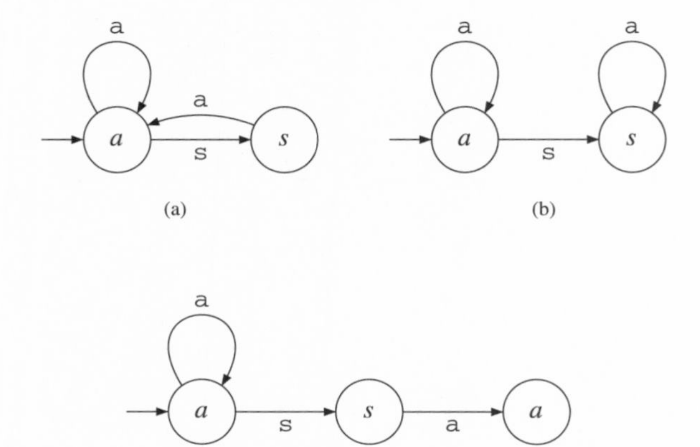

Fig. 1.9. How to parse asa $\rightarrow a$.

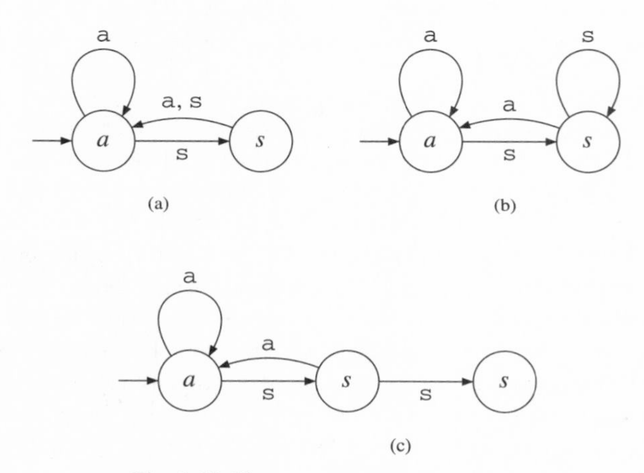

Fig. 1.10. How to parse asaass $\rightarrow s$.

would be prohibitive for a larger quantity of data. Therefore the automaton represented in Figure 1.9(b) is kept. It can be noticed that not only can *asa* be parsed, but also more strings, for instance asaa and asaas which correspond to asaa $\rightarrow a$ and asaas $\rightarrow s$ in Table 1.3. This also is a typical feature of grammatical inference: one decision can take care of more than one piece of data.

Step 5 The next string we cannot parse is asaass ($\rightarrow s$). We again have three choices and can show that the two choices which do not create states (1.10(a), 1.10(b)) fail. We keep the strategy from Figure 1.10(c).

Step 6 Let us name the states by the smallest string reaching each state: we therefore have at this point q_λ, $q_{\mathtt{s}}$ and $q_{\mathtt{ss}}$. We label each state by its name and its output. Therefore

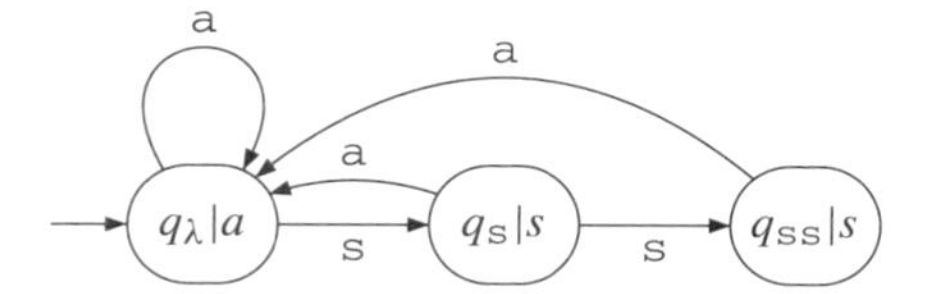

Fig. 1.11. Trying $\delta(q_{\mathrm{ss}}, \mathrm{a}) = q_\lambda$.

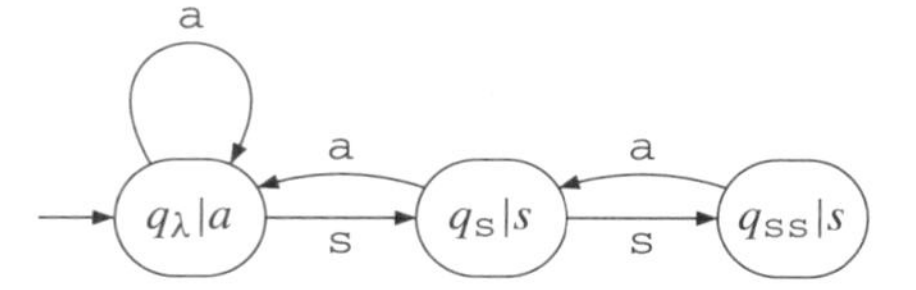

Fig. 1.12. We now have to deal with $\delta(q_{\mathrm{ss}}, \mathrm{s})$.

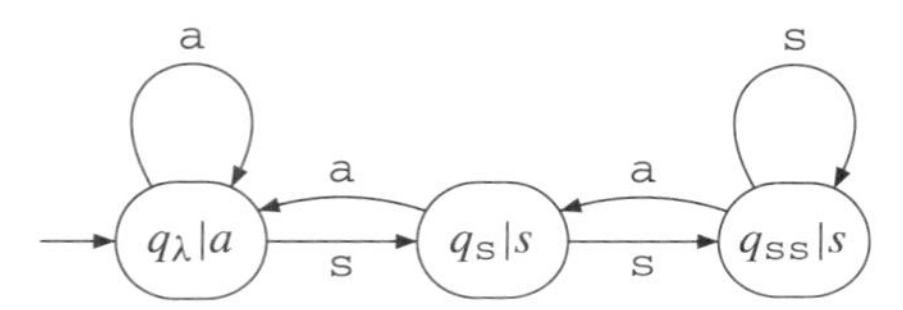

Fig. 1.13. What the algorithm returns.

$(q_{\mathrm{ss}}|s)$ corresponds to state q_{ss}, which outputs symbol s. At this point we have to decide what we do with the edge labelled a leaving state q_{ss}, which we will denote by $\delta(q_{\mathrm{ss}}, \mathrm{a})$. This is an equivalent question to asking: how do we parse an a from state q_{ss}? Again the first idea is to try ending this edge in q_λ (Figure 1.11).

As this is rejected, $\delta(q_{\mathrm{ss}}, \mathrm{a}) = q_{\mathrm{s}}$ is then tested (Figure 1.12). This decision is accepted.

After rejecting $\delta(q_{\mathrm{ss}}, \mathrm{s}) = q_\lambda$ and $\delta(q_{\mathrm{ss}}, \mathrm{s}) = q_{\mathrm{s}}$, the loop $\delta(q_{\mathrm{ss}}, \mathrm{s}) = q_{\mathrm{ss}}$ is accepted. As all the strings can now be parsed in a consistent manner, the automaton (Figure 1.13) is returned and the algorithm halts.

It should be noted that the above example is only exactly that, an example. There is certainly no guarantee that such a short game should allow us to reconstruct the strategy.

1.2.7 Some open problems

A first interesting question (albeit not necessarily for grammatical inference) is to consider what the 'better' strategy would be against an arbitrary adversary. There are a number of variants here, as 'better' could mean a number of things. Out of interest the strategy called *tit for tat* (represented in Figure 1.14) is considered the best against an arbitrary adversary.

A second open question is: How do we get hold of the learning data?

- **Through observation**. In this setting we have to do the best we can with a given set of examples. If the previous games from which we are learning are meaningful, we can hope to learn the strategy. But obviously, if the agent has only had to play some specific type of opponent, then we will only

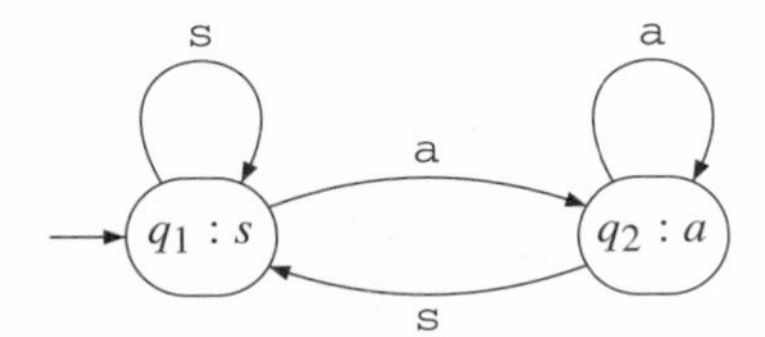

Fig. 1.14. Rational strategy *tit-for-tat*.

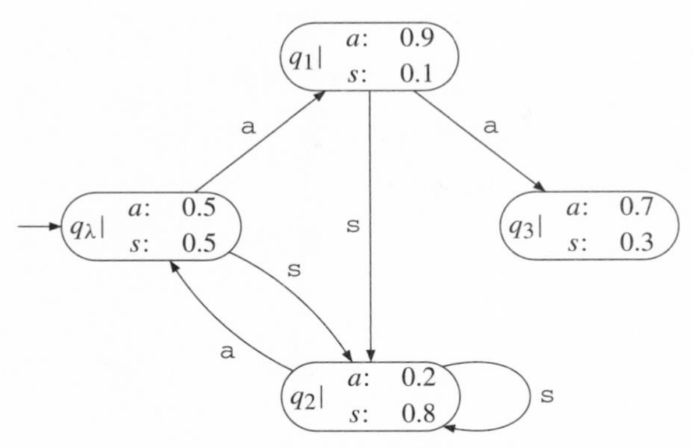

Fig. 1.15. A probabilistic strategy.

have a very partial type of information and will quite logically only find out how the player reacts in certain situations.

- **Through exploration**. If we are to play against the agent in order to gather the data, we may be in better condition to learn from him or her. But in this case there will usually be a price to pay, which can be given by the gain function. The typical situation will be the one where at some point we have learnt a strategy which, when countered, is giving us an advantage. Suppose for instance we believe the agent adopts strategy *tit for tat* (represented by Figure 1.14); then we understand that the best answer is to play s systematically. But there may be an entirely unsuspected better strategy that can only be discovered if we are prepared to consistently lose for some long period of time. For example, after playing a 1000 times in a row, perhaps the player may *resign* and consistently play s in all cases allowing us to get the maximum score in the limit by playing a. The question is then: *how do we know, after 999 losing moves of* a, *that we should do it once more?*

Another open problem is to consider that the strategy is probabilistic. The answer in each state is the result of some stochastic process. This leads to questions related with *Markov decision processes*, as discussed in Chapter 16. Just to get the idea we propose a possible probabilistic adversary in Figure 1.15. Other questions that arise through the algorithms concern validation: what can we say about the learnt automaton? How good is it? Is it provably good? If we can't say for sure that the automaton is good, perhaps can we say something about the process with which it was constructed? Could it be good *provided the data was good*?

All these are questions that will be addressed in this book.

1.3 Why is learning grammars hard?

It is one thing to build learning algorithms, and another to be able to state that these work. There may be objective reasons in given applications to learn a context-free grammar, an automaton or a pattern, but before being able to use the proposed algorithms we will be faced with the problems of validation and comparison. To solve these we are also going to have to answer a certain number of questions.

1.3.1 Some possible questions

Does this algorithm work? Does it always work? When does it not work? What does 'to work' mean?

Do we have enough learning data? Can we put a higher bound on the quantity of data needed to have some guarantee in a bad case? Or some lower bound on the quantity of data without which learning is impossible? If we are working with data that is sampled following some given distribution, can we measure the probability that the data we have is sufficient for learning?

Do we need some extra bias? If what we are looking for is just going to be impossible to find, like searching for a needle in a haystack, is there some way we can artificially (or by making extra hypotheses relevant to the application) reduce the search space in order to have better hopes of finding something sufficiently good?

Is this algorithm better than the other? In the case where we are given two algorithms that solve the same grammatical inference problem, how do they compare? What criteria should be used to compare learning algorithms?

Is this problem easier than the other? If we cannot solve a learning problem, then proving that it is unsolvable is usually difficult. An alternative is to prove that a problem is at least as difficult as another. This will call for specific reduction techniques for grammatical inference.

1.3.2 Some possible answers

There are alternatives as to how to answer the above questions:

Use well-admitted benchmarks. In machine learning, the development of shared tasks and benchmarks has allowed researchers in the field to make a lot of progress. Algorithms can be compared against each other and hopefully these comparisons allow engineers to choose the better algorithm for a given task.

Build your own benchmarks. There are many possible parameters in a grammar learning problem: the presentation, the class of grammars, the criteria for success... This means that there are in some sense more problems than available benchmarks. For a simple initial example, consider the problem of learning finite state automata from positive and negative data. There certainly is one task which consists of trying to learn automata of 500 states from as few strings as possible, but there are many variants of at least as much importance and that will not be tackled by the same techniques. Two examples are: learn when some of the strings are noisy, and learn smaller automata but with an algorithm that never fails.

Solve a real problem. In all cases, solving a new unsolved problem on *true* data is meaningful. It will convince researchers and engineers that your techniques are capable of working on something else than toy examples. But a specific issue in this case will be that of preprocessing the data, and more importantly, the number of parameters to the problem make it difficult to measure if the proposed technique can be transferred to another problem, or even get close.

Mathematical proofs. Obtaining mathematical convergence results is not only going to be an alternative to the above benchmarking problems. You may also want to be able to say something more than 'seems to work in practice'. Being able to characterise the mathematical properties of a method will allow us to pose conditions, which, if fulfilled, guarantee success in new tasks.

1.4 About identification, approximation and in a general sense convergence

Suppose that, inspired by the previous discussion, you come up with some technique that seems to fit a particular problem. That might mean, for example, that you have considered that linear grammars were exactly what you wanted and that the data available consisted of strings. Now what? Is the problem solved? Perhaps you have considered computing the complexity or the run time for your algorithm and you have reached the conclusion that it was correct, and even if you were given a particularly large set of new strings, your algorithm would perform correctly. But what would 'correctly' mean? Notice that if the problem is just about finding some grammar that could fit the data then we could argue that the linear grammar that generates all possible strings is just as good as the candidate you have come up with.

Consider for a more detailed example the learning sample represented in Figure 1.16(a). It is composed of two sets: S_+ contains the examples, i.e. the strings supposed to be in the language, and S_- contains counter-examples or negative examples, that is, strings that are not in the language. A learning algorithm for deterministic automata could return the machine represented in Figure 1.16(b) or 1.16(c) between the infinity of possibilities, even if we restrict the algorithm to accept the consistent ones only.

What arguments could we come up with to say that the second grammar is better than the first?

S_+	{aa, b, bb}
S_-	{a, ab, ba}

(a)

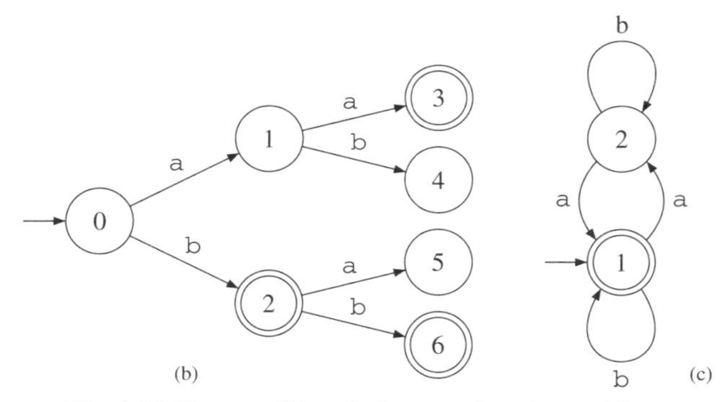

Fig. 1.16. Two possible solutions to a learning problem.

We may be tempted to say that 'in practice' the second one is 'better'. Presumably 'better' could measure the number of mistakes we could make, when parsing.

An alternative is to come up with some form of 'the better grammar is the one which best explains the data'. This would mean that we would have to avoid a grammar that would only explain the data seen so far, so we would be able to generalise, but we would at the same time need to give short descriptions (i.e. derivations) of the data.

Another alternative taking into account both ideas is generally called the minimum description length (MDL) principle: a solution is measured by summing both the size of the grammar and the size of the explanations. A good grammar is therefore one which is small in size and allows us to encode the data in a short way.

Notice that in all these arguments we have decided to measure the quality of the grammar and this should also somehow measure the quality of the learning process, i.e. the capacity to repeat this on a new set of unseen data. This is obviously arguable: why should our algorithm perform well in general if it has done well once? On the other hand, if our algorithm did not do well on one particular task, could it not be because the task was indeed particular or the specific data bad?

All these questions are to be discussed in detail in this book and one way to try to answer these questions involves introducing a target.

1.4.1 A target: from induction to inference

By saying that there is a target, and this target is what we are after, we have replaced an induction problem by an altogether different problem, of searching for a target, approximating it or identifying it. But many things have changed and new questions can be raised, as the target itself now is part of the parameters! Indeed the hardness of the learning task can depend on the complexity of the target, and questions linked with teaching (is this particular target teachable?) now make sense.

1.4.2 Identification or approximation

Now we have a target, and we are given, in some way still to be made precise, some data to work from. So the question is: how are we going to measure success?

A first school of thought is that success only takes place when the actual target is found. This leads to the elegant concept of *identification in the limit*: a language is identified when the hypothesis made is perfect. There is going to be room for discussion, because many questions remain:

- Were we lucky? Would the same thing happen if we were given different examples?
- Are we in any sense robust: if we are given more examples after identification, are we not going to change our mind?
- How do we know that we are correct? Or incorrect?
- As infinite languages can certainly not be represented *in extenso*, some sort of representation scheme will have to be chosen. How does the identification depend on this choice?

A second school of thought consists of saying that we could accept to be a little wrong. As matters stand, we could be a little wrong always or nearly always. Now this is going to oblige us to define some new measure: that of there being a distribution over the strings, and that this distribution is going to allow us to define 'the probability of being wrong'. Again an elegant setting, called PAC learning has been introduced to match this idea.

Both settings have opponents, some of the questions raised by these being: Why should the distribution be stable in time? The fact that we are using our model to classify but probably also to generate could probably change the actual distribution. Is language stable? When a child starts learning a language, this language is going to change with time.

Our feeling is that no paradigm is best. Some fit some situations better; some fit others. The study of a variety of settings is therefore a necessity.

1.4.3 Grammars and languages

As languages can be infinite, there is no way we can manipulate them directly: we can only use some representation of languages, typically called grammars. And there is going to be a point in choosing between several possible ways of defining languages.

Let us consider what is often described as the simplest case, that of regular languages. It is well known that these can be generated (by left linear or regular grammars), recognised (by non-deterministic or deterministic finite state automata), and described (by rational expressions).

Another related issue is that if we only have information about a language, then the learning algorithm itself somehow defines (or is related to) some canonical form of grammar.

More generally, these remarks make us define grammatical inference as learning grammars given information about the languages.

1.4.4 About being efficient

Once it is admitted that some sort of convergence is needed, we will want to converge efficiently. If we are talking about induction then the issue will be that of taking an appropriate amount of time, when considering the amount of data from which learning should take place.

But if identification or approximation is required then it is reasonable that the harder the language is (or the more complex a grammar is for the given language), the more time (and eventually space) we should be allowed for our task.

There are a number of options for defining variants of fast learning, neither of which should be considered as definitive. The discussion about these will take place in Chapter 7:

- Just asking bluntly for learnability to be achieved from any data in time polynomial in the size of the information is not good enough. There is no way we are going to be sure to have only very useful information.
- In some cases the basic information needed to learn can be too expensive!

1.5 Organisation of the manuscript

This book is organised into three parts: the first part describes the objects (strings, automata, grammars...), in the second part, we study the convergence criteria, and the third part contains a survey of many algorithms and techniques that can be used.

Chapter 1 is the present *Introduction*; we have described the field and those surrounding it. The key questions are raised: why prove things, what does grammatical inference mean?

What are the typical data for grammatical inference? A number of examples and applications are presented in Chapter 2.

There is an enormous literature on strings. Even if it would have been possible to just refer to the normal and usual text books, it seemed reasonable for the manuscript to be self-contained (as much as possible), to revisit the key definitions and unify the presentation of those results most needed in grammatical inference in our Chapter 3, called *Basic stringology*. There will be a special focus on topological issues: how do we order strings, how do we compare them? Moreover, the links between string distances and string kernels are explored.

In Chapter 4, called *Representing languages*, we present the two first levels of the Chomsky hierarchy: regular languages and context-free languages. In both cases we consider various representations and discuss questions of sizes, problems of parsing and of normal forms.

In Chapter 5 we introduce *Representing distributions over strings*: probabilistic finite automata (and probabilistic grammars) generate distributions over strings. Not only do we give different definitions, but we study in detail the questions of parsing, eliminating λ-transitions and computing distances between distributions. Again, the algorithmic questions will be of utmost importance.

In Chapter 6 (*About combinatorics*) we group a certain number of combinatorial elements needed when wanting to write a new algorithm or to analyse an existing one: the VC dimension of the main classes of automata and some important $\mathcal{NP}$-hardness proofs like the fact that the problem of finding the minimum consistent DFA is $\mathcal{NP}$-hard. Some strange automata constructions are also given that can help to understand why learning automata is hard: transforming clauses into automata, automata whose shortest 'important' string is of exponential size...

The second part concerns the introduction and analysis of learning models: what language learning and efficient learning are about, and what can't be done.

When learning in a non-probabilistic setting, we will need to talk about the convergence of the learning setting. This typically concerns identification in the limit and its resource-bounded versions. Thus non-probabilistic learning paradigms are introduced, commented on and analysed in Chapter 7 (*Identifying languages*).

Chapter 8 is about *Learning from text*: we survey the results and techniques allowing us to learn when presented with unique data from the target language (text). What is learnable from text and why is a class learnable or not?

Active learning or query learning is dealt with in Chapter 9. We will talk about active learning when the learner can in some way interact with its environment. We will recall proofs that learning is impossible given less help from the environment.

Chapter 10 (*Learning distributions over strings*) deals with the case where there is a distribution over the set of all strings. The first possibility is that we rely on this distribution to learn a probably approximately correct grammar, one that will parse correctly most unseen strings (most referring to the distribution), unless the sampling process during the learning phase has gone particularly badly. The second option is to suppose that the distribution itself is generated by a grammar (or an automaton). In that case the convergence can be obtained through a small error or with probability one. Both settings are analysed (and negative results are given) in Chapter 10.

The third part is a systematic survey of language learning algorithms and techniques.

Chapter 11 is about *Text learners*: we mainly study two classes of languages learnable from text, one where a window of fixed size describes acceptable strings, and the second where a look-ahead can be used to deal with ambiguity problems. We also study a simple algorithm for learning pattern languages. Finally, we survey here the ideas allowing us to learn planar languages.

Informed learners concerns the case where not only are examples given, but also counter-examples. This is an issue that has attracted a lot of interest, especially when what one wants to learn is the class of the DFA. In Chapter 12 we describe two important algorithms, GOLD and RPNI.

Learning with queries is dealt with in Chapter 13. The most important algorithm in this setting is called LSTAR, and is used to learn deterministic finite automata from membership and strong equivalence queries.

Grammatical inference is linked with cognitive questions; furthermore the combinatorial problems associated with the usual questions are intractable. This justifies that *Artificial intelligence techniques* (Chapter 14) have been used in many cases: genetic algorithms, heuristic searches, tabu methods, minimum description length ideas are some ideas that have been tried and that we briefly present here.

Learning context-free grammars has been an essential topic in grammatical inference since the beginning. We study the learnability of context-free grammars (but also of subclasses) and show some algorithms and heuristics in Chapter 15.

When what we want is to learn a distribution over strings, then it may be the case that this distribution is represented by an automaton or a grammar. Learning the automaton or the grammar is then the problem. One line of research consists of supposing the formal syntactic model given and concentrating on estimating the probabilities. Grammatical inference has provided algorithms to learn the structure as well as the probabilities. Some of these algorithms are described in Chapter 16 about *Learning probabilistic finite automata*.

One special case we need to analyse is when the structure of the automata is known and the task consists of *Estimating the probabilities* of an automaton or of a grammar. This is the theme of Chapter 17.

There are many cases where learning one language is not enough, and where we have to manipulate two languages at the same time. Automatic translation is just one of such tasks. A very simple translation device is a transducer, which is a finite state machine. We extend the results on learning finite automata to *Learning transducers* in Chapter 18.

In each chapter we end by providing a small list of exercises, of variable difficulty. Their purpose is to allow anyone wanting to check how much they have understood to do so and also to assist anyone wanting to build a course using grammatical inference. The exercises also give an idea of some difficulties that are not discussed in the text.

A discussion in three parts ends each chapter: the first is a bibliographical survey of the theme of the chapter, in which the references corresponding to the main results are given. The second part concerns some alternative routes to similar problems that have been taken by researchers. In the third part we comment upon some open problems and research directions one could take.

1.6 Conclusions of the chapter and further reading

We summarise, recall the key issues and discuss the bibliographical background.

In Section 1.1 we discussed some of the important historical landmarks. Ray Solomonoff (Solomonoff, 1960, 1964) is considered as the inspiring figure for work in inductive inference, whereas the key questions concerning discovering grammars were first posed by Noam Chomsky (Chomsky, 1955, 1957).

Mark Gold's work (Gold, 1967, 1978) around the central question of identification in the limit is essential. He put down the definitions of identification in the limit and then proved some critical results which we will describe in more detail in Chapter 7.

Jim Horning's PhD in 1969 (Horning, 1969), called *A Study on Grammatical Inference* is important for a number of reasons: it not only states that there is a field called grammatical inference, but in his PhD one can find several questions still very active today, like the importance of probabilistic grammars.

When mentioning simplicity and the different principles that learning specialists use for the introduction of convergence criteria, one should consult (Chaitin, 1966, Kolmogorov, 1967, Rissanen, 1978, Solomonoff, 1964, Wallace & Ball, 1968). A full analysis can be found in (Li & Vitanyi, 1993). The point of view is also defended from a cognitive perspective in (Wolf, 2006).

The specific field of structural and syntactic pattern recognition (Bunke & Sanfeliu, 1990, Fu, 1982, Gonzalez & Thomason, 1978, Miclet, 1990, Ney, 1992) rose using ideas from Kim Sung Fu (Fu, 1974, 1975), relayed by researchers like Laurent Miclet (Miclet, 1986) who developed many new ideas in grammatical inference (which is usually called grammar induction in this context).

Probabilistic grammars were studied in pattern recognition by Kim Sung Fu (Fu, 1974, Fu & Booth, 1975), and have been worked on in speech recognition by researchers like Hermann Ney (Ney, 1992), Enrique Vidal and Francisco Casacuberta (Casacuberta & Vidal, 2004). As will be shown in Chapter 16 the first algorithms to learn probabilistic automata were ALERGIA (Carrasco & Oncina, 1994b) and DSAI (Ron, Singer & Tishby, 1994); similar algorithms using hidden Markov models (HMM) can be found in (Stolcke, 1994, Stolcke & Omohundro, 1994).

Machine learning scientists have added a lot of extra knowledge to the question of grammatical inference. Pioneering work by Dana Angluin (Angluin, 1978, 1980, 1981, 1982, 1987b) allowed us to introduce new algorithms (with queries, for pattern languages or for reversible languages), but also to visit the intrinsic limits of the field. The introduction of the PAC learning model (Valiant, 1984) led to research on grammars and automata in this model. Negative proofs relied on cryptographic limitations (Kearns & Valiant, 1989). Research also dealt with combinatoric issues following early theoretical work by Boris Trakhtenbrot and Ya Bardzin (Trakhtenbrot & Bardzin, 1973) and experimental work by Kevin Lang (Lang, 1992). Other work of specialised interest is that by Ron Rivest and Robert Schapire (Rivest & Schapire, 1993) on learning from unique (but very long) *homing sequences* and different works on learning distributions (Abe & Warmuth, 1992). Lenny Pitt and Manfred Warmuth worked on proving the intractability of learning deterministic finite state automata from informed presentations: the results are combinatorial and close to complexity theory. Through reductions one measures the hardness of the issue (Pitt, 1989,

Pitt & Warmuth, 1988, 1993). For a survey one can read the books on learning theory (Kearns & Vazirani, 1994, Natarajan, 1991).

Computational biology has offered challenging problems to grammatical inference scientists. There are certainly different reasons for this: the importance of the data, and the difficulty to humanly manipulate it. Yasubumi Sakakibara's survey (Sakakibara, 1997) is a good starting point for researchers interested in these questions. The problems to be solved involve prediction (Abe & Mamitsuka, 1997) and modelling (Jagota, Lyngsø & Pedersen, 2001, Sakakibara *et al.*, 1994). The idea of combining probabilistic context-free grammars and n-gram models (Salvador & Benedí, 2002) helps to deal with long-term and short-term dependencies. Alvis Brazma's work on regular expressions and pattern languages is another direction (Brazma, 1997, Brazma & Cerans, 1994, Brazma *et al.*, 1998).

Computational linguistics has been an important field for grammatical inference. Not only did Noam Chomsky introduce the key notions for formal language theory but he started asking the questions in terms of discovering grammars. For an introduction of grammatical inference for linguists, see (Adriaans & van Zaanen, 2004). But syntactic structures themselves don't convince most linguists who have been attempting to use more complex constructions, with richer possibilities for the manipulation of semantics. Makoto Kanazawa's book (Kanazawa, 1998) can be used as a starting point.

Another issue, closer to statistics, is that of language modelling (Goodman, 2001), with some theoretical questions presented in (McAllester & Schapire, 2002). Language models arise whenever disambiguation is needed. As such the task of building a language model (or the dual task of using the model) is thus to be seen as a sub-task, or tool in important fields like automatic translation (Ferrer, Casacuberta & Juan-Císcar, 2008), speech recognition or natural language processing. The possibilities of grammatical inference here were guessed long ago (Baker, 1979, Feldman, 1972), but the more statistical approaches (Charniak, 1993, 1996) have so far been the better ones. Nevertheless work building language models for speech recognition (Chodorowski & Miclet, 1998, García *et al.*, 1994, Thollard, 2001, Thollard, Dupont & de la Higuera, 2000) has allowed the introduction of several important ideas. Some grammatical inference systems are actually proposed to build context-free grammars for computational linguists (Adriaans & Vervoort, 2002, van Zaanen, 2000).

Grammatical inference, as an independent field, has developed through the organisation of the international colloquium on grammatical inference (Adriaans, Fernau & van Zaannen, 2002, Carrasco & Oncina, 1994a, de Oliveira, 2000, Honavar & Slutski, 1998, Miclet & de la Higuera, 1996, Paliouras & Sakakibara, 2004, Sakakibara *et al.*, 2006). Between the surveys we include (de la Higuera, 2005, Sakakibara, 1997).

The introductory example we propose throughout Section 1.2 is adapted from work by David Carmel and Shaul Markovich (Carmel & Markovitch, 1998a, 1999), who in a series of papers try to learn such strategies and also analyse the exploration-exploitation issues.

When we described the different data grammatical inference had been tested on, we mentioned graphs. The very few works about learning graph grammars (López & Sempere, 1998, Oates, Doshi & Huang, 2003) use formalisms that can be found in theoretical computer science literature (Courcelle, 1991, Mosbah, 1996).

The ideas developed in the section about the hardness of grammatical inference have been partially introduced in (de la Higuera, 2006a). For the discussion about the relationship between identification and teaching, one can read (de la Higuera, 1997, Goldman & Kearns, 1995, Goldman & Mathias, 1996).

2

The data and some applications

> Errors using inadequate data are much less than those using no data at all.
>
> ***Charles Babbage***
>
> It is a capital mistake to theorise before one has data.
>
> ***Sir Arthur Conan Doyle***, *Scandal in Bohemia*

Strings are a very natural way to encode information: they appear directly with linguistic data (they will then be words or sentences), or with biological data. Computer scientists have for a long time organised information into tree-like data structures. It is reasonable, therefore, that trees arise in a context where the data has been preprocessed. Typical examples are the parse trees of a program or the parse trees of natural language sentences.

Graphs will appear in settings where the information is more complex: images will be encoded into graphs, and first-order logical formulae also require graphs when one wants to associate a semantic.

Grammatical inference is a task where the goal is to learn or infer a grammar (or some device that can generate, recognise or describe strings) for a language and from all sorts of information about this language.

Grammatical inference consists of finding the grammar or automaton for a language of which we are given an indirect presentation through strings, sequences, trees, terms or graphs.

As what characterises grammatical inference is at least as much the data from which we are asked to learn, as the sort of result, we turn to presenting some possible examples of data.

2.1 Linguistic data and applications

In natural language processing tasks, huge corpora are constructed from texts that have appeared in the press, been published on the web or been transcribed from conversations. A small but representative example is as follows:

> *En un lugar de la Mancha, de cuyo nombre no quiero acordarme, no ha mucho tiempo que vivía un hidalgo de los de lanza en astillero, adarga antigua, rocín flaco y galgo corredor. Una olla de algo más vaca que carnero, salpicón las más noches, duelos y quebrantos los sábados, lantejas los viernes, algún palomino de añadidura los domingos, consumían las tres partes de su hacienda. El resto della concluían sayo de velarte, calzas de velludo para las fiestas, con sus pantuflos de lo mesmo, y los días de entresemana se honraba con su vellorí de lo más fino.*
> **Primera parte del ingenioso hidalgo don Quijote de la Mancha** (The beginning of *Don Quijote*, by Cervantes)

The text can sometimes be tagged with various degrees of information. Sometimes, the tagging actually gives us the structure of the text, in which case, instead of learning from strings the task will more probably consist of learning from bracketed strings which correspond to trees. A typical parse tree is represented in Figure 2.1. A corresponding tagged sentence might be (NP `John`)(VP (V `hit`) (NP (Det `the`) (N `ball`))). Trees found in tree banks are transcriptions of natural text that has been tagged and archived in view of better analysis. The syntactic structure is supposed to help the semantic manipulation.

A number of tasks can be described in computational linguistics. These include the following, in the case where what we are given is raw text.

- A first task can consist of associating with each word in a sentence its grammatical role. This is called *part-of-speech tagging*.
- In order to get a better bracketing of a sentence, the brackets are named, corresponding to their type (noun phrase, verb phrase). The goal is to find the border of these *chunks*. The task is usually called *chunking* or *shallow parsing*.
- The task called *named entity recognition* consists of labelling words into categories, for example `Person` or `Location`.
- *Semantic role labelling* aims to give a semantic role to the syntactic constituents of a sentence. It will therefore go one step further than chunking.

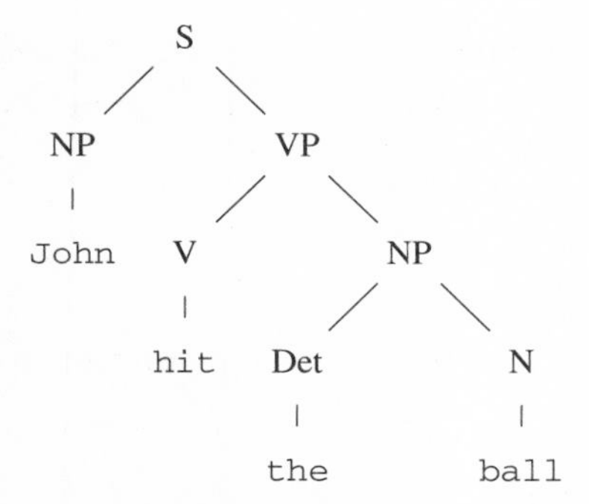

Fig. 2.1. Parse tree for the sentence `John hit the ball`.

- *Language modelling* is another task: the goal is to build a model with which the probability of the next word in a sentence can be estimated.
- A more complex task is that of predicting if two words are *semantically related* (synonyms, holonyms, hypernyms...)

2.1.1 Language model learning

In several tasks related to natural language, it is necessary to choose between different plausible (lexically, or even grammatically) phrases. For example, these may be transcriptions of some acoustic sequence in a speech recognition task or candidate translations of a text. In order to choose between these candidates one would like to propose the sentence which seems to be most likely to belong to the intended language. Obviously, there would also be an acoustic score, or a word-by-word score, and perhaps other elements to take into account.

For example, when using an automatic translation system, we may have to translate the sentence: *'Il fait beau et nous voulons nous baigner.'* Candidate translations may be:

- It makes beautiful and we want us swim
- It is beautiful and we want to swim
- He is beautiful and we want to swim

Each of these translations may come with a score corresponding to the individual translations of each individual word. What is needed now is to compute a score reflecting whether the sentence seems to be English or not.

A language model can give us a score of likeliness of transcription. It allows us to somehow know the score of each sequence inside a language. This can be obtained through having an estimation of the distribution of all sentences in the chosen language.

A very successful technique is to use a window over a corpus to count which sequences of length n occur often. By simply counting these occurrences we obtain n-grams. These can be then used to predict the next word or to give a weight to a sentence.

An interesting alternative is to learn probabilistic automata. This line of research has been followed by a number of researchers. Some algorithms are described in Chapter 16.

Natural language processing (NLP) is a very active research field. The task is to build models that can be used at the semantic level in order to understand and create natural language. In this field a natural cooperation between linguists and computer scientists can take place. Between the issues addressed here we have the important question of language acquisition: How does a young child learn his or her first language? The basic model studied is often that of learning from text and will be explored in Chapter 8.

Another specific point is that it is never of use to learn regular languages as they do not correspond to good models for natural language. Even if there are arguments against context-free languages, these are usually preferred. We will explore this question in Chapter 15.

2.1.2 Automatic translation

The goal is to translate from one language to the other. In a corpus-based approach, corpora for both the initial and the target language will be given. But another corpus containing pairs of strings, one in each language, can also be used.

A typical example might be a pair with a sentence in Gaelic and its translation to English: (*Tha thu cho duaichnidh ri èarr àirde de a coisich deas damh*, You are as ugly as the north end of a southward travelling ox.)

The problems raised in this setting are multiple: to build alignments between the sentences, to learn transducers, but also to have statistical language models not only for both languages, but also for the pair of languages.

In the case of less frequently spoken languages, one has to face the problem of having less data and not being able to rely on statistics. In that case, one can consider the alternatives of learning through a third language (which may typically be English or a more abstract language) or considering an active learning approach: an expert might be interrogated in order to help translate specific sentences.

The case where the two languages use altogether different alphabets is called transliteration. A typical issue is that of working on texts written with different alphabets and attempting to make proper nouns correspond.

Automatic translation has become an increasingly important task. Translation using rule-based systems is interesting but development of the translation rules by experts is a tedious and far too expensive task. Therefore constructing these rules automatically from corpora is an attractive alternative. These automatic translators can take a variety of forms. The one we are interested in here, being closer to the formal grammar approach, makes use of transducers. Learning restricted types of transducers is a task for grammatical inference, as will be seen in Chapter 18.

An interesting point to note is that whereas in natural language there are now very large corpora for the main languages, this is not the case for less common languages or artificial languages. In this case the creation of corpora for translations is expensive, and syntactic methods (like grammatical inference) may hope to have an edge over purely statistical ones.

2.2 Biological data and applications

Following Wikipedia, 'deoxyribonucleic acid (DNA) is a nucleic acid that contains the genetic instructions used in the development and functioning of all known living organisms and some viruses'. In the DNA is stored the genetic information. The basic units are the nucleotides, each nucleotide being tagged by one of four types of molecule, called bases. More abstractly DNA is composed over a four-letter alphabet $\{A, C, T, G\}$ where A stands for adenine, C for cytosine, T for thymine and G for guanine. Strings of DNA can measure up to millions in length, and there are a number of databanks in which such sequences can be found. See Figure 2.2 for an example. The number of possible problems in which some form of prediction clustering or classification is needed is ever increasing.

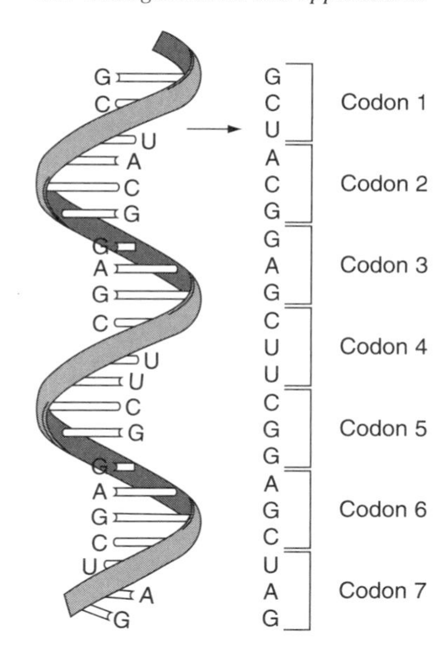

Fig. 2.2. DNA. (Courtesy of National Human Genome Research Institute)

Proteins are macromolecules made of amino acids. They have most of the active roles in the cell. Abstractly, proteins are composed over a 20-letter alphabet, each letter being an amino acid itself made of the concatenation of 3 nucleotides. In the first example below we present some DNA. The second case corresponds to a protein sequence.

```
ATGAAGGCTCCCACCGTGCTGGCACCTGGCATTCTG
GTGCTGCTGCTTGTCCTTGGTGCAG
```

```
MKAPTVLAPGILVLLLSLVQRSHGECKEALVKSEMNVNM
KYQLPNFTAETP
```

Trees are used in biology for many purposes. Between the many sorts of trees one can find in computational biology, phylogenetic (or evolutionary) trees are used to describe the ways the species have been modified during the ages. They show the evolutionary interrelationships among various species or other entities. Distances over strings (see Section 3.4) can be used to decide which species are close to one another and to find the genes that might have mutated over time. An example is represented in Figure 2.3.

Typical tasks in computational biology are to find encoding regions, to classify sequences, to searching for encoding sections, to align, to discover distance parameters...

An important issue in this context is to be able to propose intelligible solutions to the biologists; this means that algorithms returning just one blackbox classifier might not quite solve their problem.

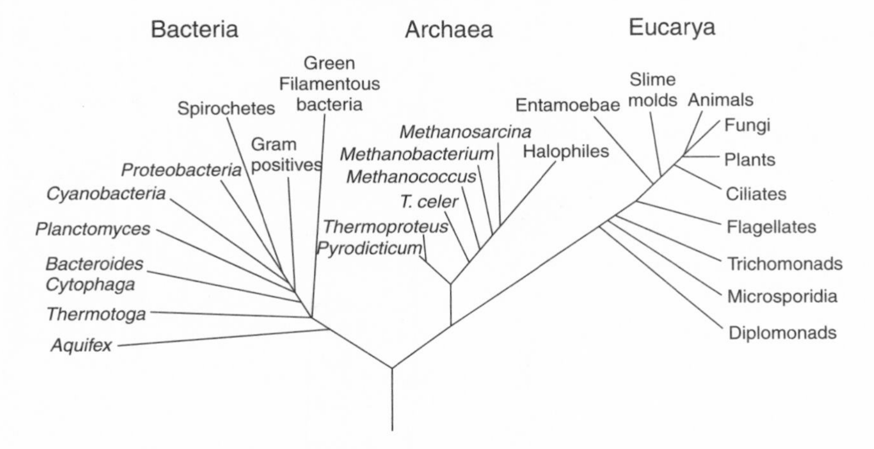

Fig. 2.3. Phylogenetic tree. (Courtesy of NASA)

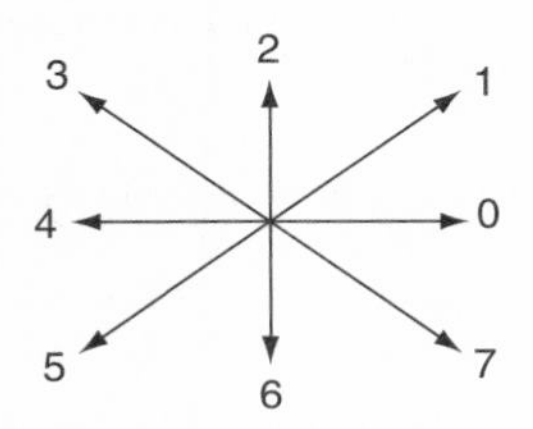

(a) An eight letter alphabet for images.

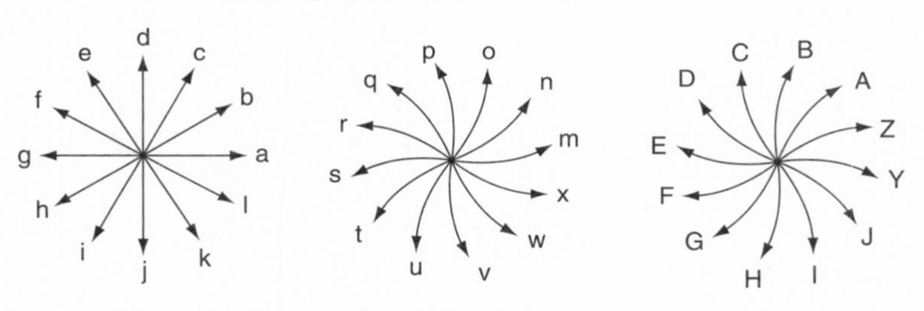

(b) A more elaborate alphabet.

Fig. 2.4. Images.

A specificity of the applications in this field is that the strings under consideration are very long and therefore the algorithms have to be fast.

2.3 Data and applications in pattern recognition

Images, and more specifically contours of images, can be made by providing either a four- or an eight-letter alphabet describing direction in the plane. We represent in Figure 2.4(a) a way to encode a contour with an eight-symbol alphabet, and in Figure 2.4(b) typical digits and encodings of these that have been used in several applications of hand-writing character recognition by means of grammatical inference.

Typical tasks will include automatically extracting patterns for each digit or letter, and then using these patterns for classification. Automata and grammars can be used as class representatives.

Graphs are used to represent complex data. Intricate relationships can be described by graphs when trees or strings are insufficient. The drawback is that many problems that were easy to solve in the case of trees cease to be so for graphs. This is the case when testing if two graphs are isomorphic, finding the largest common subgraph, computing a distance or aligning two graphs. The development of the World Wide Web and the need for richer semantic descriptions nevertheless makes the use of graphs as data from which

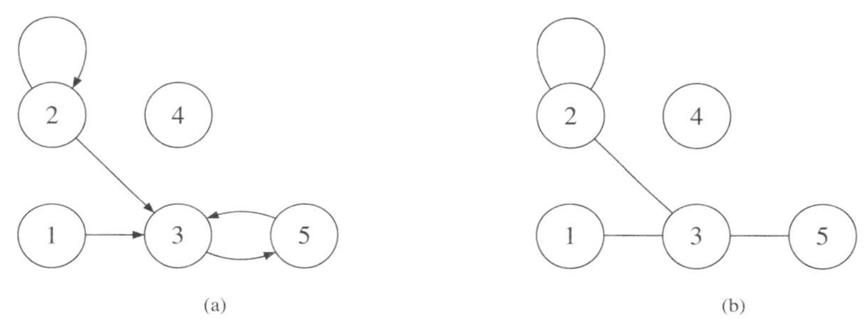

Fig. 2.5. Two graphs.

Fig. 2.6. Delaunay triangulation.

to learn important. The hardness of grammatical inference makes the learnability of graph grammars a nearly untouched problem.

Two very simple graphs are represented in Figure 2.5. In the context of image processing, graphs can appear through Delaunay triangulation, that is by selecting a number of nodes (that can correspond either to zones or to important points in the image), and then relating the nodes by edges by making use of the topological information. A simple example is represented in Figure 2.6.

2.4 Data in computer science applications

Parse trees for programs are obtained through compiling a program written in some programming language as a result of syntactic analysis. In different tasks of engineering these parse trees are the basis for learning. An example of a parse tree for an arithmetic expression is provided in Figure 2.7. A very simple fragment of the parse tree for a program in C is given in Figure 2.8.

A related task can be to infer the correct syntax of some unknown (or forgotten) programming language given some instances of programs, and perhaps some additional information, like the grammar of a language supposed to be close to the target. Automatic programme synthesis is a field concerned with (re)building the program given its traces.

2.4.1 Inductive logic programming

SLD-trees appear when studying the semantics of a logic program. Given a logic program, the way the resolution takes place is best described through Selected Literal Definite (SLD) clause resolution. The trees representing these computations can be used as a basis for learning in Inductive Logic Programming (ILP). See Figure 2.9 for a trivial example.

Inductive logic programming is an active subfield in machine learning. The goal is to discover, from data and background knowledge both described in a logical programming language, the program that could account for the data.

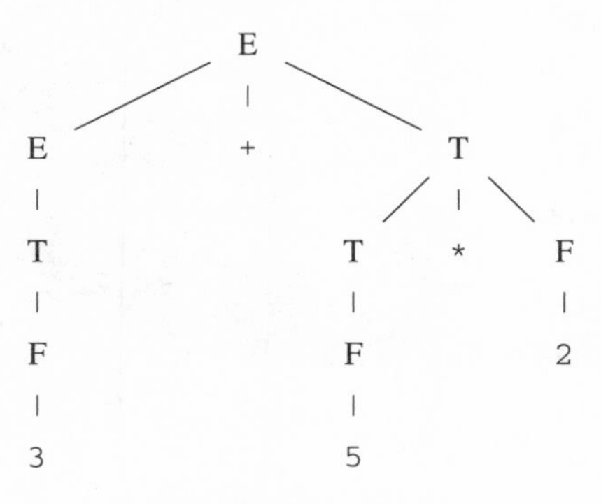

Fig. 2.7. Parse tree for 3+5*2.

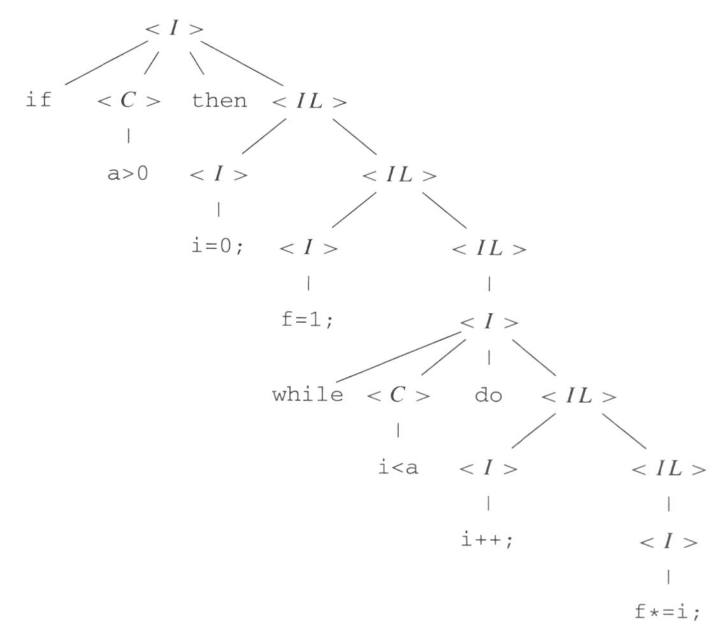

Fig. 2.8. Parse tree for a small program.

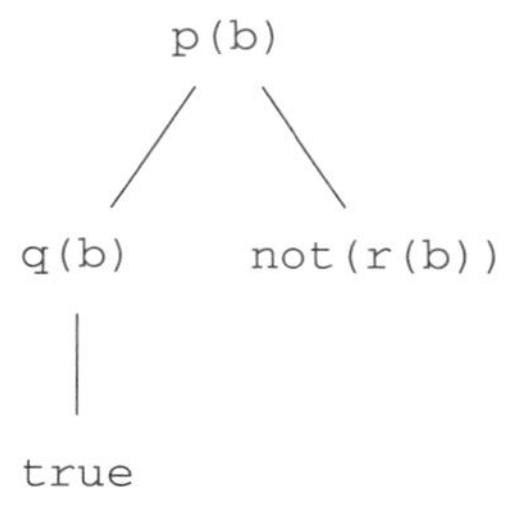

Fig. 2.9. SLD-tree.

There are many difficulties in the task: the encoding of the data is one, the usual noise one can find is another. But, more than anything else, one should notice that the richness of representation of the logic programs is much stronger than that of regular (or even context-free) languages. A number of added biases have therefore to be used if something interesting is wanted.

2.4.2 Information extraction: automatic wrapper generation

The quantity of structured data available today due to the exponential growth of the World Wide Web introduces a number of challenges to scientists interested in grammatical inference. HTML and XML data appear as text, but the text is well bracketed through a number of tags that are either syntactic (HTML) and give indications as to how the file should be represented, or semantic (XML).

Here is a small piece from an XML file:

```
<book>
 <chapter>
  <name>Introduction</name>
  <length>25 pages</length>
  <description>
    Motivations about the book
  </description>
  <exercises>0</exercises>
 </chapter>
 <chapter>
  <name>The Data</name>
  <length>18 pages</length>
  <description>
    Describe some cases where the data is made of strings
  </description>
  <exercises>0</exercises>
 </chapter>
<chapter>
  <name>Strings and Languages</name>
  <length>35 pages</length>
  <description>
    Definitions of strings and stringology
  </description>
  <exercises>23</exercises>
 </chapter>
</book>
```

Between the many problems when working with these files, one can aim to find the grammar corresponding to a set of XML files.

One very nice application in which grammatical inference has been helpful is that of building a wrapper automatically (or semi-automatically). A wrapper is supposed to take a web page and extract from it the information for which it has been designed. For instance, if we need to build a mailing list, the wrapper would find in a web page the information that is needed. Obviously, the wrapper will work on the code of the web page: the HTML or XML file. Therefore, grammatical inference of tree automata is an obvious candidate.

Another feature of the task is that labelling the examples is cumbersome and can be noisy. The proposal is to do this on the fly, through an interaction between the system and the user. This will justify in part the rising interest in *active learning* methods.

2.5 Other situations

2.5.1 Time series

Time series (Figure 2.10) describe (numerical) data over time. As these data are usually not symbolic, an effort of discretisation has to be made before using them as strings. In many cases, the time series depends on a variety of factors. This interdependence usually makes grammatical inference techniques less reliable for problems linked with such data.

2.5.2 Music pieces and partitions

Music (Figure 2.11) can be represented by strings in many ways, depending on the sort of encoding one uses, in knowing if one is directly transcribing analogue signals or just re-encoding a partition. The pitches and lengths of the notes will be the symbols of the alphabet.

Strings allow us to structure the information along just one axis. An extension used in many fields inside computer science is by means of trees. We mathematically define trees and terms in Section 3.3. Music can be seen as a language with each piece being represented by notes, the encoding of which corresponds to the characters in the alphabet.

A number of problems are of interest: creating models for a composer or a style, clustering composers and classifying new pieces of music.

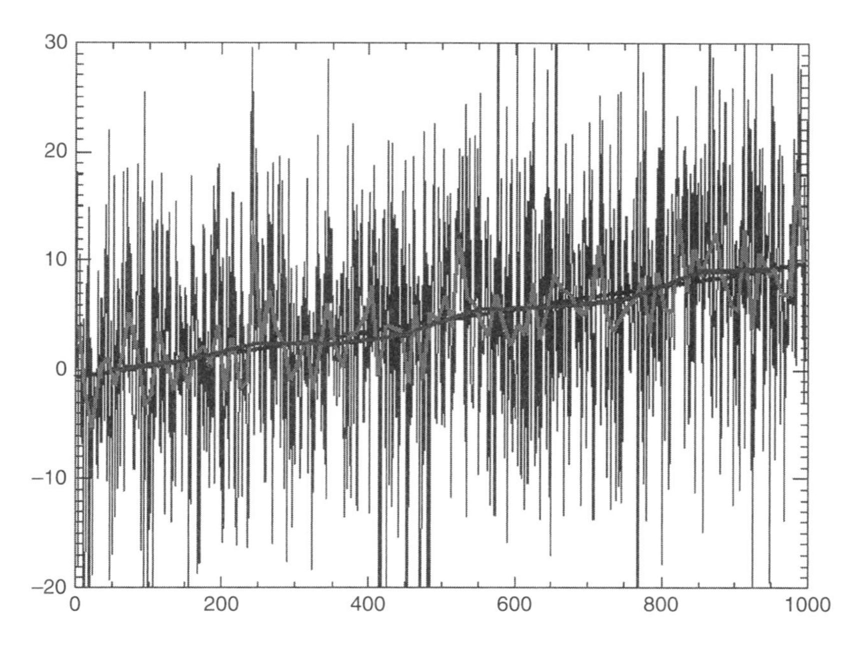

Fig. 2.10. Time series. (Courtesy of Wikipedia. Used under the GNU Free Documentation License)

Fig. 2.11. Music.

The question of polyphony is also crucial as the melody on its own does not give us all the information we may want.

This is a field where, typically, probabilistic automata have been learnt, not so much because of their capacity to adapt as because of the fact that they can then be used to compose new pieces. Probabilistic automata are studied in Chapter 5, and some learning algorithms that have been used on musical style recognition tasks are presented in Chapter 16.

2.5.3 Robotics

An autonomous robot may need to construct a representation of the space in which it is moving. For this purpose it may move around the room and use its captors. One should note that there usually is no reset: the robot will not suddenly get lifted and put back into its original position to start a new experiment! But we are in a setting where all the prefixes of the string are meaningful, so the situation is somehow close to the preliminary example from game theory we have explored in this chapter.

The learning is usually said to be active. This setting is explored in Chapter 9.

2.5.4 Chemical data

Chemical data are usually represented as graphs. But for many reasons, chemists have proposed linear encodings of these graphs. Two formal chemistry languages can be seen as examples.

The first, *simplified molecular line entry specification* (SMILES) is defined as a context-free language. The strings corresponding to molecules are mostly unambiguous.

The second language is called the IUPAC international identifier. It allows us to define in the same string various phenomena within the molecule. An example is drawn in Figure 2.12.

In both cases, three-dimensional information corresponding to the molecules is encoded into the strings.

2.5.5 User modelling

One may consider the question of trying to anticipate a user's needs by comparing the sequence of steps the user is following with prototypes that have been previously learned. This enables the system to anticipate the user's needs and desires, which may be a good commercial strategy.

A typical case corresponds to following a web user's navigation patterns. The actual graph of the website is known, and the goal is to predict the next page the user will require. Cache management or commercial strategy can be two reasons to want to do this.

The data will be strings, but an enormous amount of noise may be present: following an individual user who may be using proxies and caches of all types or even having a cup of coffee in the middle of his or her session is not a simple task, especially if privacy issues (which are essential to consider here) are taken into account.

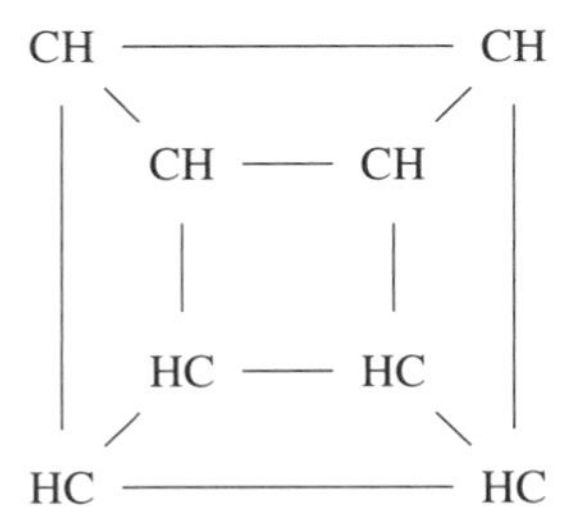

Fig. 2.12. SMILES string corresponding to cubane: C12C3C4C1C5C4C3C25.

2.5.6 Ethology and others

Along with the other tasks on which grammatical inference techniques have been tried, we can mention some strange ones like ethology (bird-songs, but also mating behaviour of flies), learning computer programming language dialects for reverse engineering, or analysis of a driver's fuel consumption!

2.6 Conclusions of the chapter and further reading

The use of grammatical inference for computational linguistic tasks can be researched independently; the historic links can be found in Noam Chomsky's PhD (Chomsky, 1955), and, between the surveys linking the fields of grammatical inference and computational linguistics one can find (Adriaans & van Zaanen, 2004, Nowak, Komarova & Niyogi, 2002). The use of transducers (Mohri, 1997), of Markov models (Saul & Pereira, 1997) or of categorial grammars (Adriaans, 1992, Costa Florêncio, 2003, Kanazawa, 1998, Tellier, 2005) can be studied elsewhere. Neural networks have also been applied in this setting (Collobert & Weston, 2008).

Tree bank grammars have been used for some time (Charniak, 1996). Language acquisition is studied mainly by linguists. A point of view closer to that of grammatical inference can be found in (Becerra-Bonache, 2006, Becerra-Bonache & Yokomori, 2004). Language modelling is a task studied by many authors. Andres Stolcke's approach (Stolcke, 1994) is probably the first using probabilistic finite state machines for this.

Automatic translation has used automata called transducers for some time now (Alshawi, Bangalore & Douglas, 2000a, 2000b). Complex systems have attempted to use syntactic methods relying on transducers instead of just statistical information (Amengual *et al.*, 2001, Casacuberta & Vidal, 2004).

Many algorithms for the task of learning transducers for automatic translation have been developed in Spain (Castellanos, Galiano & Vidal, 1994, Castellanos *et al.*, 1998, Oncina, 1998, Oncina & Varó, 1996, Vilar, 2000). Transliteration has been investigated via learning of edit distance weights (Pouliquen, 2008).

Other questions relating to language technologies in which grammar learning issues are raised include acoustics (Castro & Casacuberta, 1996, Vidal *et al.*, 1988) and morphology (Roark & Sproat, 2007, Wang & Acero, 2002).

The number of works dealing with formal language approaches to computational linguistics is high. To keep the list short, let us mention the work of Alvis Brazma (Brazma, 1997, Brazma & Cerans, 1994) on learning regular expressions and pattern discovery in biosequences (Brazma *et al.*, 1998). Work on the many questions that lead from this can be found in (Lyngsø, Pedersen & Nielsen, 1999, Sakakibara, 1997, Sakakibara *et al.*, 1994, Salvador & Benedí, 2002, Wang *et al.*, 1999).

In pattern recognition the work on strings and trees and grammatical representations has taken place in subcommunities (typically the structural and syntaxic pattern recognition working groups) (Bunke & Sanfeliu, 1990, Fu, 1974, Fu & Booth, 1975, García & Vidal,

1990, Gonzalez & Thomason, 1978, Lucas *et al.*, 1994). Extensions to graphs have been studied in (Qiu & Hancock, 1996, Yu & Hancock, 1996).

Grammatical inference has also been used to find programming language dialects: the tasks consist of building, from programs written in some forgotten dialect, and some background knowledge about similar languages, the grammar for that program (Dubey, Jalote & Aggarwal, 2006). Other software engineering tasks have been looked into, such as of synthesising the behaviour of software (Dupont *et al.*, 2008).

ILP is concerned with finding a first-order logic representation explaining some data and possibly using a background theory. As this representation can be a program written in some logic programming language, the question of learning recursive programs arises. Work in the field has either used SLD resolutions as data to learn from (Boström, 1996, 1998) or a transformation of the ground data into terms as data (Bernard & de la Higuera, 2001, Bernard & Habrard, 2001, de la Higuera & Bernard, 2001).

The fact that structured document analysis could be done through grammatical inference has been studied for some time (Ahonen, Mannila & Nikunen, 1994, Young-Lai & Tompa, 2000). The next step was to attempt to extract information from the World Wide Web (Hong & Clark, 2001, Kosala *et al.*, 2003).

Also linked with the World Wide Web, research has attempted to learn something from the navigation of a user, in order perhaps to predict the next page he or she will access or to better adapt the site to the current user. This is a task for which different grammatical inference algorithms have been tried (Korfiatis & Paliouras, 2008). Manipulating XML files has made researchers introduce different types of pattern over semi-structured data (Arimura, Sakamoto & Arikawa, 2001).

Boris Chidlovskii (Chidlovskii, 2001) extracts schema from XML, as also proposed by Henning Fernau (Fernau, 2001), whereas an alternative is DTDs (Bex *et al.*, 2006).

One task in which grammatical inference is proving to be particularly useful is that of wrapper induction: the idea is to find in a web page (or something of the same type) all arguments of a special sort (Carme *et al.*, 2005, Chidlovskii, 2000, Chidlovskii, Ragetli & de Rijke, 2000, Crescenzi & Merialdo, 2008, Kosala *et al.*, 2003).

Little work has been done in order to model time series and to predict with grammatical inference techniques. One notable exception is (Giles, Lawrence & Tsoi, 2001) in which the value of different currencies are predicted by means of automata learnt by neural network training methods. The links between grammatical inference and compression have been studied either in a paradigmatic way (Wolf, 1978), or in order to develop a compression tool that at the same time discovers the structure of the document to be compressed (Nevill-Manning & Witten, 1997a).

Different representations of music can be used (Lerdahl & Jackendoff, 1983); attempts to classify musical styles and to create music using grammatical inference have taken place (Cruz & Vidal, 1998, Cruz-Alcázar & Vidal, 2008, de la Higuera, 2005).

There have been some attempts to link grammatical inference with robotics (Dean *et al.*, 1992, Reutenauer & Shützenberger, 1995, Ron & Rubinfeld, 1993). The goal has been to get a robot to construct a map of its environment. The work has actually been more interesting towards the theoretical setting (no *reset*: the robot does not magically come back to its starting point when needed).

The conversions from graphs representing molecules to strings are described in (James, 2007) for SMILES (James, 2007)) and in (Stein *et al.*, 2006) for IUPAC.

In (Murgue, 2005, Murgue & de la Higuera, 2004), Thierry Murgue tries to learn probabilistic finite automata for web user modelling. Other user navigation pattern extraction techniques have been used (Borges & Levene, 2000, Castro & Gavaldá, 2008, Korfiatis & Paliouras, 2008). An alternative approach based again on finite state machines (Poggi *et al.*, 2007) attempts to anticipate the needs of customers of a travel agency.

Other and future applications include testing systems: testing can be a very expensive matter and the specifications of these systems can often be described through one type of automaton or anther. Then grammatical inference can be used to develop a test-set (Bréhélin, Gascuel & Caraux, 2001), or to check through interaction with the actual chips used as an Oracle that the specifications are met (Berg, Jonsson & Raffelt, 2006, Raffelt & Steffen, 2006).

Other security issues have been studied by modelling the systems as automata for infinite strings (Alpern, Demers & Schneider, 1985, de la Higuera & Janodet, 2004, Saoudi & Yokomori, 1993): indeed since the hope is that the system never crashes, the natural assumption is that the positive examples are infinitely long.

Let us end this survey of examples by noting that the example presented in Chapter 1 corresponded to a problem where strategies were discovered by learning (Carmel & Markovitch, 1998a, 1999).

PART I

The Tools

3
Basic stringology

The biscuit tree. This remarkable vegetable production has never yet been described or delineated.

Edward Lear, *Flora Nonsensica*

Man muss immer generalisieren.

Carl Jacobi

Formal language theory has been developed and studied consistently over the past 50 years. Because of their importance in so many fields, strings and tools to manipulate them have been studied with special care, leading to the specific topic of *stringology*. Usual definitions and results can therefore be found in several text books.

We re-visit these objects here from a pragmatic point of view: grammatical inference is about learning grammars and automata which are then supposed to be used by programs to deal with strings. Their advantage is that we can parse with them, compare them, compute distances... Therefore, we are primarily interested in studying how the strings are organised: knowing that a string is **in** a language (or perhaps more importantly **out** of the language) is not enough. We will also want to know how it belongs or why it doesn't belong. Other questions might be about finding close strings or building a kernel taking into account their properties. The goal is therefore going to be to organise the strings, to put some topology over them.

3.1 Notations

We start by introducing here some general notations used throughout the book.

In a definition if_{def} is a *definition* 'if'.

The main mathematical objects used in this book are letters and strings. The simplest notations will consist of using *italics* for variables denoting letters and strings with the letters from the beginning of the alphabet $(a, b, c, \ldots)$ reserved for the symbols, and the rest of the alphabet $u, v, \ldots, z$ used for the strings. When using examples, reference alphabets will be used. The letters intervening in the examples will be denoted as $\mathtt{a}, \mathtt{b}, \mathtt{c}, \ldots$

Finally when we manipulate letters from two alphabets, one being ***final*** (the observable alphabet) and the other being ***auxiliary*** or composed of auxiliary symbols, letters of the final alphabet will be indicated by $\mathtt{a}, \mathtt{b}, \mathtt{c}, \ldots$, and letters from the auxiliary alphabet by capital letters $A, B, C, \ldots$ Strings over the union of the two alphabets will be denoted by Greek symbols $\alpha, \beta, \gamma \ldots$

3.1.1 Mathematical preliminaries

- $\mathbb{Q}$ is the set of all rational numbers.
- $\mathbb{N}$ is the set of positive or null integers.
- $\mathbb{R}$ is the set of real numbers.
- $|X|$ is the ***cardinality*** of any finite set X.
- $\max(X)$ is the maximum value in a set X (if it exists) ($\min(X)$ being the minimum value in X).
- $\text{argmax}_{x \in X}\{\theta(x)\}$ denotes one value in X that yields the maximal value over the set X for the function θ. Note that whereas unicity is often compulsory, we will not follow this line here.
- $\text{argmin}_{x \in X}\{\theta(x)\}$ conversely denotes one value in X that yields the minimal value over the set X for the function θ.
- $[n] = \{1, \ldots, n\}$ (and for $n = 0$, $[0] = \emptyset$).
- We denote $A \subseteq B$ and $A \subset B$ for (respectively) 'A is a subset of B' and 'A is a proper subset of B'.
- When given a finite set E, a ***partition*** of E is a set of subsets of E denoted by $\Pi = \{E_1, \ldots E_k\}$ such that on one hand $\bigcup_{i \in [k]} E_i = E$ and on the other $\forall i, j \in [k]$, $E_i \cap E_j = \emptyset$. A partition defines an equivalence relation over E: two elements are ***equivalent*** for Π (denoted by $x \equiv_\Pi y$) if they are in the same ***class*** or ***block*** of the partition. A partition Π is ***finer*** than a partition Π' if $x \equiv_\Pi y \implies x \equiv_{\Pi'} y$.
- A ***multiset*** is a pair $M =< A, \text{cnt}_M >$ where $\text{cnt}_M : A \to \mathbb{N} \setminus \{0\}$ returns the number of occurrences of each element of A in M in M. A is called the ***support*** of the multiset: $A = \text{support}(M)$.
- We shall denote in extension a multiset as $\{(\mathtt{a}, 2), (\mathtt{b}, 3), (\mathtt{c}, 1)\}$ or alternatively as $\ll \mathtt{a}, \mathtt{a}, \mathtt{b}, \mathtt{b}, \mathtt{b}, \mathtt{c} \gg$.
- The cardinality of a multiset written $|M|$ is $\sum_{a \in A} \text{cnt}_M(a)$. Alternatively $\lceil M \rceil$=|support(M)| is the number of different symbols in M.

Example 3.1.1 Let $\sin : \mathbb{R} \to \mathbb{R}$ be the sine function. Then we obtain $\max\{\sin(x) : x \in [0; 2\pi]\} = 1$ whereas $\text{argmax}_{x \in [0;2\pi]}\{\sin(x)\} = \frac{\pi}{2}$.

For $M = \{(\mathtt{a}, 2), (\mathtt{b}, 3), (\mathtt{c}, 1)\}$ we have $\text{support}(M) = \{\mathtt{a}, \mathtt{b}, \mathtt{c}\}$. Also, $|M| = 6$ but $\lceil M \rceil$=3.

3.1.2 Notation in algorithms

The notation $\mathcal{O}$ corresponds to the asymptotic class. $\mathcal{O}(f(n))$ stands for the set of all functions dominated asymptotically by function $n \to f(n)$. That means: $g \in \mathcal{O}(f(n)) \iff \lim_{n\to\infty} \frac{g(n)}{f(n)} < +\infty$.

$\mathcal{P}$ is the class of all problems solvable in polynomial time by a deterministic Turing machine.

$\mathcal{NP}$ is the class of all decision problems solvable in polynomial time by a non-deterministic Turing machine, with the $\mathcal{NP}$-complete problems those hardest in the class and thus, if $\mathcal{P} \neq \mathcal{NP}$, not solvable in polynomial time. Problems that are at least as hard as any $\mathcal{NP}$-complete problem are $\mathcal{NP}$-hard. This concerns not only decision problems, but any problem (optimisation, search) as hard as any 'hardest' problem in $\mathcal{NP}$ (for instance SAT, the satisfiability problem).

Tables used in an algorithm will be denoted as $\mathrm{T}[x][y]$ when two dimensional. This allows us to consider a row in the table as $\mathrm{T}[x]$.

3.1.3 Distances and metrics

Let X be the set over which a distance is to be defined. We will primarily be dealing with the case where X is a set of strings, but this of course need not necessarily be the case.

Definition 3.1.1 A **metric** over some set X is a function $X^2 \rightarrow \mathbb{R}^+$ which has the following properties:

$$\begin{aligned} d(x, y) = 0 &\iff x = y \\ d(x, y) &= d(y, x) \\ d(x, y) + d(y, z) &\geq d(x, z) \end{aligned}$$

The last condition is known as the triangular inequality, and it is interesting to notice how many so called *distances* do not comply with it. We choose to use the term ***distance*** for any mapping associating a value to a pair of elements with the general idea that the closer the elements are, the smaller the distance is. When the properties of Definition 3.1.1 hold, we will use the term *metric*.

In $\mathbb{R}^n$ a number of distances have been well studied. Let $x = [x_1, \ldots, x_n]$ and $y = [y_1, \ldots, y_n]$ be two vectors; then the Minkovski distance of order k, or k-norm distance, is denoted by $\mathbf{L}_k$:

$$\mathbf{L}_k(x, y) = \left(\sum_{i \in [n]} |x_i - y_i|^k\right)^{\frac{1}{k}}.$$

If now $x = [x_1, \ldots, x_n]$ and $y = [y_1, \ldots, y_n]$,

$$\mathbf{L}_\infty(x, y) = \max_{i \leq n} \left\{|x_i - y_i|\right\}.$$

The notation $\mathbf{L}_\infty$ is well founded since $\lim_{k\rightarrow\infty} \mathbf{L}_k(x, y) = \mathbf{L}_\infty(x, y)$.

Note the two important special cases where $k = 1$ and $k = 2$:

- The **variation metric**: $\mathbf{L}_1(x, y) = \sum_{i \in [n]} |x_i - y_i|$, sometimes also called the **Manhattan metric**.
- The **Euclidean metric**: $\mathbf{L}_2(x, y) = \sqrt{\sum_{i \in [n]} (x_i - y_i)^2}$.

3.2 Alphabets, strings and languages

3.2.1 Alphabets

An ***alphabet*** Σ is a finite non-empty set of symbols called ***letters***. It should be noted that even if elementary examples are typically given on alphabets of size 2, the practical issues addressed by grammatical inference in fields like computational linguistics involve alphabets that can be huge, and in many cases of unknown and unbounded size.

3.2.2 Strings

A ***string*** x over Σ is a finite sequence $x = a_1 \cdots a_n$ of letters. Let $|x|$ denote the length of x. In this case we have $|x| = |a_1 \cdots a_n| = n$.

The ***empty string*** is denoted by λ (in certain books the notation ϵ is used for the empty string).

Alternatively a string x of length n can be used as a mapping $x : [n] \rightarrow \Sigma$: If $x = a_1 \cdots a_n$ we have $x(i) = a_i$, $\forall i \in [n]$.

Given $a \in \Sigma$, and x a string over Σ, $|x|_a$ denotes the number of occurrences of the letter a in x.

Given two strings u and v we will denote by $u \cdot v$ the concatenation of strings u and v: $u \cdot v$: $\big[|u|+|v|\big] \rightarrow \Sigma$ with $\forall i \in [|u|]$, $(u \cdot v)(i) = u(i)$ and $\forall i \in [|v|]$, $(u \cdot v)(|u|+i) = v(i)$. When the context allows it $u \cdot v$ shall be simply written uv.

We write $u^R = u(n) \cdots u(1)$ for the ***reversal*** of u.

Let $\Sigma^\star$ be the set of all finite strings over alphabet Σ. We also define:

$$\Sigma^+ = \{x \in \Sigma^\star : |x| > 0\}$$
$$\Sigma^{<n} = \{x \in \Sigma^\star : |x| < n\}$$
$$\Sigma^{\leq n} = \{x \in \Sigma^\star : |x| \leq n\}$$

Given a string x, u is a ***substring*** (or a ***factor***) of x if_{def} there are two strings l and r such that $x = lur$. In that case we will also say that x is a ***superstring*** of u.

We can count the number of occurrences of a given string u as a substring of a string x and denote this value by $|x|_u = \big|\{l \in \Sigma^\star : \exists r \in \Sigma^\star \wedge x = lur\}\big|$.

u is a ***subsequence*** of x if_{def} it can be obtained from x by erasing letters from x. Alternatively: $\forall x, y, z, x_1, x_2 \in \Sigma^\star, \forall a \in \Sigma$:

- x is a subsequence of x,
- x_1x_2 is a subsequence of x_1ax_2,
- if x is a subsequence of y and y is a subsequence of z then x is a subsequence of z.

Example 3.2.1 Consider x = abbababa; then abb is a prefix of x, and aba is a suffix of x. Both are substrings of x and so is bab. bbbb is a subsequence of x, but not a substring.

Note that:

- Finding the longest common subsequence between u and v is in $\mathcal{O}(|u| \cdot |v|)$ time.
- Finding the longest common subsequence of a set of strings is $\mathcal{NP}$-hard.
- Checking if string u is a subsequence (or substring) of string x requires $\mathcal{O}(|x|)$ time.

For $u, x \in \Sigma^\star$, u is a subsequence of x *if$_{def}$* there exist indices $\mathbf{i} = (i_1, \ldots i_{|\mathbf{i}|})$ with $1 \leq i_1 < \cdots < i_{|u|} \leq |x|$, such that $u_j = x_{i_j}$ for $j \in [|u|]$. If u is a subsequence of x in the positions given by $\mathbf{i}$, we will use $u = x(\mathbf{i})$ as notation. The length of the subsequence $l(\mathbf{i})$ is $i_{|u|} - i_1 + 1$, and corresponds not to the length of $x(\mathbf{i})$ but to the difference (+1) between the two extreme positions.

Example 3.2.2 Consider the alphabet $\Sigma = \{\texttt{a}, \texttt{b}, \texttt{c}\}$ and let x be the string abbabacac. We have $|x| = 9$. The string $x(2, 4, 6, 8) =$ baaa is a subsequence of length 4, but whose $l(2, 4, 6, 8)$ in x is 7.

3.2.3 *Ordering strings*

Suppose we have a total order relation over the letters of an alphabet Σ. We denote by $\leq_{alpha}$ this order, which is usually called the alphabetical order.

Different orders can be defined over $\Sigma^\star$:

- The ***prefix order***: $x \leq_{pref} y$ *if$_{def}$* $\exists w \in \Sigma^\star : y = xw$.
- The ***lexicographic order***: $x \leq_{lex} y$ *if$_{def}$* $x \leq_{pref} y \vee \big[x = uaw : y = ubz \wedge a \leq_{alpha} b\big]$.
- The ***subsequence order***: $x \leq_{subseq} y$ *if$_{def}$* x is a subsequence of y.

A more interesting order is the ***length-lexicographic*** order (also sometimes called the ***hierarchical*** or ***length-lex*** order). We will order the strings totally according to the hierarchical order by: if x and y belong to $\Sigma^\star$, $x \leq_{length\text{-}lex} y$ *if$_{def}$*$|x| < |y| \vee \big(|x| = |y| \wedge x \leq_{lex} y\big)$.

The first strings, according to the hierarchical order, with $\Sigma = \{\texttt{a}, \texttt{b}\}$ are λ, a, b, aa, ab, ba, bb, aaa, ...

In all cases we associate with each of these orders the *strict* orders $<_{alpha}$, $<_{pref}$, $<_{lex}$, and $<_{length\text{-}lex}$. We should pay special attention to the above definitions: although the lexicographic order seems the most natural, common and therefore interesting one, in practice it suffers from several flaws (see for example Exercise 3.5).

Example 3.2.3 Let $\Sigma = \{\texttt{a}, \texttt{b}, \texttt{c}\}$ with a $<_{alpha}$ b $<_{alpha}$ c. Then aab $\leq_{lex}$ ab, but ab $\leq_{lex\text{-}length}$ aab. And the two strings are incomparable for $\leq_{pref}$.

Note that, if $<_{alpha}$ is a total order, then so are $<_{lex}$ and $<_{length\text{-}lex}$.

3.2.4 *Languages*

A ***language*** is any set of strings, so therefore a subset of $\Sigma^\star$. Operations over languages include:

- Set operations (union and intersection)
- Product: $L_1 \cdot L_2 = \{uv : u \in L_1, v \in L_2\}$
- Powerset: $L^0 = \{\lambda\},\ L^{n+1} = L^n \cdot L = L \cdot L^n$
- Star: $L^* = \bigcup_{i \in \mathbb{N}} L^i$

The ***complement*** of language L is taken, if no precision is given, with respect to $\Sigma^\star$: $\overline{L} = \{w \in \Sigma^\star : w \notin L\}$. But notation $L_1 \setminus L_2$ will be used for the complement of L_2 in L_1, thus:

$$L_1 \setminus L_2 = L_1 \cap \overline{L_2} = \{x \in L_1 : x \notin L_2\}.$$

The ***symmetric difference*** between two languages L_1 and L_2 is denoted by $L_1 \oplus L_2 = L_1 \setminus L_2 \cup L_2 \setminus L_1 = \{x \in \Sigma^\star : (x \in L_1 \wedge x \notin L_2) \vee (x \in L_2 \wedge x \notin L_1)\}$.

Example 3.2.4 Let $\Sigma = \{\mathtt{a}, \mathtt{b}\}$. We consider languages $L_1 = \{\mathtt{a}, \mathtt{ab}\}$, $L_2 = \{\lambda, \mathtt{b}, \mathtt{bb}\}$ and $L_3 = \{\mathtt{a}, \mathtt{b}\}$. Then $L_1 \cdot L_2 = \{\mathtt{a}, \mathtt{ab}, \mathtt{abb}, \mathtt{abbb}\}$ and $L_2^* = \{\mathtt{b}^n : n \in \mathbb{N}\}$. Also $L_1 \setminus L_3 = \{\mathtt{ab}\}$ whereas $L_1 \oplus L_3 = \{\mathtt{ab}, \mathtt{b}\}$.

We denote by $\boldsymbol{\mathcal{L}}$ a class of languages. These will be indexed, in order to be well defined, by a particular alphabet. When not mentioned, we will suppose the size of this alphabet to be at least 2.

3.2.5 Prefixes, suffixes and quotients

Let $u, v \in \Sigma^*, u^{-1}v = w$ such that $v = uw$ (undefined if u is not a prefix of v) and $uv^{-1} = w$ such that $u = wv$ (undefined if v is not a suffix of u). Let L be a language and $u \in \Sigma^*$. Then $u^{-1}L = \{v \in \Sigma^* : uv \in L\}$ and $Lu^{-1} = \{v \in \Sigma^* : vu \in L\}$.

Generalising the above notation to languages, if L and M are languages, $L^{-1}M = \{v : \exists u \in L \wedge uv \in M\}$ and $LM^{-1} = \{v : \exists\, u \in M \wedge vu \in L\}$.

Let L be a language. The ***prefix set*** of L is $\text{PREF}(L) = \{u \in \Sigma^\star : uv \in L\}$ and the ***suffix set*** of L is $\text{SUFF}(L) = \{v \in \Sigma^\star : uv \in L\}$.

A language is ***prefix-closed*** if_{def} $\forall u, v \in \Sigma^\star, uv \in L \Rightarrow u \in L$. It is ***suffix-closed*** if_{def} $uv \in L \Rightarrow v \in L$.

The ***longest common suffix*** $\text{lcs}(L)$ of L is the longest string u such that each string in L accepts u as a prefix. Technically, $(Lu^{-1})u = L$. The ***longest common prefix*** $\text{lcp}(L)$ of L is the longest string u such that $u(u^{-1}L) = L$.

Example 3.2.5 Let $\Sigma = \{\mathtt{a}, \mathtt{b}\}$.
$\text{PREF}(\Sigma^\star) = \Sigma^\star$.
$\mathtt{ab}^{-1}\,\Sigma^\star = \Sigma^\star$.
Let $L = \{\mathtt{a}, \mathtt{aba}, \mathtt{abba}, \mathtt{bba}, \mathtt{bab}\}$ and $u = \mathtt{a}$. Then $u^{-1}L = \{\lambda, \mathtt{ba}, \mathtt{bba}\}$ and $Lu^{-1} = \{\lambda, \mathtt{ab}, \mathtt{bb}, \mathtt{abb}\}$.

3.3 Trees and terms

There are several ways of defining trees, depending on the point of view and the use: in graph theory trees are just a special case of graphs, in data structures trees will often be binary, and in formal language theory they are usually ranked. We choose here to introduce only ranked trees: the labels of the nodes are chosen from a ranked alphabet, each symbol having a unique arity, in which case the order of the subtrees matters.

Well-formed trees are built from a ranked alphabet Σ. In a ranked alphabet each f in Σ has an associated rank given by the arity function $\rho : \Sigma \to \mathbb{N}$. The maximum value the arity takes is denoted by *rmax*= $\max\{\rho(f) : f \in \Sigma\}$. This allows us to partition Σ into $\Sigma_0 \cup \Sigma_1 \cup \cdots \cup \Sigma_{rmax}$.

A Dewey tree (or empty tree, or tree domain) is a (finite) language (L_T) over an alphabet $[rmax]$, which verifies, for $\boldsymbol{u}, \boldsymbol{v} \in [rmax]^*$, $\boldsymbol{a}, \boldsymbol{b} \in [rmax]$:

- $\boldsymbol{u} \cdot \boldsymbol{v} \in L_T \implies \boldsymbol{u} \in L_T$;
- $\boldsymbol{u} \cdot \boldsymbol{a} \in L_T$ and $\boldsymbol{b} < \boldsymbol{a} \implies \boldsymbol{u} \cdot \boldsymbol{b} \in L_T$.

In order to better separate the addresses (elements of the Dewey tree) and the labels, the former are represented in bold characters: **112** is an address, whereas a is a label.

Definition 3.3.1 A **tree** t over a (ranked) alphabet Σ is a total mapping $L_T \to \Sigma$ where L_T is an empty tree and $\forall \boldsymbol{u} \in L_T$, $\left|\{\boldsymbol{a} \in \mathbb{N} : \boldsymbol{ua} \in L_T\}\right| = \rho(t(\boldsymbol{u}))$. L_T is called the **domain** of tree t and is denoted by $\mathrm{Dom}(t)$.

We use typical notations and terms for trees: nodes, leaves,... Examples are shown in Figures 3.1(a) and 3.1(b). In the first case, what is depicted is the skeleton with as labels, wrongfully, the addresses of the nodes.

Definition 3.3.2 A tree t is a **subtree** of t' at node $\boldsymbol{u}$ if_{def} $\forall \boldsymbol{w} \in \mathrm{Dom}(t)$, $t(\boldsymbol{w}) = t'(\boldsymbol{uw})$.

Definition 3.3.3 A **context** $C[\,]$ is a tree where exactly one leaf is labelled with a special symbol \$ that does not belong to Σ. If $C[\,]$ is a context and t is a tree, the tree $C[t]$ is

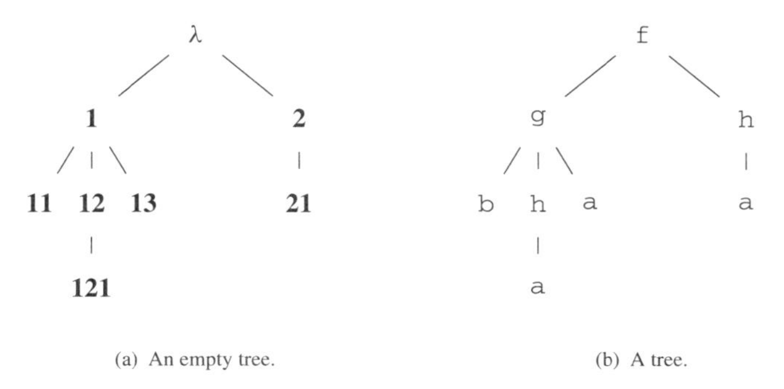

(a) An empty tree. (b) A tree.

Fig. 3.1. A tree over alphabet {a, b, f, g, h} with its tree domain.

obtained by substituting the node \$ by the tree t. Formally, $C[t](\boldsymbol{u}) = C[\,](\boldsymbol{u})$ if_{def} $\boldsymbol{u}$ belongs to Dom($C[\,]$), and $C[t](\boldsymbol{uv}) = t(\boldsymbol{v})$ if $C[\,](\boldsymbol{u}) =\$$.

Example 3.3.1

Other definitions are classical and are not recalled technically here:

- a ***node*** in the tree comprises an ***address*** and a ***label***;
- the ***root*** of a tree is the node of address λ;
- a ***leaf*** of t is a node whose address is not a proper prefix of another string in Dom(t);
- an ***internal node*** is a node that is not a leaf;
- the ***frontier*** of a tree is the string composed by concatenating all leaves of the tree when read in prefixial order;
- the ***height*** of a tree is the length of the longest string in the domain of the tree.

Definition 3.3.4 A tree t' is a **subtree** of t at node $\boldsymbol{u}$ if_{def} $\forall \boldsymbol{w} \in \text{Dom}(t), t'(\boldsymbol{w}) = t(\boldsymbol{uw})$.

In Example 3.3.1, tree t (Figure 3.2(b)) is a subtree of the tree represented in Figure 3.2(c).

3.3.1 Defining trees as strings

To manipulate trees there are a number of possibilities: one is to convert them to strings, and another is to transform them into binary trees, by means of the 'first son, right brother' encoding.

A linear description of a tree is obtained with the help of the following notation:

- a is a tree of height 0 (just a leaf, labelled by a);
- $f(t_1 \dots t_k)$ is a tree with root labelled by f and with k subtrees $t_1 \dots t_k$.

For example tree (a) from Figure 3.3 will be denoted by f(g(a(h(b) g(a b)))).

Another possibility is to describe a tree corresponding to a bracketed string. If you consider string (b(a(a b))), then an alternative representation is given in Figure 3.3(b). We will prefer the first encoding.

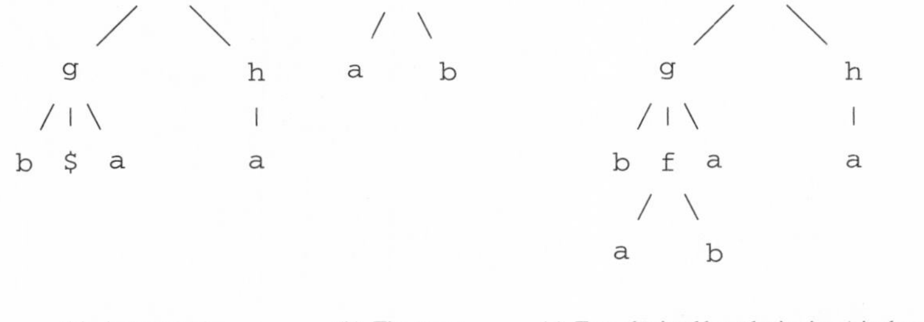

(a) A context C. (b) The tree t. (c) Tree obtained by substituting \$ in the context C by the tree t.

Fig. 3.2. Contexts and substitutions.

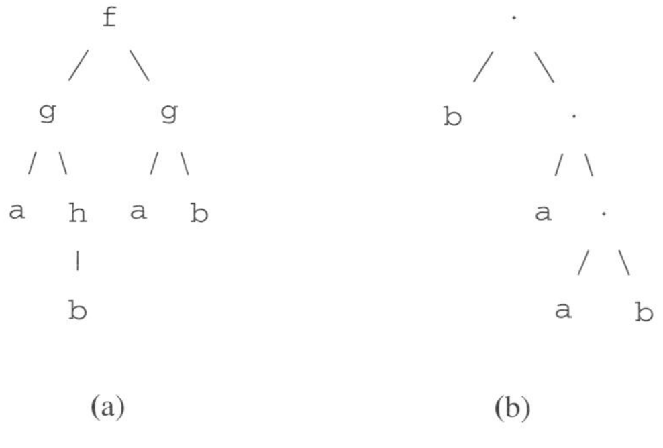

(a) (b)

Fig. 3.3. Trees.

3.3.2 Associating binary trees to trees

There are many reasons for limiting oneself to binary trees: notations are simpler, and programs may seem fitted to that case. It is therefore of interest to be able to transform general trees into binary ones. This is always possible through the following operation:

$t \to t'$: $\forall \boldsymbol{u} \in \mathrm{Dom}(t)$

$$\begin{aligned}\phi(\lambda) &= \lambda\\ \phi(\mathbf{0}\boldsymbol{u}) &= \mathbf{0}\phi(\boldsymbol{u})\\ \phi(\boldsymbol{a}\boldsymbol{u}) &= \mathbf{1}^{\boldsymbol{a}}\phi(\boldsymbol{u})\\ t'(\boldsymbol{u}) &= t(\phi(\boldsymbol{u}))\end{aligned}$$

Note that ψ is the inverse function which transforms a string in $\mathrm{Dom}(t')$ into the corresponding string $t \to t'$: $\forall \boldsymbol{u} \in \mathrm{Dom}(t)$

$$\begin{aligned}\psi(\lambda) &= \lambda\\ \psi(\mathbf{1}^{\boldsymbol{n}}\mathbf{0}\boldsymbol{u}) &= \boldsymbol{n}\psi(\boldsymbol{u})\\ t(\boldsymbol{u}) &= t'(\psi(\boldsymbol{u})).\end{aligned}$$

3.3.3 Extensions

Extensions of strings and trees include:

- Graphs and hypergraphs are even more complex to define by generative means and grammars that produce them have been studied in specific fields only. Typically a graph is composed of a finite set of nodes or vertices X and a subset of X^2 of edges, but there are variants where the graphs can be undirected or have multiple edges or hyper-edges.
- Infinite strings are used to model situations in reactive systems. We denote by Σ^∞ the set of all infinite strings defined as total functions $\mathbb{N} \to \Sigma$. If L is a subset of Σ^∞ it will be an ***infinitary*** language. In the case where L is an infinitary language, the set of finite prefixes of L is $\mathrm{PREF}(L) = \{x \in \Sigma^\star : xy \in L\}$.

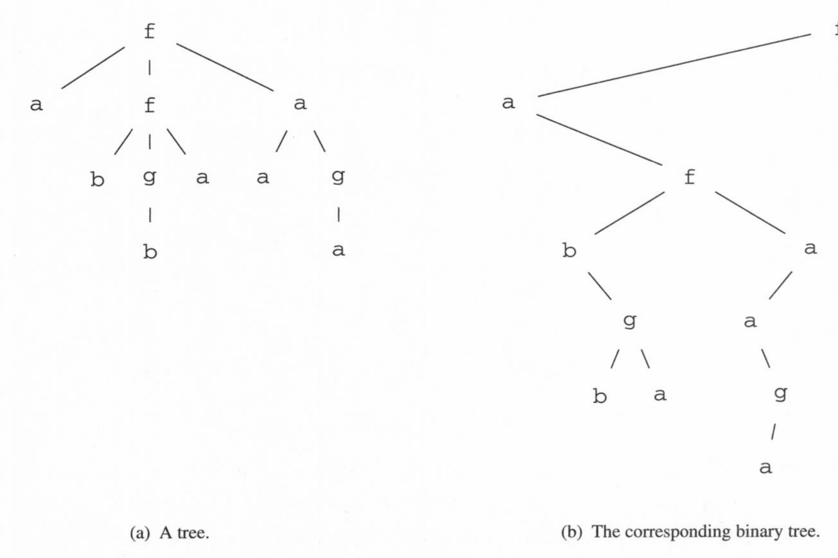

(a) A tree.

(b) The corresponding binary tree.

Fig. 3.4. Trees.

3.4 Distances between strings

There are many good reasons to define distances between strings. One may want to capture the notion that some string is only a slight deformation of another, or want to use this distance for classification purposes: given a set of strings, find some representative of this set, i.e. a string to which all the other ones are close. We survey here some of these distances and discuss their tractability.

Distances over strings can be defined in a number of ways, either by selecting features which are then measured, and using a distance over $\mathbb{R}^n$, or by computing directly the distance over the strings themselves.

3.4.1 Feature distances

A typical (and very successful) approach is to transform a string into a vector in a (usually high-dimensional) space by extracting some features (or measurable information) about the string. You can then use a conventional numerical distance over the vectors. The construction goes as follows:

- Find a finite set of measurable features.
- Compute a numerical vector for x and y ($\overrightarrow{x}$ and $\overrightarrow{y}$). These vectors are elements of some set $\mathbb{R}^n$.
- Use some distance $d_{\rightarrow}$ over $\mathbb{R}^n$. Then $d(x, y) = d_{\rightarrow}(\overrightarrow{x}, \overrightarrow{y})$.

Example 3.4.1 Let $|\Sigma| = n$. Suppose we order the symbols in Σ as $\mathtt{a}_1, \mathtt{a}_2, \ldots, \mathtt{a}_k$. With each string x over Σ we associate the vector that counts the number of occurrences of each symbol in x: $\overrightarrow{x} = [|x|_{a_1}, |x|_{a_2}, \dots, |x|_{a_k}]$. This is known as the Parikh mapping. With this, it is possible to use any distance over $\mathbb{R}^k$ to compare strings.

An alternative is, using the fact that the number of strings is infinite, to do the same but in an infinite dimension:

- Find an infinite (enumerable) set of measurable features.
- Compute a numerical vector for x and y ($\overrightarrow{x}$ and $\overrightarrow{y}$). These vectors are elements of $\mathbb{R}^\infty$.
- Use some distance $d_\rightarrow$ over $\mathbb{R}^\infty$.
- $d(x, y) = d_\rightarrow(\vec{x}, \vec{y})$.

The difference between the two cases is that the set of values associated with a string is usually unbounded, but not infinite (only a finite number of values is going to be used for the computation of the distance).

Example 3.4.2 For strings x and u in $\Sigma^\star$ denote by $|x|_u$ the number of occurrences of string u as a substring of string x. Now order all strings of $\Sigma^\star$ using the hierarchical order. We can now associate with each string x the infinite vector $\vec{x}$ where the coordinate corresponding to u is $|x|_u$.

For instance, string ababaa is represented by $(7, 4, 2, 2, 2, 0, 0, 0, 2, 0, \dots)$. Note that the value of $|x|_\lambda$ is arbitrarily set to 1.

We now can use the $\mathbf{L}_\infty$ distance. Thus, in this case, $d(\texttt{ababaa}, \texttt{bba}) = 3$.

3.4.2 Modification distances

This class of distances is broadly defined as follows:

- Compute the number of modifications of some type allowing us to change string x to string y.
- Perhaps *normalise* this distance according to the sizes of x and y or to the number of possible paths.

Typically, the edit distance (see Section 3.4.4) is of this sort, but when the strings are of identical length, the Hamming distance is also suitable:

Definition 3.4.1 (Hamming distance) Let x and y be two strings of identical length.

$$d_{Hamming}(x, y) = \sum_{i \in [|x|]} \left|\{i : x(i) \neq y(i)\}\right|$$

Example 3.4.3 $d_{Hamming}(\texttt{aaba}, \texttt{abab}) = 3$. Notice that a simple 'rotation' of the string (the first symbol becomes the last) can have dramatic effects: $d_{Hamming}(\texttt{ababababab}, \texttt{bababababa}) = 10$. A rotation corresponds to imagining that a string is circular, and that then, somehow, the first symbol in the string becomes the last.

3.4.3 Similarity distances

Another type of distance can be obtained from similarities instead of from differences. The general idea goes as follows:

We first compute a similarity between x and y. This will be a positive value $S(x, y) \in \mathbb{R}^+$. The idea is that the higher the value, the closer the strings.

We then have at least two options to build a distance:

(i) • If $x = y$, then $d(x, y) = 0$.
• If $x \neq y$, then $d(x, y) = 2^{-S(x,y)}$.

This typically is the case for the ***prefix distance***, or the distance on trees defined by the following similarity:

$$S(t_1, t_2) = \min\{|x| : x \in \mathrm{Dom}(t_1) \cap \mathrm{Dom}(t_2) \wedge t_1(x) \neq t_2(x)\}$$

(ii) The second idea is to define $d(x, y) = S(x, x)^2 - 2S(x, y)S(y, x) + S(y, y)^2$. This sort of distance is typically used when we want to transform a kernel into a distance (see Section 3.5, page 60 about kernels).

3.4.4 The edit distance

Defined by Levenshtein in 1966, there have been many variants, studies and extensions since then. The idea is to define some simple operations to get from one string to the other. The basic operations are those corresponding to typographic mistakes (insertion, deletion and substitution) and we measure the minimum number of these operations needed to transform one string into another.

Definition 3.4.2 Given two strings x and y in $\Sigma^\star$, x rewrites into y in one step if_{def} one of the following correction rules holds:

- $x = uav,\ y = uv$ and $u, v \in \Sigma^\star, a \in \Sigma$ (single-symbol deletion);
- $x = uv,\ y = uav$ and $u, v \in \Sigma^\star, a \in \Sigma$ (single-symbol insertion);
- $x = uav,\ y = ubv$ and $u, v \in \Sigma^\star, a, b \in \Sigma$, (single-symbol substitution).

Example 3.4.4
- abc $\rightarrow$ ac (one deletion);
- ac $\rightarrow$ abc (one insertion);
- abc $\rightarrow$ aec (one substitution).

We will consider the reflexive and transitive closure of this derivation, and $x \xrightarrow{k} y$ if_{def} x rewrites into y by k operations of single-symbol deletion, single-symbol insertion and single-symbol substitution.

Given two strings x and y, the Levenshtein or edit distance between x and y denoted by $d_{edit}(x, y)$ is the smallest k such that $x \xrightarrow{k} y$.

Table 3.1. *A cost matrix.*

	λ	a	b
λ	0	0.5	0.7
a	0.5	0	1.3
b	0.7	1.3	0

Example 3.4.5 $d_{edit}(\texttt{abaa}, \texttt{aab}) = 2$. abaa rewrites into aab via (for instance) a deletion of the 'b' and a substitution of the last 'a' by a 'b'.

In the most general case the cost of each operation may not be 1. The cost is then given by a ***cost matrix***. An example of such a matrix can be found in Table 3.1. Cost matrices are an important part of the edit distance: if one does understand the point of counting the number of errors in a text, clearly, in many applications, some errors are more probable (and thus have less cost) that others. For instance, in biology, certain mutations are easier, and in copying a text one might make more errors through using one key instead of its neighbour on the keyboard.

Certain conditions have to be met for the edit distance with costs still to be a metric: the cost matrix has itself to be symmetric.

With the matrix in Table 3.1 this is the case, but the cost matrix does not respect the triangular inequality as $C[\texttt{b}][\texttt{a}] = 1.3 > C[\texttt{a}][\lambda] + C[\lambda][\texttt{b}] = 0.5 + 0.7$. We leave the consequences of this remark as an exercise (Exercise 3.8, page 66).

General algorithm. The definition of the edit distance does not provide us with a direct algorithm. Indeed using something like 'recursively trying out all possible operations' will end up with an exponential amount of work.

The correct (yet inefficient) formula is:

$$d_{edit}(xa, yb) = \min \begin{cases} d_{edit}(xa, y) + C[\lambda][b] \\ d_{edit}(x, y) + C[a][b] \\ d_{edit}(x, yb) + C[a][\lambda] \end{cases}$$

In the above, $C[a][b]$ refers to the cost of substituting symbol a by symbol b, $C[a][\lambda]$ to the cost of erasing symbol a, and $C[\lambda][b]$ to the cost of inserting symbol b. One way to compute this efficiently is through dynamic programming. This is what Algorithm 3.1 does.

We show in Table 3.2 an example computation of the edit distance between strings abbaa and abaacabb for the unit cost matrix (all operations are worth 1). It should be noticed that the same table can easily be interpreted to find the number of optimal alignments or paths from one string to the other, or to explore the different optimal paths.

Algorithm 3.1: Computing the edit distance.

Data: A cost matrix $C[|\Sigma|+1][|\Sigma|+1]$, two strings $x = a_1 \cdots a_m$ and $y = b_1 \cdots b_n$
Result: $d[m][n] = d_{edit}(x, y)$
$d[0][0] \leftarrow 0$;
for $i : 1 \leq i \leq m$ **do** $d[i][0] \leftarrow d[i-1][0] + C[a_i][\lambda]$;
for $j : 1 \leq j \leq n$ **do** $d[0][j] \leftarrow d[0][j-1] + C[\lambda][b_j]$;
for $i : 1 \leq i \leq m$ **do**
  for $j : 1 \leq j \leq n$ **do** $d[i][j] \leftarrow$
  $\min\big(d[i-1][j] + C[a_i][\lambda],\ d[i][j-1] + C[\lambda][b_j],\ d[i-1][j-1] + C[a_i][b_j]\big)$
end
return $d[m][n]$

Table 3.2. *An example computation of the edit distance for the unitary cost matrix.*

a	5	4	3	2	1	2	3	4	5
a	4	3	2	1	2	3	4	5	6
b	3	2	1	2	3	4	5	5	5
b	2	1	1	2	3	4	5	5	6
a	1	1	2	2	3	4	5	6	7
λ	0	1	2	3	4	5	6	7	8
	λ	b	b	a	a	c	a	b	b

Complexity. Time and space of Algorithm 3.1 are in $\mathcal{O}(|x|.|y|)$. But an alternative implementation by one-dimensional vectors is possible, reducing the space complexity to $\mathcal{O}(min(|x|, |y|))$.

Extensions. The general idea behind the computation of the edit distance can be adapted to other cases and settings. This may be of interest in particular applications. But in all these cases the problems of the distance (the fact that it is not easy to compute and in the case of very long strings it actually becomes extremely complex) remain. Some of the extensions are as follows:

- We can add other operations such as the inversion of two contiguous symbols: $uabv \rightarrow ubav$; usually the algorithms and results can be adapted.
- Similar algorithms can work (with some extra cost) on circular strings: circular strings can be used in pattern analysis to represent contours.
- We can compute how far a string is from a set or a language. The definition would then be: $d_{edit}(x, L) = \min\{d_{edit}(x, y) : y \in L\}$. The next step consists of defining the distance between two languages: $d_{edit}(L_1, L_2) = \min\{d(x, y) : x \in L_1 \wedge y \in L_2\}$.
- In practice there are arguments in favour of using a normalised distance and being able to say that string aaabababa**a**bababba is closer to aaabababa**b**bababba than a is to b. That is

	C	S	T	P	A	G	N	D	E	Q	H	R	K	M	I	L	V	F	Y	W
C	9																			
S	-1	4																		
T	-1	1	5																	
P	-3	-1	-1	7																
A	0	1	0	-1	4															
G	-3	0	-2	-2	0	6														
N	-3	1	0	-2	-2	0	6													
D	-3	0	-1	-1	-2	-1	1	6												
E	-4	0	-1	-1	-1	-2	0	2	5											
Q	-3	0	-1	-1	-1	-2	0	0	2	5										
H	-3	-1	-2	-2	-2	-2	1	-1	0	0	8									
R	-3	-1	-1	-2	-1	-2	0	-2	0	1	0	5								
K	-3	0	-1	-1	-1	-2	0	-1	1	1	-1	2	5							
M	-1	-1	-1	-2	-1	-3	-2	-3	-2	0	-2	-1	-1	5						
I	-1	-2	-1	-3	-1	-4	-3	-3	-3	-3	-3	-3	-3	1	4					
L	-1	-2	-1	-3	-1	-4	-3	-4	-3	-2	-3	-2	-2	2	2	4				
V	-1	-2	0	-2	0	-3	-3	-3	-2	-2	-3	-3	-2	1	3	1	4			
F	-2	-2	-2	-4	-2	-3	-3	-3	-3	-3	-1	-3	-3	0	0	0	-1	6		
Y	-2	-2	-2	-3	-2	-3	-2	-3	-2	-1	2	-2	-2	-1	-1	-1	-1	3	7	
W	-2	-3	-2	-4	-3	-2	-4	-4	-3	-2	-2	-3	-3	-1	-3	-2	-3	1	2	11

Fig. 3.5. Typical cost matrix used for proteins.

why, in many applications variants of this normalised distance are used. But these normalised distances are problematic: note for instance that if normalising by dividing by the sum of lengths $d_N(x, y) = \frac{d_{edit}(x,y)}{|x|+|y|}$ you end up with something that is not a metric:

- $d_N(\mathtt{ab}, \mathtt{aba}) = 0.2$
- $d_N(\mathtt{aba}, \mathtt{ba}) = 0.2$
- $d_N(\mathtt{ab}, \mathtt{ba}) = 0.5$

Therefore we have $d_N(\mathtt{ab}, \mathtt{ba}) > d_N(\mathtt{ab}, \mathtt{aba}) + d_N(\mathtt{aba}, \mathtt{ba})$ and the triangular inequality no longer holds.

- Extending the edit distance to trees and graphs is not straightforward.

Other questions related to the distance are of interest when trying to define, from a given set of strings, a specific string as representative of the set. We could in this case consider computing:

- the string median: $u = \text{argmin}_{c\in\Sigma^\star}\{\sum_{x\in S} d_{edit}(c, x)\}$ or
- the string centre: $u = \text{argmin}_{c\in\Sigma^\star}\{\max_{x\in S} d_{edit}(c, x)\}$.

But finding any of these values is intractable and corresponds to attempting to solve $\mathcal{NP}$-hard problems.

It should be added that if the edit distance can (with many technical complications) extend to trees, the extension to graphs poses a problem. Indeed, if we want it to be a metric, then it would enable us to solve the graph isomorphism problem, which is believed not to be in $\mathcal{P}$. That is why the computation of the edit distance for graphs is only done by heuristics.

3.5 String kernels

Kernels provide us with another way to associate a numerical value to a pair of objects (here strings). In a certain sense, the value associated does not indicate a closeness between the objects, but how much they share.

A kernel is a function κ: $A \times A \rightarrow \mathbb{R}$ such that there exists a feature mapping ϕ: $A \rightarrow \mathbb{R}^n$, and $\kappa(x, y) =< \phi(x), \phi(y) >$.

$< \phi(x), \phi(y) >$ represents the dot product (also known as the scalar product): $< \phi(x), \phi(y) >= \phi_1(x) \cdot \phi_1(y) + \phi_2(x) \cdot \phi_2(y) + \ldots + \phi_n(x) \cdot \phi_n(y)$.

Some important points that we should make are:

- The κ function is explicit, whereas the feature mapping ϕ may only be implicit. This means that we need an algorithm to compute κ, whereas this is not necessary for the feature mapping.
- Instead of $\mathbb{R}^n$ some other Hilbert space can be used.
- If we build the kernel directly from the feature mapping ϕ, this one respects the kernel conditions (known as Mercer conditions: the associated matrix should be semi-definite positive).
- If a semantic can be attached to the function ϕ, then interpretation of whatever learning results we obtain is possible. If not, in general, we are left with a *black box*.
- Another essential issue is algorithmic: the computation of $\kappa(x, y)$ must be cheap since this is going to be repeated many times. A typical complexity of $\mathcal{O}(|x|+|y|)$ or $\mathcal{O}(|x|\cdot|y|)$ is considered correct in most cases.

  Notice that this point is not as trivial as it may seem since the actual definition is $\kappa(x, y) = \sum_{i\in[n]} = \phi_i(x)\phi_i(y)$, and we do not want this value n to intervene in the complexity. Actually, for some kernels, n can be infinite! Avoiding computing the kernel through the feature function is called the *kernel trick*.

3.5.1 *Basic definitions*

We give the elements necessary to construct string kernels, used to compare strings in $\Sigma^\star$. Adaptations to trees are sometimes possible.

Definition 3.5.1 (String kernel function) A **string kernel** is a function κ that, for all $x, y \in \Sigma^\star$, satisfies

$$\kappa(x, y) = \langle\phi(x), \phi(y)\rangle,$$

where ϕ is a mapping from Σ^* to an (inner product) feature space F

$$\phi : x \rightarrow \phi(x) \in F.$$

Note that kernels can be combined:

Proposition 3.5.1 *Let κ_1 and κ_2 be kernels. Then the following function is a kernel:*

$$\kappa(x, y) = \kappa_1(x, y) + \kappa_2(x, y).$$

We now introduce some string kernels.

3.5.2 The Parikh kernel

The Parikh map associates to each string a vector of natural numbers where each component is the number of occurrences of a symbol of the alphabet in the string.

Definition 3.5.2 (Parikh kernel) The feature space associated with the Parikh kernel is indexed by Σ, with the feature mapping:

$$\phi_a^P(x) = |\{i : x(i) = a\}| = |w|_a, a \in \Sigma.$$

The associated kernel is defined as

$$\kappa^P(x, y) = \langle \phi^P(x), \phi^P(y) \rangle = \sum_{a \in \Sigma} \phi_a^P(x) \phi_a^P(y).$$

Example 3.5.1 Consider the strings bac, baa, cab and bad. Their Parikh kernels are given in the following table:

ϕ^P	a	b	c	d
bac	1	1	1	0
baa	2	1	0	0
cab	1	1	1	0
bad	1	1	0	1

So the Parikh kernel between bad and cab has value

$$\kappa^P(\texttt{bad}, \texttt{cab}) = 1 * 1 + 1 * 1 + 0 * 1 + 1 * 0 = 2.$$

3.5.3 The spectrum kernel

This corresponds to counting contiguous subsequences, or substrings, of a given length. The idea is similar to that of the n-grams or the k-testable languages (these are presented in Section 11.1, page 217).

Definition 3.5.3 (Spectrum kernel) The feature space associated with the spectrum kernel is indexed by Σ^k, with the feature mapping:

$$\phi_u^{S,k}(x) = |x|_u, u \in \Sigma^k.$$

The associated kernel is defined as

$$\kappa^{S,k}(x, y) = \langle \phi^{S,k}(x), \phi^{S,k}(y) \rangle = \sum_{a \in \Sigma} \phi_u^{S,k}(x) \phi_u^{S,k}(y).$$

Note that of course for k=1 we obtain the Parikh kernel. As an example $\kappa^{S,2}(\texttt{aabaaac}, \texttt{baaa}) = 8 + 1$ (for $k = 2$).
Complexity of the computation of $\kappa^{S,k}(x, y)$ is in $\mathcal{O}(k \cdot |x| \cdot |y|)$.

3.5.4 The all k-subsequences kernel

The ***all k-subsequences*** kernel computes the dot product of two vectors of which each component is the number of occurrences of contiguous or non-contiguous subsequences of length at most k in a string.

We consider all subsequences of length at most k. The goal of this kernel is to grow with the number of shared subsequences of bounded length. Alternatively, one counts all the alignments of at most k symbols between the two strings.

Definition 3.5.4 (All k-subsequences kernel) The feature space associated with the embedding of the all k-subsequences kernel is indexed by $I = \Sigma^{\leq k}$, with the component labelled by u given by

$$\phi_u^{U,k}(x) = |\{\mathbf{i} : u = x(\mathbf{i})\}|, u \in I,$$

that is, the count of the number of times the indexing string u occurs as a subsequence in the string x. The associated kernel is defined by:

$$\kappa^{U,k}(x, y) = \langle \phi^{U,k}(x), \phi^{U,k}(y) \rangle = \sum_{u \in \Sigma^{\leq k}} \phi_u^{U,k}(x)\phi_u^{U,k}(y).$$

We will also write: $\forall x \in \Sigma^\star$, $\phi_\lambda^{U,k}(x) = 1$.

Example 3.5.2 All the 2-subsequences in the strings bac, baa, cab and bad are given in the following table:

$\phi^{U,2}$	λ	a	b	c	d	aa	ab	ac	ad	ba	bc	bd	ca	cb
bac	1	1	1	1	0	0	0	1	0	1	1	0	0	0
baa	1	2	1	0	0	1	0	0	0	2	0	0	0	0
cab	1	1	1	1	0	0	1	0	0	0	0	0	1	1
bad	1	1	1	0	1	0	0	0	1	1	0	1	0	0

with all other dimensions indexed by other strings of length 2 having value zero. So the all 2-subsequences kernel between baa and bad has value

$$\kappa^{U,2}(\text{baa}, \text{bad}) = 6.$$

3.5.5 The all-subsequences kernel

We now extend the previous kernel to the case where we do not bound *a priori* the size of the subsequences. The ***all-subsequences*** kernel associates to each string a vector where each component is the number of occurrences of contiguous or non-contiguous subsequences in a string.

This also corresponds to counting the number of alignments between two strings.

Definition 3.5.5 (All-subsequences kernel) The feature space associated with the embedding of the all-subsequences kernel is indexed by $I = \Sigma^\star$, with the component labelled by u given by

$$\phi_u^U(x) = |\{\mathbf{i} : u = x(\mathbf{i})\}|, u \in I,$$

that is, the count of the number of times the indexing string u occurs as a subsequence in the string x. The associated kernel is defined by:

$$\kappa^U(x, y) = \langle \phi^U(x), \phi^U(y) \rangle = \sum_{u \in \Sigma^\star} \phi_u^U(x)\phi_u^U(y).$$

Note that the number of empty alignments is 1, so $\forall x \in \Sigma^\star$, $\phi_\lambda^U(x) = 1$.

Example 3.5.3 All the subsequences in the strings bac, baa, cab are given in the following table:

ϕ^U	λ	a	b	c	aa	ab	ac	ba	bc	ca	cb	bac	baa	cab
bac	1	1	1	1	0	0	1	1	1	0	0	1	0	0
baa	1	2	1	0	1	0	0	2	0	0	0	0	1	0
cab	1	1	1	1	0	1	0	0	0	1	1	0	0	1

with all other dimensions indexed by the (infinitely many!) other strings of $\Sigma^\star$ having value zero.
So the all-subsequences kernel values for the pairs baa and bac, and aaba and abac, are

$$\kappa^U(\text{baa}, \text{bac}) = 1{+}3{+}2{=}6$$
$$\kappa^U(\text{aaba}, \text{abac}) = 1 + 7 + 6 + 2 = 16.$$

3.5.6 *The gap-weighted subsequences kernel*

The idea of the ***gap-weighted subsequences*** kernel is to associate to each string a vector where each component is a weight which measures the degree of contiguity of the subsequence of length k (k fixed) in the string. The goal is to penalise those alignments that use letters far apart.

Definition 3.5.6 (Gap-weighted subsequences kernel) We consider all contiguous and non-contiguous subsequences of length k where the gaps are weighted by γ. The feature space associated with the gap-weighted subsequences kernel of length k is indexed by $I = \Sigma^k$, with the embedding given by

$$\phi_u^G(x) = \sum_{\mathbf{i}:\, u = x(\mathbf{i})} \gamma^{l(\mathbf{i})}, u \in \Sigma^k.$$

The associated kernel is defined as

$$\kappa^{G,k}(x, y) = \langle \phi^{G,k}(x), \phi^{G,k}(y) \rangle = \sum_{u \in \Sigma^k} \phi_u^{G,k}(x) \phi_u^{G,k}(y).$$

Example 3.5.4 Consider the simple strings `bac`, `baa`, `cab` and `bad`. Fixing $k = 2$, the strings are mapped as follows:

$\phi^{G,2}$	aa	ab	ac	ad	ba	bc	bd	ca	cb
bac	0	0	γ^2	0	γ^2	γ^3	0	0	0
baa	γ^2	0	0	0	$\gamma^2 + \gamma^3$	0	0	0	0
cab	0	γ^2	0	0	0	0	0	γ^2	γ^3
bad	0	0	0	γ^2	γ^2	0	γ^3	0	0

with all other coordinates having value zero.

So the gap-weighted subsequences kernel between `baa` and `bad` has value

$$\kappa^{G,2}(\texttt{baa}, \texttt{bad}) = \gamma^4 + \gamma^5.$$

If practically the idea is that the value of γ is less than 1, for theoretical reasons an interesting value for γ is 2. Complexity of the computation of this kernel is $\mathcal{O}(k \cdot |x| \cdot |y|)$.

3.5.7 *How do we compute a kernel?*

The goal in most cases is obviously to avoid computing all the $\phi_u(x)$ and then using the dot product. This requires computing the kernel directly. Dynamic programming is typically used for this. Algorithms for the different kernels presented in this section can be found in the literature. Just to give the spirit, we present in Algorithm 3.2 the computation of the all-subsequences kernel.

Note that $\kappa(a_1 \cdots a_m, b_1 \cdots b_n)$ counts the number of good alignments between $x = a_1 \cdots a_m$ and $y = b_1 \cdots b_n$. And this count can be decomposed into on one hand those alignments where a_m is not used (this is then $\kappa(a_1 \cdots a_{m-1}, b_1 \cdots b_n)$) and on the other those alignments where a_m is used: in this case a_m is aligned with one of $b_1 \cdots b_n$ (for example b_j with $b_j = a_m$). We count this in an auxiliary variable, denoted by Aux[j].

3.6 Some simple classes of languages

Languages are sets of strings and can be defined in many ways. In the next chapter we will use grammars and automata to generate or recognise strings, but it is possible to define simple classes of languages through topological operations.

$\mathcal{FIN}(\Sigma)$ is the class of all finite languages over an alphabet Σ.

Algorithm 3.2: Computing the all-subsequences kernel.

```
Input: a_1 ··· a_m, b_1 ··· b_n
Output: K[m][n] = κ(a_1 ··· a_m, b_1 ··· b_n)
for j ∈ [n] do                                   /* Only the empty alignment */
 | K[0][j] ← 1
end
for i ∈ [m] do
 | last ← 0;
 | Aux[0] ← 0;
 | for j ∈ [n] do                                 /* a_1 ··· a_i, b_1 ··· b_j */
 |  | Aux[j] ← Aux[last]
 | end
 | if a_i = b_j then Aux[j] ←Aux[last]+K[i − 1][j − 1];
 | last ← j;
 | for j ∈ [n] do K[i][j] ← K[i − 1][j]+Aux[j]
end
return K[m][n]
```

3.6.1 Balls

Let u be a string over an alphabet Σ. $B_r(u) = \{x \in \Sigma^\star : d_{edit}(u, x) \leq r\}$ is a ***ball of strings***. $r \in \mathbb{N}$ is the ***radius*** of the ball, and u is the ***centre*** of the ball.

We denote by $\mathcal{BALL}(\Sigma)$ the class of all balls over an alphabet Σ. Those balls that have the radius at most as long as the length of the centre are called ***good balls***. The class of all good balls over an alphabet Σ is denoted by $\mathcal{GB}(\Sigma)$.

3.6.2 Cones

Let u be a string over an alphabet Σ. $K(u) = \{x \in \Sigma^\star : x \leq_{subseq} u\}$ is a **cone**.

We denote by $\mathcal{CONE}(\Sigma)$ the class of all cones over an alphabet Σ.

3.6.3 Co-cones

Let u be a string over an alphabet Σ. $KK(u) = \{x \in \Sigma^\star : u \leq_{subseq} x\}$ is a **cocone**.

We denote by $\mathcal{COCONE}(\Sigma)$ the class of all cocones over an alphabet Σ.

3.7 Exercises

3.1 Prove that if L is a finite set, then so is PREF(L).

3.2 Compute $u^{-1}L$ and Lu^{-1} for:

- $L = \{\texttt{a}, \texttt{aba}, \texttt{abba}\}$ and $u = \texttt{a}$,
- $L = \{\texttt{a}, \texttt{aba}, \texttt{abba}\}$ and $u = \texttt{ab}$,
- $L = \Sigma^\star$ and $u = \texttt{a}$,
- $L = \Sigma^\star$ and $u = \lambda$.

3.3 Order the following strings for the orders $\leq_{pref}$, $\trianglelefteq_{lex}$, $\preceq_{subseq}$ and $\leq_{lex\text{-}length}$: aabab, bbabacaa, cc, acabaab, baaa.

3.4 Compute the Hamming distance (when possible) and the edit distance (with unit costs) between the three following strings: ababa, baaaba and babab.

3.5 Prove that the lexicographic order is not Noetherian, i.e. that you can construct a non-empty set with no minimal element. Alternatively, that there is a strictly decreasing infinite sequence.

3.6 Let $\Sigma = \{\texttt{a}, \texttt{b}\}$, prove that $\leq_{lex\text{-}length}$ is a good ordering, i.e. that every non-empty set has a minimal element.

3.7 Prove that the induction principle holds for $\leq_{lex\text{-}length}$, i.e. that given any property P, if we can prove $\forall y\big[\forall x \leq_{lex\text{-}length} y \; P(x) \implies P(y)\big]$ then the property P is true on every element of $\Sigma^\star$.

3.8 Prove that if the cost matrix does not respect the triangular inequality, Algorithm 3.1 (page 58) may not return the correct value.

3.9 Give an example of a distance over $\Sigma^\star$ for which the triangular inequality does not hold.

3.10 Prove that the Hamming distance is indeed a metric over each Σ^n.

3.11 Suppose we normalise the Hamming distance by dividing the distance by the length of x (and with $d(\lambda, \lambda) = 0$). Is this still a metric?

3.12 Extend the Hamming distance to $\Sigma^\star$ by first extending it to the alphabet $(\Sigma \cup \{\$\})^\infty$ as follows: given any strings u and v in $\Sigma^\star$, complete u and v to the right by an infinity of symbols \$ and 'compute' the Hamming distance between these two strings. Is this a metric?

3.13 Prove that the prefix distance is indeed a metric. The prefix distance between two strings is defined as

- if $x = y$, then $d_{pref}(x, y) = 0$,
- if $x \neq y$ $d_{pref}(x, y) = 2^{|\text{lcp}(x,y)|}$.

Recall that the lcp of a set is the longest common prefix of the set.

3.14 Prove that the prefix distance is an ultra-metric, i.e. that the following conditions hold:

- $\forall x, y \in \Sigma^\star, d_{pref}(x, y) \leq 1$,
- $\forall x, y, z \in \Sigma^\star, d_{pref}(x, z) \leq \max\{d_{pref}(x, y), d_{pref}(y, z)\}$.

3.15 Suppose we normalise the prefix distance by dividing the distance between x and y by $|x| + |y|$ (and with $d(\lambda, \lambda) = 0$). Is this still a metric?

3.16 Give an example of a distance over $\Sigma^\star$ which uses substrings. Is it a metric?

3.17 Consider using the edit distance normalised by dividing by the length of the smallest of the two strings: $d(x, y) = \frac{d_{edit}(x,y)}{1+\min\{|x|,|y|\}}$. Is this a metric?

3.18 Another possible idea to take into account the lengths is to divide by the product of both lengths. In order to deal with the special case where one of the strings is the empty string we have to add one. $d(x, y) = \frac{d_{edit}(x,y)}{1+|x||y|}$. Is this a metric?

3.19 Build an algorithm that computes the longest common subsequence of two strings u and v, and modify this algorithm to check if the longest common sequence is unique or not.

3.8 Conclusions of the chapter and further reading

3.8.1 Bibliographical background

Most of the definitions and techniques from this chapter can be found in typical textbooks for formal language theory (Harrison, 1978, Hopcroft & Ullman, 1979, Simon, 1999, Sakarovich, 2004). More specifically, the book by Maxime Crochemore *et al.* (Crochemore, Hancart & Lecroq, 2007) offers most of what should be known in stringology.

The mathematical notations and algorithmic conventions we use are standard. The definitions concerning alphabets, strings and languages are also well known. Between the different choices made, we have preferred to write the empty string as λ rather than ϵ, because of possible confusion in the context of distribution-free learning. One might also point out that the λ refers to the German ***leer***, which means *empty*.

We have very briefly given, as extensions, definitions for the infinitary case because of some very specific grammatical inference in this setting (de la Higuera & Janodet, 2004, Maler & Pnueli, 1991, Saoudi & Yokomori, 1993).

In the same sense we have proposed some basic definitions concerning trees.

The study of algorithms for finding patterns in strings is important. Several research papers deal with the question of finding the longest common subsequence or checking if a string is a substring of another. A good starting point is (Aho, 1990).

The definitions concerning distances over strings come from a number of places. If the early definition was by Vladimir Levenshtein (Levenshtein, 1965), the algorithm was introduced in (Wagner & Fisher, 1974). Discussions concerning the variants can be found in a number of places including (Crochemore, Hancart & Lecroq, 2001, Gusfield, 1997, Luzeaux, 1992, Miclet, 1986). The complexity of the algorithm is an issue, so, for applicative reasons, heuristics have been proposed (Rico-Juan & Micó, 2003). The cited problem that the median and centre string problems were intractable was proved in (de la Higuera & Casacuberta, 2000), and there have been several heuristics to try to solve the problem, albeit in an approximative way. But a number of authors argue that in practice, having to rewrite twice on a string of length 2 is not the same as having to rewrite twice on a string of length 200 (Navarro, 2002). For such reasons, several ideas have been proposed to try to relate (in an inverse way) the lengths of the strings with the distance (de la Higuera & Micó, 2008, Yujian & Bo, 2007).

The advantages of the distance function being a metric are that alternative algorithms and data structures can be used for nearest neighbour algorithms (Chávez *et al.*, 2001, Marzal & Vidal, 1993, Micó, Oncina & Vidal, 1994). The triangular inequality can be used to avoid certain computations, resulting in more cost-effective algorithms (Rico-Juan & Micó, 2003).

Section 3.5 closely follows work by Alexander Clark *et al.* (Clark *et al.*, 2006, Clark, Florêncio & Watkins, 2006), and the book by John Shawe-Taylor and Nello Christianini (Shawe-Taylor & Christianini, 2004).

There has been an increasing amount of literature dealing with string kernels. An association of kernels based on automata has been introduced by Corinna Cortes *et al.* (Cortes, Kontorovich & Mohri, 2007).

From a formal language point of view it would seem that a kernel taking into account the size of the best alignment (and not the number of alignments) would be better. But it remains to be seen if such a kernel would comply with Mercer's conditions and if the computation can be efficient enough. In the same way, counting through an automaton would be of real help.

3.8.2 Some alternative lines of research

Kernels are an alternative to distances; the links between the two still deserve more attention. The idea here is to count some similarities. More importantly certain mathematical properties have to be defined. We discuss these questions briefly in Section 3.5. The relationship between the edit distance and grammatical inference is not new, and there has been work on algorithms that try to use the distance as a generalisation criterion (Rulot, Prieto & Vidal, 1989), with applications to speech recognition (Chodorowski & Miclet, 1998). Another relationship can be found through parsing and smoothing (Dupont & Amengual, 2000). Using the edit distance to model noise (Tantini, de la Higuera & Janodet, 2006), or to deal with corrections (Becerra-Bonache *et al.*, 2007) in an active learning setting have also been investigated.

Rough sets also represent an interesting alternative to represent strings more-or-less in or more-or-less out of a language. Some studies have been done (Yokomori & Kobayashi, 1994).

3.8.3 Open problems and possible new lines of research

There are still several open questions of research dealing with the topological issues for strings. For instance, the notions of compactness or of convexity remain to be properly defined, and proposing rational expressions that somehow preserve this compactness would be of great help when dealing with noisy settings.

The edit distance has been adapted for trees and graphs, but if in the first case they just introduce complex but tractable definitions, this is not true for graphs. Also, taking into account the gaps in between the symbols that are not aligned is an important question.

Finding new distances and kernels is of interest. Again, as in the kernel discussion, one has to combine expressive power (the distance and kernel need to measure something worth measuring) and fast computation. If there are no fast algorithms, at least fast approximations are required.

A specific question is that of learning the weights of the edit distance. Experimental approaches (Dupont & Amengual, 2000) and approaches where the weights are probabilities (Oncina & Sebban, 2006) have been looked into. But there is room for more algorithms for this problem.

4

Representing languages

Structures are the weapons of the mathematician.

Bourbaki

It is no coincidence that in no known language does the phrase 'As pretty as an airport' appear.

Douglas Adams

Learning languages requires, for the process to be of any practical value, agreement on a representation of these languages. We turn to formal language theory to provide us with such meaningful representations, and adapt these classical definitions to the particular task of grammatical inference only when needed.

4.1 Automata and finite state machines

Automata are finite state machines used to recognise strings. They correspond to a simplified and limited version of Turing machines: a string is written on the input tape, the string is then read from left to right and, at each step, the next state of the system is chosen depending only on the previous state and the letter or symbol being read. The fact that this is the only information that can be used to parse the string makes the system powerful enough to accept just a limited class of languages called regular languages. The recognition procedure can be made deterministic by allowing only one action to be possible at each step (therefore for each state and each symbol). It is usually nicer and easier to manipulate these deterministic machines (called deterministic finite automata) because parsing is then performed in a much more convenient and economic way, and also because a number of theoretical results only apply to these. On the other hand, non-determinism may be better suited to model certain phenomena and could also be a partial solution to the difficulties one has when facing noisy data when learning. There are several definitions of automata in the literature, so we will use a variant which is of reasonable use for us. In this variant we admit three sorts of states: ***accepting*** states, ***rejecting*** states and ***neutral*** states, those for which we cannot decide on acceptance or rejection. The two first types of states are also called ***final***.

4.1.1 Non-deterministic finite automata

Definition 4.1.1 (NFA) A **non-deterministic finite automaton** (NFA) is a sextuple $\mathcal{A} = \langle \Sigma, Q, \mathbb{I}, \mathbb{F}_{\mathbb{A}}, \mathbb{F}_{\mathbb{R}}, \delta_N \rangle$ where Σ is an alphabet, Q is a finite set of **states**, $\mathbb{I} \subseteq Q$ is the set of **initial** states, $\mathbb{F}_{\mathbb{A}}$ and $\mathbb{F}_{\mathbb{R}}$ are the sets of **final** states, respectively **accepting** and **rejecting**, and $\delta_N : Q \times (\Sigma \cup \{\lambda\}) \to 2^Q$ is a transition function.

We will depict the automaton as a graph where the states are nodes labelled by q_s where s is a subscript, usually an integer or a string, and where there is an edge (q_1, q_2) labelled by $\mathtt{a}$ if $q_2 \in \delta_N(q_1, \mathtt{a})$. Notice that $\mathtt{a}$ can be either a symbol from alphabet Σ or the empty string λ. The initial states are labelled with an entering arrow, and the final states are marked with a double circle, when from $\mathbb{F}_{\mathbb{A}}$, and with a thicker grey line when from $\mathbb{F}_{\mathbb{R}}$.

Example 4.1.1 In Figure 4.1 a 4-state NFA is depicted, with two initial states (q_1 and q_2), two final accepting states q_2 and q_3, and one final rejecting state, q_4.

Definition 4.1.2 ($\mathbb{L}_{\mathbb{F}_{\mathbb{A}}}$) The language $\mathbb{L}_{\mathbb{F}_{\mathbb{A}}}(\mathcal{A})$ **recognised** by the automaton $\mathcal{A}$ is the set of all strings $x = a_1 \cdots a_n$ (with $a_i \in \Sigma \cup \{\lambda\}$) for which there exists a sequence $q_{i_0}, q_{i_1}, \ldots, q_{i_m}$ of states and a sequence $b_{i_1}, \ldots, b_{i_m}$ of symbols in $\Sigma \cup \{\lambda\}$ and $\forall j \in [m]$, $q_{i_j} \in \delta_N(q_{i_{j-1}}, b_{i_j})$ with $b_{i_1} \cdots b_{i_m} = a_1 \cdots a_n$, $q_{i_0} \in \mathbb{I}$ and $q_{i_n} \in \mathbb{F}_{\mathbb{A}}$.

Definition 4.1.3 ($\mathbb{L}_{\mathbb{F}_{\mathbb{R}}}$) The language $\mathbb{L}_{\mathbb{F}_{\mathbb{R}}}(\mathcal{A})$ **recognised by rejection** in the automaton $\mathcal{A}$ is the set of strings rejected by $\mathcal{A}$. $\mathbb{L}_{\mathbb{F}_{\mathbb{R}}}(\mathcal{A})$ is the set of all strings $x = a_1 \cdots a_n$ (with $a_i \in \Sigma \cup \{\lambda\}$) for which there exists a sequence $q_{i_0}, q_{i_1}, \ldots, q_{i_m}$ of states and a sequence $b_{i_1}, \ldots, b_{i_m}$ of symbols in $\Sigma \cup \{\lambda\}$ and $\forall j \in [m]$, $q_{i_j} \in \delta_N(q_{i_{j-1}}, b_{i_j})$ with $b_{i_1} \cdots b_{i_m} = a_1 \cdots a_n$, $q_{i_0} \in \mathbb{I}$ and $q_{i_n} \in \mathbb{F}_{\mathbb{R}}$.

We extend the function δ_N to $(Q \times \Sigma^\star) \to 2^Q$ by defining $\delta_N(q, a_1 \cdots a_n)$ as the set of all $q' \in Q$ such that there exists a sequence $q_{i_0}, q_{i_1}, \ldots, q_{i_m}$ and a sequence $b_{i_1}, \ldots, b_{i_m}$ of symbols in $\Sigma \cup \{\lambda\}$ and $\forall j \in [m]$, $q_{i_j} \in \delta_N(q_{i_{j-1}}, b_{i_j})$ with $b_{i_1} \cdots b_{i_m} = a_1 \cdots a_n$, and $q_{i_0} = q$ and $q_{i_m} = q'$.

We will denote by $\mathbb{L}(\mathcal{A})$ the language $\mathbb{L}_{\mathbb{F}_{\mathbb{A}}}(\mathcal{A})$. This emphasises the fact that the definition, even if mathematically symmetrical, is often used in an asymmetrical way: anything accepted belongs to the language; everything else doesn't belong to the language.

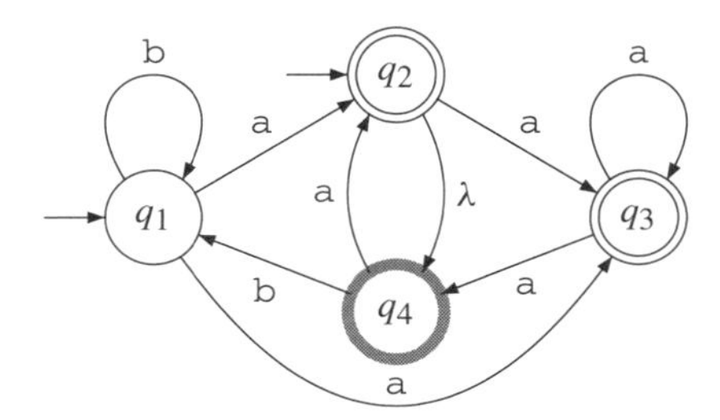

Fig. 4.1. Graphical representation of an NFA.

Remarks

- In the above definition, λ-transitions are allowed. These are used to move freely from one state to another. It will be important to make clear, when using these automata, if such empty string transitions are allowed or not. These (λ-transitions) are often cumbersome and should, whenever possible, be avoided. This can be noted in the above definitions, where, in order to define a path which reads a string x, several levels of subscripts are necessary. This also has algorithmic consequences.
- The usual theory for NFA has only one type of accepting state, and a string belongs to the language as soon as there is one path leading from an initial state to a final state, which reads the string. We have introduced a definition with two types of final state, because of the particularities of grammatical inference. A very typical situation arises when we are given some positive examples (strings that lead to a state in $\mathbb{F}_{\mathbb{A}}$), and some negative examples, that lead to some state in $\mathbb{F}_{\mathbb{R}}$, but in that case a lot of states remain uncertain in the sense that at least for the available data, both of the labels for these unreached states would be consistent. There are even arguments in favour of considering more than two types of state; this will be discussed at the end of the chapter.

It should be noted that even if $\mathbb{F}_{\mathbb{A}}$ and $\mathbb{F}_{\mathbb{R}}$ are disjoint there may be strings belonging to both $\mathbb{L}_{\mathbb{F}_{\mathbb{A}}}(\mathcal{A})$ and $\mathbb{L}_{\mathbb{F}_{\mathbb{R}}}(\mathcal{A})$, which will create a situation of inconsistency.

Definition 4.1.4 (Consistent NFA**)** An NFA $\mathcal{A} = \langle \Sigma, Q, \mathbb{I}, \mathbb{F}_{\mathbb{A}}, \mathbb{F}_{\mathbb{R}}, \delta_N \rangle$ is **consistent** *if*$_{def}$ $\mathbb{L}_{\mathbb{F}_{\mathbb{A}}}(\mathcal{A}) \cap \mathbb{L}_{\mathbb{F}_{\mathbb{R}}}(\mathcal{A}) = \emptyset$.

Example 4.1.2 The automaton from Figure 4.1 is not consistent, since string a belongs both to $\mathbb{L}_{\mathbb{F}_{\mathbb{A}}}(\mathcal{A})$ and to $\mathbb{L}_{\mathbb{F}_{\mathbb{R}}}(\mathcal{A})$. An example of a consistent NFA is depicted in Figure 4.2.

The following results are well known:

- Computing whether a string x belongs to $\mathbb{L}_{\mathbb{F}_{\mathbb{A}}}(\mathcal{A})$ is in $\mathcal{O}(|x| \cdot |Q|)$. This can be done through Algorithm 4.2 which first requires the λ-closure of the NFA to be computed (by Algorithm 4.1). This closure can be computed off-line instead of during parsing. The algorithm (4.1) corresponds to finding the transitive closure of a graph. Parsing is done by dynamic programming: the set S of states reachable by the current prefix is updated with each new symbol. Adapted data structures allow us to compute in linear time the update of set S.
- Checking if two NFAs are equivalent is $\mathcal{P}$-space complete. $\mathcal{A}$ and $\mathcal{B}$ are equivalent *if*$_{def}$ they recognise the same language, i.e. if $\mathbb{L}_{\mathbb{F}_{\mathbb{A}}}(\mathcal{A}) = \mathbb{L}_{\mathbb{F}_{\mathbb{A}}}(\mathcal{B})$ and $\mathbb{L}_{\mathbb{F}_{\mathbb{R}}}(\mathcal{A}) = \mathbb{L}_{\mathbb{F}_{\mathbb{R}}}(\mathcal{B})$.

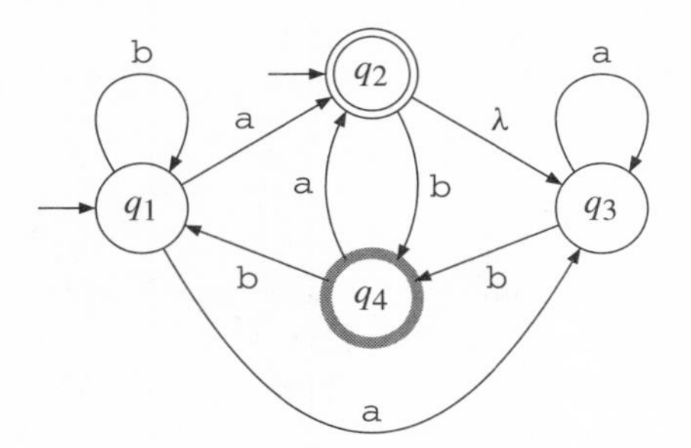

Fig. 4.2. A consistent NFA.

Algorithm 4.1: λ-CLOSURE of an NFA.

Data: An NFA : $\mathcal{A} = \langle \Sigma, Q, \mathbb{I}, \mathbb{F}_{\mathbb{A}}, \mathbb{F}_{\mathbb{R}}, \delta_N \rangle$, with $Q = \{q_1, \ldots, q_{|Q|}\}$
Result: A Boolean matrix E = $[|Q|], [|Q|]$, with E$[i][j]$ = **true** if $q_j \in \delta_N(q_i, \lambda)$, **false** if not.

```
for i ∈ [|Q|] do
    for j ∈ [|Q|] do
        if i = j then
            E[i][j] ← true
        else
            if q_j ∈ δ_N(q_i, λ) then E[i][j] ← true
            else E[i][j] ← false
        end
    end
end
for k ∈ [|Q|] do
    for i ∈ [|Q|] do
        for j ∈ [|Q|] do  E[i][j] ← E[i][j] ∨ (E[i][k] ∧ E[k][j])
    end
end
return E
```

Algorithm 4.2: Parsing with an NFA.

Data: An NFA : $\mathcal{A} = \langle \Sigma, Q, \mathbb{I}, \mathbb{F}_{\mathbb{A}}, \mathbb{F}_{\mathbb{R}}, \delta_N \rangle$, a string $x = a_1 a_2 \cdots a_n$
Result: A set of states S reachable from $\mathbb{I}$ by reading x

```
S← I;
E← λ-CLOSURE(A);
for 1 ≤ i ≤ n do
    S← {q_k ∈ Q : ∃q_j ∈ S ∧ E[j][k]};
    S← {q_k ∈ Q : ∃q_j ∈ S ∧ q_k ∈ δ_N(q_j, a_i)}
end
S← {q_k ∈ Q : ∃q_j ∈ S ∧ E[j][k]};
return S
```

- Minimising an NFA is an $\mathcal{NP}$-hard problem. This is linked with the fact that there is no natural tractable normal or canonical form for an NFA. One should be careful with this statement: obviously there exists a minimum NFA; what is intractable is to build it from an existing NFA. Here minimality refers to any definition which would associate a canonical and thus unique (up to isomorphisms, possibly) minimum form to two equivalent automata.

4.1.2 Deterministic finite automata

A ***deterministic finite automaton*** (DFA) is an NFA in which the two non-deterministic liberties have been eliminated: there is only one initial state and, in each state, reading a symbol brings us to a unique state. Therefore $\forall q \in Q, \forall a \in \Sigma, |\delta_N(q,a)| \leq 1$.

The above definition also excludes the use of λ-transitions.

This makes the set notation for the result of the transitions cumbersome, so we will denote transitions in a slightly different manner:

Definition 4.1.5 (DFA) A **deterministic finite automaton** (DFA) is a sextuple $\mathcal{A} = \langle \Sigma, Q, q_\lambda, \mathbb{F}_\mathbb{A}, \mathbb{F}_\mathbb{R}, \delta \rangle$ where Σ is an alphabet, Q is a finite set of states, $q_\lambda \in Q$ is the initial state, $\delta : Q \times \Sigma \to Q$ is a transition function, and $\mathbb{F}_\mathbb{A} \subseteq Q$ and $\mathbb{F}_\mathbb{R} \subseteq Q$ are sets of marked states, called the **final accepting** and **rejecting** states (respectively).

It is usual to recursively extend δ to $Q \times \Sigma^\star \to Q$: $\delta(q,\lambda) = q$ and $\delta(q, a.w) = \delta(\delta(q,a), w)$ for all $q \in Q, a \in \Sigma, w \in \Sigma^*$. Let $\mathbb{L}_{\mathbb{F}_\mathbb{A}}(\mathcal{A})$ denote the language recognised by automaton $\mathcal{A}$:

$$\mathbb{L}_{\mathbb{F}_\mathbb{A}}(\mathcal{A}) = \{w \in \Sigma^* \mid \delta(q_\lambda, w) \in \mathbb{F}_\mathbb{A}\}.$$

In the same way:

$$\mathbb{L}_{\mathbb{F}_\mathbb{R}}(\mathcal{A}) = \{w \in \Sigma^* \mid \delta(q_\lambda, w) \in \mathbb{F}_\mathbb{R}\}.$$

We will usually denote by $\mathbb{L}(\mathcal{A})$ the language $\mathbb{L}_{\mathbb{F}_\mathbb{A}}(\mathcal{A})$ and in a general way denote by $\mathbb{L}$ the ***naming*** function, i.e. the function associating with a finite automaton the language it recognises.

Definition 4.1.6 A language is **regular** *if*$_{def}$ it is recognised by a DFA.

It is well-known result that DFA and NFA have an identical expressive power: any language recognised by an NFA can also be recognised by a DFA. The result is constructive: a determinisation algorithm exists, but the resulting DFA can be much larger than the equivalent NFA.

Definition 4.1.7 We denote by $\mathcal{DFA}(\Sigma)$ the class of all DFAs over the alphabet Σ and by $\mathcal{NFA}(\Sigma)$ the class of all NFAs over the alphabet Σ. $\mathcal{REG}(\Sigma)$ is the family of all regular languages over the alphabet Σ.

There are alternative ways of defining regular languages: regular expressions, regular grammars or systems. Because of the interest in certain fields in manipulating regular expressions (for instance X-paths and *grep* expressions) it may be of interest to learn and manipulate these in a grammatical inference setting.

Definition 4.1.8 Given a DFA $\mathcal{A} = \langle \Sigma, Q, q_\lambda, \mathbb{F}_\mathbb{A}, \mathbb{F}_\mathbb{R}, \delta \rangle$ and a state q in Q, we call ***marker*** for q the shortest string (for the lex-length order) that reaches state q: $\text{marker}(q) = \min_{\leq_{lex\text{-}length}} \{w \in \Sigma^\star : \delta(q_\lambda, w) = q\}$.

Markers will allow us to index the states not by numbers, but by strings.

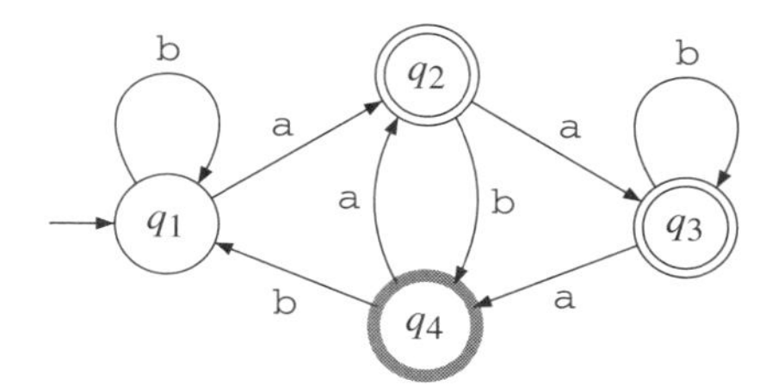

Fig. 4.3. Graphical representation of a DFA.

Algorithm 4.3: Parsing with a DFA.

Data: A DFA : $\mathcal{A} = \langle \Sigma, Q, q_\lambda, \mathbb{F}_\mathbb{A}, \mathbb{F}_\mathbb{R}, \delta \rangle$, a string $x = a_1 a_2 \cdots a_n$
Result: The state q_{curr} (if any) reached from q_λ by reading x

$q_{curr} \leftarrow q_\lambda$;
for $1 \leq i \leq n$ **do**
| $q_{curr} \leftarrow \delta(q_{curr}, a_i)$
end
return q_{curr}

Example 4.1.3 In Figure 4.3 for states q_1, q_2, q_3 and q_4, the markers are respectively λ, a, aa and ab. The states can be renamed q_λ, $q_{\texttt{a}}$, $q_{\texttt{aa}}$ and $q_{\texttt{ab}}$.

Some properties of DFAs are as follows:

- There exist languages for which the smallest NFA is exponentially smaller than the smallest equivalent DFA. A typical well-known example is the language L_n over alphabet {a, b} in which the nth letter before the end is an 'a'. L_n is recognised by an NFA with n+1 states but the smallest DFA recognising L_n needs 2^n states. We will discuss these questions of sizes of representations when defining the learning models, in Section 7.3, page 152.
- Every regular language is recognised by a specific automaton called the minimum canonical automaton, which is the complete automaton with a minimum number of states. In some cases this automaton may have one ***sink*** state. This is a state q_s that is neither accepting nor rejecting and such that $\forall a \in \Sigma$, $\delta(q_s, a) = q_s$. The construction of this unique (up to state isomorphism, i.e. renaming of the states) automaton can be done in time $\mathcal{O}(n \log n)$ where n is the number of states of the initial DFA. What really is constructed is known as the *Myhill-Nerode equivalence*:

$$\forall u, v \in \Sigma^\star, u \equiv v \iff [\forall w \in \Sigma^\star\, uw \in L \Leftrightarrow vw \in L]$$

  Note that in the above the language L defines the equivalence classes.
- An issue linked with this is that of the equivalence of automata. If these are DFAs, the problem is simple: minimise both DFAs, then compare the results. Hence equivalence of DFAs is in $\mathcal{P}$.

4.1.3 Regular expressions

We introduce briefly ***regular expressions***, also sometimes called *rational expressions*. They allow us to *define* languages (as opposed to 'recognise' or 'generate').

Definition 4.1.9 Let Σ be a finite alphabet. We define recursively a regular expression as:

- $\emptyset$, Λ, and a ($\forall a \in \Sigma$) are regular expressions over Σ;
- if e_1 and e_2 are regular expressions over Σ then so are $(e_1) \cdot (e_2)$, $(e_1) + (e_2)$ and $(e_1)^*$.

The cumbersome brackets will be disposed of whenever possible. The concatenation dot $(\cdot)$ is also only used when necessary. Therefore $(a) \cdot (b)$ is denoted more conveniently by ab. Usual priorities are that '$*$' has higher priority than '$\cdot$' and (then) '$+$'.

Definition 4.1.10 Let Σ be a finite alphabet. The value of a regular expression is the language defined recursively as follows:

- $\text{value}(\emptyset) = \emptyset$,
- $\text{value}(\Lambda) = \{\lambda\}$,
- $\forall a \in \Sigma$, $\text{value}(a) = \{a\}$,
- $\text{value}((e)) = \text{value}(e)$,
- $\text{value}(e_1 \cdot e_2) = \text{value}(e_1) \cdot \text{value}(e_2)$,
- $\text{value}(e_1 + e_2) = \text{value}(e_1) \cup \text{value}(e_2)$,
- $\text{value}(e^*) = \text{value}(e)^*$.

Example 4.1.4 $\texttt{ab}^*(\texttt{a} + \texttt{b}^*)^*$ is a regular expression over $\{\texttt{a}, \texttt{b}\}$ having as value the language composed of all strings starting with the symbol 'a'. It is easy to notice that many equivalent regular expressions exist.

We denote by $\mathcal{REGEXP}(\Sigma)$ the set of all regular expressions over the alphabet Σ.

There exist algorithms allowing us to transform a regular expression into an NFA. These usually introduce λ-transitions in the process. One can also extract a regular expression from an automaton. The complexity classes for the usual problems of parsing, minimisation and equivalence are the same as those for the NFA.

It is known that the class of regular languages contains all languages that can either be described by a regular expression, or recognised by a finite automaton, which can be either deterministic or non-deterministic.

4.1.4 A sample as an automaton

A ***sample*** is a finite set of data. The sample can be ***informed*** in which case it is a pair $\langle S_+, S_- \rangle$ of finite sets of strings. S_+ contains the ***positive examples*** and S_- contains the ***negative examples*** (also called the counter-examples). A sample is ***non-conflicting*** when $S_+ \cap S_- = \emptyset$. We will only deal with non-conflicting samples.

We may denote a sample $\langle S_+, S_- \rangle$ as a set of ***labelled*** strings, the labels being either '1' for the examples or '0' for the counter-examples.

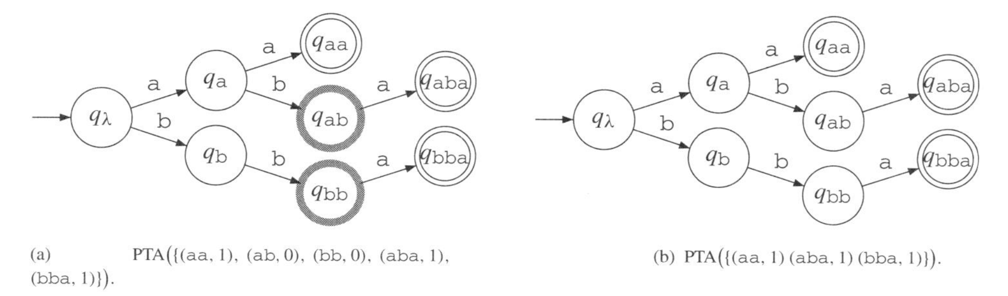

(a) PTA({(aa, 1), (ab, 0), (bb, 0), (aba, 1), (bba, 1)}).

(b) PTA({(aa, 1) (aba, 1) (bba, 1)}).

Fig. 4.4. Two PTAs.

Example 4.1.5 The sample $\langle S_+, S_- \rangle$ with $S_+ = \{\texttt{aa}, \texttt{aba}, \texttt{bba}\}$ and $S_+ = \{\texttt{ab}, \texttt{bb}\}$ can also be denoted by $\{(\texttt{aa}, 1), (\texttt{ab}, 0), (\texttt{bb}, 0), (\texttt{aba}, 1), (\texttt{bba}, 1)\}$.

A sample is made of ***text*** if it contains no counter-example.

The ***prefix tree acceptor*** of S_+ is the DFA$\langle \Sigma, Q, q_\lambda, \mathbb{F}_\mathbb{A}, \mathbb{F}_\mathbb{R}, \delta \rangle$ denoted PTA(S_+) such that:

- $Q = \{q_u : u \in \text{PREF}(S_+)\}$,
- $\forall ua \in \text{PREF}(S_+) : \ \delta(q_u, a) = q_{ua}$,
- $\mathbb{F}_\mathbb{A} = \{q_u : u \in S_+\}$,
- $\mathbb{F}_\mathbb{R} = \emptyset$.

The PTA is therefore the smallest DFA which recognises S_+, has minimal size $\mathbb{F}_\mathbb{A}$ and $\mathbb{F}_\mathbb{R}$ for which the transition function δ is injective (i.e. if $\delta(q, a) = \delta(q', a)$ then $q = q'$). In fact, in PTA(S_+), $\mathbb{F}_\mathbb{R} = \emptyset$. Notice that in this case the DFA is a tree, each state being the successor of exactly one state (except for the initial state, which has no predecessor).

If furthermore we are given a set S_+ and a set S_-, PTA(S_+, S_-) is

- $Q = \{q_u : u \in \text{PREF}(S_+ \cup S_-)\}$,
- $\forall ua \in \text{PREF}(S_+ \cup S_-) : \ \delta(q_u, a) = q_{ua}$,
- $\mathbb{F}_\mathbb{A} = \{q_u : u \in S_+\}$,
- $\mathbb{F}_\mathbb{R} = \{q_u : u \in S_-\}$.

In the case where the PTA is built from both positive and negative strings, then $\mathcal{A} =$ PTA (S_+, S_-) is again a tree, the smallest such that $S_+ \subseteq \mathbb{L}_{\mathbb{F}_\mathbb{A}}(\mathcal{A})$ and $S_- \subseteq \mathbb{L}_{\mathbb{F}_\mathbb{R}}(\mathcal{A})$.

In Section 12.1 an algorithm building the PTA is given and studied. But it can easily be deduced from the above definitions. We depict in Figure 4.4 the PTA built from a sample $\langle S_+, S_- \rangle$, and another one corresponding to only a positive sample.

4.2 Grammars

Automata are devices used to recognise strings. The other classical way of defining languages is through generative devices called grammars. Parsing is then an essential issue.

4.2.1 Context-free grammars

Definition 4.2.1 A **context-free grammar** is a quadruple $< \Sigma, V, R, N_1 >$ where Σ is a finite alphabet (of terminal symbols), V is a finite alphabet (of variables or non-terminals, denoted by $N_1, \ldots, N_i$), $R \subset V \times (\Sigma \cup V)^*$ is a finite set of production rules, and N_1 ($\in V$) is the axiom.

We will write $N \rightarrow \beta$ for the rule $(N, \beta) \in R$. If $\alpha, \beta, \gamma \in (\Sigma \cup V)^*$ and $(N, \beta) \in R$ we have $\alpha N \gamma \Rightarrow \alpha\beta\gamma$. This means that the string $\alpha N \gamma$ derives (in one step) into the string $\alpha\beta\gamma$.

$\stackrel{*}{\Longrightarrow}$ is the reflexive and transitive closure of $\Rightarrow$. If there exist $\alpha_0, \ldots, \alpha_k$ such that $\alpha_0 \Rightarrow \ldots \Rightarrow \alpha_k$ we will write $\alpha_0 \stackrel{k}{\Rightarrow} \alpha_k$. We denote by $\mathbb{L}(G, N)$ the language $\{x \in \Sigma^\star : N \stackrel{*}{\Longrightarrow} x\}$. If N_1 is the axiom, we shall denote $\mathbb{L}(G, N_1)$ by $\mathbb{L}(G)$ and call this the language generated by G (from N). When different rules share a same left-hand side it will be possible to write these in the shortened way $N \rightarrow \alpha + \beta + \gamma$ (for rules (N, α), (N, β) and (N, γ), for instance).

Definition 4.2.2 Two grammars are equivalent *if$_{def}$* they generate the same language.

We denote by $\mathcal{CFG}(\Sigma)$ the class of all context-free grammars using the alphabet Σ, whereas $\mathcal{CFL}(\Sigma)$ is the set of all context-free languages over Σ.

A derivation $T = \alpha_0 \Rightarrow \ldots \Rightarrow \alpha_k = u$ can be represented by a ***derivation tree*** where the root is the non-terminal T and the frontier of the tree reads the string u.

We will not provide here all the definitions concerning derivations and derivation trees, but we summarise some of the typical notions and questions through an example.

Example 4.2.1 Let $G = < \{\mathtt{a}, \mathtt{b}\}, \{N\}, R, N >$ with $N \rightarrow \mathtt{a}NN + \mathtt{b}$. The language generated by G starting from N is the Lukaciewitz language:

$$\left\{ w \in \{\mathtt{a}, \mathtt{b}\}^* : \forall p,\ s \in \{\mathtt{a}, \mathtt{b}\}^*,\ w = p \cdot s \begin{pmatrix} s = \lambda \implies |p|_{\mathtt{a}} + 1 = |p|_{\mathtt{b}} \\ s \neq \lambda \implies |p|_{\mathtt{a}} \geq |p|_{\mathtt{b}} \end{pmatrix} \right\}.$$

A derivation is $N \Rightarrow \mathtt{a}NN \Rightarrow \mathtt{aa}NNN \Rightarrow \mathtt{aa}N\mathtt{a}NNN \Rightarrow \mathtt{aaba}NNN \stackrel{*}{\Longrightarrow} \mathtt{aababbb}$. We depict the corresponding derivation tree in Figure 4.5.

Definition 4.2.3 A context-free grammar $G =< \Sigma, V, R, N_1 >$ is in **reduced** normal form *if$_{def}$* all non-terminals are **useful**:

$$\forall N \in V\ \exists u \in \Sigma^\star : N \stackrel{*}{\Longrightarrow} u$$
$$\forall N \in V\ \exists u, v \in \Sigma^\star : N_1 \stackrel{*}{\Longrightarrow} uNv.$$

Reduced normal forms correspond to the idea that each non-terminal must be reachable from the axiom and can produce at least one string. It is clear that non-terminals that do not obey these rules can be eliminated from the grammar without changing the language. Another extension is to make sure that applying a rule gets us closer to the target string. That can be done by considering proper normal forms:

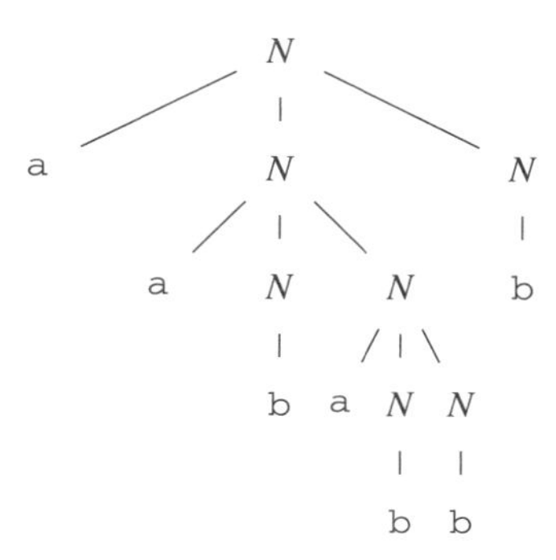

Fig. 4.5. Derivation tree for string aababbb.

Definition 4.2.4 A context-free grammar $G =< \Sigma, V, R, N_1 >$ is in **proper** normal form if_{def} all non-terminals (with the possible exception of the axiom) are **constructive**:

$$(N, \lambda) \in R \text{ or } (N, N') \in R \implies N = N_1.$$

In this case the advantage is that there is a relation between the length of the derivation and the size of the string:

Lemma 4.2.1 *If $G =< \Sigma, V, R, N_1 >$ is in proper and reduced normal form then*

$$w \in \mathbb{L}(G) \Longleftrightarrow \big[N_1 \stackrel{k}{\Longrightarrow} w \text{ with } k \leq 2|w| - 1\big].$$

Two other important normal forms are:

Definition 4.2.5 A context-free grammar $G =< \Sigma, V, R, N_1 >$ is in **quadratic** or **Chomsky** normal form if_{def} $R \subseteq V \times \big(V^2 \cup \Sigma \cup \{\lambda\}\big)$.

A context-free grammar $G =< \Sigma, V, R, N_1 >$ is in **Greibach** normal form if_{def} $R \subset V \times \Sigma V^*$.

The above normal forms can be obtained constructively. But they are normal forms, as opposed to canonical forms: the same language may admit various incomparable normal forms.

Parsing can be done in cubic time when starting from a grammar in Chomsky normal form. We provide a general parser in Algorithm 4.4, known as the Cocke-Younger-Kasami (CYK) algorithm. Alternative algorithms can be found in the literature. The algorithm computes $\text{CYK}[i][j][k] = \textbf{true} \iff N_i \stackrel{*}{\Longrightarrow} a_j \cdots a_k$. The algorithm can be adapted to context-free grammars not in quadratic normal form, but with more complicated notations.

We summarise the other important results concerning context-free languages and grammars:

Proposition 4.2.2 *The following problems are undecidable:*

- *Does a given context-free grammar generate $\Sigma^\star$?*
- *Does a given context-free grammar generate a regular language?*
- *Are two context-free grammars equivalent?*

Algorithm 4.4: CYK algorithm.

```
Data: A context-free grammar G =< Σ, V, R, N_1 > in quadratic normal form, with
      V = {N_1, N_2, ..., N_|V|}, a string x = a_1 a_2 ··· a_n.
Result: CYK[i][j][k] = true ⟺ a_j ··· a_k ∈ L(G, N_i).
for k ∈ [|x|] do
    for j ∈ [k] do
        for i ∈ [|V|]/ do CYK[i][j][k] ← false
    end
end
for j ∈ [|x|] do
    for i ∈ [|V|] do
        if (N_i, a_j) ∈ R then CYK[i][j][j] ← true
    end
end
for m : 1 ≤ m ≤ |x| − 1 do
    for j : 1 ≤ j ≤ |x| − m do
        for k : j ≤ k < j + m do
            for (N_i, N_{i1} N_{i2}) ∈ R do
                if CYK[i1][j][k] ∧ CYK[i2][k + 1][j + m] then
                    CYK[i][j][j + m] ← true
                end
            end
        end
    end
end
return CYK[1][1][n]
```

We do not provide the classical proofs here but comment upon their meaning for the problem of inferring or learning grammars. The fact that the equivalence problem is undecidable means that the shortest string separating two context-free grammars is of unbounded length; if not we would only have to check strings of length at most the bound and thus be able to decide equivalence. This in turn means that when intending to learn one grammar (and not the other) from only short strings, we are facing a problem: we can be sure not to encounter a small significant string in a reasonable sized learning sample. Let us call this the ***curse of expansiveness***, which can be illustrated through the following example:

Example 4.2.2 Let $G_n = < \{\mathtt{a}\}, \{N_i : i \leq n\}, R_n, N_1 >$ with, for each $i < n$, the following rule in R_n: $N_i \rightarrow N_{i+1}N_{i+1}$. We also have the rule $N_n \rightarrow \mathtt{a}$.

It is easy to show that $\mathbb{L}(G_n) = \mathtt{a}^{2^{n-1}}$. This is therefore a language whose shortest (and only) string is of length exponential in the size of the grammar!

4.2.2 Linear grammars

To avoid the expansiveness problems one may want to only consider *linear grammars*:

Definition 4.2.6 A context-free grammar $G = <\Sigma, V, R, N_1>$ is **linear** if_{def} $R \subset V \times (\Sigma^\star\ V\ \Sigma^\star \cup \Sigma^\star)$.

A context-free language is ***linear*** if it is generated by a linear grammar. Not all context-free languages are linear. This is the case, for example, of the following product of two linear languages: $\{\mathtt{a}^m\mathtt{b}^m\mathtt{c}^n\mathtt{d}^n : m, n \in \mathbb{N}\}$. Moreover it is undecidable to know whether a language is linear or not, just as it is to know whether two linear languages are equivalent or not.

These results justify considering even more restricted classes of linear languages:

Definition 4.2.7 A linear grammar $G = <\Sigma, V, R, N_1>$ is **even linear** if_{def} $(N, uN'v) \in R \implies |u| = |v|$.

A linear grammar $G = <\Sigma, V, R, N_1>$ is **left linear** (respectively right linear) if_{def} $R \subseteq V \times (V\ \Sigma^\star \cup \Sigma^\star)$ (respectively $R \subseteq V \times (\Sigma^\star\ V \cup \Sigma^\star)$).

It is well known that all regular languages can be generated by both even linear and left (or right) linear languages. The converse is true for the right and left linear languages but not for the even linear ones: consider the language $\{\mathtt{a}^n\mathtt{b}^n : n \in \mathbb{N}\}$ which is even linear but not regular.

We denote by $\mathcal{LING}(\Sigma)$ the class of linear grammars over the alphabet Σ and by $\mathcal{LIN}(\Sigma)$ the class of linear languages over the alphabet Σ.

Algorithm CYK (4.4) can be adapted to parse strings with a linear grammar in $\mathcal{O}(n^2)$ time.

4.2.3 Deterministic linear grammars

A variety of definitions have been given capturing the ideas of determinism and linearity.

Definition 4.2.8 (Deterministic linear grammars) A linear grammar $G = <\Sigma, V, R, N_1>$ is a **deterministic linear grammar** if_{def}

- all rules are of the form $(N, aN'u)$ or (N, λ) and
- $(N, au), (N, av) \in R \Longrightarrow u = v$ with $u, v \in V\ \Sigma^\star$.

This definition induces the determinism on the first symbol of the right-hand side of the rules. This extends by easy induction to the derivations: let $N \overset{*}{\Longrightarrow} uN'v$ and $N \overset{*}{\Longrightarrow} uN''w$ then $N' = N''$ and $v = w$.

We denote by $\mathcal{DL}(\Sigma)$ the set of all languages generated by a deterministic linear grammar.

4.2.4 Pushdown automata

Another way of defining context-free languages is through *pushdown automata* (PDAs).

Informally a PDA is a one-way finite state machine with a stack. Criteria for recognition can be by empty stack or by accepting states, but in both cases the class of languages is that of the context-free ones. If the machine is deterministic (in a given configuration of the stack, with a certain symbol to be read, only one rule is possible) the class of languages is that of the *deterministic languages*, denoted as $\mathcal{DETL}(\Sigma)$. Given a computation of a PDA, a ***turn*** in the computations is a move that decreases the height of the stack and is preceded by a move that did not decrease it. A PDA is said to be ***one-turn*** if in any computation there is at most one turn.

Theorem 4.2.3 *A language is linear* if and only if *if it is recognised by a one-turn* PDA.

4.3 Exercises

4.1 Prove that checking if two NFAs are inconsistent (as in Definition 4.1.4, page 72) is a $\mathcal{P}$-space complete problem. Remember that testing the inequivalence of two NFAs is $\mathcal{P}$-space complete.

4.2 Prove Lemma 4.2.1 (page 79).

4.3 How many states are needed in a DFA (or an NFA) recognising all strings that do not have repeated symbols?

4.4 Prove that DFAs are always consistent, i.e. that a string cannot be accepted and rejected.

4.5 Propose an algorithm that checks if a grammar is in reduced normal form.

4.6 Propose an algorithm that checks if a grammar is in proper normal form.

4.7 Propose an algorithm that computes a grammar in quadratic normal form equivalent to a given grammar. How much larger is the new grammar?

4.8 Find a deterministic linear grammar for the palindrome language, which is the set of all strings w that read the same from left to right as from right to left, i.e. such that $\forall i \in [|w|], w(i) = w(n+1-i)$.

4.9 Prove that every regular language is even linear.

4.10 Prove that the union of two deterministic linear languages may not be deterministic linear.

4.4 Conclusions of the chapter and further reading

4.4.1 Bibliographical background

Automata and finite state machines (Section 4.1) have been studied since the 1950s, where variants were sometimes known as Moore or Mealy machines. The theory is now well documented and can be found in different textbooks (Harrison, 1978, Hopcroft & Ullman, 1977, 1979, Sakarovich, 2004, Simon, 1999). The central questions concerning equivalence between DFAs and NFAs were solved by Michael Rabin and Dana Scott (Rabin & Scott, 1959). The links between regular expressions and automata were explored in a number of places, with Ken Thompson's method (Thompson, 1968) widely used, even if it does introduce many λ-transitions.

We have followed here the usual definitions with the exception that we are using two types of final states, better suited for grammatical inference (even if sometimes cumbersome). A similar path was followed by François Coste (Coste & Fredouille, 2003) who argues that automata are classifying functions and can thus have values other than 0 and 1. The point we make is that when learning DFAs the current formalism always allows us to decide that 'anything that is not accepting is rejecting' after the learning process, but that if background knowledge tells us otherwise we can just as well decide the opposite.

Between the different sorts of context-free grammars (introduced in Section 4.2), even linear grammars (Mäkinen, 1996, Sempere & García, 1994, Takada, 1988) and deterministic linear grammars have been more extensively studied in the context of grammatical inference. Deterministic linear grammars were introduced in grammatical inference in (de la Higuera & Oncina, 2002) (and (de la Higuera & Oncina, 2003) for the probabilistic version), but alternative definitions (Ibarra, Jiang & Ravikumar, 1988, Nasu & Honda, 1969) were introduced earlier. Pushdown automata have not been used that much in grammatical inference, but they define the important class of the deterministic languages. A specific study on these machines can be found in (Autebert, Berstel & Boasson, 1997).

4.4.2 Some alternative lines of research

We have described here some classical ways of describing languages, i.e. sets of strings. But there are several alternative ways of doing this, with interesting work dealing with learning in different cases.

Categorical grammars (Costa Florêncio, 2003, Kanazawa, 1998, Tellier, 1998) are used to associate semantics to syntax directly, and follow very often work by researchers in logic and in computational linguistics. The idea is to associate the rules with the lexicon: each word can be used, depending on its type, following some specific pattern. The intricate mechanisms used to parse and the computational costs involved make these grammars out of the main line of results presented in this book. Relating these elegant mechanisms with acute complexity models (MDL principle, Kolmogorov complexity) is a research direction followed by some (Tellier, 2005).

Regular expressions of different sorts have been elaborated and studied in grammatical inference (Fernau, 2005), but, in order to avoid facing the difficulties of learning NFAs, structural restrictions (on the type of operators) have been imposed (Brazma *et al.*, 1998, Kinber, 2008). There are of course several ways to restrict or enrich the types of regular expressions we may use. One important idea has consisted of measuring the *star height* of an expression as the number of embedded stars one has to use. Adding a complementary operation allows us to manipulate infinite languages of star height 0. There have been separate learnability results for these (Brazma, 1997).

Extensions of the above mechanisms (automata, grammars) to deal with trees and graphs have been proposed. For the case of tree automata a general survey is (Comon *et al.*, 1997) and for graph grammars there are a number of possible sources (Courcelle, 1991).

Other extensions consist of not having halting states but states through which the run must (or must not) pass infinitely often: these define infinitary languages (Büchi, 1960), and have also been considered in learning settings (de la Higuera & Janodet, 2004, Maler & Pnueli, 1991, Saoudi & Yokomori, 1993).

In Chapter 18 we will discuss the extension of the final state models to deal with transducers.

4.4.3 Open problems and possible new lines of research

Between the many interesting lines of research corresponding to studying language models of use in grammatical inference tasks, we propose one presented originally in (de la Higuera, 2006b).

We propose to consider the case where the alphabet is ranked, i.e. there is a partial order over the symbols in the alphabet. The situation arises in a number of settings:

- either when the alphabet is naturally ordered as can be the case in music;
- if the original data are numeric, the normal discretisation loses the proximity/topological issues that should help and that are contained in the data;
- sometimes the alphabet can consist of subsets of strings, in which case we can also have a relation which may be a generalisation or subsumption.

We introduce k-edge deterministic finite automata to deal with such languages.

A ranked alphabet $\langle\Sigma, \leq\rangle$ is an alphabet Σ with a relation $\leq_\Sigma$ which is a partial order (reflexive, antisymmetric and transitive) over Σ. Given a ranked alphabet $\langle\Sigma, \leq\rangle$ and two symbols a and b in Σ, with $a \leq b$, we denote by $[a, b]$ the set $\{c \in \Sigma : a \leq c \leq b\}$.

Example 4.4.1 Here are two possible relations:

- $\langle\Sigma, \leq\rangle$ where $\Sigma = \{\mathtt{0}, \mathtt{1}, \mathtt{2}, \mathtt{3}\}$ and $\mathtt{0} < \mathtt{1} < \mathtt{2} < \mathtt{3}$. This is the case in music or working with numerical data.
- $\langle\Sigma, \leq\rangle$ where $\Sigma = \{\mathtt{00}, \mathtt{01}, \mathtt{10}, \mathtt{11}\}$ and $\mathtt{00} \leq \mathtt{01} \leq \mathtt{11}$ and $\mathtt{00} \leq \mathtt{10} \leq \mathtt{11}$. This means that the automaton may not need to be based on a total order.

Definition 4.4.1 (k-edge deterministic finite automaton) A ***k-edge deterministic finite automaton (k-*DFA*)*** $\mathcal{A}$ is a tuple $\langle\Sigma, Q, q_\lambda, \mathbb{F}_\mathbb{A}, \mathbb{F}_\mathbb{R}, \delta\rangle$ where Σ is a finite alphabet, Q is a finite set of *states*, $q_1 \in Q$ is the initial state, $\mathbb{F}_\mathbb{A} \subseteq Q$ is the set of final accepting states, $\mathbb{F}_\mathbb{R} \subseteq Q$ is the set of final rejecting states, and $\delta : Q \times \Sigma \times \Sigma \rightarrow Q$ is the total transition function verifying: $\forall q \in Q$, $|\{a, b : \delta(q, a, b) \text{ is defined}\}| \leq k$, and if $\delta(q, a_1, b_1) \neq \delta(q, a_2, b_2)$ then $[a_1, b_1] \cap [a_2, b_2] = \emptyset$.

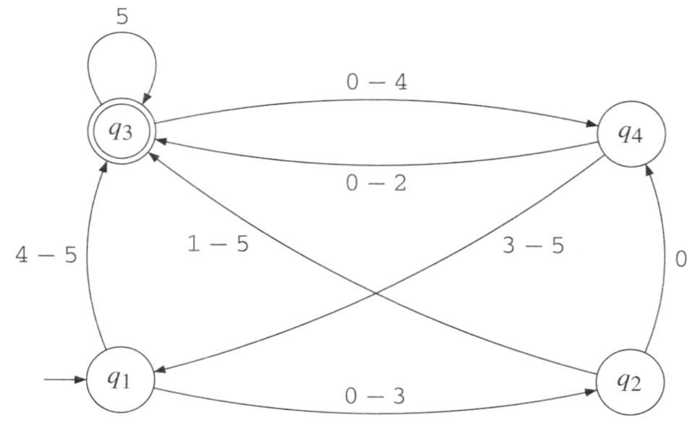

Fig. 4.6. A 2-edge DFA.

A string $x = c_1 \cdots c_{|x|}$ belongs to $\mathbb{L}(\mathcal{A})$ *if*$_{def}$ there is a sequence of states $q_{i_0}, q_{i_1}, \ldots, q_{i_{|x|}}$ with $\delta(q_{i_j}, a_j, b_j) = q_{i_{j+1}}$ and $a_j \leq c_j \leq b_j$. And of course $q_{i_{|x|}}$ has to be a final accepting state.

The extension of δ is done as usual and the language recognised by $\mathcal{A}$, $\mathbb{L}(\mathcal{A})$, is $\{a \in \Sigma^* : \delta(q_\lambda, w) \in \mathbb{F}_{\mathbb{A}}\}$.

Example 4.4.2 We represent in Figure 4.6 a 2-edge automaton. Notice that the same language can also be represented by a 3-edge automaton, but not by a 1-edge automaton. Here, `102` $\in \mathbb{L}(\mathcal{A})$.

Clearly any k-edge DFA is also a DFA but the converse is not true. Moreover if k is fixed, one can easily build (for an alphabet of size more than k) a regular language that cannot be represented by a k-edge DFA. Also it should be pointed out that the case where k is the size of the alphabet is of no new interest at all, as it corresponds to a normal DFA.

There are then a number of possible problems concerning these automata: to decide if a given sample allows at least one k-edge DFA consistent with it, to find some pumping lemma or to study the closure properties. These characterisations would possibly allow learning algorithms to be designed in the setting where the alphabets are huge but structured.

5
Representing distributions over strings with automata and grammars

If your experiment needs statistics, you ought to have done a better experiment.
Ernest Rutherford

'I think you're begging the question,' said Haydock, 'and I can see looming ahead one of those terrible exercises in probability where six men have white hats and six men have black hats and you have to work it out by mathematics how likely it is that the hats will get mixed up and in what proportion. If you start thinking about things like that, you would go round the bend. Let me assure you of that!'
***Agatha Christie**, The Mirror Crack'd from Side to Side (1962)*

Instead of defining a language as a set of strings, there are good reasons to consider the seemingly more complex idea of defining a distribution over strings. The distribution can be regular, in which case the strings are then generated by a probabilistic regular grammar or a probabilistic finite automaton. We are also interested in the special case where the automaton is deterministic.

Once distributions are defined, distances between the distributions and the syntactic objects they represent can be defined and in some cases they can be conveniently computed.

5.1 Distributions over strings

Given a finite alphabet Σ, the set $\Sigma^\star$ of all strings over Σ is enumerable, and therefore a distribution can be defined.

5.1.1 Distributions

A ***probabilistic language*** $\mathcal{D}$ is a probability distribution over $\Sigma^\star$.

The probability of a string $x \in \Sigma^\star$ under the distribution $\mathcal{D}$ is denoted by a positive value $Pr_{\mathcal{D}}(x)$ and the distribution $\mathcal{D}$ must verify

$$\sum_{x\in\Sigma^\star} Pr_{\mathcal{D}}(x) = 1.$$

If the distribution is modelled by some syntactic machine $\mathcal{A}$, the probability of x according to the probability distribution defined by $\mathcal{A}$ is denoted $Pr_{\mathcal{A}}(x)$. The distribution modelled by a machine $\mathcal{A}$ will be denoted $\mathcal{D}_{\mathcal{A}}$ and simplified to $\mathcal{D}$ if the context is not ambiguous.

If L is a language (a subset of $\Sigma^\star$), and $\mathcal{D}$ a distribution over $\Sigma^\star$, $Pr_{\mathcal{D}}(L) = \sum_{x \in L} Pr_{\mathcal{D}}(x)$.

Two distributions $\mathcal{D}$ and $\mathcal{D}'$ are equal (denoted by $\mathcal{D} = \mathcal{D}'$) *if*$_{def}$ $\forall\, w \in \Sigma^\star$, $Pr_{\mathcal{D}}(w) = Pr_{\mathcal{D}'}(w)$.

In order to avoid confusion, even if we are defining languages (albeit probabilistic), we will reserve the notation L for sets of strings and denote the probabilistic languages as distributions, thus with symbol $\mathcal{D}$.

5.1.2 About the probabilities

In theory all the probabilistic objects we are studying could take arbitrary values. There are even cases where negative (or even complex) values are of interest! Nevertheless to fix things and to make sure that computational issues are taken into account we will take the simplified view that all probabilities will be rational numbers between 0 and 1, described by fractions. Their encoding then depends on the encoding of the two integers composing the fraction.

5.2 Probabilistic automata

We are concerned here with generating strings following distributions. If, in the non-probabilistic context, automata are used for parsing and recognising, they will be considered here as *generative* devices.

5.2.1 Probabilistic finite automata (PFA)

The first thing one can do is add probabilities to the non-deterministic finite automata:

Definition 5.2.1 (Probabilistic finite automaton (PFA)) A **probabilistic finite automaton (PFA)** is a tuple $\mathcal{A}=\langle \Sigma, Q, \mathbb{I}_{\mathbb{P}}, \mathbb{F}_{\mathbb{P}}, \delta_{\mathbb{P}} \rangle$, where:

- Q is a finite set of **states**; these will be labelled $q_1, \ldots, q_{|Q|}$ unless otherwise stated,
- Σ is the alphabet,
- $\mathbb{I}_{\mathbb{P}} : Q \to \mathbb{Q}^+ \cap [0, 1]$ (initial-state probabilities),
- $\mathbb{F}_{\mathbb{P}} : Q \to \mathbb{Q}^+ \cap [0, 1]$ (final-state probabilities),
- $\delta_{\mathbb{P}} : Q \times (\Sigma \cup \{\lambda\}) \times Q \to \mathbb{Q}^+$ is a transition function; the function is complete: $\delta_{\mathbb{P}}(q, a, q') = 0$ can be interpreted as 'no transition from q to q' labelled with a'. We will also denote (q, a, q', P) instead of $\delta_{\mathbb{P}}(q, a, q') = P$ where P is a probability.

$\mathbb{I}_{\mathbb{P}}$, $\delta_{\mathbb{P}}$ and $\mathbb{F}_{\mathbb{P}}$ are functions such that:

$$\sum_{q \in Q} \mathbb{I}_{\mathbb{P}}(q) = 1,$$

and $\forall q \in Q$,

$$\mathbb{F}_{\mathbb{P}}(q) + \sum_{a \in \Sigma \cup \{\lambda\},\ q' \in Q} \delta_{\mathbb{P}}(q, a, q') = 1.$$

In contrast with the definition of NFA, there are no accepting (or rejecting) states. As will be seen in the next section, the above definition of automata describes models that are generative in nature. Yet in the classical setting automata are usually introduced in order to parse strings rather than to generate them, so there may be some confusion here. The advantages of the definition are nevertheless that there is a clear correspondence with probabilistic left (or right) context-free grammars. But on the other hand if we were requiring a finite state machine to parse a string and somehow come up with some probability that the string belongs or not to a given language, we would have to turn to another type of machine.

Figure 5.2 shows a ***graphical representation*** of a PFA with four states, $Q = \{q_1, q_2, q_3, q_4\}$, two initial states, q_1 and q_2, with respective probabilities of being chosen of 0.4 and 0.6, and a two-symbol alphabet, $\Sigma = \{\mathtt{a}, \mathtt{b}\}$. The numbers on the edges are the transition probabilities. The initial and final probabilities are drawn inside the state (as in Figure 5.1) before and after the name of the state. The transitions whose weights are 0 are not drawn. We will introduce this formally later, but we will also say that the automaton ***respects the following set of constraints***: $\{(q_1, \mathtt{a}, q_2), (q_1, \mathtt{b}, q_1), (q_2, \lambda, q_3), (q_2, \mathtt{b}, q_4), (q_3, \mathtt{b}, q_3), (q_3, \mathtt{b}, q_4), (q_4, \mathtt{a}, q_1), (q_4, \mathtt{a}, q_2)\}$. This means that these are the only transitions with non-null weight in the automaton.

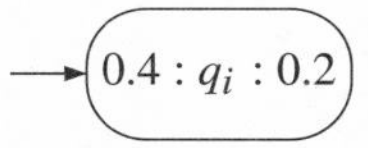

$0.4 : q_i : 0.2$

Fig. 5.1. Graphical representation of a state q_i. $\mathbb{I}_{\mathbb{P}}(q_i) = 0.4$, $\mathbb{F}_{\mathbb{P}}(q_i) = 0.2$.

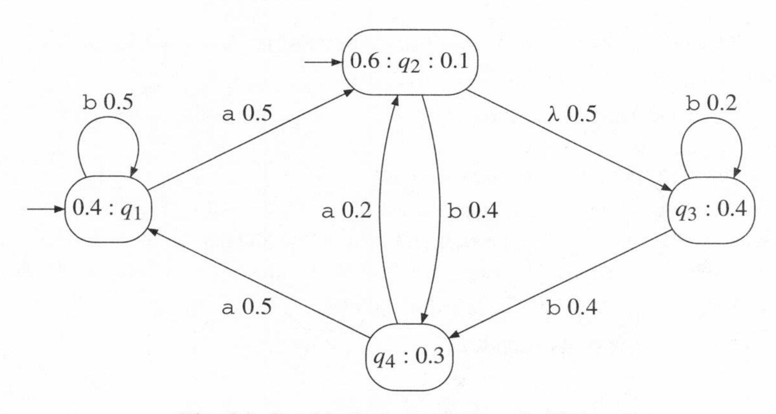

Fig. 5.2. Graphical representation of a PFA.

The above definition allows λ-transitions and λ-loops. In Figure 5.2, there is a λ-transition between state q_2 and state q_3. Such λ-transitions make things difficult for parsing. We show in Section 5.2.4 that λ-transitions and λ-loops can actually be removed in polynomial time without changing the distribution. When needing to differentiate we shall call the former λ-PFA, and the latter (when λ-transitions are forbidden) λ-free PFA.

Definition 5.2.2 For any λ-PFA $\mathcal{A} = \langle \Sigma, Q, \mathbb{I}_{\mathbb{P}}, \mathbb{F}_{\mathbb{P}}, \delta_{\mathbb{P}} \rangle$:

- a **λ-transition** is any transition labelled by λ,
- a **λ-loop** is a transition of the form (q, λ, q, P),
- a **λ-cycle** is a sequence of λ-transitions from $\delta_{\mathbb{P}}$ with the same starting and ending state: $(q_{i_1}, \lambda, q_{i_2}, P_1) \ldots (q_{i_j}, \lambda, q_{i_{j+1}}, P_j) \ldots (q_{i_k}, \lambda, q_{i_1}, P_k)$.

5.2.2 *Deterministic probabilistic finite-state automata* (DPFA)

As in the non-probabilistic case we can restrict the definitions in order to make parsing deterministic. In this case 'deterministic' means that there is only one way to generate each string at each moment, i.e. that in each state, for each symbol, the next state is unique.

The main tool we will use to generate distributions is deterministic probabilistic finite (state) automata. These will be the probabilistic counterparts of the deterministic finite automata.

Definition 5.2.3 (Deterministic probabilistic finite automata) A PFA $\mathcal{A} = \langle \Sigma, Q, \mathbb{I}_{\mathbb{P}}, \mathbb{F}_{\mathbb{P}}, \delta_{\mathbb{P}} \rangle$ is a **deterministic probabilistic finite automaton (DPFA)** if_{def}

- $\exists q_1 \in Q$ (**unique initial state**) such that $\mathbb{I}_{\mathbb{P}}(q_1) = 1$;
- $\delta_{\mathbb{P}} \subseteq Q \times \Sigma \times Q$ (no λ-transitions);
- $\forall q \in Q,\ \forall a \in \Sigma,\ |\{q' : \delta_{\mathbb{P}}(q, a, q') > 0\}| \leq 1$.

In a DPFA, a transition (q, a, q', P) is completely defined by q and a. The above definition is cumbersome in this case and we will associate with a DPFA a non-probabilistic transition function:

Definition 5.2.4 Let $\mathcal{A}$ be a DPFA. $\delta_{\mathcal{A}} : Q \times \Sigma \to Q$ is **the transition function** with $\delta_{\mathcal{A}}(q, a) = q' : \delta_{\mathbb{P}}(q, a, q') \neq 0$.

This function is extended (as in Chapter 4) in a natural way to strings:

$$\delta_{\mathcal{A}}(q, \lambda) = q$$
$$\delta_{\mathcal{A}}(q, a \cdot u) = \delta_{\mathcal{A}}(\delta_{\mathcal{A}}(q, a), u).$$

The probability function also extends easily in the deterministic case to strings:

$$\delta_{\mathbb{P}}(q, \lambda, q) = 1$$
$$\delta_{\mathbb{P}}(q, a \cdot u, q') = \delta_{\mathbb{P}}(q, a, \delta_{\mathcal{A}}(q, a)) \cdot \delta_{\mathbb{P}}(\delta_{\mathcal{A}}(q, a), u, q').$$

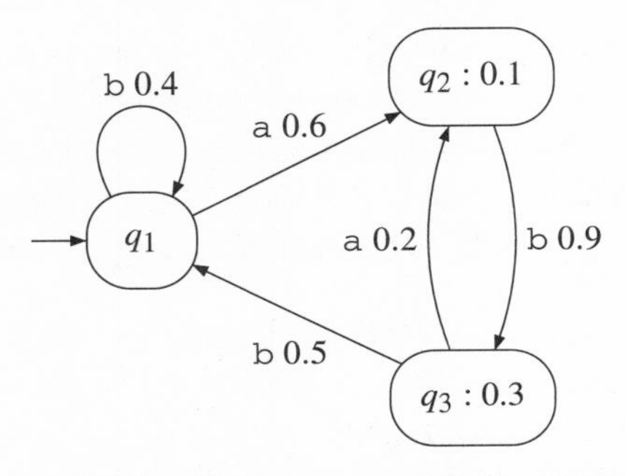

Fig. 5.3. Graphical representation of a DPFA.

But, in the deterministic case, one should note that $\delta_{\mathbb{P}}(q, \lambda, q) = 1$ is not referring to the introduction of a λ-loop with weight 1 (and therefore the loss of determinism), but to the fact that the probability of having λ as a prefix is 1.

The size of a DPFA depends on the number n of states, the size $|\Sigma|$ of the alphabet and the number of bits needed to encode the probabilities. This is also the case when dealing with a PFA or λ-PFA.

In the DPFA depicted in Figure 5.3, the initial probabilities need not be represented: there is a unique initial state (here q_1) with probability of being initial equal to 1.

5.2.3 Parsing with a PFA

PFAs are adequate probabilistic machines to generate strings of finite length. Given a PFA (or λ-PFA) $\mathcal{A}$, the process is described in Algorithm 5.1. For this we suppose we have two functions:

Algorithm 5.1: Generating a string with a PFA.

Data: a PFA $\mathcal{A} = \langle \Sigma, Q, \mathbb{I}_{\mathbb{P}}, \mathbb{F}_{\mathbb{P}}, \delta_{\mathbb{P}} \rangle$, $\mathbf{Random}_Q$, $\mathbf{Random}_{\mathbb{I}_{\mathbb{P}}}$
Result: a string x
$x \leftarrow \lambda$;
$q \leftarrow \mathbf{Random}_{\mathbb{I}_{\mathbb{P}}}(\mathcal{A})$;
$(a, q') \leftarrow \mathbf{Random}_Q(q, \mathcal{A})$;
while $a \neq H$ **do** /* generate another symbol */
  $x \leftarrow x \cdot a$;
  $q \leftarrow q'$;
  $(a, q') \leftarrow \mathbf{Random}_Q(q, \mathcal{A})$
end
return x

- **Random**$_Q(q, \mathcal{A})$ takes a state q from a PFA $\mathcal{A}$ and returns a pair (a, q') drawn according to the distribution at state q in PFA $\mathcal{A}$ from all pairs in $\Sigma \times Q$ including the pair (H, q), where H is a special character not in Σ, whose meaning is 'halt'.
- **Random**$_{\mathbb{I}_\mathbb{P}}(\mathcal{A})$, which returns an initial state depending on the distribution defined by $\mathbb{I}_\mathbb{P}$.

Inversely, given a string and a PFA, how do we compute the probability of the string? We will in this section only work with λ-free automata. If the automaton contains λ-transitions, parsing requires us to compute the probabilities of moving freely from one state to another. But in that case, as this in itself is expensive, the elimination of the λ-transitions (for example with the algorithm from Section 5.2.4) should be considered. There are two classical algorithms that compute the probability of a string by dynamic programming.

The FORWARD algorithm (Algorithm 5.2) computes $Pr_\mathcal{A}(q_s, w)$, the probability of generating string w and being in state q_s after having done so. By then multiplying by the final probabilities (Algorithm 5.3) and summing up, we get the actual probability of the string. The BACKWARD algorithm (Algorithm 5.4) works in a symmetrical way. It computes Table B, whose entry B[0][s] is $Pr_\mathcal{A}(w|q_s)$, the probability of generating string

---

Algorithm 5.2: FORWARD.

Data: a PFA$A = \langle \Sigma, Q, \mathbb{I}_\mathbb{P}, \mathbb{F}_\mathbb{P}, \delta_\mathbb{P} \rangle$, a string $x = a_1 a_2 \cdots a_n$
Result: the probabilities $\mathrm{F}[n][s] = Pr_\mathcal{A}(x, q_s)$
for $j : 1 \le j \le |Q|$ **do** `/* Initialise */`
  $\mathrm{F}[0][j] \leftarrow \mathbb{I}_\mathbb{P}(q_j)$;
  for $i : 1 \le i \le n$ **do** $\mathrm{F}[i][j] \leftarrow 0$
end
for $i : 1 \le i \le n$ **do**
  for $j : 1 \le j \le |Q|$ **do**
    for $k : 1 \le k \le |Q|$ **do**
      $\mathrm{F}[i][j] \leftarrow \mathrm{F}[i][j] + \mathrm{F}[i-1][k] \cdot \delta_\mathbb{P}(q_k, a_i, q_j)$
    end
  end
end
return F

Algorithm 5.3: Computing the probability of a string with FORWARD.

Data: the probabilities $\mathrm{F}[n][s] = Pr_\mathcal{A}(x, q_s)$, a PFA $\mathcal{A}$
Result: the probability $\mathrm{T} = Pr_\mathcal{A}(x)$
$\mathrm{T} \leftarrow 0$;
for $j : 1 \le j \le |Q|$ **do** $\mathrm{T} \leftarrow \mathrm{T} + \mathrm{F}[n][j] \cdot \mathbb{F}_\mathbb{P}[j]$;
return T

Algorithm 5.4: BACKWARD

Data: a PFA $\mathcal{A} = \langle \Sigma, Q, \mathbb{I}_{\mathbb{P}}, \mathbb{F}_{\mathbb{P}}, \delta_{\mathbb{P}} \rangle$, a string $x = a_1 a_2 \cdots a_n$
Result: the probabilities B[0][s] $= Pr_{\mathcal{A}}(x|q_s)$
for $j : 1 \leq j \leq |Q|$ **do** /* Initialise */
  B[n][j] $\leftarrow \mathbb{F}_{\mathbb{P}}[j]$;
  for $i : 1 \leq i \leq n$ **do** B[i][j] $\leftarrow 0$
end
for $i : n-1 \geq i \geq 0$ **do**
  for $j : 1 \leq j \leq |Q|$ **do**
    for $k : 1 \leq k \leq |Q|$ **do**
      B[i][j] $\leftarrow$B[i][j]+B[$i+1$][k] $\cdot \delta_{\mathbb{P}}(q_j, a_i, q_k)$
    end
  end
end
return B

Algorithm 5.5: Computing the probability of a string with BACKWARD.

Data: the probabilities B[0][s] $= Pr_{\mathcal{A}}(x|q_s)$, a PFA $\mathcal{A}$
Result: the probability $Pr_{\mathcal{A}}(x)$
T$\leftarrow$ 0;
for $j : 1 \leq j \leq |Q|$ **do** T$\leftarrow$ T+B[0][j] $\cdot \mathbb{I}_{\mathbb{P}}[j]$;
return T

w when starting in state q_s (which has initial probability 1). If we then run Algorithm 5.5 to this we also get the value $Pr_{\mathcal{A}}(w)$.

These probabilities will be of use when dealing with parameter estimation questions in Chapter 17.

Another parsing problem, that of finding the most probable path in the PFA that generates a string w, is computed through the VITERBI algorithm (Algorithm 5.6).

The computation of Algorithms 5.2, 5.4 and 5.6 have a time complexity in $\mathcal{O}(|x| \cdot |Q|^2)$. A slightly different implementation allows us to reduce this complexity by only visiting the states that are really successors of a given state. A vector representation allows us to make the space complexity linear.

In the case of DPFAs, the algorithms are simpler than for non-deterministic PFAs. In this case, the computation cost of Algorithm 5.6 is in $\mathcal{O}(|x|)$, that is, the computational cost does not depend on the number of states since at each step the only possible next state is computed with a cost in $\mathcal{O}(1)$.

Algorithm 5.6: VITERBI.

Data: a PFA $\mathcal{A} = \langle \Sigma, Q, \mathbb{I}_\mathbb{P}, \mathbb{F}_\mathbb{P}, \delta_\mathbb{P} \rangle$, a string $x = a_1 a_2 \cdots a_n$
Result: a sequence of states bestpath= $q_{p_0} \dots q_{p_n}$ reading x and maximising:
$\mathbb{I}_\mathbb{P}(q_{p_0}) \cdot \delta_\mathbb{P}(q_{p_0}, a_1, q_{p_1}) \cdot \delta_\mathbb{P}(q_{p_1}, a_2, q_{p_2}) \dots \delta_\mathbb{P}(q_{p_{n-1}}, a_n, q_{p_n}) \cdot \mathbb{F}_\mathbb{P}(q_{p_n})$

```
for j : 1 ≤ j ≤ |Q| do                                          /* Initialise */
    V[0][j] ← I_P(q_j);
    Vpath[0][j] ← λ;
    for i : 1 ≤ i ≤ n do
        V[i][j] ← 0;
        Vpath[i][j] ← λ
    end
end
for i : 1 ≤ i ≤ n do
    for j : 1 ≤ j ≤ |Q| do
        for k : 1 ≤ k ≤ |Q| do
            if V[i][j] < V[i − 1][k] · δ_P(q_k, a_i, q_j) then
                V[i][j] ← V[i − 1][k] · δ_P(q_k, a_i, q_j);
                Vpath[i][j] ← Vpath[i − 1][j] · q_j
            end
        end
    end
end
bestscore← 0;             /* Multiply by the halting probabilities */
for j : 1 ≤ j ≤ |Q| do
    if V[n][j] · F_P[j] >bestscore then
        bestscore← V[n][j] · F_P[j];
        bestpath←Vpath[n][j]
    end
end
return bestpath
```

We compute the probability of string ab in the automaton of Figure 5.4:

$$\begin{aligned}
Pr_\mathcal{A}(\mathtt{ab}) &= \mathbb{I}_\mathbb{P}(q_1) \cdot \delta_\mathbb{P}(q_1, \mathtt{a}, q_2) \cdot \delta_\mathbb{P}(q_2, \mathtt{b}, q_3) \cdot \mathbb{F}_\mathbb{P}(q_3) \\
&\quad + \mathbb{I}_\mathbb{P}(q_2) \cdot \delta_\mathbb{P}(q_2, \mathtt{a}, q_4) \cdot \delta_\mathbb{P}(q_4, \mathtt{b}, q_2) \cdot \mathbb{F}_\mathbb{P}(q_2) \\
&\quad + \mathbb{I}_\mathbb{P}(q_2) \cdot \delta_\mathbb{P}(q_2, \mathtt{a}, q_4) \cdot \delta_\mathbb{P}(q_4, \mathtt{b}, q_1) \cdot \mathbb{F}_\mathbb{P}(q_1) \\
&= 0.4 \cdot 0.5 \cdot 0.2 \cdot 0.5 + 0.6 \cdot 0.2 \cdot 0.2 \cdot 0.6 + 0.6 \cdot 0.2 \cdot 0.5 \cdot 0 \\
&= 0.0344.
\end{aligned}$$

Table 5.1. *Details of the computation of B and F for string* ab *on Automaton 5.4 by Algorithms* BACKWARD *and* FORWARD.

State	B	F
q_1	B[2][1] $= Pr(\lambda\|0) = 0$	F[0][1] $= Pr(\lambda, 1) = 0.4$
q_2	B[2][2] $= Pr(\lambda\|1) = 0.6$	F[0][2] $= Pr(\lambda, 2) = 0.6$
q_3	B[2][3] $= Pr(\lambda\|2) = 0.5$	F[0][3] $= Pr(\lambda, 3) = 0$
q_4	B[2][4] $= Pr(\lambda\|3) = 0.3$	F[0][4] $= Pr(\lambda, 4) = 0$
q_1	B[1][1] $= Pr(\mathrm{b}\|0) = 0$	F[1][1] $= Pr(\mathrm{a}, 1) = 0$
q_2	B[1][2] $= Pr(\mathrm{b}\|1) = 0.1$	F[1][2] $= Pr(\mathrm{a}, 2) = 0.2$
q_3	B[1][3] $= Pr(\mathrm{b}\|2) = 0.06$	F[1][3] $= Pr(\mathrm{a}, 3) = 0$
q_4	B[1][4] $= Pr(\mathrm{b}\|3) = 0.12$	F[1][4] $= Pr(\mathrm{a}, 4) = 0.12$
q_1	B[0][1] $= Pr(\mathrm{ab}\|0) = 0.05$	F[2][1] $= Pr(\mathrm{ab}, 1) = 0.06$
q_2	B[0][2] $= Pr(\mathrm{ab}\|1) = 0.024$	F[2][2] $= Pr(\mathrm{ab}, 2) = 0.024$
q_3	B[0][3] $= Pr(\mathrm{ab}\|2) = 0.018$	F[2][3] $= Pr(\mathrm{ab}, 3) = 0.04$
q_4	B[0][4] $= Pr(\mathrm{ab}\|3) = 0$	F[2][4] $= Pr(\mathrm{ab}, 4) = 0$

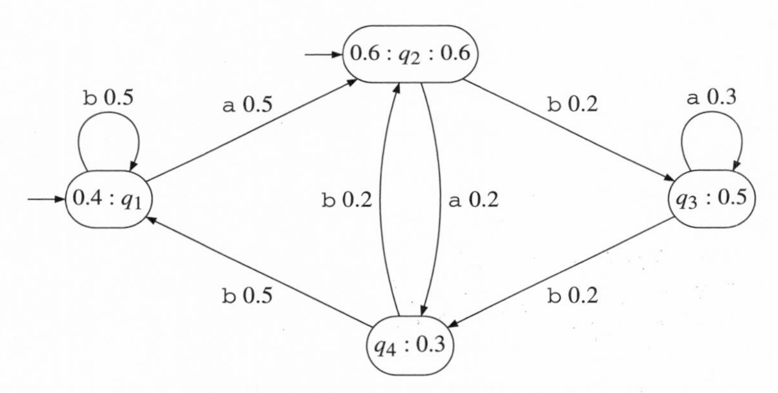

Fig. 5.4. A PFA $\mathcal{A}$, $Pr_{\mathcal{A}}(\mathrm{ab}) = 0.0344$.

We can also trace the BACKWARD and FORWARD computations in Table 5.1. We obtain:

$$0.05 \cdot 0.4 + 0.024 \cdot 0.6 + 0.018 \cdot 0 + 0 \cdot 0 =$$
$$0.06 \cdot 0 + 0.024 \cdot 0.6 + 0.04 \cdot 0.5 + 0 \cdot 0.3 = 0.0264.$$

We conclude this section by defining classes of string distributions on the basis of the corresponding generating automata.

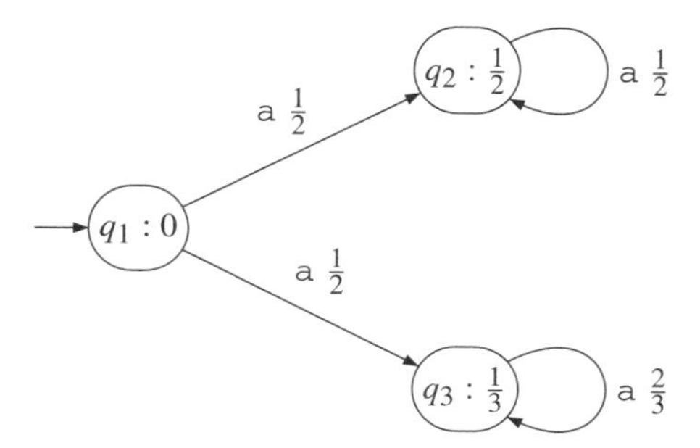

Fig. 5.5. A regular distribution that is not deterministic.

Definition 5.2.5 A distribution is **regular** *if$_{def}$* it can be generated by some PFA. We denote by $\mathcal{REG}_{\mathcal{P}}(\Sigma)$ the class of all regular distributions (over alphabet Σ).

Definition 5.2.6 A distribution is **regular deterministic** *if$_{def}$* it can be generated by some DPFA. We denote by $\mathcal{DET}_{\mathcal{P}}(\Sigma)$ the class of all regular deterministic distributions (over alphabet Σ).

Definition 5.2.7 Two PFAs are **equivalent** *if$_{def}$* they generate the same distribution.

From the definitions of PFA and DPFA the following hierarchy is obtained:

Proposition 5.2.1 *A regular deterministic distribution is also a regular distribution. There exist distributions which are regular but not regular deterministic.*

Proof It can be checked that the distribution defined by the PFA from Figure 5.5 is not regular deterministic. □

5.2.4 *Eliminating λ-transitions in a* PFA

The algorithms from the previous section were able to parse whenever the PFA contained no λ-transitions. But even if a specific parser using λ-transitions can be built, it is tedious to compute with it or to generate strings in a situation where the λ-transitions may even be more frequent than the other transitions. We prove in this section that λ-transitions are not necessary in probabilistic finite automata and that there is no added power in automata that use λ-transitions compared to those that don't. To do so we propose an algorithm that takes a λ-PFA as input and returns an equivalent PFA with no more states, but where the (rational) probabilities may require a polynomial number only of extra bits for encoding.

Remember that a λ-PFA can not only have λ-transitions (including λ-loops), but also various initial states. Thus $\mathbb{I}_{\mathbb{P}}$ is a function $Q \to \mathbb{Q}^+$, such that $\sum_{q\in Q} \mathbb{I}_{\mathbb{P}}(q) = 1$.

Theorem 5.2.2 *Given a* λ-PFA $\mathcal{A}$ *representing distribution* $\mathcal{D}_{\mathcal{A}}$*, there exists a* λ*-free* PFA $\mathcal{B}$ *such that* $\mathcal{D}_{\mathcal{A}} = \mathcal{D}_{\mathcal{B}}$*. Moreover* $\mathcal{B}$ *is of total size at most n times the total size of* $\mathcal{A}$*, where n is the number of states in* $\mathcal{A}$*.*

Algorithm 5.7: Transforming the λ-PFA into a λ-PFA with just one initial state.

Input: a λ-PFA : $\langle \Sigma, Q, \mathbb{I}_\mathbb{P}, \mathbb{F}_\mathbb{P}, \delta_\mathbb{P} \rangle$
Output: a λ-PFA : $\langle \Sigma, Q \cup \{q_{new}\}, \mathbb{I}_\mathbb{P}', \mathbb{F}_\mathbb{P}', \delta_\mathbb{P}' \rangle$ with one initial state
$Q' \leftarrow Q \cup \{q_{new}\}$;
for $q \in Q$ **do**
  $\delta_\mathbb{P}'(q_{new}, \lambda, q) \leftarrow \mathbb{I}_\mathbb{P}(q)$;
  $\mathbb{I}_\mathbb{P}'(q) \leftarrow 0$
end
$\mathbb{I}_\mathbb{P}'(q_{new}) \leftarrow 1$;
$\mathbb{F}_\mathbb{P}'(q_{new}) \leftarrow 0$;
return $\langle \Sigma, Q \cup \{q_{new}\}, \mathbb{I}_\mathbb{P}', \mathbb{F}_\mathbb{P}', \delta_\mathbb{P}' \rangle$

Proof To convert $\mathcal{A}$ into an equivalent PFA $\mathcal{B}$ there are two steps. The starting point is a PFA $\mathcal{A} = \langle \Sigma, Q, \mathbb{I}_\mathbb{P}, \mathbb{F}_\mathbb{P}, \delta_\mathbb{P} \rangle$ where the set of states Q is labelled by numbers. We will update this numbering as we add more states.

Step 1: If there is more than one initial state, add a new initial state and λ-transitions from this state to each of the previous initial states, with probability equal to that of the state being initial. Notice that this step can be done in polynomial time and that the resulting automaton contains at most one more state than the initial one. This is described by Algorithm 5.7.

Step 2: Algorithm 5.8 iteratively removes a λ-loop if there is one, and if not, the λ-transition with maximal extremity. In a finite number of steps, the algorithm terminates.

To prove that in a finite number of steps all λ-transitions are removed, we first associate with any automaton $\mathcal{A}$ the value $\mu(\mathcal{A})$ corresponding to the largest number of a state in which a λ-transition ends.

$$\mu(\mathcal{A}) = \max\{i : q_i \in Q \text{ and } \delta_\mathbb{P}(q, \lambda, q_i) \neq 0\}.$$

Now let $\rho(\mathcal{A})$ be the number of λ-transitions ending in $q_{\mu(\mathcal{A})}$.

$$\rho(\mathcal{A}) = \left|\{q \in Q : \delta_\mathbb{P}(q, \lambda, q_{\mu(\mathcal{A})}) \neq 0\}\right|.$$

Finally, the value $v(\mathcal{A}) = \langle \mu(\mathcal{A}), \rho(\mathcal{A}) \rangle$ will decrease at each step of the algorithm, thus ensuring the termination and the convergence. In Figure 5.6, we have $\mu(\mathcal{A}) = 3$, $\rho(\mathcal{A}) = 2$ and $v(\mathcal{A}) = \langle 3, 2 \rangle$.

Algorithm 5.8 takes the current PFA and eliminates a λ-loop if there is one. If not it chooses a λ-transition ending in the state with largest number and eliminates it.

One can notice that in Algorithm 5.8, new λ-loops can appear when such a λ-transition is eliminated, but that is no problem because they will only appear in states of smaller index than the current state that is being considered.

Algorithm 5.8: Eliminating λ-transitions.

Input: a λ − PFA : $\langle \Sigma, Q, \{q_1\}, \mathbb{F}_{\mathbb{P}}, \delta_{\mathbb{P}} \rangle$ with only one initial state
Output: a λ-free PFA : $\langle \Sigma, Q, \{q_1\}, \mathbb{F}_{\mathbb{P}}, \delta_{\mathbb{P}} \rangle$
while *there still are λ-transitions* **do**
  if *there exists a λ-loop (q, λ, q, P)* **then**
    for *all transitions (q, a, q') $((a, q') \neq (\lambda, q))$* **do**
      $\delta_{\mathbb{P}}(q, a, q') \leftarrow \delta_{\mathbb{P}}(q, a, q') \cdot \frac{1}{1-\delta_{\mathbb{P}}(q,\lambda,q)}$
    end
    $\mathbb{F}_{\mathbb{P}}(q) \leftarrow \mathbb{F}_{\mathbb{P}}(q) \cdot \frac{1}{1-\delta_{\mathbb{P}}(q,\lambda,q)}$;
    $\delta_{\mathbb{P}}(q, \lambda, q) \leftarrow 0$
  else /* there are no λ-loops */
    let (q, λ, q_m) be a λ-transition with m maximal;
    foreach $(q_m, \lambda, q_n, P_\lambda)$ **do** /* $n < m$ */
      $\delta_{\mathbb{P}}(q, \lambda, q') \leftarrow \delta_{\mathbb{P}}(q, \lambda, q_n) + \delta_{\mathbb{P}}(q, \lambda, q_m) \cdot \delta_{\mathbb{P}}(q_m, \lambda, q_n)$
    end
    foreach (q_m, a, q_n, P_a) **do** /* $a \in \Sigma$ */
      $\delta_{\mathbb{P}}(q, a, q_n) \leftarrow \delta_{\mathbb{P}}(q, a, q_n) + \delta_{\mathbb{P}}(q, \lambda, q_m) \cdot \delta_{\mathbb{P}}(q_m, a, q_n)$
    end
    $\mathbb{F}_{\mathbb{P}}(q) \leftarrow \mathbb{F}_{\mathbb{P}}(q) + \delta_{\mathbb{P}}(q, \lambda, q_m) \cdot \mathbb{F}_{\mathbb{P}}(q_m)$;
    $\delta_{\mathbb{P}}(q, \lambda, q_m) \leftarrow 0$
  end
end
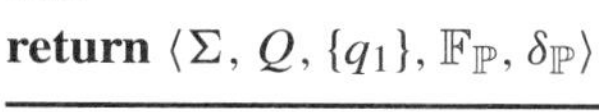
return $\langle \Sigma, Q, \{q_1\}, \mathbb{F}_{\mathbb{P}}, \delta_{\mathbb{P}} \rangle$

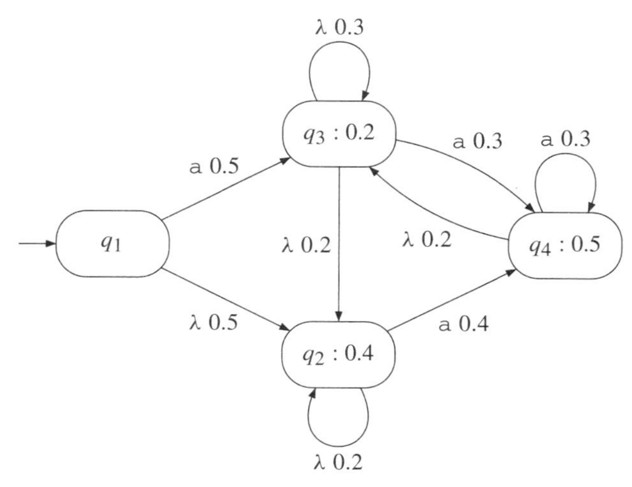

Fig. 5.6. A PFA.

Therefore the quantity $v(\mathcal{A})$ decreases (for the lexicographic order over $\mathbb{N}^2$) at each round of the algorithm, since a new λ-transition only appears as a combination of a (q, λ, q_m) and a (q_m, λ, q_n) and $n < m$ with $m = \mu(\mathcal{A})$.

As clearly this can only take place a finite number of times, the algorithm converges.

Then we prove that at each step $v(\mathcal{A})$ does not increase (for the lexicographic order), and that there can only be a finite number of consecutive steps where $v(\mathcal{A})$ remains equal. Summarising, at each step one of the following holds:

- A λ-loop is erased. Then $v(\mathcal{A})$ is left untouched because no new λ-transition is introduced, but the number of λ-loops is bounded by the number of states of the PFA. So only a finite number of λ-loop elimination steps can be performed before having no λ-loops left.
- A λ-transition (that is not a loop) is replaced. This transition is a (q, λ, q_m, P) with $\mu(\mathcal{A}) = m$. Therefore only λ-transitions with terminal vertex of index smaller than m can be introduced. So $v(\mathcal{A})$ diminishes. □

Also, clearly, if the probabilities are rational, they remain so.

A run of the algorithm. We suppose the λ-PFA contains just one initial state and is represented in Figure 5.7(a). First, the λ-loops at states q_2 and q_3 are eliminated and several transitions are updated (Figure 5.7(b)).

The value of $v(\mathcal{A})$ is $\langle 3, 1\rangle$. At that point the algorithm eliminates transition (q_4, λ, q_3) because $\mu(\mathcal{A}) = 3$. The resulting PFA is represented in Figure 5.8(a). The new value of $v(\mathcal{A})$ is $\langle 2, 3\rangle$ which, for the lexicographic order, is less than the previous value.

The PFA is represented in Figure 5.8(b). The new value of $v(\mathcal{A})$ is $\langle 2, 2\rangle$. Then transition (q_3, λ, q_2) is eliminated resulting in the PFA represented in Figure 5.9(a) whose value

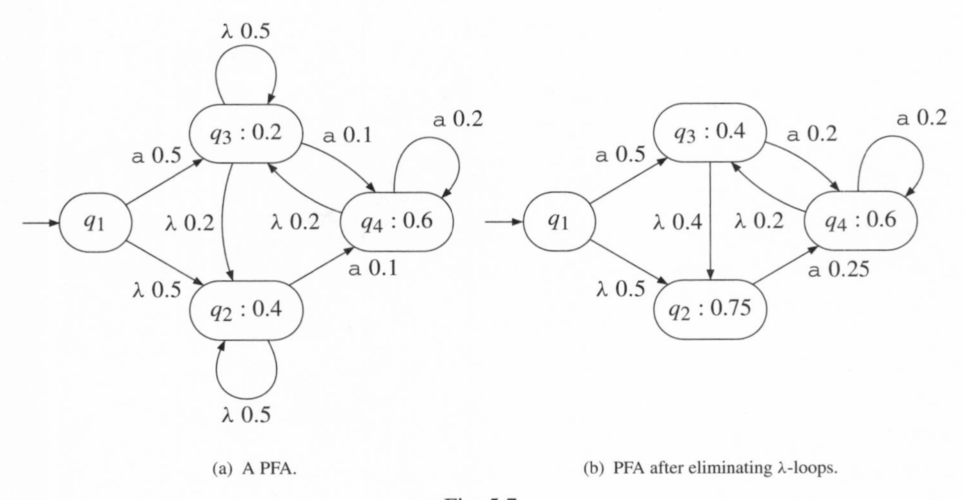

(a) A PFA. (b) PFA after eliminating λ-loops.

Fig. 5.7.

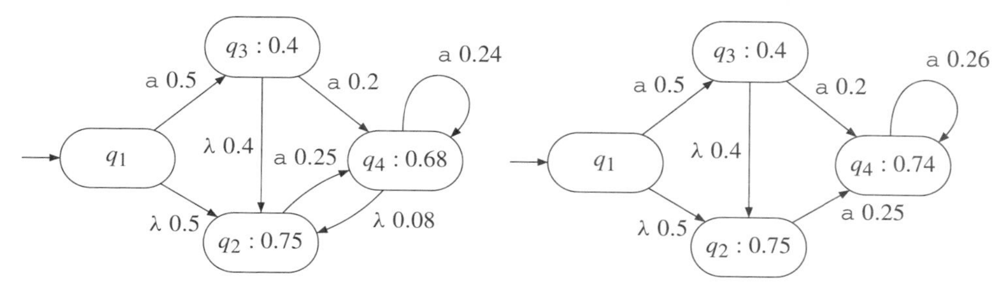

(a) PFA after eliminating transition (q_4, λ, q_3).

(b) PFA after eliminating transition (q_4, λ, q_2).

Fig. 5.8.

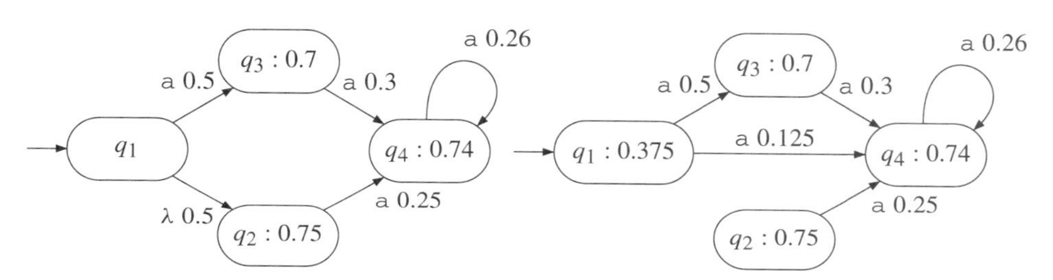

(a) PFA after eliminating transition (q_3, λ, q_2).

(b) PFA after eliminating transition (q_1, λ, q_2).

Fig. 5.9.

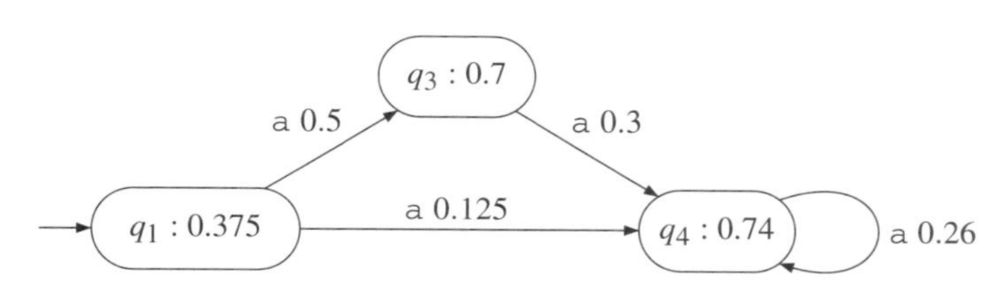

Fig. 5.10. PFA after simplification.

$v(\mathcal{A})$ is $\langle 2, 1\rangle$. Finally the last λ-transition is removed and the result is represented in Figure 5.9(b). At this point, since state q_2 is no longer reachable, the automaton can be pruned as represented in Figure 5.10.

Complexity issues. The number of states in the resulting automaton is identical to the initial one. The number of transitions is bounded by $|Q|^2 \cdot |\Sigma|$. But each multiplication of probabilities can result in the summing of the number of bits needed to encode each probability. Since the number of λ-loops to be removed on one path is bounded by n, this also gives a polynomial bound to the size of the resulting automaton.

Conclusion. Automatic construction of a PFA (through union operations for example) will lead to the introduction of λ-transitions or multiple initial states.

Algorithms 5.7 and 5.8 remove λ-transitions, and an equivalent PFA with just one initial state can be constructed for a relatively small cost.

It should be remembered that not all non-determinism can be eliminated, as it is known that PFAs are strictly more powerful than their deterministic counterparts (see Proposition 5.2.1, page 95).

5.3 Probabilistic context-free grammars

One can also add probabilities to grammars, but the complexity increases. We therefore only survey here some of the more elementary definitions.

Definition 5.3.1 A **probabilistic context-free grammar (PCFG)** G is a quintuple $< \Sigma, V, R, P, N_1 >$ where Σ is a finite alphabet (of terminal symbols), V is a finite alphabet (of variables or non-terminals), $R \subset V \times (\Sigma \cup V)^*$ is a finite set of production rules, $P : R \to \mathbb{R}^+$ is the probability function, and N_1 $(\in V)$ is the axiom.

As in the case of automata, we will restrict ourselves to the case where the probabilities are rational (in $\mathbb{Q} \cap [0, 1]$).

Given a PCFG G, a string w, and a left-most derivation $d = \alpha_0 \Rightarrow \ldots \Rightarrow \alpha_k$ where

$$\begin{aligned}
&\alpha_0 = N_{i_0} = N_1,\\
&\alpha_k = w,\\
&\alpha_j = l_j N_{i_j} \gamma \text{ with } l_j \in \Sigma^\star, \gamma \in (\Sigma \cup V)^*, \text{and } j < k,\\
&\alpha_{j+1} = l_j \beta_{i_j} \gamma \text{ with } (N_{i_j}, \beta_{i_j}) \in R
\end{aligned}$$

we define the weight of derivation d as: $Pr_{G,d}(w) = \prod_{0 \le j < k} P(N_{i_j}, \beta_{i_j})$ where (N_{i_j}, β_{i_j}) is the rule used to rewrite α_j into α_{j+1}. Notice that it is essential to count only the left-most derivations in order not to count the same derivation various times. The quantity $Pr_{G,d}(w)$ is the probability of generating string w **and** of doing this using derivation d.

Now we sum over all (left-most) derivations: $Pr_G(w) = \sum_d Pr_{G,d}(w)$.

Example 5.3.1 Consider the grammar $G = <\{\mathtt{a}, \mathtt{b}\}, \{N_1\}, R, P, N_1>$ with $R = \{(N_1, \mathtt{a}N_1\mathtt{b}), (N_1, \mathtt{a}N_1), (N_1, \lambda)\}$ and the probabilities of the rules given by $P(N_1, \lambda) = \frac{1}{6}$, $P(N_1, \mathtt{a}N_1\mathtt{b}) = \frac{1}{2}$ and $P(N_1, \mathtt{a}N_1) = \frac{1}{3}$.

Then string $\mathtt{aab}$ can be generated through $d_1 = N_1 \Rightarrow \mathtt{a}N_1\mathtt{b} \Rightarrow \mathtt{aa}N_1\mathtt{b} \Rightarrow \mathtt{aab}$ and $d_2 = N_1 \Rightarrow \mathtt{a}N_1 \Rightarrow \mathtt{aa}N_1\mathtt{b} \Rightarrow \mathtt{aab}$. Both derivations have probability $Pr_{G,d_i}(\mathtt{aab}) = \frac{1}{36}$, and therefore $Pr_G(\mathtt{aab}) = \frac{1}{18}$.

As with PFAs, we can give the algorithms allowing us by dynamic programming to effectively sum over all the paths. Instead of using BACKWARD and FORWARD, we define INSIDE and OUTSIDE, two algorithms that compute respectively $Pr_G(w|N)$, the

Algorithm 5.9: INSIDE.

Data: a PCFG $G =< \Sigma, V, R, P, N_1 >$ in quadratic normal form, with
$V = \{N_1, \ldots, N_{|V|}\}$, a string $x = a_1 a_2 \cdots a_n$,
Result: The probabilities $\mathrm{I}[i][j][k] = Pr_G(a_j \cdots a_k | N_i)$
for $j : 1 \leq j \leq n$ **do** /* Initialise */
  for $k : j \leq k \leq n$ **do**
    for $i : 1 \leq i \leq |V|$ **do** $\mathrm{I}[i][j][k] \leftarrow 0$
  end
end
for $(N_i, b) \in R$ **do**
  for $j : 1 \leq j \leq n$ **do**
    if $a_j = b$ **then** $\mathrm{I}[i][j][j] \leftarrow P(N_i, b)$
  end
end
for $m : 1 \leq m \leq n - 1$ **do**
  for $j : 1 \leq j \leq n - m$ **do**
    for $k : j \leq k \leq j + m - 1$ **do**
      for $(N_i, N_{i_1} N_{i_2}) \in R$ **do** $\mathrm{I}[i][j][j + m] \leftarrow$
      $\mathrm{I}[i][j][j + m] + \mathrm{I}[i_1][j][k] \cdot \mathrm{I}[i_2][k + 1][j + m - 1] \cdot P(N_i, N_{i_1} N_{i_2})$
    end
  end
end
return I

probability of generating string w from non-terminal N, and $Pr_G(uNv)$, the probability of generating uNv from the axiom.

Notice that the OUTSIDE algorithm needs the computation of the INSIDE one.

One should be careful when using PCFGs to generate strings: the process can diverge. If we take for example grammar $G= < \{\mathtt{a}, \mathtt{b}\}, \{N_1\}, R, P, N_1 >$ with rules $N_1 \rightarrow N_1 N_1$ (probability $\frac{1}{2}$) and $N_1 \rightarrow \mathtt{a}$ (probability $\frac{1}{2}$), then if generating with this grammar, although everything looks fine (it is easy to check that $Pr(\mathtt{a}) = \frac{1}{2}$, $Pr(\mathtt{aa}) = \frac{1}{8}$ and $Pr(\mathtt{aaa}) = \frac{1}{16}$), the generation process diverges. Let x be the estimated length of a string generated by G. The following recursive relation must hold:

$$x = \frac{1}{2} \cdot 2x + \frac{1}{2} \cdot 1$$

and it does not accept any solution.

In other words, as soon as there are more than two occurrences of the symbol 'N_1', then in this example, the odds favour an exploding and non-terminating process.

Algorithm 5.10: OUTSIDE.

Data: a PCFG $G =< \Sigma, V, R, P, N_1 >$ in quadratic normal form, with
$V = \{N_1, \dots, N_{|V|}\}$, string $u = a_1 a_2 \cdots a_n$
Result: the probabilities $\mathrm{O}[i][j][k] = Pr_G(a_1 \cdots a_j N_i a_k \cdots a_n)$
for $j : 0 \le k \le n+1$ **do**
  for $k : j+1 \le k \le n+1$ **do**
    | **for** $i : 1 \le i \le |V|$ **do** $\mathrm{O}[i][j][k] \leftarrow 0$
  end
end
$\mathrm{O}[1][0][n+1] \leftarrow 1$;
for $e : 0 \le e \le n$ **do**
  for $j : 1 \le j \le n-e$ **do**
    $k \leftarrow n+1-j+e$;
    for $(N_i, N_{i_1} N_{i_2}) \in R$ **do**
      for $s : j \le s \le k$ **do**
        $\mathrm{O}[i_1][j][s] \leftarrow \mathrm{O}[i_1][j][s] + P(N_i, N_{i_1} N_{i_2}) \cdot \mathrm{O}[i][j][k] \cdot \mathrm{I}[i_2][s][k-1]$;
        $\mathrm{O}[i_2][s][k] \leftarrow \mathrm{O}[i_2][s][k] + P(N_i, N_{i_1} N_{i_2}) \cdot \mathrm{I}[i_1][j+1][s] \cdot \mathrm{O}[i][j][k]$
    end
  end
end
return O

5.4 Distances between two distributions

There are a number of reasons for wanting to measure a distance between two distributions:

- In a testing situation where we want to compare two resulting PFAs obtained by two different algorithms, we may want to measure the distance towards some ideal target.
- In a learning algorithm, one option may be to decide upon merging two nodes of an automaton. For this to make sense we want to be able to say that the distributions at the nodes (taking each state as the initial state of the automaton) are close.
- There may be situations where we are faced by various candidate models, each describing a particular distribution. Some sort of nearest neighbour approach may then be of use, if we can measure a distance between distributions.
- The relationship between distances and kernels can also be explored. An attractive idea is to relate distances and kernels over distributions.

Defining similarity measures between distributions is the most natural way of comparing them. Even if the question of exact equivalence (discussed in Section 5.4.1) is of interest, in practical cases we wish to know if the distributions are close or not. In tasks involving the learning of PFAs or DPFAs we may want to measure the quality of the result or of the learning process. For example, when learning takes place from a sample, measuring how far the learnt automaton is from the sample can also be done by comparing distributions since a sample can be encoded as a DPFA.

5.4.1 Equivalence questions

We defined equality between two distributions earlier; equivalence between two models is true when the underlying distributions are equal. But suppose now that the distributions are represented by PFAs, what more can we say?

Theorem 5.4.1 *Let $\mathcal{A}$ and $\mathcal{B}$ be two* PFA*s. We can decide in polynomial time if $\mathcal{A}$ and $\mathcal{B}$ are equivalent.*

Proof If $\mathcal{A}$ and $\mathcal{B}$ are DPFAs the proof can rely on the fact that the Myhill-Nerode equivalence can be redefined in an elegant way over regular deterministic distributions, which in turn ensures there is a canonical automaton for each regular deterministic distribution. It being canonical means that for any other DPFA generating the same distribution, the states can be seen as members of a partition block, each block corresponding to state in the canonical automaton.

The general case is more complex, but contrarily to the case of non-probabilistic objects, the equivalence is polynomially decidable. One way to prove this is to find a polynomially computable metric. Having a null distance (or not) will then correspond exactly to the PFAs being equivalent. Such a polynomially computable metric exists (the $\mathbf{L}_2$ norm, as proved in the Proposition 5.5.3, page 110) and the algorithm is provided in Section 5.5.3. □

But in practice we will often be confronted with either a sample and a PFA or two samples and we have to decide this equivalence upon incomplete knowledge.

5.4.2 Samples as automata

If we are given a sample S drawn from a (regular) distribution, there is an easy way to represent this empirical distribution as a DPFA. This automaton will be called the ***probabilistic prefix tree acceptor*** (PPTA(S)).

The first important thing to note is that a sample drawn from a probabilistic language is not a set of strings but a multiset of strings. Indeed, if a given string has a very high probability, we can expect this string to be generated various times. The total number of *different* strings in a sample S is denoted by $\lceil S \rceil$, but we will denote by $|S|$ the total number of strings, including repetitions. Given a language L, we can count how many occurrences of strings in S belong to L and denote this quantity by $|S|_L$. To count the number of occurrences of a string x in a sample S we will write $\mathrm{cnt}_S(x)$, or $|S|_x$. We will use the same notation when counting occurrences of prefixes, suffixes and substrings in a string: for example $|S|_{\Sigma^\star \mathtt{ab} \Sigma^\star}$ counts the number of strings in S containing ab as a substring.

The empirical distribution associated with a sample S drawn from a distribution $\mathcal{D}$ is denoted $\widehat{S}$, with $Pr_{\widehat{S}}(x) = \frac{\mathrm{cnt}_S(x)}{|S|}$.

Example 5.4.1 Let $S = \{\mathtt{a}(3), \mathtt{aba}(1), \mathtt{bb}(2), \mathtt{babb}(4), \mathtt{bbb}(1)\}$ where babb(4) means that there are 4 occurrences of string babb. Then $\lceil S \rceil = 5$, but $|S| = 11$ and $|S|_{b^*} = 3$.

Definition 5.4.1 Let S be a multiset of strings from $\Sigma^\star$. The **probabilistic prefix tree acceptor** PPTA(S) is the DPFA $\langle \Sigma, Q, q_\lambda, \mathbb{F}_\mathbb{P}, \delta_\mathbb{P} \rangle$ where

- $Q = \{q_u : u \in \mathrm{PREF}(S)\}$.
- $\forall ua \in \mathrm{PREF}(S) : \ \delta_\mathbb{P}(q_u, a, q_{ua}) = \frac{|S|_{ua\,\Sigma^\star}}{|S|_{u\,\Sigma^\star}}$.
- $\forall u \in \mathrm{PREF}(S) : \ \mathbb{F}_\mathbb{P}(q_u) = \frac{|S|_u}{|S|_{u\,\Sigma^\star}}$.

Example 5.4.2 Let $S = \{\mathtt{a}(3), \mathtt{aba}(1), \mathtt{bb}(2), \mathtt{babb}(4), \mathtt{bbb}(1)\}$. Then the corresponding probabilistic prefix tree acceptor PPTA(S) is depicted in Figure 5.11. The fractions are represented in a simplified way (1 instead of $\frac{4}{4}$). This nevertheless corresponds to a loss of information: $\frac{4}{4}$ is 'different' statistically than $\frac{400}{400}$; the issue will be dealt with in Chapter 16.

The above transformation allows us to define in a unique way the distances between regular distributions over $\Sigma^\star$. In doing so they implicitly define distances between automata, but also between automata and samples, or even between samples.

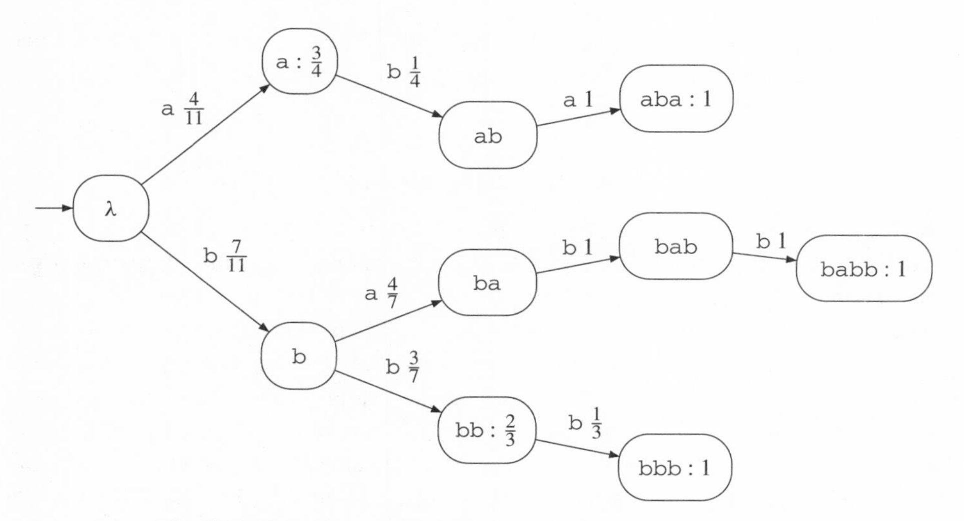

Fig. 5.11. PPTA for $S = \{\mathtt{a}(3), \mathtt{aba}(1), \mathtt{bb}(2), \mathtt{babb}(4), \mathtt{bbb}(1)\}$.

5.4.3 Distances between automata

- The most general family of distances is referred to as the $\mathbf{L}_n$ distances or distances for the norm $\mathbf{L}_n$. The general definition goes as follows:

$$\mathbf{L}_n(\mathcal{D}, \mathcal{D}') = \left(\sum_{x \in \Sigma^\star} \left| Pr_{\mathcal{D}}(x) - Pr_{\mathcal{D}'}(x) \right|^n \right)^{\frac{1}{n}}.$$

- For n=1 we get a natural distance also known as the $\mathbf{L}_1$ distance, the ***variation*** or ***Manhattan*** distance, or distance for the norm $\mathbf{L}_1$.

$$\mathbf{L}_1(\mathcal{D}, \mathcal{D}') = \sum_{x \in \Sigma^\star} \left| Pr_{\mathcal{D}}(x) - Pr_{\mathcal{D}'}(x) \right|.$$

- And in the special case where $n = 2$ we obtain:

$$\mathbf{L}_2(\mathcal{D}, \mathcal{D}') = \sqrt{\sum_{x \in \Sigma^\star} (Pr_{\mathcal{D}}(x) - Pr_{\mathcal{D}'}(x))^2}.$$

  This is also known as the quadratic distance or the Euclidean distance. Notice that with $n = 2$ the absolute values of the definition of $\mathbf{L}_n$ vanish, which will allow computation to be easier.
- The following distance (sometimes denoted also as d_* or $d_{\max}$) for the $\mathbf{L}_\infty$ norm is the limit when $n \to \infty$ of the $\mathbf{L}_n$:

$$\mathbf{L}_\infty(\mathcal{D}, \mathcal{D}') = \max_{x \in \Sigma^\star} \left| Pr_{\mathcal{D}}(x) - Pr_{\mathcal{D}'}(x) \right|.$$

When concerned with very small probabilities such as those that may arise when an infinite number of strings have non-null probability, it may be more useful to use logarithmic probabilities. In this way a string with very small probability may influence the distance because the relative probabilities in both distributions are very different. Suppose $Pr_1(x) = 10^{-7}$ and $Pr_2(x) = 10^{-9}$, then the effect for $\mathbf{L}_1$ of this particular string will be of $99 \cdot 10^{-9}$ whereas for the logarithmic distance the difference will be the same as if the probabilities had been 10^{-1} and 10^{-3}.

The *logarithmic distance* is

$$d_{\log}(\mathcal{D}, \mathcal{D}') = \max_{x \in \Sigma^\star} \left| \log Pr_{\mathcal{D}}(x) - \log Pr_{\mathcal{D}'}(x) \right|.$$

But the logarithmic distance erases all the differences between the large and the less large probabilities. A good compromise is the ***Kullback-Leibler divergence***:

$$d_{\mathrm{KL}}(\mathcal{D}, \mathcal{D}') = \sum_{x \in \Sigma^\star} Pr_{\mathcal{D}}(x) \cdot \log \frac{Pr_{\mathcal{D}}(x)}{Pr_{\mathcal{D}'}(x)}.$$

We set in a standard way that $0 \log 0 = 0$ and $\frac{0}{0} = 1$.

The ***Kullback-Leibler divergence*** is the sum over all strings of the logarithmic loss weighted down by the actual probability of a string. The divergence is an asymmetrical measure: the distribution $\mathcal{D}$ is used twice as it clearly is the 'real' distribution, the one that assigns the weights in the sum.

It should be noticed that in the case where some string has a null probability in $\mathcal{D}'$, but not in $\mathcal{D}$, the denominator $Pr_{\mathcal{D}'}(x)$ is null. As it is supposed to divide the expression, the Kullback-Leibler divergence is infinite. This is obviously a crucial issue:

- A model that assigns a null probability to a possible string is bad. The KL-measure tells us it is worse than bad, as no matter what the rest of the distribution looks like, the penalty is infinite and cannot be compensated.
- A first way to deal with this serious problem is through *smoothing*: the model is generalised in order to assign (small) probabilities to events that were not regarded as possible during the learning phase. The crucial question is how to do this and is a research field in itself.
- The second way consists of using a metric that does not have the inconveniences of the KL-divergence. This is theoretically sound, but at the same time does not address the practical issues of assigning a null probability to a possible event.

Rewriting the Kullback-Leibler divergence as

$$d_{\mathrm{KL}}(\mathcal{D}, \mathcal{D}') = \sum_{x\in\Sigma^\star} Pr_{\mathcal{D}}(x) \cdot \log Pr_{\mathcal{D}}(x) - \sum_{x\in\Sigma^\star} Pr_{\mathcal{D}}(x) \cdot \log Pr_{\mathcal{D}'}(x), \tag{5.1}$$

one can note the first term is the entropy of $\mathcal{D}$ and does not depend on $\mathcal{D}'$, and the second term is the cross-entropy of $\mathcal{D}$ given $\mathcal{D}'$. One interpretation (from information theory) is that we are measuring the difference between the optimal number of bits needed to encode a random message (the left-hand side) and the average number of bits when using distribution $\mathcal{D}'$ to encode the data.

5.4.4 Some properties

- $\mathbf{L}_n$ ($\forall n$), $\mathbf{L}_\infty$ are metrics, i.e. they comply with the usual properties: $\forall \mathcal{D}, \mathcal{D}',\ \forall d \in \{\mathbf{L}_n, \mathbf{L}_\infty\}$,
  1. $d(\mathcal{D}, \mathcal{D}') = 0 \iff \mathcal{D} = \mathcal{D}'$,
  2. $d(\mathcal{D}, \mathcal{D}') = d(\mathcal{D}', \mathcal{D})$,
  3. $d(\mathcal{D}, \mathcal{D}') + d(\mathcal{D}', \mathcal{D}'') \geq d(\mathcal{D}, \mathcal{D}'')$.
- d_{KL} is not a metric (Definition 3.1.1, page 47) but can be adapted to comply with those conditions. This is usually not done because the asymmetrical aspects of d_{KL} are of interest in practical and theoretical settings.
  It nevertheless verifies the following properties: $\forall \mathcal{D}, \mathcal{D}'$,
  1. $d_{\mathrm{KL}}(\mathcal{D}, \mathcal{D}') \geq 0$,
  2. $d_{\mathrm{KL}}(\mathcal{D}, \mathcal{D}') = 0 \iff \mathcal{D} = \mathcal{D}'$.
- Obviously $\forall \mathcal{D}, \mathcal{D}', \mathbf{L}_\infty(\mathcal{D}, \mathcal{D}') \leq \mathbf{L}_1(\mathcal{D}, \mathcal{D}')$.
- Pinsker's inequality states that:

$$d_{\mathrm{KL}}(\mathcal{D}, \mathcal{D}') \geq \frac{1}{2\ln 2}\mathbf{L}_1(\mathcal{D}, \mathcal{D}')^2.$$

These inequalities help us to classify the distances over distributions in the following (informal) way:

$$d_{\log} \precsim D_{\mathrm{KL}} \precsim \mathbf{L}_1 \precsim \cdots \precsim \mathbf{L}_n \precsim \mathbf{L}_{n+1} \precsim \mathbf{L}_\infty.$$

This can be read as:

- From left to right we find the distances that attach most to least importance to the relative differences between the probabilities. The further to the right, the more important are the absolute differences.
- If you want to measure closeness of two distributions by the idea that you want to be close on the important probabilities, then you should turn to the right-hand side of the expression above.
- If you want all the probabilities to count and to count this in a relative way, you should use measures from the left-hand side of the expression above.

Alternatively, we can consider a sample as a *random variable*, in which case we can consider the probability that a sample of a given size has such or such property. And the notation $Pr_{\mathcal{D}}(f(S_n))$ will be used to indicate the probability that property f holds over a sample of size n sampled following $\mathcal{D}$.

Lemma 5.4.2 *Let $\mathcal{D}$ be any distribution on $\Sigma^\star$. Then $\forall a > 1$, the probability that a sample S of size n has*

$$\mathbf{L}_\infty(\mathcal{D}, \widehat{S}) \leq \sqrt{6a(\log n)/n}$$

is at least $4n^{-a}$.

Essentially the lemma states that the empirical distribution converges (for the $\mathbf{L}_\infty$ distance) to the true distance.

5.5 Computing distances

When we have access to the models (the automata) for the distributions, an exact computation of the distance is possible in certain cases. This can also solve two other problems:

- If the distance we are computing respects the conditions $d(x, y) = 0 \iff x = y$ and $d(x, y) \geq 0$, computing the distance allows us to solve the equivalence problem.
- Since samples can easily be represented by PFAs, we can also compute the distance between two samples or between a sample and a generator through these techniques. Error bounds corresponding to different statistical risks can also be computed.

5.5.1 Computing prefixial distances between states

Let $\mathcal{A}$ and $\mathcal{B}$ be two PFAs (without λ-transitions) with $\mathcal{A} = \langle \Sigma, Q_{\mathcal{A}}, \mathbb{I}_{\mathbb{P}\mathcal{A}}, \mathbb{F}_{\mathbb{P}\mathcal{A}}, \delta_{\mathbb{P}\mathcal{A}} \rangle$ and $\mathcal{B} = \langle \Sigma, Q_{\mathcal{B}}, \mathbb{I}_{\mathbb{P}\mathcal{B}}, \mathbb{F}_{\mathbb{P}\mathcal{B}}, \delta_{\mathbb{P}\mathcal{B}} \rangle$ with associated probability functions $Pr_{\mathcal{A}}$ and $Pr_{\mathcal{B}}$.

We will denote in a standard fashion by $\mathcal{A}_q$ the PFA obtained from $\mathcal{A}$ when taking state q as the unique initial state. Alternatively it corresponds to using function $\mathbb{I}_{\mathbb{P}\mathcal{A}}(q) = 1$ and $\forall q' \neq q,\ \mathbb{I}_{\mathbb{P}\mathcal{A}}(q') = 0$.

We define now η_q as the probability of reaching state q in $\mathcal{A}$. Computing η_q may seem complex but is not:

$$\eta_q = \mathbb{I}_{\mathbb{P}\mathcal{A}}(q) + \sum_{s\in Q_\mathcal{A}} \sum_{a\in\Sigma} \eta_s \cdot \delta_{\mathbb{P}\mathcal{A}}(s, a, q).$$

In a similar way, we define $\eta_{qq'}$ as the probability of jointly reaching (with the same string) state q in $\mathcal{A}$ and state q' in $\mathcal{B}$. This means summing over all the possible strings and paths. Luckily, the recursive definition is much simpler and allows a polynomial implementation by dynamic programming.

By considering on one hand the only zero-length prefix λ and on the other hand the other strings which are obtained by reading all but the last character, and then the last one, $\forall q, q' \in Q_\mathcal{A} \times Q_\mathcal{B}$,

$$\eta_{qq'} = \mathbb{I}_{\mathbb{P}\mathcal{A}}(q)\mathbb{I}_{\mathbb{P}\mathcal{B}}(q') + \sum_{s\in Q_\mathcal{A}} \sum_{s'\in Q_\mathcal{B}} \sum_{a\in\Sigma} \eta_{s,s'} \cdot \delta_{\mathbb{P}\mathcal{A}}(s, a, q) \cdot \delta_{\mathbb{P}\mathcal{B}}(s', a, q'). \tag{5.2}$$

The above equation is one inside a system of linear equations. A cubic algorithm (in the number of equations and variables) can solve this, even if specific care has to be taken with manipulation of the fractions. In certain cases, where the PFAs have reasonably short probable strings, it can be noticed that convergence is very fast, and a fixed number of iterations is sufficient to approximate the different values closely.

We will use this result in Section 5.5.3 to compute $\mathbf{L}_2$.

5.5.2 Computing the KL-divergence

We recall the second definition of the KL-divergence and adapt it for two DPFAs (Equation 5.1):

$$d_{\text{KL}}(\mathcal{A}, \mathcal{B}) = \sum_{x\in\Sigma^\star} Pr_\mathcal{A}(x) \cdot \log Pr_\mathcal{A}(x) - \sum_{x\in\Sigma^\star} Pr_\mathcal{A}(x) \cdot \log Pr_\mathcal{B}(x).$$

It follows that we need to know how to compute $\sum_{x\in\Sigma^\star} Pr_\mathcal{A}(x) \cdot \log Pr_\mathcal{B}(x)$. Then the first part of the formula can be computed by simply taking $\mathcal{B} = \mathcal{A}$. We denote by $Pr_\mathcal{A}(a|x)$ the probability of reading a after having read x and by $Pr_\mathcal{A}(a|x)$ the probability of halting after having read x.

Let us, in the case of DPFAs, show that we can compute $\sum_{x\in\Sigma^\star} Pr_\mathcal{A}(x) \cdot \log Pr_\mathcal{B}(x)$. If we take any string $x = a_1 \cdots a_n$, we have $\log Pr_\mathcal{B}(x) = \sum_{i\in[n]} \log Pr_\mathcal{B}(a_i|a_1 \cdots a_{i-1}) + \log Pr_\mathcal{B}(\lambda|x)$. Each term $\log Pr_\mathcal{B}(a_i)$ therefore appears in every string (and every time) containing letter a_i. We can therefore factorise:

$$\sum_{x\in\Sigma^\star} Pr_\mathcal{A}(x) \cdot \log Pr_\mathcal{B}(x) = \sum_{x\in\Sigma^\star} \sum_{a\in\Sigma} Pr_\mathcal{A}(xa\,\Sigma^\star) \cdot \log Pr_\mathcal{B}(a|x) + \sum_{x\in\Sigma^\star} Pr_\mathcal{A}(x) \cdot \log Pr_\mathcal{B}(\lambda|x).$$

Since $\mathcal{A}$ and $\mathcal{B}$ are DPFAs, another factorisation is possible, by taking together all the strings x finishing in state q in $\mathcal{A}$ and in state q' in $\mathcal{B}$. We recall that $\eta_{q,q'}$ is the probability of reaching simultaneously state q in $\mathcal{A}$ and state q' in $\mathcal{B}$. The value of $\sum_{x\in\Sigma^\star} Pr_{\mathcal{A}}(x) \cdot \log Pr_{\mathcal{B}}(x)$ is:

$$\sum_{q\in Q_{\mathcal{A}}} \sum_{q'\in Q_{\mathcal{B}}} \sum_{a\in\Sigma} \eta_{q,q'} Pr_{\mathcal{A}}(a|q) \cdot \log Pr_{\mathcal{B}}(a|q') + \sum_{q\in Q_{\mathcal{A}}} \sum_{q'\in Q_{\mathcal{B}}} \eta_{q,q'} \mathbb{F}_{\mathbb{P}_{\mathcal{A}}}(q) \cdot \log \mathbb{F}_{\mathbb{P}_{\mathcal{B}}}(q').$$

Since the $\eta_{q,q'}$ can be computed thanks to Equation 5.2, we can state:

Proposition 5.5.1 *Let $\mathcal{A}$, $\mathcal{B}$ be two* DPFA*s. Then $d_{\mathrm{KL}}(\mathcal{A}, \mathcal{B})$ can be computed in polynomial time.*

This proposition will be the key to understanding algorithm MDI, given in Section 16.7.

5.5.3 Co-emission

If a direct computation of a distance between two automata is difficult, one can compute instead the probability that both automata generate the same string at the same time. We define the co-emission between $\mathcal{A}$ and $\mathcal{B}$:

Definition 5.5.1 (Co-emission probability) The **co-emission probability** of $\mathcal{A}$ and $\mathcal{B}$ is

$$\mathbf{coem}(\mathcal{A}, \mathcal{B}) = \sum_{w\in\Sigma^\star} (Pr_{\mathcal{A}}(w) \cdot Pr_{\mathcal{B}}(w)).$$

This is the probability of emitting w simultaneously from automata $\mathcal{A}$ and $\mathcal{B}$. The main interest in being able to compute the co-emission is that it can be used to compute the $\mathbf{L}_2$ distance between two distributions:

Definition 5.5.2 ($\mathbf{L}_2$ distance between two models) The **distance for the $\mathbf{L}_2$ norm** is defined as:

$$\mathbf{L}_2(\mathcal{A}, \mathcal{B}) = \sqrt{\sum_{w\in\Sigma^\star} (Pr_{\mathcal{A}}(w) - Pr_{\mathcal{B}}(w))^2}.$$

This can be computed easily by developing the formula:

$$\mathbf{L}_2(\mathcal{A}, \mathcal{B}) = \sqrt{\mathbf{coem}(\mathcal{A}, \mathcal{A}) + \mathbf{coem}(\mathcal{B}, \mathcal{B}) - 2\,\mathbf{coem}(\mathcal{A}, \mathcal{B})}.$$

If we use $\eta_{qq'}$ (the probability of jointly reaching state q in $\mathcal{A}$ and state q' in $\mathcal{B}$),

$$\mathbf{coem}(\mathcal{A}, \mathcal{B}) = \sum_{q\in Q} \sum_{q'\in Q'} \eta_{qq'} \cdot \mathbb{F}_{\mathbb{P}_{\mathcal{A}}}(q) \cdot \mathbb{F}_{\mathbb{P}_{\mathcal{B}}}(q').$$

Since the computation of the $\eta_{qq'}$ could be done in polynomial time (by Equation 5.2, page 108), this in turns means that the computation of the co-emission and of the $\mathbf{L}_2$ distance is polynomial.

Concluding:

Theorem 5.5.2 *If $\mathcal{D}$ and $\mathcal{D}'$ are given by* PFAs*, $\mathbf{L}_2(\mathcal{D},\mathcal{D}')$ can be computed in polynomial time.*

Theorem 5.5.3 *If $\mathcal{D}$ and $\mathcal{D}'$ are given by* DPFAs*, $d_{\mathrm{KL}}(\mathcal{D},\mathcal{D}')$ can be computed in polynomial time.*

5.5.4 Some practical issues

In practical settings we are probably going to be given a sample (drawn from the unknown distribution $\mathcal{D}$) which we have divided into two subsamples in a random way. The first sample has been used to learn a probabilistic model ($\mathcal{A}$) which is supposed to generate strings in the same way as the (unknown) target sample does.

The second sample is usually called the ***test*** sample (let us denote it by S) and is going to be used to measure the quality of the learnt model $\mathcal{A}$.

There are a number of options using the tools studied up to now:

- We could use the $\mathbf{L}_2$ distance between $\mathcal{A}$ and $\widehat{S}$. The empirical distribution $\widehat{S}$ can be represented by a PPTA and we can compute the $\mathbf{L}_2$ distance between this PPTA and $\mathcal{A}$ using Proposition 5.5.1.
- Perhaps the $\mathbf{L}_1$ distance is better indicated and we would like to try to do the same with this. But an exact computation of this distance (between two DPFAs) cannot be done in polynomial time (unless $\mathcal{P}=\mathcal{NP}$), so only an approximation can be obtained.
- The Kullback-Leibler divergence is an alternative candidate. But as $\mathcal{D}$ is unknown, again the distance has to be computed over the empirical distribution $\widehat{S}$ instead of $\mathcal{D}$. In this case we may notice that in Equation 5.1 the first part is constant and does not depend on $\mathcal{A}$, so we are really only interested in minimising the second part (the relative entropy).
- The alternative is to use *perplexity* instead.

5.5.5 Perplexity

If we take the equation for the KL-divergence, we can measure the distance between a hypothesis distribution H and the real distribution $\mathcal{D}$. If we only have a sample S of the real distribution, then it may still be possible to use this sample (or the corresponding empirical distribution $\widehat{S}$) instead of the real distribution $\mathcal{D}$. We would then obtain (from Equation 5.1):

$$d_{KL}(\widehat{S},H)=\sum_{x\in\Sigma^\star}Pr_{\widehat{S}}(x)\cdot\log Pr_{\widehat{S}}(x)-\sum_{x\in\Sigma^\star}Pr_{\widehat{S}}(x)\cdot\log Pr_H(x), \tag{5.3}$$

In the above equation the first term doesn't depend on the hypothesis. We can therefore remove it when what we want is to compare different hypotheses. For those strings not in the sample, the probability in $\widehat{S}$ is 0, so they need not be counted either.

Simplifying, we get the estimation of the ***divergence*** (DIV) between S and the hypothesis H:

$$\mathrm{DIV}(S|H) = -\frac{1}{|S|}\sum_{x\in S} \mathrm{cnt}_S(x)\log(Pr_H(x)). \tag{5.4}$$

This in turn helps us define the ***perplexity*** of H with respect to the sample:

$$\begin{aligned} PP(S|H) &= \left[\prod_{x\in S} Pr_H(x)\right]^{-\frac{\mathrm{cnt}_S(x)}{|S|}} \\ &= \sqrt[n]{\prod_{x\in S} Pr_H(x)^{\mathrm{cnt}_S(x)}}. \end{aligned} \tag{5.5}$$

The properties of the perplexity can be summarised as follows:

- Equation 5.4 tells us that the perplexity measures the average number of bits one must 'pay' by using the model H instead of $\mathcal{D}$ while coding the sample S.
- From Equation 5.5, the perplexity can be seen as the inverse of the geometric mean of the probabilities – according to the model – of the sample words.

In practical situations, the following issues must be carefully taken into account:

- Perplexity and entropy are infinite whenever there is a string in the sample for which $Pr_H(x) = 0$. In practice, this implies that the perplexity must be used on a *smoothed* model, i.e. a model that provides a non-null probability for any string of $\Sigma^\star$. One should add that this is not a defect of using perplexity. This is a normal situation: a model that cannot account for a possible string is a bad model.
- The perplexity can compare models only when using the same sample S.

5.6 Exercises

5.1 Compute the PPTA corresponding to the sample $S = \{\lambda(6)$, a(2), aba(1), baba(4), bbbb(1)$\}$.

5.2 Consider the PFA from Figure 5.12. What is the probability of string aba? Of string bbaba? Which is the most probable string (in $\Sigma^\star$)?

5.3 Write a polynomial time algorithm which, given a DPFA $\mathcal{A}$, returns the most probable string in $\mathcal{D}_\mathcal{A}$.

5.4 Write a polynomial time algorithm which, given a DPFA $\mathcal{A}$ and an integer n, returns $\mathrm{mps}(\mathcal{A}, n)$, the n most probable string in $\mathcal{D}_\mathcal{A}$.

5.5 Prove that the probabilistic language from Figure 5.13 is not deterministic.

5.6 Build two DPFAs $\mathcal{A}$ and $\mathcal{B}$ with n states, each such that $\{w \in \Sigma^\star : Pr_\mathcal{A}(w) > 0 \wedge Pr_\mathcal{B}(w) > 0\} = \emptyset$ yet $\mathbf{L}_2(\mathcal{D}_\mathcal{A}, \mathcal{D}_\mathcal{B}) \leq 2^{-n}$.

5.7 Let $\mathcal{D}$ and $\mathcal{D}'$ be two distributions. Prove that $\forall\alpha, \beta > 0 : \alpha + \beta = 1, \alpha\mathcal{D} + \beta\mathcal{D}'$ is a distribution.

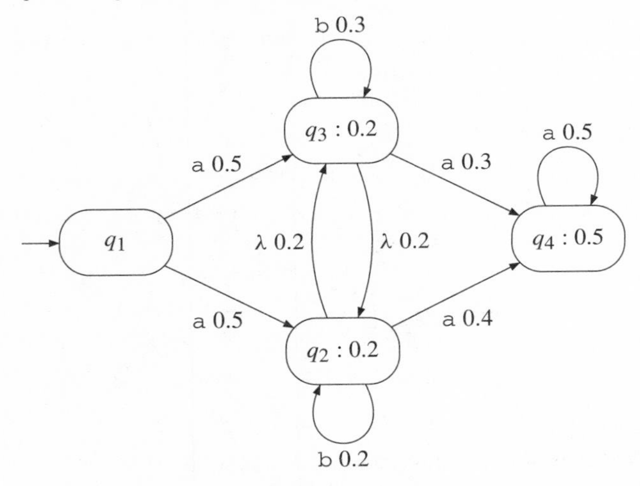

Fig. 5.12. A PFA.

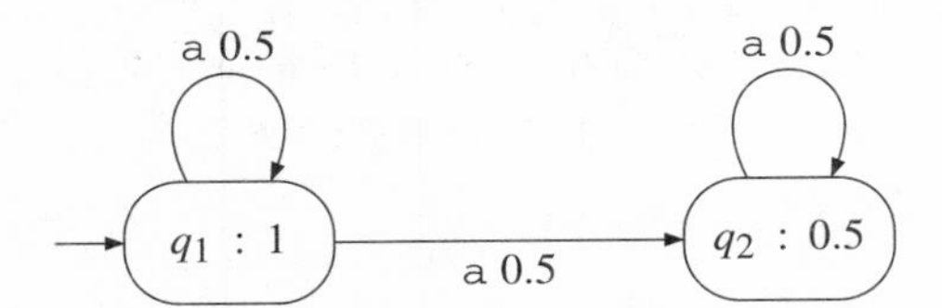

Fig. 5.13. A very simple distribution.

5.8 Consider the PFA from Figure 5.12. Eliminate the λ-transitions from it and obtain an equivalent λ-free PFA.
5.9 Let $S = \{\mathrm{a}(3), \mathrm{aba}(1), \mathrm{bb}(2), \mathrm{babb}(4), \mathrm{bbb}(1)\}$ and call $\mathcal{A}$ the PFA from Figure 5.12. Compute $\mathbf{L}_1(\mathcal{D}_\mathcal{A}, \mathcal{D}_\S)$ and $\mathbf{L}_2(\mathcal{D}_\mathcal{A}, \mathcal{D}_\S)$.
5.10 Prove that $\mathbf{L}_1(\mathcal{D}_\mathcal{A}, \mathcal{D}_\S)$ can be computed, given any sample S and any PFA $\mathcal{A}$.
5.11 Build a PFA $\mathcal{A}_n$ and a string x_n for which the VITERBI score of x is less than any non-null polynomial fraction of $Pr_\mathcal{A}(x)$.
5.12 Write a randomised algorithm which, given a PFA $\mathcal{A}$ and a value $\alpha > 0$ tells us whether there exists a string w whose probability is at least α.

5.7 Conclusions of the chapter and further reading

5.7.1 Bibliographical background

The main reference about probabilistic finite automata is Azaria Paz's book (Paz, 1971); two articles by Enrique Vidal *et al.*, present these automata from an engineering point of view (Vidal *et al.*, 2005a, 2005b). Extended proofs of some of the results presented here, like Proposition 5.2.1 (page 95), can be found there.

We have chosen here to use the term 'probabilistic automata' uniquely whereas a number of authors have sometimes used 'stochastic automata' for exactly the same objects. This can create confusion and if one wants to access the different papers written on the subject, it has to be kept in mind.

Among the other probabilistic finite state machines we can find hidden Markov models (HMMs) (Jelinek, 1998, Rabiner, 1989), probabilistic regular grammars (Carrasco & Oncina, 1994b), Markov chains (Saul & Pereira, 1997), n-grams (Jelinek, 1998, Ney, Martin & Wessel, 1997), probabilistic suffix trees (Ron, Singer & Tishby, 1994), deterministic probabilistic automata (Carrasco & Oncina, 1994b) and weighted automata (Mohri, 1997). These are some names of syntactic objects which have been used to model distributions over sets of strings of possibly infinite cardinality, sequences, words, phrases but also terms and trees. Pierre Dupont *et al.* prove the links between HMMs and PFAs in (Dupont, Denis & Esposito, 2005), with an alternative proof in (Vidal *et al.*, 2005a, 2005b).

The parsing algorithms are now well known; the FORWARD algorithm is described by Leonard Baum *et al.* (Baum *et al.*, 1970). The VITERBI algorithm is named after Andrew Viterbi (Viterbi, 1967).

Another problem related to parsing is the computation of the probability of all strings sharing a given prefix, suffix or substring in a PFA (Fred, 2000).

More complicated is the question of finding the most probable string in a distribution defined by a PFA. In the general case the problem is intractable (Casacuberta & de la Higuera, 2000), with some associated problems undecidable (Blondel & Canterini, 2003), but in the deterministic case a polynomial algorithm can be written using dynamic programming (see Exercise 5.3). Curiously enough, randomised algorithms solve this question very nicely with small error.

Several other interesting questions are raised in Omri Guttman's PhD thesis (Guttman, 2006), for instance the question of knowing how reasonable it can be to attempt to approximate an unknown distribution with a regular one. He proves that for a fixed bounded number of states n, the best PFA with at most n states can be arbitrarily bad.

In Section 5.2.4 we consider the problem of eliminating λ-transitions; λ-PFAs appear as natural objects when combining distributions. λ-PFAs introduce specific problems, in particular, when sequences of transitions labelled with λ are considered. Some authors have provided alternative parsing algorithms to deal with λ-PFAs (Picó & Casacuberta, 2001). In that case parsing takes time that is no longer linear in the size of the string to be parsed but in the size of this string multiplied by the number of states of the PFA. In (Mohri, Pereira & Riley, 2000) Mehryar Mohri proposes to eliminate λ-transitions by means of first running the Floyd-Warshall algorithm in order to compute the λ-transition distance between pairs of edges before removal of these edges. The algorithm we give here can be found in (Hanneforth & de la Higuera, 2009). Thomas Hanneforth points out (Hanneforth, 2008) that when dealing with $\mathcal{A}$ obtained automatically, once the λ-transitions have been eliminated, there is pruning to be done, as some states can be no longer accessible.

We have only given some very elementary results concerning probabilistic context-free grammars (Section 5.3). The parsing algorithms INSIDE and OUTSIDE were introduced by Karim Lari and Steve Young and can be found (with non-trivial differences) in several places (Casacuberta, 1994, Lari & Young, 1990).

Several researchers have worked on distances between distributions (Section 5.4). The first important results were found by Rafael Carrasco *et al.* (Calera-Rubio & Carrasco, 1998, Carrasco, 1997). In the context of HMMs and with intended bio-informatics applications, intractability results were given by Rune Lyngsø *et al.* (Lyngsø & Pedersen, 2001, Lyngsø, Pedersen & Nielsen, 1999). Other results are those by Michael Kearns *et al.* (Kearns *et al.*, 1994). Properties about the distances between automata have been published in various places. Thomas Cover and Jay Thomas' book (Cover & Thomas, 1991) is a good place for these.

A first algorithm for the equivalence of PFAs was given by Vijay Balasubramanian (Balasubramanian, 1993). Then Corinna Cortes *et al.* (Cortes, Mohri & Rastogi, 2006) noticed that being able to compute the $\mathbf{L}_2$ distance in polynomial time ensures that the equivalence is also testable.

The algorithms presented here to compute distances have first appeared in work initiated by Rafael Carrasco *et al.* (Calera-Rubio & Carrasco, 1998, Carrasco, 1997) for the KL-divergence (for string languages and tree languages) and for the $\mathbf{L}_2$ between trees and Thierry Murgue (Murgue & de la Higuera, 2004) for the $\mathbf{L}_2$ between strings. Rune Lyngsø *et al.* (Lyngsø & Pedersen, 2001, Lyngsø, Pedersen & Nielsen, 1999) introduced the co-emission probability, which is an idea also used in kernels. Corinna Cortes *et al.* (Cortes, Mohri & Rastogi, 2006) proved that computing the $\mathbf{L}_1$ distance was intractable; this is also the case for each $\mathbf{L}_n$ with n odd. The same authors better the complexity of (Carrasco, 1997) in a special case and present further results for computations of distances.

Practical issues with distances correspond to topics in speech recognition. In applications such as language modelling (Goodman, 2001) or statistical clustering (Brown *et al.*, 1992, Kneser & Ney, 1993), perplexity is introduced.

5.7.2 Some alternative lines of research

If comparing samples co-emission may be null when large vocabulary and long strings are used. An alternative idea is to compare not only the whole strings, but all their prefixes (Murgue & de la Higuera, 2004):

Definition 5.7.1 (Prefixial co-emission probability) The **prefixial co-emission probability** of $\mathcal{A}$ and $\mathcal{B}$ is

$$\mathbf{coempr}(\mathcal{A}, \mathcal{B}) = \sum_{w \in \Sigma^\star} \left(Pr_{\mathcal{A}}(w\, \Sigma^\star) \cdot Pr_{\mathcal{B}}(w\, \Sigma^\star) \right).$$

Definition 5.7.2 ($\mathbf{L}_2 pref$ distance between two models) The **prefixial distance for the $\mathbf{L}_2$ norm**, denoted by $\mathbf{L}_{2pref}$, is defined as:

$$\mathbf{L}_{2pref}(\mathcal{A}, \mathcal{B}) = \sqrt{\sum_{w \in \Sigma^\star} (Pr_{\mathcal{A}}(w\, \Sigma^\star) - Pr_{\mathcal{B}}(w\, \Sigma^\star))^2}$$

which can be computed easily using:

$$\mathbf{L}_{2pref}(\mathcal{A}, \mathcal{B}) = \sqrt{\mathbf{coempr}(\mathcal{A}, \mathcal{A}) + \mathbf{coempr}(\mathcal{B}, \mathcal{B}) - 2\mathbf{coempr}(\mathcal{A}, \mathcal{B})}.$$

Theorem 5.7.1 $\mathbf{L}_{2pref}$ *is a metric over* $\Sigma^\star$.

Proof These proofs are developed in (Murgue & de la Higuera, 2004). □

5.7.3 Open problems and possible new lines of research

Probabilistic acceptors are defined in (Fu, 1982), but they have only seldom been considered in syntactic pattern recognition or in (probabilistic) formal language theory.

One should also look into what happens when looking at approximation issues. Let $\boldsymbol{\mathcal{D}}_1$ and $\boldsymbol{\mathcal{D}}_2$ be two classes of distributions, $\epsilon > 0$ and d a distance. Then $\boldsymbol{\mathcal{D}}_1$ d-ϵ approximates $\boldsymbol{\mathcal{D}}_2$ whenever given any distribution $\mathcal{D}$ from $\boldsymbol{\mathcal{D}}_1$, $\exists \mathcal{D} \in \boldsymbol{\mathcal{D}}_2$ such that $d(\mathcal{D}, \mathcal{D}') < \epsilon$. The question goes as follows: are regular distributions approximable by deterministic regular ones? For what value of ϵ? Along these lines, one should consider distributions represented by automata of up to some fixed size.

Analysing (in the spirit of (Guttman, 2006)) the way distributions can or cannot be approximated can certainly help us better understand the difficulties of the task of learning them.

6
About combinatorics

All generalisations, with the possible exception of this one, are false.

Kurt Gödel

A child of five would understand this. Send someone to fetch a child of five.

Groucho Marx

In order to get a better grasp of the problem of learning grammars, we need to understand both how the individual objects we are trying to infer are shaped and how the set of these objects is structured. This will enable us to formally state learnability and non-learnability results, but also to identify and study the search space and the operators to move around this space; in turn, this will enable us to develop new heuristics.

The first results are mainly negative: if to learn a grammar you have to solve some intractable combinatorial problem, then only wild guessing is going to allow you to identify or infer correctly, but then you are relying on luck, not on convergence probability. This sort of result is usually obtained by *reductions*: typically, we show that if a well-known hard problem can be solved via a learning algorithm (perhaps with some error, so it may depend on a randomised version), it will mean that something is wrong. The learning algorithm cannot do what it *promises* to do.

But working on the combinatorics of automata and grammars also helps us to build intuitions that contribute to designing learning processes. Being able to describe the learning space, or the space where solutions to our learning problems are to be searched for, or the operators that permit us to modify one automaton into another, will allow us to use searching methods effectively, and more importantly to resort to artificial techniques to look for items in this space.

6.1 About VC-dimensions

In learning problems, a number of combinatorial dimensions can be associated with the objects one wants to learn. These dimensions may then provide lower (sometimes upper) bounds on the number of examples or queries needed to learn. Among the several dimensions that have been introduced for learning, the best known is the *Vapnik-Chernovakis dimension* (also called VC-dimension). This measures the size of the maximum set of strings that can be *shattered* by functions of the studied class. A set of size n is said to be shattered when there is a function for every subset (there are 2^n of them) which accepts the elements of the subset and none from the complement. In the case where the number of functions is infinite and where every subset can be exactly accepted by a function, one may want to index these functions by the size of the functions. Since in the case of grammatical inference the PTA can be used to exactly represent any set of strings, this will be an issue here.

6.1.1 Why VC-dimensions?

The VC-dimension measures how descriptive and complex the class is. It can be used as a better parameter to measure the minimum size of a sample needed to guarantee that a consistent solution is approximately correct. We study here the VC-dimension of deterministic and non-deterministic automata.

Definition 6.1.1 Given a class of grammars $\mathcal{G}$, VC($\mathcal{G}$) is the size of the largest subset of strings shattered by $\mathcal{G}$, where a set $X = \{x_1, x_2, \ldots, x_n\}$ is shattered by $\mathcal{G}$ if_{def} for every subset $Y \subset X$ there exists a grammar $G \in \mathcal{G}$ such that $Y \subset \mathbb{L}(G)$ and $(X \setminus Y) \cap \mathbb{L}(G) = \emptyset$.

The VC-dimension is usually used to obtain bounds for PAC-learning, and is also indicative of the hardness of the learning task in other settings. A finite (or slowly increasing) VC-dimension can be used to derive PAC-bounds.

6.1.2 VC-dimension of DFA

Proposition 6.1.1 *The* VC*-dimension of the class* $\mathcal{DFA}(\Sigma)$ *is infinite. But if we denote by* $\mathcal{DFA}_n(\Sigma)$ *the set of all* DFA*s with at most* n *states, then* $\text{VC}(\mathcal{DFA}_n(\Sigma)) \in \mathcal{O}(n \log n)$.

Proof The infinity of VC($\mathcal{DFA}(\Sigma)$) follows from the fact that any finite language is regular. Hence given any $Y \subseteq X$, with X as large as one wants, PTA(Y) can be constructed recognising exactly Y.

Now consider the class of all DFAs with at most n states, and suppose for simplicity that the alphabet is of size two. Then one can check that $|\mathcal{DFA}_n(\Sigma)| \leq n^{2n} \cdot 2^{n^2}$. This bound is obtained by considering all the ways one can fill a transition table with n rows and 2 columns, and all the ways one can label the states as final. By Stirling's formula that means that $|\mathcal{DFA}_n(\Sigma)| \in \mathcal{O}(2^{n \log n})$. So the largest set that can be shattered is of size in $\mathcal{O}(n \log n)$.

Conversely we show that this bound can be met by providing such a class of strings/automata. For simplicity consider the case where $n = 2^k$. Let $\Sigma = \{\mathtt{a}, \mathtt{b}\}$ and

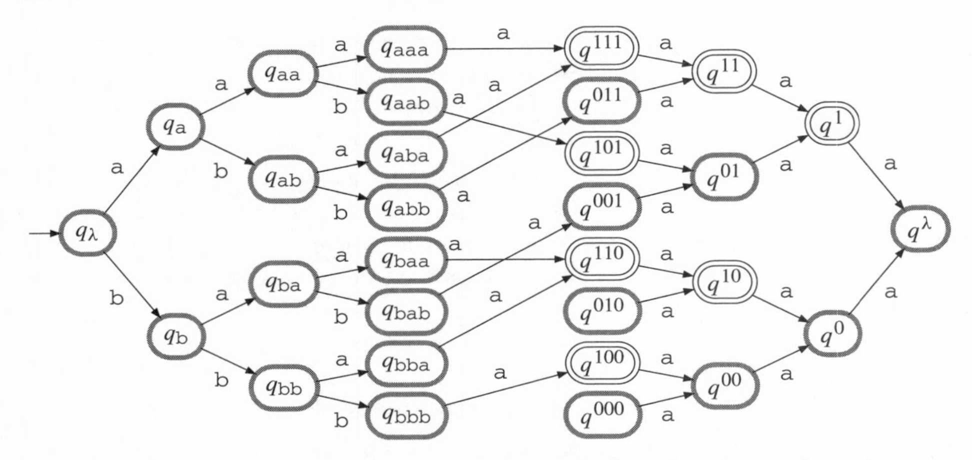

Fig. 6.1. The DFA to show that the VC-dimension is in $\mathcal{O}(n \log n)$: It recognises strings $\mathtt{a}^3 \cdot \mathtt{a} \cdot \lambda$, $\mathtt{a}^3 \cdot \mathtt{a} \cdot \mathtt{a}$, $\mathtt{a}^3 \cdot \mathtt{a} \cdot \mathtt{aa}$, $\mathtt{aab} \cdot \mathtt{a} \cdot \lambda$, $\mathtt{aab} \cdot \mathtt{a} \cdot \mathtt{aa}, \ldots$

let X be the set of all strings $u\mathtt{a}v$ with $u \in \Sigma^k$ and v in $\{\mathtt{a}\}^{<k}$. This set contains strings of length between $k+1$ and $2k$, the first k $(= \log n)$ symbols being $\mathtt{a}$ or $\mathtt{b}$, the next ones all being $\mathtt{a}$. The total size of X is therefore $2^k k$ or in other words $n \log n$.

Then any subset Y of X can be recognised by the automaton $\mathcal{A}_Y = \langle \Sigma, Q, q_\lambda, \mathbb{F}_\mathbb{A}, \mathbb{F}_\mathbb{R}, \delta \rangle$ where

- $Q = \{q_u : u \in \Sigma^{\leq k}\} \cup \{q^v : v \in \{0,1\}^{\leq k}\}$,
- $\mathbb{F}_\mathbb{A} = \{q^{1u} : u \in \{0,1\}^*\}$,
- $\mathbb{F}_\mathbb{R} = Q \setminus \mathbb{F}_\mathbb{A}$,
- $\forall u \in \Sigma^{<k}$, $\delta(q_u, \mathtt{a}) = q_{u\mathtt{a}}$, $\delta(q_u, \mathtt{b}) = q_{u\mathtt{b}}$,
- $\forall u \in \{0,1\}^{<k}$, $\delta(q^{u1}, \mathtt{a}) = q^u$, $\delta(q^{u0}, \mathtt{a}) = q^u$,
- $\forall u \in \Sigma^k$, let $w_u = c_1 c_2 \cdots c_n$ with $c_i = 1$ if $u\mathtt{a}^i \in Y$, 0 if not. Then $\delta(q_u, \mathtt{a}) = q^{w_u}$.

We illustrate the proposed construction for $k = 3$ (and so $n = 8$) and a specific subset $Y = \{\mathtt{a}^4, \mathtt{a}^5, \mathtt{a}^6, \mathtt{a}^2\mathtt{ba}, \mathtt{a}^2\mathtt{ba}^3, \mathtt{aba}, \mathtt{aba}^2, \mathtt{aba}^3, \mathtt{ab}^2\mathtt{a}^2, \mathtt{ab}^2\mathtt{a}^3, \mathtt{ba}^3, \mathtt{ba}^4, \mathtt{b}^2\mathtt{a}^2, \mathtt{b}^2\mathtt{a}^3, \mathtt{b}^3\mathtt{a}\}$ in Figure 6.1. □

The DFA represented in Figure 6.1 is used to show that the VC-dimension of the DFA is in $\mathcal{O}(n \log n)$. It is easy to construct, with the same states, an automaton for every subset of $\Sigma^n \times \{\mathtt{a}\} \times \Sigma^k$. Only the central transitions have to be changed to recognise a different language. Remember that $|X| = n \log n$ and the automaton has at most $4n - 2$ states.

6.1.3 VC-dimension of NFAs

Proposition 6.1.2 *The* VC*-dimension of* $\mathcal{NFA}(\Sigma)$ *is infinite. But if we call* $\mathcal{NFA}_n(\Sigma)$ *the set of all languages accepted by automata of at most n states, then we have:* $\mathrm{VC}(\mathcal{NFA}_n(\Sigma)) \in \mathcal{O}(n^2)$.

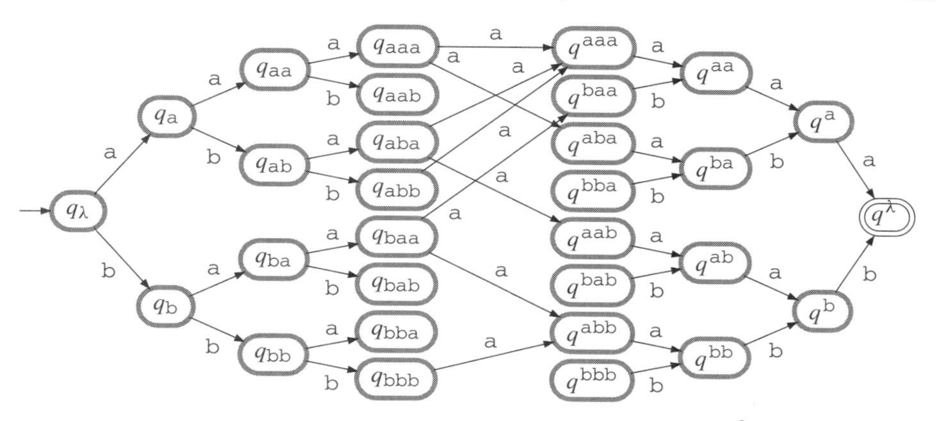

Fig. 6.2. The NFA to prove that the VC-dimension is in $\mathcal{O}(n^2)$.

Proof The first item follows as above from the fact that any finite language is regular.

Now consider the class of all NFAs with at most n states, and suppose again for simplicity that the alphabet is of size two. Then $|\mathcal{NFA}_n(\Sigma)| \leq 2^{2n^2} \cdot (2^n)^2$. So the largest set that can be shattered is in $\mathcal{O}(n^2)$.

Conversely we show that this bound can be met by providing such a class of strings/automata. Let $\Sigma = \{\mathtt{a}, \mathtt{b}\}$ and consider the set X of all strings of length $2k + 1$ with a as a centre symbol. This set is clearly of size n^2 when $n = 2^k$.

Then any subset Y of X can be recognised with the automaton $\mathcal{A} = \langle \Sigma, Q, \mathbb{I}, \mathbb{F}_{\mathbb{A}}, \mathbb{F}_{\mathbb{R}}, \delta_N \rangle$ where:

- $Q = \{q_u : u \in \Sigma^{\leq n}\} \cup \{q^v : v \in \Sigma^{\leq n}\}$,
- $\mathbb{I} = \{q_\lambda\}$,
- $\mathbb{F}_{\mathbb{A}} = \{q^\lambda\}$,
- $\mathbb{F}_{\mathbb{R}} = Q \setminus \mathbb{F}_{\mathbb{A}}$,
- $\forall u \in \Sigma^{<k}$, $q_{u\mathtt{a}} \in \delta_N(q_u, \mathtt{a})$, $q_{u\mathtt{b}} \in \delta_N(q_u, \mathtt{b})$,
- $\forall v \in \Sigma^{<k}$, $q^v \in \delta_N(q^{v\mathtt{a}}, \mathtt{a})$, $q^v \in \delta_N(q^{v\mathtt{b}}, \mathtt{b})$,
- $\forall u\mathtt{a}v \in Y$ with $|u| = k$, $|v| = k$, $q^v \in \delta_N(q_u, \mathtt{a})$.

With the above construction only the strings in Y (among all strings in $\Sigma^k \times \{\mathtt{a}\} \times \Sigma^k$) are recognised. We therefore have constructed an automaton with $4n - 2$ states for each $Y \subseteq X$, with $|Y| = n^2$. □

The NFA shown in Figure 6.2 recognises strings $\mathtt{aaa} \cdot \mathtt{a} \cdot \mathtt{aaa}$, $\mathtt{aaa} \cdot \mathtt{a} \cdot \mathtt{aba}$, $\mathtt{aba} \cdot \mathtt{a} \cdot \mathtt{aaa}, \ldots$

6.2 About consistency

In most other learning tasks (from examples and counter-examples), just finding a consistent solution is usually a good way to learn. Indeed many negative results are obtained through proving that since the consistency problem is $\mathcal{NP}$-hard, learning is as hard as solving this hard problem by a randomised algorithm.

This is why the complexity status of the consistency problem is of interest for learning.

6.2.1 The problem of finding the minimum consistent automaton

In the case of grammatical inference, finding a consistent grammar or automaton is not difficult: the consistency problem is usually trivial. For example, in the case of the DFA, the PTA is a perfect separator between two non-contradictory sets of strings.

The really interesting problem is therefore not of finding a consistent automaton but of finding the smallest consistent solution. In this case such an algorithm would be considered an Occam algorithm (and thus would allow polynomial PAC-learnability results) and would also identify in the limit.

A simple example that can be given to illustrate this point is the problem of learning rectangles in a two-dimensional space. If we fit the rectangle tightly to the positive data, then the error will be small unless the sampling was very bad. And this can easily be controlled by drawing enough examples. If furthermore the coordinates are integers, it is easy to see that just solving the consistency problem will identify in the limit.

6.2.2 NP-completeness of the problem of finding the minimum consistent automaton

Finding the minimum consistent DFA can be no easier than the corresponding decision problem.

Theorem 6.2.1 *The following problem is $\mathcal{NP}$-complete:*

> ***Name:*** *Smallest consistent* DFA *(*SCD*)*
> ***Instance:*** *Two sets S_+ and S_- of strings over some alphabet Σ, $n \in \mathbb{N}$*
> ***Question:*** *Does there exist a* DFA *$\mathcal{A}$ with at most n states such that $S_+ \subseteq \mathbb{L}(\mathcal{A})$ and $S_- \cap \mathbb{L}(\mathcal{A}) = \emptyset$?*

Proof The fact the problem is in $\mathcal{NP}$ is clear enough: An automaton with at most n states can be checked against S_+ and S_- in polynomial time.

The hardness will be shown in the following. □

To prove that the problem is $\mathcal{NP}$-hard we reduce SAT to SCD:

> **Name:** SAT
> **Instance:** A set of variables $U = \{x_1, \ldots, x_k\}$, a collection $\mathbb{C} = \{C_1, \ldots, C_m\}$ of clauses over U.
> **Question:** Does there exist an evaluation ϕ of U that satisfies all clauses?

We start by reducing SAT to a constrained version (called SAT_0) where there is the same number of variables as clauses and where each clause is either purely positive (all literals are positive) or purely negative (all literals are negative). A clause is ***pure*** if it is either positive (each literal l_j is positive: $C_i \in \mathbb{C}_P \iff \forall l_j \in C_i, l_j = x_k$) or negative (each literal l_j is negative: $C_i \in \mathbb{C}_N \iff \forall l_j \in C_i, l_j = \overline{x_k}$).

> **Name:** SAT_0
> **Instance:** A set of variables $U = \{x_1, \ldots, x_n\}$, a collection $\mathbb{C} = \{C_1, \ldots, C_m\}$ of pure clauses over U, with $\mathbb{C} = \mathbb{C}_P \cup \mathbb{C}_N$.
> **Question:** Does there exists an evaluation ϕ of U that satisfies all clauses?

It is straightforward to transform a normal clause into two pure clauses by introducing a new variable: $C_i = x_{i_1} \vee \ldots \vee x_{i_j} \vee \overline{x_{i_{j+1}}} \vee \ldots \vee \overline{x_{i_m}}$ becomes $x_{i_1} \vee \ldots \vee x_{i_j} \vee x_{i_{m+1}}$ and $\overline{x_{i_{j+1}}} \vee \ldots \vee \overline{x_{i_m}} \vee \overline{x_{i_{m+1}}}$ and C_i is satisfied if and only if the two new clauses are both satisfied.

The obtained set of clauses is equivalent to the initial one in the sense that they are either both satisfiable or neither one is.

We now reduce SAT_0 to SCD.

We construct from U and C as above the following sample over the alphabet $\Sigma = \{\mathtt{a}, \mathtt{b}\}$:

- $\mathtt{a}^n \in S_+$,
- $\forall k : 1 \leq k < n,\ \mathtt{a}^k \in S_-,\ \mathtt{a}^{n+k} \in S_-$,
- $\forall C_i \in \mathbb{C}_P$, $\mathtt{a}^{i-1}\mathtt{b}\mathtt{a}^n\mathtt{b} \in S_+$ and $\forall x_j \notin C_i,\ \mathtt{a}^{i-1}\mathtt{b}\mathtt{a}^{n-j+1} \in S_-$,
- $\forall C_i \in \mathbb{C}_N$, $\mathtt{a}^{i-1}\mathtt{b}\mathtt{a}^n\mathtt{b} \in S_-$ and $\forall \overline{x_j} \notin C_i,\ \mathtt{a}^{i-1}\mathtt{b}\mathtt{a}^{n-j+1} \in S_-$.

Claim 1 If $\mathcal{A}$ is consistent with $\langle S_+, S_- \rangle$, $\mathcal{A}$ has at least $n+1$ states. Indeed it is impossible to have $\mathtt{a}^i$ and $\mathtt{a}^j$ leading to the same state (if $i \neq j$) because $\mathtt{a}^{i+n-i} \in S_+$, whereas $\mathtt{a}^{j+n-i} \in S_-$. So $\mathcal{A}$ is certain to have at least states $q_{\mathtt{a}}, q_{\mathtt{a}^2}, \ldots, q_{\mathtt{a}^n} (= q_\lambda)$.

Claim 2 If the set C of clauses is satisfiable, it is satisfied by a valuation ϕ. Take clause C_i, and suppose without loss of generality that it is true thanks to the literal l_j which is either x_j or $\overline{x_j}$. Then we merge state $q_{\mathtt{a}^{i-1}\mathtt{b}}$ with state $q_{\mathtt{a}^{j-1}}$, and we have $\delta(q_\lambda, \mathtt{a}^{i-1}\mathtt{b}\mathtt{a}^{n-j+1}) \in \mathbb{F}_{\mathbb{A}}$. We therefore (since C is satisfiable) can merge each $q_{\mathtt{a}^{i-1}\mathtt{b}}$ with at least one $q_{\mathtt{a}^{j-1}}$ corresponding to the variable whose value of x_j satisfies clause C_i.

Claim 3 If there is an n-state automaton consistent with the sample, since each prefix $\mathtt{a}^k$ corresponds to at least one state, this means that for each C_i, $q_{\mathtt{a}^{i-1}\mathtt{b}}$ is merged with a state $q_{\mathtt{a}^i - 1}$. Therefore we have $\delta(q_{\mathtt{a}^{i-1}\mathtt{b}}, \mathtt{a}^{n-j+1}) \in \mathbb{F}_{\mathbb{A}}$, so we can choose $x_j = 1$ if $C_i \in \mathbb{C}_P$, $x_j = 0$ if $C_i \in \mathbb{C}_N$.

Notice that clashes between two merges due to the same variable but corresponding to opposite values are impossible: if $C_i \in \mathbb{C}_P$, $\mathtt{a}^{i-1}\mathtt{b}\mathtt{a}^n\mathtt{b} \in S_+$, and if $C_k \in \mathbb{C}_N$, $\mathtt{a}^{k-1}\mathtt{b}\mathtt{a}^n\mathtt{b} \in S_-$. So $q_{\mathtt{a}^{i-1}\mathtt{b}}$ and $q_{\mathtt{a}^{k-1}\mathtt{b}}$ cannot be merged.

Claim 4 The reduction is polynomial: the size of $\langle S_+, S_- \rangle$ is at most n^2, and it is simple to construct the learning sample from U and $\mathbb{C}$.

We conclude that SCD is $\mathcal{NP}$-complete.

Example 6.2.1 In order to illustrate the construction let us consider the following set of clauses: $x_1 \vee \overline{x_2} \vee x_3$ and $x_2 \vee \overline{x_3}$. The first step consists of building an equivalent system containing only pure clauses. For this the two new variables x_4 and x_5 are introduced. The new (and this time pure) system is:

$x_1 \vee x_3 \vee x_4$
$\overline{x_2} \vee \overline{x_4}$
$x_2 \vee x_5$
$\overline{x_3} \vee \overline{x_5}$

This is converted into the following sample:

$$\begin{aligned}
S_+ &= \{\mathtt{a}^5, \mathtt{ba}^5\mathtt{b}, \mathtt{a}^2\mathtt{ba}^5\mathtt{b}\} \\
S_- &= \{\mathtt{a}^i : 1 \le i < 5\} \cup \{\mathtt{a}^i : 5 < i < 10\} \cup \\
&\quad \{\mathtt{ba}, \mathtt{ba}^4, \mathtt{aba}, \mathtt{aba}^3, \mathtt{aba}^5, \mathtt{a}^2\mathtt{ba}^2, \mathtt{a}^2\mathtt{ba}^3, \mathtt{a}^2\mathtt{ba}^5, \mathtt{a}^3\mathtt{ba}^2, \mathtt{a}^3\mathtt{ba}^4, \mathtt{a}^3\mathtt{ba}^5\} \cup \\
&\quad \{\mathtt{aba}^5\mathtt{b}, \mathtt{a}^3\mathtt{ba}^5\mathtt{b}\}.
\end{aligned}$$

The corresponding PTA is represented in Figure 6.3. The question therefore is: can the states of this PTA be merged in some consistent way so as to give a 6-state DFA? In other terms, is there a 6-state DFA consistent with the data in the sample?

Using the assignment $x_1 = 1$, $x_2 = 1$, $x_3 = 0$, $x_4 = 0$, $x_5 = 1$, we can build the DFA represented in Figure 6.4. Notice that state $q_{\mathtt{b}}$ is merged with state q_λ, using $x_1 = 1$, state $q_{\mathtt{ab}}$ is merged with state $q_{\mathtt{a}^4}$, using $x_4 = 0$, state $q_{\mathtt{a}^2\mathtt{b}}$ is merged with state $q_{\mathtt{a}^2}$, using $x_2 = 1$, and state $q_{\mathtt{a}^3\mathtt{b}}$ is merged with state $q_{\mathtt{a}^3}$, using $x_3 = 0$.

6.2.3 Related questions

But the previous result itself is insufficient to conclude that learning DFAs is impossible. A typical approach in learning theory consists of compressing the data in some way. This does not require finding the smallest consistent solution but finding **a** small consistent solution. Yet this is also going to prove difficult since there are non-approximation results: finding a DFA (only) polynomially larger than the smallest consistent DFA is $\mathcal{NP}$-hard (given the polynomial in advance). Even the seemingly much simpler problem of deciding if, given two sets of strings S_+ and $S-$, there exists a DFA $\mathcal{A}$ with just two states such that $S_+ \subseteq \mathbb{L}(\mathcal{A})$ and $S_- \cap \mathbb{L}(\mathcal{A}) = \emptyset$ is $\mathcal{NP}$-complete. Note here that for this last result to hold, the size of the alphabet should not be fixed and is a parameter to the problem.

Another line of very negative results concerns linking learning DFAs with deciphering. In a public key setting the deciphering problem can be thought of as having access to the encoding scheme (as a black box) and therefore being able to, in some way, query the solution through membership queries. PAC predicting DFA can be reduced from the problem of inverting RSA encryption functions: an algorithm capable of obtaining some representation of a DFA whose error can be bounded in a PAC way could be transformed to decrypt RSA.

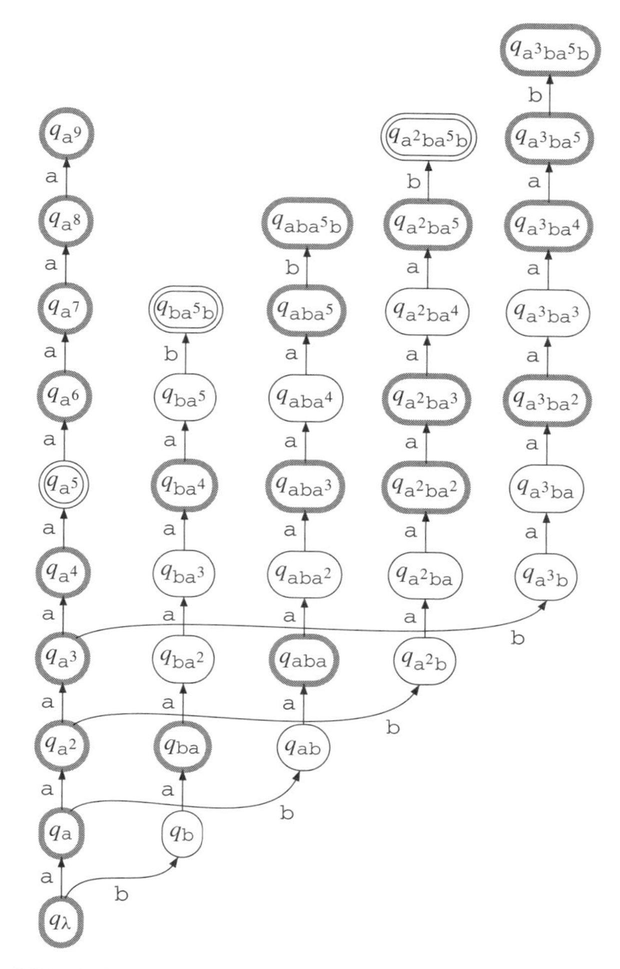

Fig. 6.3. The PTA obtained by the proposed construction from clauses $x_1 \vee \overline{x_2} \vee x_3$ and $x_2 \vee \overline{x_3}$.

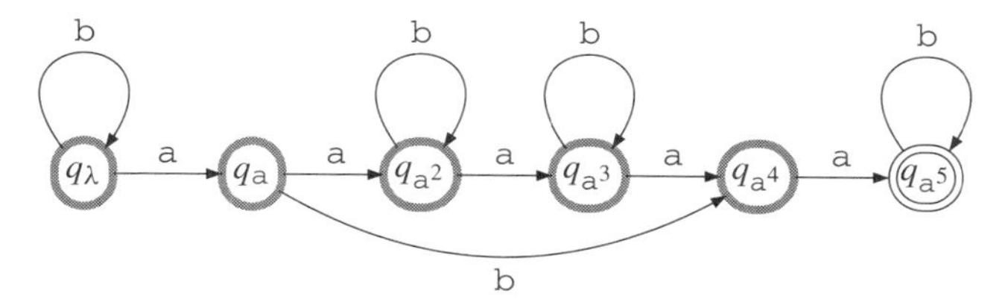

Fig. 6.4. A DFA with six states accepting S_+ and rejecting S_-.

6.3 The search space for the DFA learning problem

A successful way to rely on the many heuristic methods introduced in artificial intelligence is to describe the space in which the goal lies, propose operators to move around the space and an objective (or fitness) function that will tell us in some way how good the current solution is. There are then many alternative ways to search this space: some of these are described in Chapter 14.

We explore here some attempts to define the search space for the automata learning problem.

6.3.1 *Typical approach*

We will not argue here the fact that given a learning sample $< S_+, S_- >$, a learner should return a consistent solution. There may be cases where this would not be the right approach, namely when the data is suspected of being noisy. We mention works in this setting elsewhere (see Section 7.6, page 166).

So if we are interested in talking about the different possible solutions, we want to answer a first question:

Given two sets of strings S_+ and S_-, how many consistent DFAs can we find?

Let us analyse this question with the following example:

Example 6.3.1 Suppose we are given the following sample with alphabet $\Sigma = \{\texttt{a}, \texttt{b}\}$:
$S_+ = \{\texttt{b}, \texttt{aab}, \texttt{aaaba}, \texttt{bbaba}\}$
$S_- = \{\texttt{ba}, \texttt{abba}, \texttt{baaa}\}$
What are the correct solutions, if by correct we mean those consistent with the sample?

The answer to the above question is that any regular language L such that $\{\texttt{b}, \texttt{aab}, \texttt{aaaba}, \texttt{bbaba}\} \subseteq L \subseteq \Sigma^\star \setminus \{\texttt{ba}, \texttt{abba}, \texttt{baaa}\}$ will do!

The first answer to our question is therefore disappointing as the number of possible candidates is infinite.

In order to reduce this search space we need to add an extra bias.

6.3.2 *The partition lattice*

We first define an equivalence relation between automata: two states will be equivalent when they are in the same block. Let $\mathcal{A} = \langle \Sigma, Q, \mathbb{I}, \mathbb{F}_{\mathbb{A}}, \mathbb{F}_{\mathbb{R}}, \delta_N \rangle$ be a non-deterministic finite automaton and Π be a partition over the set of states Q. This partition induces an equivalence relation on the states, denoted by $\equiv_\Pi$.

Definition 6.3.1 The quotient automaton $\mathcal{A}/\Pi = \langle \Sigma, \overline{Q}, \overline{\mathbb{I}}, \overline{\mathbb{F}_{\mathbb{A}}}, \overline{\mathbb{F}_{\mathbb{R}}}, \overline{\delta_N} \rangle$ is defined as:

- $\overline{Q} = Q/\Pi$ is the set of equivalence classes defined by the partition Π,
- $\overline{\delta_N}$ is a function $\overline{Q} \times \Sigma \to 2^{\overline{Q}}$ such that $\forall \overline{q}, \overline{q}' \in \overline{Q}$, $\forall a \in \Sigma$, $\overline{q}' \in \overline{\delta_N}(\overline{q}, a)$ *if*$_{def}$ $\exists q \in \overline{q}\ \exists q' \in \overline{q}' : q' \in \delta_N(q, s)$,
- $\overline{\mathbb{I}}$ is the set of equivalence classes to which belongs at least one state in $\mathbb{I}$,
- $\overline{\mathbb{F}_{\mathbb{A}}}$ is the set of equivalence classes to which belongs at least one state q in $\mathbb{F}_{\mathbb{A}}$,
- $\overline{\mathbb{F}_{\mathbb{R}}}$ is the set of equivalence classes to which belongs at least one state q in $\mathbb{F}_{\mathbb{R}}$.

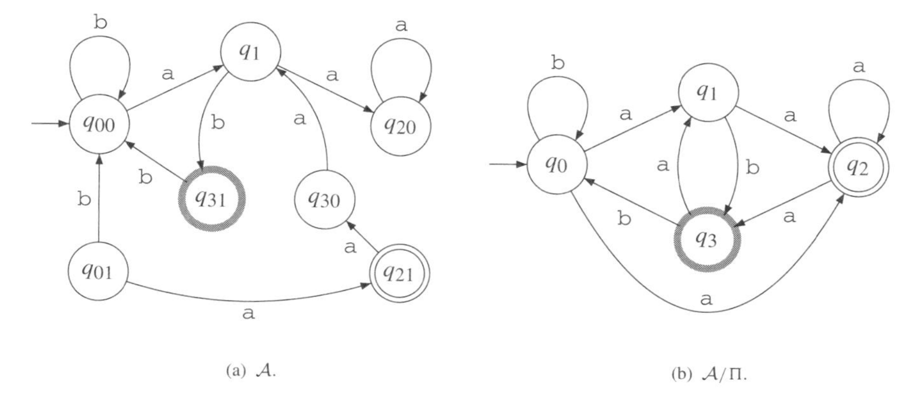

(a) $\mathcal{A}$. (b) $\mathcal{A}/\Pi$.

Fig. 6.5. $\Pi = \big\{\{q_{00}, q_{01}\}, \{q_1\}, \{q_{20}, q_{21}\}, \{q_{30}, q_{31}\}\big\}$.

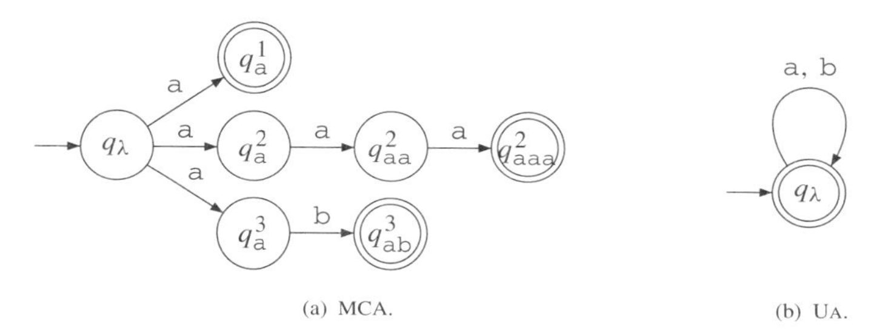

(a) MCA. (b) UA.

Fig. 6.6. A maximal canonical automaton and a universal automaton for $S_+ = \{\texttt{a}, \texttt{ab}, \texttt{aaa}\}$.

Example 6.3.2 We have represented in Figure 6.5(a) a finite automaton, and in Figure 6.5(b) the quotient automaton $\mathcal{A}/\Pi$.

In a complete lattice there are two extreme elements. In this case these are the maximal canonical automaton and the universal automaton.

Definition 6.3.2 (Maximal canonical automaton) Given a sample $S_+ = \{x_1, \dots x_n\}$, MCA(S_+) is the NFA $\mathcal{A} = \langle \Sigma, Q, \mathbb{I}, \mathbb{F}_{\mathbb{A}}, \mathbb{F}_{\mathbb{R}}, \delta_N \rangle$, where:

- $Q \leftarrow \{q_u^i : u \in \text{PREF}(x_i) \wedge u \neq \lambda\} \cup \{q_\lambda\}$,
- $\delta_N(q_u^i, a) = \{q_{ua}^i : ua \in \text{PREF}(x_i)\}$, $\forall i \in [n], \forall a \in \Sigma$,
- $\delta_N(q_\lambda, a) = \{q_a^i : a \in \text{PREF}(x_i)\}$, $\forall i \in [n], \forall a \in \Sigma$,
- $\forall i \in [n], q_{x_i}^i \in \mathbb{F}_{\mathbb{A}}$,
- $\mathbb{F}_{\mathbb{R}} = Q \setminus \mathbb{F}_{\mathbb{A}}$,
- If $\lambda \in S_+$, $q_\lambda \in \mathbb{F}_{\mathbb{A}}$ else $q_\lambda \in \mathbb{F}_{\mathbb{R}}$,
- $\mathbb{I} = \{q_\lambda\}$.

An MCA is therefore a star-shaped NFA with one branch per string in S_+, except for the empty string (if it belongs to S_+). We represent the MCA for sample $S_+ = \{\texttt{a}, \texttt{aaa}, \texttt{ab}\}$ in Figure 6.6(a).

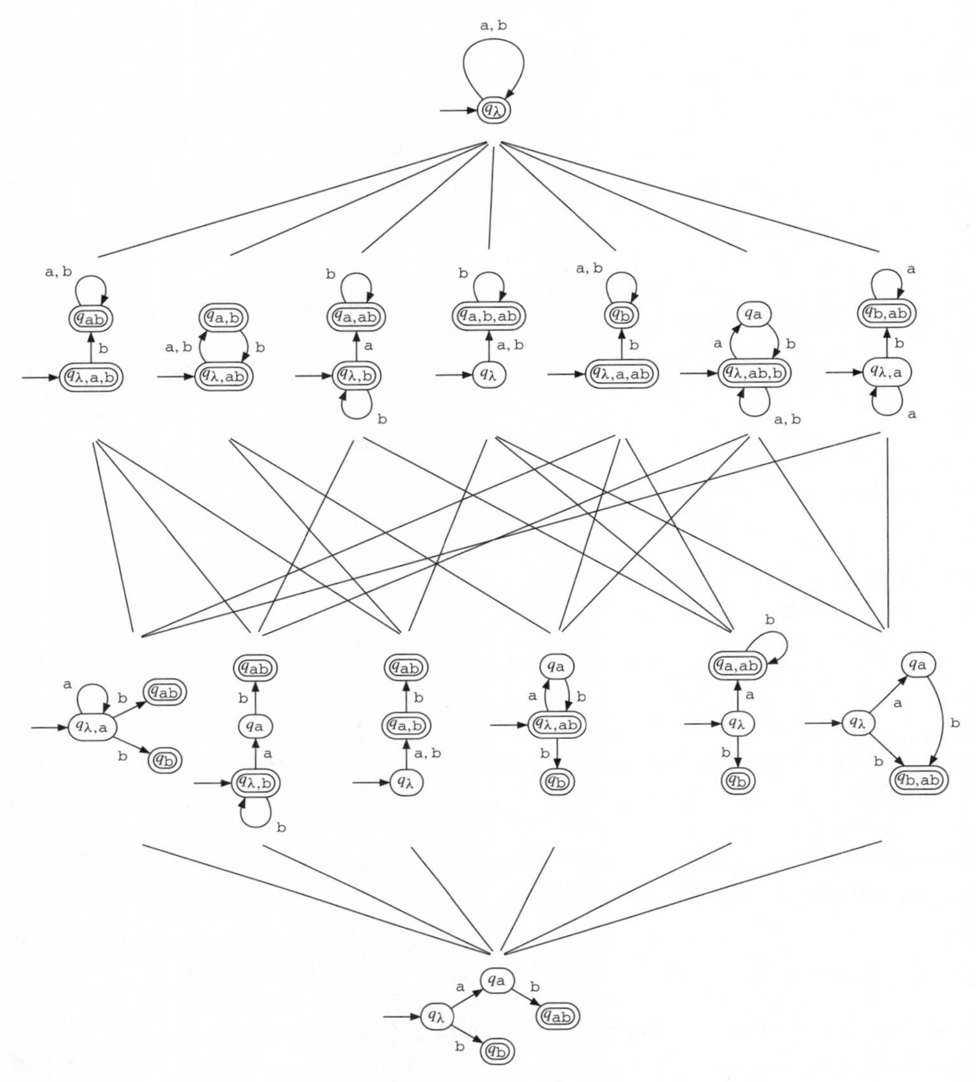

Fig. 6.7. The lattice for the derived automata built on MCA({b, ab}).

Definition 6.3.3 (Universal automaton) The universal automaton (UA) over an alphabet Σ is the smallest automaton accepting all strings of $\Sigma^\star$.

The universal automaton is the simplest of automata: it is deterministic, with just one state, and accepts everything. There is, of course, no rejecting state. The UA for alphabet {a, b} is represented in Figure 6.6(b).

We define the following relation on the set of all possible quotient automata of an automaton:

Definition 6.3.4 Let $\mathcal{A} = \langle \Sigma, Q, \mathbb{I}, \mathbb{F}_{\mathbb{A}}, \mathbb{F}_{\mathbb{R}}, \delta_N \rangle$ be an automaton and Π_1 and Π_2 be two partitions of Q. We say that $\mathcal{A}/\Pi_2$ **derives from** $\mathcal{A}/\Pi_1$, denoted $\mathcal{A}/\Pi_2 \preceq \mathcal{A}/\Pi_1$, if_{def} Π_1 is finer than Π_2.

A partition is finer than another if it has more classes and if being equivalent in the finer partition implies the same in the coarser one. This relation is a partial order on the set of all possible quotient automata of a given automaton and we have the following property:

Proposition 6.3.1 *If an automaton $\mathcal{A}/\Pi_j$ derives from an automaton $\mathcal{A}/\Pi_i$ then $\mathbb{L}(\mathcal{A}/\Pi_i) \subseteq \mathbb{L}(\mathcal{A}/\Pi_j)$.*

Proof Let $u = a_1 \cdots a_n$ in $\mathbb{L}_{\mathbb{F}_{\mathbb{A}}}(\mathcal{A})$. There is a path $q_{i_0}, a_1, q_{i_1}, \ldots, q_{i_{n-1}}, a_n, q_{i_n}$ in $\mathcal{A}$. By construction $\overline{q_{i_0}} a_1 \ldots a_n \overline{q_{i_n}}$ is a path in $\overline{\mathcal{A}}$. Furthermore $\overline{q_{i_0}} \in \overline{\mathbb{I}}$ and $\overline{q_{i_0}} \in \overline{\mathbb{F}}_{\mathbb{A}}$. □

Proposition 6.3.2 *The set of all non-deterministic quotient automata of an* NFA $\mathcal{A}$ *together with the partial order $\preceq$ is a complete lattice denoted by* ***LAT***$(\mathcal{A})$ *for which $\mathcal{A}$ and the universal automaton are respectively the null element and the universal element.*

Proof Each partition of the states of $\mathcal{A}$ defines a point in the lattice. Note also that we can have different points and automata corresponding to an identical language. □

6.3.3 Structural completeness

If the number of automata consistent with a given sample is infinite (as seen in Example 6.3.1, page 124), one might wonder if all these automata make sense. Take for instance the case where the alphabet consists of `a`, `b` and `c` but in the positive examples only letters `a` and `b` are seen. Unless there is some external additional knowledge to tell us that extra generalisation to unseen letters (and specifically to `c`) is a good idea, one can legitimately take the bias that only those letters seen in the learning sample should be mentioned in the learnt solution.

Not only does this bias make sense, but it is necessary in order to achieve identification.

We can be more precise in the case of DFAs by stating that we are only interested in automata that:

- use each transition at least once when parsing the learning sample,
- are able to parse the learning sample,
- have at least one string of the learning sample ending in each final state.

Notice that although the definition is of structural completeness of an automaton with respect to a sample, the definition can be written in a symmetrical way:

Definition 6.3.5 (Structural completeness for a DFA) Let $\mathcal{A}$ be a deterministic finite automaton, $\mathcal{A} = \langle \Sigma, Q, q_\lambda, \mathbb{F}_{\mathbb{A}}, \mathbb{F}_{\mathbb{R}}, \delta \rangle$, and S_+ be a finite set of strings over Σ.
$\mathcal{A}$ is **structurally complete** with respect to S_+ (and S_+ is **structurally complete** with respect to $\mathcal{A}$) if_{def}

- $\forall q \in \mathbb{F}_{\mathbb{A}},\ \exists w \in S_+ : \delta(q_\lambda, w) = q$,
- $\forall w \in S_+,\ w \in \mathbb{L}(\mathcal{A})$,
- $\forall q \in Q,\ \forall a \in \Sigma$ such that $\delta(q, a)$ is defined, $\exists w \in S_+ : w = uav$ and $\delta(q_\lambda, u) = q$.

The bias we were considering therefore consists of only searching for DFAs for which the sample is structurally complete! Notice that the set of final rejecting strings $\mathbb{F}_{\mathbb{R}}$ plays no part in this construction. An alternative definition incorporating this set is of course possible:

Definition 6.3.6 (Symmetrical structural completeness for a DFA) Let $\mathcal{A}$ be a deterministic finite automaton, $\mathcal{A} = \langle \Sigma, Q, q_\lambda, \mathbb{F}_{\mathbb{A}}, \mathbb{F}_{\mathbb{R}}, \delta \rangle$, and $\langle S_+, S_- \rangle$ be a finite sample over Σ.
$\mathcal{A}$ is **symmetrically structurally complete** with respect to $\langle S_+, S_- \rangle$ (and $\langle S_+, S_- \rangle$ is **symmetrically structurally complete** with respect to $\mathcal{A}$) *if*$_{def}$

- $\forall q \in \mathbb{F}_{\mathbb{A}},\ \exists w \in S_+ : \delta(q_\lambda, w) = q$,
- $\forall q \in \mathbb{F}_{\mathbb{R}},\ \exists w \in S_- : \delta(q_\lambda, w) = q$,
- $\forall w \in S_+,\ w \in \mathbb{L}_{\mathbb{F}_{\mathbb{A}}}(\mathcal{A})$,
- $\forall w \in S_-,\ w \in \mathbb{L}_{\mathbb{F}_{\mathbb{R}}}(\mathcal{A})$,
- $\forall q \in Q,\ \forall a \in \Sigma$ such that $\delta(q, a)$ is defined, $\exists w \in S_+ \cup S_- : w = uav$ and $\delta(q_\lambda, u) = q$.

Example 6.3.3 Suppose our learning sample is:
$S_+ = \{\texttt{ab}, \texttt{bbab}\}$
$S_- = \{\lambda, \texttt{b}, \texttt{aba}, \texttt{abb}, \texttt{aaba}\}$.
Then the automaton depicted in Figure 6.8 is not structurally complete: notice that the transition $(q_{\texttt{b}}, \texttt{a}, q_\lambda)$ is not exercised and the final state $q_{\texttt{a}}$ is never used as such. If the sample also contained string abaa then this automaton would be structurally complete.

Structural completeness, as defined above, can be extended with care to the non-deterministic setting. But this cannot be done directly, as shown in Figure 6.10(b). To make things slightly simpler we choose to restrict the NFAs to those with a unique initial state. The idea is that each transition and each accepting state in the NFA are validated by at least one string in the sample, but that a string is used in only one path!

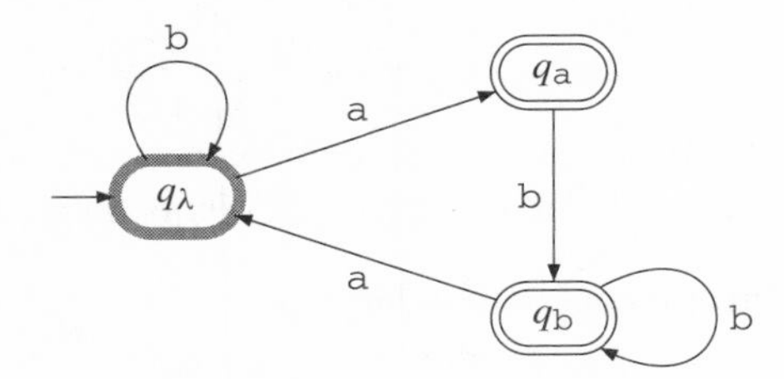

Fig. 6.8. A DFA that is not structurally complete with respect to the sample $\{(\texttt{ab}, 1), (\texttt{bbab}, 1), (\lambda, 0), (\texttt{b}, 0), (\texttt{aba}, 0), (\texttt{aaba}, 0)\}$.

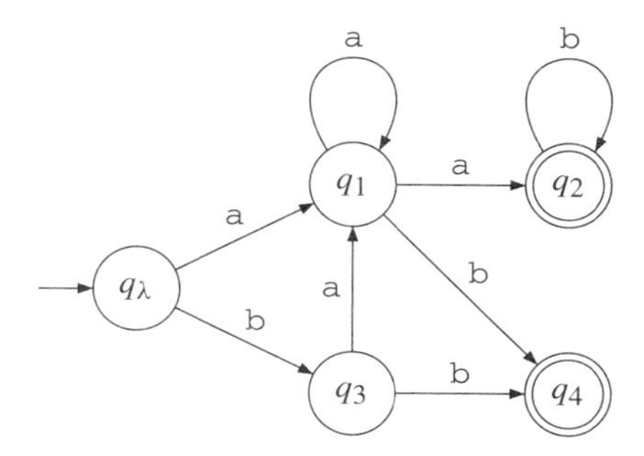

Fig. 6.9. This NFA is strongly structurally complete for the sample $S_+ = \{\mathtt{aab}, \mathtt{baab}, \mathtt{bb}\}$.

Definition 6.3.7 (Strong structural completeness)
Let $\mathcal{A}$ be a non-deterministic finite automaton with a unique initial state q_λ, $\mathcal{A} = \langle \Sigma, Q, \{q_\lambda\}, \mathbb{F}_\mathbb{A}, \mathbb{F}_\mathbb{R}, \delta_N \rangle$, and $S_+ = \{x_1, \ldots, x_n\}$ a finite set of strings over Σ.
$\mathcal{A}$ is **strongly structurally complete** with respect to S_+ (and we will say that S_+ is **strongly structurally complete** with respect to $\mathcal{A}$) if_{def} there exists a surjective function $\phi : \text{PREF}(S_+) \times [n] \to Q$ such that $\phi(ua, i) \in \delta_N(\phi(u, i), a)$, and

- $\forall q \in \mathbb{F}_\mathbb{A}, \ \exists x_i \in S_+ : \ \phi(x_i, i) = q$,
- $\forall (q, a, q') \in \delta_N : \exists x_i = uav \in S_+$ and $\phi(u, i) = q$.

Example 6.3.4 The following sample is strongly structurally complete for the NFA represented in Figure 6.9. $S_+ = \{\mathtt{aab}, \mathtt{baab}, \mathtt{bb}, \}$, with ϕ given as follows:

- $\phi(\lambda, 1) = \phi(\lambda, 2) = \phi(\lambda, 3) = q_\lambda$,
- $\phi(\mathtt{a}, 1) = q_1, \phi(\mathtt{aa}, 1) = q_1, \phi(\mathtt{aab}, 2) = q_4$,
- $\phi(\mathtt{b}, 2) = q_3, \phi(\mathtt{ba}, 2) = q_1, \phi(\mathtt{baa}, 2) = q_2, \phi(\mathtt{baab}, 2) = q_2$,
- $\phi(\mathtt{b}, 3) = q_3, \phi(\mathtt{bb}, 3) = q_4$.

And ϕ complies with the rules from Definition 6.3.7. Notice that what ϕ is doing is showing that each element of the NFA is justified by some separate piece of data.

In the above definition it is essential that a parse validates each state of an NFA. This will avoid cases like the one represented in Figure 6.10(b), in which the automaton is not strongly structurally complete for the sample $S_+ = \{\mathtt{aa}, \mathtt{bb}, \mathtt{cc}\}$ (see Proposition 6.3.4 for details). This may also mean that if strings are present in the set more than once (these are then multisets), different automata become structurally complete.

6.3.4 *Defining the search space by structural completeness*

Once we have admitted that structural completeness could be our added bias, the one that could restrict our search space, we will then only consider *biased solutions* to our problem, that is DFAs for which the structural completeness condition holds. But the space consisting only of DFAs does not have the combinatorial qualities we are searching for, so we

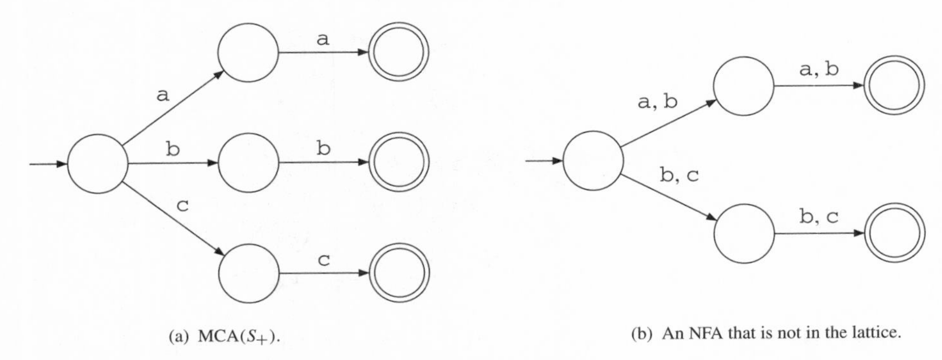

(a) MCA(S_+). (b) An NFA that is not in the lattice.

Fig. 6.10. The MCA cannot be the correct starting point for structural completeness.

consider the more general space of all NFAs (with one initial state) for which the strong structural completeness holds. This search space is then a partition lattice and we have the following theorem:

Theorem 6.3.3 *Every* DFA *in* ***LAT***(MCA(S_+)) *is structurally complete for* S_+. *Every* NFA *in* ***LAT***(MCA(S_+)) *is strongly structurally complete for* S_+.

Proof It is easy to see that any NFA derived from MCA(S_+) is strongly structurally complete as the function ϕ is constructed easily during the merging process. □

In Figure 6.7 the search space corresponding to a sample $S_+ = \{\texttt{a}, \texttt{ab}\}$ is represented. One should be careful with the converse:

Proposition 6.3.4 *There exists a sample* S_+ *and an* NFA $\mathcal{A}$ *(with just one state) such that* $\mathcal{A}$ *is structurally complete with* S_+ *yet does not belong to* ***LAT***(MCA(S_+)).

Proof One can build a sample S_+ and an NFA $\mathcal{A}$ such that:

- $\forall q \in \mathbb{F}_{\mathbb{A}},\ \exists x_i \in S_+ : q \in \delta_N(q_\lambda, x_i)$,
- $\forall q \in Q,\ \forall a \in \Sigma$ such that $\exists x, y \in \Sigma^\star : q \in \delta_N(q_\lambda, x, q)$ and $xay \in \mathbb{L}(\mathcal{A})$, $\exists x_i \in S_+ : x_i = uav$ and $q \in \delta(q_\lambda, u)$.

and $\mathcal{A}$ is not in **LAT**(MCA(S_+)).

Let $S_+ = \{\texttt{aa}, \texttt{bb}, \texttt{cc}\}$. MCA($S_+$) is represented in Figure 6.10(a). Now consider the NFA represented in Figure 6.10(b). It is easy to see that it is *structurally complete* but cannot be obtained as a quotient of the MCA, and nor can any other automaton equivalent to it. □

Theorem 6.3.5
Every DFA *structurally complete for* S_+ *is in* ***LAT***(MCA(S_+)).
Every NFA *strongly structurally complete for* S_+ *is in* ***LAT***(MCA(S_+)).

Proof We only need to prove the second part. If $\mathcal{A}$ is strongly structurally complete for S_+ there exists a function ϕ as in Definition 6.3.7. Then the partition corresponding to ϕ is such that states q_u^i and q_w^j are in the same class if $\phi(u, i) = \phi(w, j)$. □

We are obviously defining here the search space as the set of all possible non-deterministic finite state automata. But since in most cases we are looking for a deterministic automaton we have the following corollary:

Proposition 6.3.6 *The smallest* DFA *consistent with a sample* $\langle S_+, S_-\rangle$ *is in* ***LAT***(PTA(S_+)).

This suggests strategies adopted by most algorithms that learn automata: start from PTA(S_+) and explore the corresponding lattice of solutions using the merging operation. S_- is used to control the generalisation.

Such strategies have been followed in the case of RPNI (Algorithm 12.4), but also using some of the heuristics proposed in Chapter 14.

The approach consisting of an exhaustive exploration of the lattice is going to be difficult as the size of the partition lattice grows far too fast with the size of the sample, as the following formula tells us:

Let E be a set with n elements; the number of partitions of E is given by the Bell number:

$$\begin{aligned} \omega(0) &= 1 \\ \omega(n+1) &= \textstyle\sum_{p=0}^{n} \binom{n}{p}\omega(n). \end{aligned}$$

For instance, $\omega(16) = 10{,}480{,}142{,}147$. The size of the partition lattice is therefore going to be far too big to consider possible success by systematic exploration.

6.4 About the equivalence problem and its relation to characteristic sets

The equivalence problem consists of checking if two grammars/automata are equivalent, i.e. if they recognise or generate the same language.

For intuitive reasons the problem is important to grammatical inference: indeed, a grammatical inference algorithm is essentially attempting to build a canonical grammar for a given language. Being able to do this would give us a way around the equivalence problem: one might consider generating data from two different grammars and then checking if the algorithm has inferred the same thing. Obviously, things are not quite that simple, and the intended proof is very loose. But a more technical proof can be obtained for a specific learning paradigm (POLY-CS learnability), in which a class will be called POLY-CS *learnable* if each grammar in the class admits a polynomial characteristic sample, such that once this sample is included in the learning data, the algorithm will always return the correct grammar. Being POLY-CS *learnable by an informant* means that the learning algorithm learns from positive and negative examples. The necessary definitions are given in Section 7.3.3.

Theorem 6.4.1 *If the equivalence problem for $\mathcal{G}$ is undecidable, then $\mathcal{G}$ is not* POLY − CS *learnable by an informant.*

Proof Given any polynomial $p()$ there exist two grammars G_1 and G_2 indistinguishable by strings of size less than $p(\|G_1\| + \|G_2\|)$; if not the equivalence problem would be decidable. Suppose (for contradiction) that each admits a polynomial characteristic sample: call these respectively $\langle S_{1+}, S_{1-}\rangle$ for G_1 and $\langle S_{2+}, S_{2-}\rangle$ for G_2. Then what language is inferred from $\langle S_{1+} \cup S_{2+}, S_{1-} \cup S_{2-}\rangle$? Notice that both G_1 and G_2 are consistent with $\langle S_{1+} \cup S_{2+}, S_{1-} \cup S_{2-}\rangle$. □

Corollary 6.4.2 *For any alphabet Σ of size at least 2, the classes $\mathcal{CFG}(\Sigma)$, $\mathcal{LIN}(\Sigma)$ and $\mathcal{NFA}(\Sigma)$ are not* POLY − CS *learnable.*

Proof Indeed the equivalence problem is undecidable for context-free grammars and for linear grammars. In the case of $\mathcal{NFA}(\Sigma)$ the equivalence problem is $\mathcal{P}$-space complete, which allows us to use the same argument as in the proof of Theorem 6.4.1. □

6.5 Some remarkable automata

We introduce now some constructions that are of particular use in grammatical inference: how to encode parity functions into DFAs, and SAT instances into NFAs. We also show an NFA whose 'important' string is of exponential size.

6.5.1 Parity automata

A parity function is a Boolean function $p_u : \mathbb{B}^n \rightarrow \mathbb{B}$. It is defined by a Boolean vector u (which we will interpret as a string of length n). The function p_u takes a string w as an argument and returns the parity of the number of bits equal to '1' that w has in the positions indicated by u. Formally $p_u(w) = \sum_{i=1}^{n}(u_i \cdot w_i) \mod 2$.

Informally, if the number of 1s in the specified positions is even, the result is '0', and the string is accepted. For example take n=8, and $u = 10011011$. Then string 11111111 does not belong to L since $p_u(11111111) = (1 \cdot 1) + (1 \cdot 0) + (1 \cdot 0) + (1 \cdot 1) + (1 \cdot 1) + (1 \cdot 0) + (1 \cdot 1) + (1 \cdot 1) \mod 2 = 5 \mod 2 = 1$. On the other hand $p_u(01010101) = (0 \cdot 1) + (1 \cdot 0) + (0 \cdot 0) + (1 \cdot 1) + (0 \cdot 1) + (1 \cdot 0) + (0 \cdot 1) + (1 \cdot 1) \mod 2 = 2 \mod 2 = 0$.

These functions are known to be hard to learn in noisy settings: it can be proved that above a certain level of noise, and under cryptographic assumptions, no algorithm can learn parity functions with the guarantee of only a small error.

Another way to express this is that the language recognised by a parity automaton is such that given any string in the language, there are as many chances of remaining inside the language as of getting out of it, as soon as even a bit of noise is added. The sort of noise considered can be noise on labels, on the bits in the string, noise through either the edit distance or the Hamming distance…

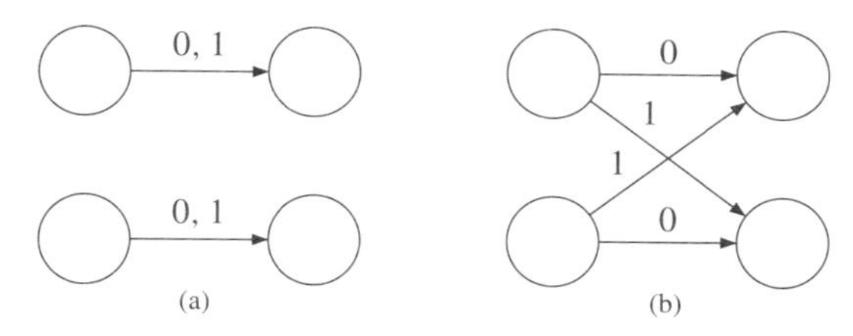

Fig. 6.11. Parity function blocks.

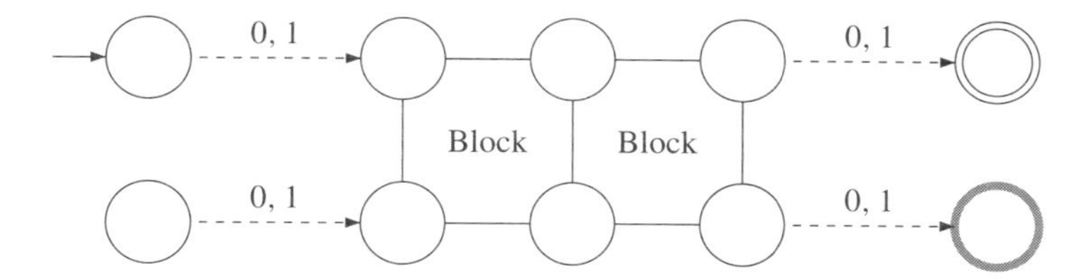

Fig. 6.12. Building a parity function automaton.

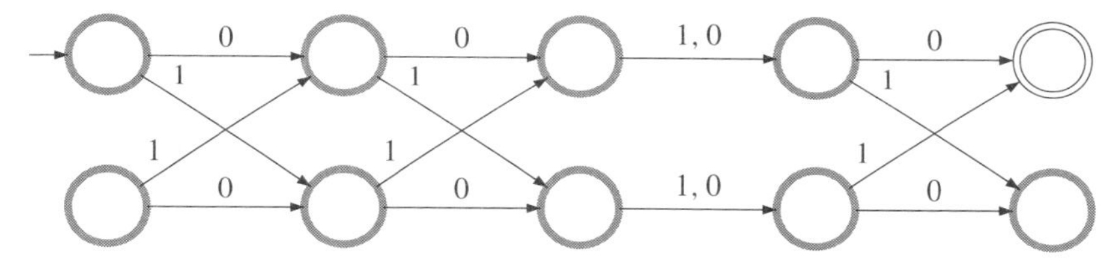

Fig. 6.13. The parity function automaton encoding function p_{1101}.

The construction of a DFA corresponding to a parity function is done by building blocks corresponding to the positions that count and those that don't. This is shown in Figure 6.11, with the block corresponding to the non-parity position in Figure 6.11(a), and the one corresponding to the parity position in Figure 6.11(b). The parity automaton when joining the blocks is of the type shown in Figure 6.12.

In Figure 6.13 we show the parity function automaton corresponding to selected first, second and fourth bits, that is, to string 1101. Notice that string 1111 is rejected whereas 1001 is accepted.

6.5.2 From SAT *to* NFA

We show here how to reduce an instance of SAT to an NFA such that there is an equivalent NFA with n states if and only if the instance of SAT is satisfiable. We first associate to each clause a sub-automaton. Recall that SAT is defined as:

> **Name:** SAT
> **Instance:** A set of variables $U = \{x_1, \ldots, x_k\}$, a collection $C_1, \ldots, C_m$ of clauses over U.
> **Question:** Does there exist an evaluation ϕ of U that satisfies all clauses?

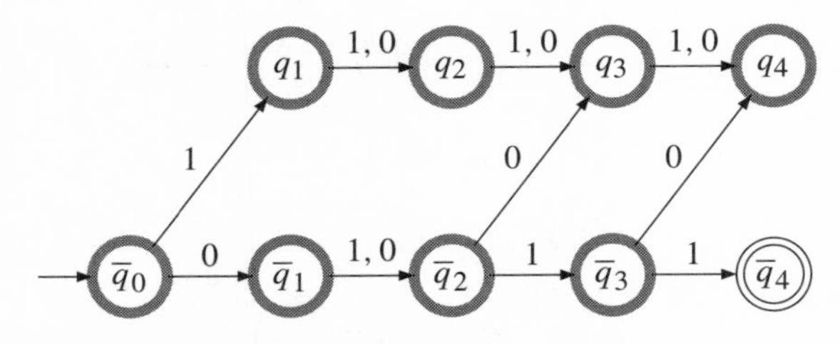

Fig. 6.14. A DFA that rejects only strings satisfying the clause $x_1 \vee \overline{x_3} \vee \overline{x_4}$. State q_0 is not drawn.

Let us first illustrate the reduction through an example: let $n = 4$ and C_i the clause $x_1 \vee \overline{x_3} \vee \overline{x_4}$. We associate with this clause the automaton depicted in Figure 6.14.

More generally, we first build for each clause C_i a DFA that rejects exactly the strings of length n corresponding to a set of values that satisfy the clause.

We construct the automaton corresponding to a clause C_i as follows: $\mathcal{A}_i = \langle \Sigma, Q^i, \overline{q}^i_0, \mathbb{F}^i_{\mathbb{A}}, \mathbb{F}^i_{\mathbb{R}}, \delta^i_N \rangle$ where

- $Q^i = \{q^i_k : k \leq n\} \cup \{\overline{q}^i_k : k \leq n\}$,
- $\mathbb{F}^i_{\mathbb{R}} = \{q^n\}$,
- $\mathbb{F}^i_{\mathbb{A}} = Q \setminus \mathbb{F}^i_{\mathbb{R}}$,
- $\forall j < n$
  - if $x_j \in C_i$, $\delta^i(\overline{q}^i_j, 1) = q^i_{j+1}$,
  - if $x_j \notin C_i$, $\delta^i(\overline{q}^i_{j-1}, 0) = \overline{q}^i_{j+1}$,
  - if $\overline{x_j} \in C_i$, $\delta^i(\overline{q}^i_j, 0) = q^i_{j+1}$,
  - if $\overline{x_j} \notin C_i$, $\delta^i(\overline{q}^i_j, 1) = \overline{q}^i_{j+1}$,
  - $\forall j < n$, $\delta^i(q^i_j, 1) = q^i_{j+1}$, $\delta^i(q^i_j, 0) = q^i_{j+1}$.

The m DFAs are then associated into a unique NFA by adding a single initial node q_I and allowing us to read a '1' from q_I to each initial node $\overline{q}^i_0$. It is straightforward to see that the new automaton does not accept a string $1w$, with w of length n, if and only if w is the string writing of a valuation that satisfies all clauses. Indeed, in that case the different parses will all end in the rejecting states of each small automaton.

From this construction, a number of results concerning NFAs follow: for instance, the equivalence problem is $\mathcal{NP}$-hard.

As a further illustration, we represent the automaton used to encode the set of clauses $(\overline{x_1} \vee \overline{x_3})$, $(x_1 \vee \overline{x_3} \vee \overline{x_4})$, $(x_1 \vee \overline{x_2} \vee x_3 \vee x_4)$ in Figure 6.15.

In this case string 1100 is rejected, so the set of clauses is satisfiable.

6.5.3 *A very long string*

From the previous constructions it may seem that the fact that SAT instances can be encoded into NFAs helps us understand why the equivalence problem for NFAs is hard.

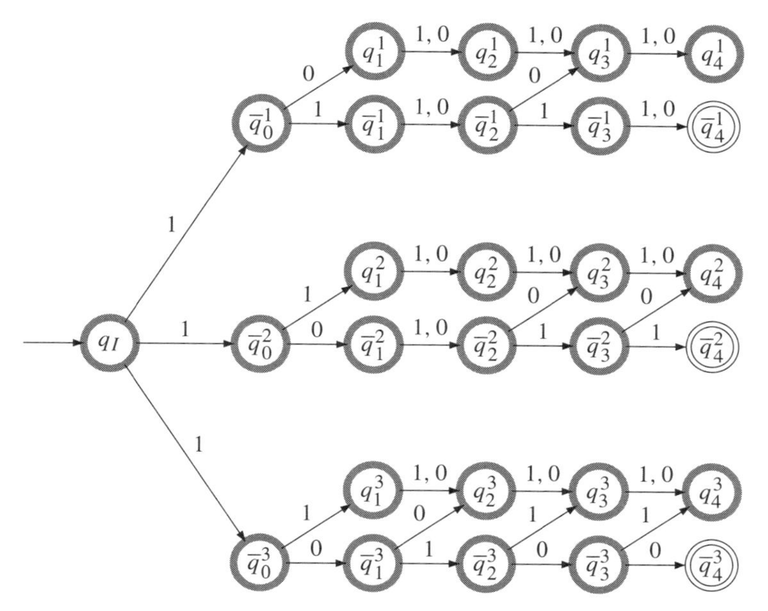

Fig. 6.15. An NFA that rejects strings satisfying clauses $(\overline{x_1} \vee \overline{x_3})$, $(x_1 \vee \overline{x_3} \vee \overline{x_4})$, $(x_1 \vee \overline{x_2} \vee x_3 \vee x_4)$.

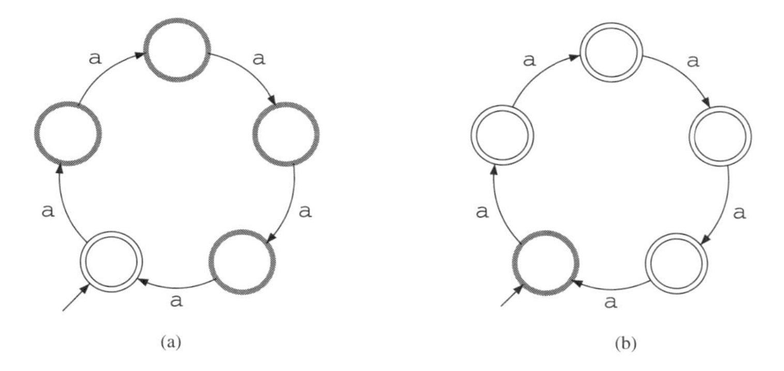

Fig. 6.16. Automata for $(\mathrm{a}^5)^*$ and $\Sigma^\star \setminus (\mathrm{a}^5)^*$.

Since the equivalence problem is in itself linked to various inference problems, this is indeed alarming.

But there is worse to come. We show now an automaton for which the smallest string **not** in the language is of exponential length with the size of the automaton. By doing this we illustrate why the equivalence problem for NFAs is not necessarily in $\mathcal{NP}$. In the following we will only even require a one-symbol alphabet.

We first note (Figure 6.16) how to build an automaton that recognises $(\mathrm{a}^p)^*$ (Figure 6.16(a)), and $\Sigma^\star \setminus (\mathrm{a}^p)^*$ (Figure 6.16(b)), for any value of p (and of particular interest are the prime numbers). We now extend this construction by taking for a set of

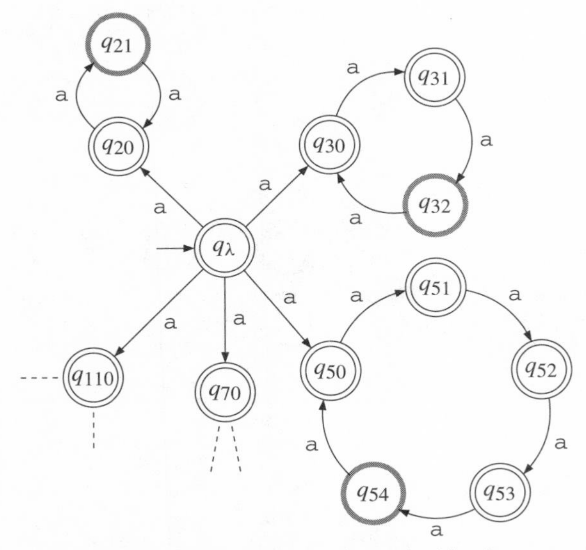

Fig. 6.17. An automaton whose smallest 'interesting string' is of exponential length.

prime numbers $\{p_1, p_2, \ldots, p_n\}$ the automaton that recognises all non-empty strings that are neither multiples of p_1, nor of $p_2 \ldots$ nor of p_n. This NFA can be constructed as proposed in Figure 6.17, and has $1 + \sum_{i=1}^{n} p_i$ states. But the smallest string **not** accepted by this automaton is $\mathtt{a}^k$ with $k = \prod_{i=1}^{n} p_i$. Indeed $\prod_{i=1}^{n} p_i$ is the smallest non-null integer which is a multiple of each p_i. This value grows exponentially with the size of the automaton.

This means that if a string is needed (in a characteristic sample) to distinguish this automaton from one that recognises $\Sigma^\star$, an exponentially long string is needed.

6.6 Exercises

6.1 Compute the size of the lattice corresponding to an MCA with 5, 10, and 20 states.

6.2 What is the VC-dimension of balls? Good balls? Cones? Co-cones?

6.3 What is the VC-dimension of linear languages?

6.4 Prove that in the lattice **LAT**(PTA(S_+)), there can be several equivalent automata. Find a sample S_+ for which there are an exponential number of NFAs but a linear number of different languages.

6.5 Build (as in Section 6.5.2) the NFA corresponding to the clause $x_1 \vee \overline{x_2} \vee x_3 \vee \overline{x_5}$.

6.6 Give conditions over NFAs such that any NFA complying with these conditions admits a strongly structurally complete sample which is a set (and not a multiset).

6.7 Prove that any ball $B_k(x)$ (corresponding to the set of all strings at edit distance at most k from string x) can be represented by an NFA which has at most $k(|x| + 1)$ states. See Sections 3.4.4 and 3.6.1 for the definitions.

6.8 Consider the PTA corresponding to the sample $S_+ = \{\texttt{a}, \texttt{aa}, \texttt{ab}, \texttt{bba}\}$. How large is the lattice $\mathbf{LAT}(\text{PTA}(S_+))$? How many elements from this lattice are deterministic?

6.9 Prove that the number of DFAs in $\mathbf{LAT}(\text{PTA}(S_+))$ can be exponential in $|\text{PTA}(S_+)|$.

6.10 The proof from Section 6.5.2 can be adapted to the probabilistic case. This allows us to prove, for instance, that finding the most probable string for a PFA is $\mathcal{NP}$-hard. Following the construction represented in Figure 6.14 (page 134), build a PFA whose most probable string is (when it exists) one that is the encoding of a satisfying assignment.

6.7 Conclusions of the chapter and further reading

6.7.1 Bibliographical background

In Section 6.1 we discussed the VC-dimension. Detailed proofs and further discussions concerning the links with learning can be found in (Ishigami & Tani, 1997, Natarajan, 1991).

The essential question of consistency (Section 6.2) corresponds to results usually reasonably well known to grammatical inference scientists, but the proofs are rarely mastered. It is clear to all that finding the minimum consistent DFA 'looks like' a variant of a graph colouring problem, but the full extent of the hardness is less clear. There were two historical proofs of this, by E. Mark Gold (Gold, 1978) and Dana Angluin (Angluin, 1978), through different techniques. The proof we give of Theorem 6.2.1 is adapted from (Gold, 1978). Since it was later proved that Occam algorithms which provided some sort of compression of the data (not just the ideal one) could also PAC-learn, it was important to have a negative result to this approach. This was obtained by Lenny Pitt and Manfred Warmuth, who proved that finding a DFA (only) polynomially larger than the smallest consistent DFA is $\mathcal{NP}$-hard (Pitt & Warmuth, 1993). This allows us to state that PAC-predicting DFAs is as difficult as inverting the RSA encryption function (Kearns & Valiant, 1989, Pitt & Warmuth, 1988).

The search space for the DFA learning problem (Section 6.3) was introduced by King-Sun Fu and Taylor Booth (Fu & Booth, 1975), and further looked into by Laurent Miclet (Miclet, 1990), and Pierre Dupont *et al.* (Dupont, Miclet & Vidal, 1994) who define an automaton to be structurally complete with respect to a sample, with a very different flavour from Definition 6.3.5 (page 127) given here. There are problems (shown here in Proposition 6.3.4) with the classical definition. We should notice that in pattern recognition the data are what matter, and therefore a point of view directed by the data would consider the automata to depend on the data. Plausibly, a researcher thinking in terms of convergence of the algorithm would consider the question of only being able to learn a target automaton if the sample is structurally complete.

Previous work by King-Sun Fu and Taylor Booth (Fu & Booth, 1975) consisted of defining the operators enabling us to move around the lattice. The special part played by the PTA in grammatical inference has led to various alternatives where for instance the

PTA may contain both the examples and the counter-examples (Coste & Nicolas, 1998a, de Oliveira & Silva, 1998). Reducing the size of the lattice (Coste *et al.*, 2004, Kermorvant & de la Higuera, 2002, Kermorvant, de la Higuera & Dupont, 2004) has been another line of research: typically the goal is to find some property prohibiting specific merges. In this way the lattice is the one of all admissible partitions, some partitions being declared illegal for all sorts of reasons: background knowledge, co-learning or just added bias. This idea was already present in Jose Oncina and Miguel Angel Varó's work on learning transducers (Oncina & Varó, 1996).

In Section 6.4 we relate the equivalence problem and the size of the characteristic samples. Theorem 6.4.1 was proved in (de la Higuera, 1997) using techniques and results from the teaching model introduced by Sally Goldman and David Mathias (Goldman & Mathias, 1996). Teaching issues correspond to exploring the idea that something difficult to teach is presumably going to be difficult to learn.

The remarkable automata described in Section 6.5 come from a variety of papers. Parity automata (Kearns *et al.*, 1994) were introduced to show a class that could not be learnt from noisy data. The SAT to NFA construction was used to prove that the decoding problem (related to finding the best translation in a stochastic transducer) was not tractable (Casacuberta & de la Higuera, 1999). The NFA that has an exponentially long string as *characteristic* was introduced in (de la Higuera, 1997). This construction has been used to show that NFAs were not identifiable from polynomial time and data. The same automaton can be adapted to the probabilistic setting to show that the most probable string for a PFA can be of exponential length.

6.7.2 Some alternative lines of research

The combinatorial properties of automata and grammars are studied in many fields and not all their results have reached researchers in grammatical inference. There certainly is much to be learnt from the specialists in formal language theory, and particularly by those applying formal models to biology.

There are also intricate issues dealing with combinatorics when looking into the edit distances and the string kernels. We have not discussed here the case of trees and graphs, for which the combinatorics are going to prove harder than in the case of strings. One of the key combinatorial problems is that of isomorphism. In the case of strings, this is of course not even an issue. For trees, if the ordering of the siblings is not important, then it becomes an interesting question, and it is supposed to be intractable in the case of graphs.

6.7.3 Open problems and possible new lines of research

There are many researchers working on alternative implementations of automata as these are essential objects in a number of fields. Yet there still are a number of open research problems in combinatorics over automata.

Techniques to explore the lattice in a faster way are needed, and so are algorithms to test consistency in a faster way. In fact, consistency is a key issue and one can argue that the representations of languages should perhaps be made around a fast consistency check.

If consistency corresponds to comparing a sample with a grammar or an automaton, compatibility concerns two non-terminals or states we want to merge. Here again, a fast check of compatibility would greatly accelerate grammatical inference algorithms.

Alternative proofs of the hardness of the consistency problem would be of help, in order to better understand what is really hard.

The size and number of characteristic samples are also good indicators of the hardness of learning; if there are many small characteristic samples, the odds may favour one of these being present in a learning sample. A formal study of these issues remains to be done.

The definition of strong structural completeness presented here is different from the one presented usually (and which would correspond to Definition 6.3.5). Yet our Definition 6.3.7 is probably not very useful in practice, as it is unclear that there is a polynomial time algorithm which, given a sample S_+ and an NFA $\mathcal{A}$, tells us if $\mathcal{A}$ is strongly structurally complete for S_+ or not. In fact, the problem can be conjectured to be $\mathcal{NP}$-complete.

A curious problem consists of relating the edit distance with automata. A ball of strings can easily be represented by a small NFA (see Exercise 6.7) but it remains unclear how large the corresponding DFA should be. Most authors (de la Higuera, Janodet & Tantini, 2008, Navarro, 2002, Schulz & Mihov, 2002) conjecture that the size of the smallest DFA recognising the language $B_k(x)$ can be exponential in k and $|x|$, but a formal proof is still needed.

PART II

What Does Learning a Language Mean?

7
Identifying languages

On ne sait prévoir que des répétitions et comprendre, c'est dégager le quelque chose qui se répète.

Antoine de Saint-Exupery, *Carnets, 1953*

Namely, in interesting identification problems, a learner cannot help but make errors due to incomplete knowledge. But, using an 'identification in the limit' guessing rule, a learner can guarantee that he will be wrong only a finite number of times.

E. Mark Gold *(Gold, 1967)*

Grammatical inference is concerned with learning language representations from information, which can be text, examples and counter-examples, or anything really that should provide us insight about the elements of the language being sought. If the problem can sometimes be stated in such an informal way and there are many papers presenting heuristics and ideas allowing us to extract some grammar or automaton from all types of information, theoretical properties of convergence of the algorithms are usually desired. This therefore requires more precise definitions.

7.1 Introductory discussion

Convergence is defined by not just finding some grammar but **the** grammar, which means considering that the problem is about searching for a hidden grammar, i.e. one defined *a priori*, or at least one equivalent to this ideal one. Obviously, in practice, such a hidden grammar may not exist, but the assumption is needed to study the convergence of the algorithms.

Yet, apart from very special cases, the information the learner gets is about the language generated, recognised or represented by the grammar, not about the grammar itself. For instance, it is usually normal to have access to strings produced by a context-free grammar but not to the parses that have produced these strings.

A further point is that there is now experimental evidence that theoretically founded algorithms work better: when competitions for learning automata, grammars or transducers have been won, the common feature is that the winning algorithm was based on some theoretically founded idea, even if tuning with powerful heuristics and careful implementation were also required.

7.1.1 The setting

There are some problems whose tractability is of great importance in grammatical inference. We present these problems in the general setting where $\mathcal{L}$ is a language class, $\mathcal{G}$ is a class of representation of objects in this class and $\mathbb{L}: \mathcal{G} \rightarrow \mathcal{L}$ is the ***naming function***, i.e. $\mathbb{L}(G)$ (with $G \in \mathcal{G}$) is the language denoted, accepted, recognised or represented by G.

For example we may be considering regular languages over an alphabet Σ, and these can be represented by DFAs, NFAs or regular expressions. In this case we would talk about the class $\mathcal{REG}(\Sigma)$ represented by $\mathcal{DFA}(\Sigma)$, $\mathcal{NFA}(\Sigma)$ or $\mathcal{REGEXP}(\Sigma)$.

It should be noted that in the above setting the language class is given with reference to some set alphabet Σ.

7.1.2 Some decision problems that are central here

The first problem concerns the fact that given some language class $\mathcal{L}$ and some representation of objects in this class $\mathcal{G}$, the following **membership problem** is decidable:

Given $w \in \Sigma^\star$ and $G \in \mathcal{G}$, is $w \in \mathbb{L}(G)$?

For example, if we are working on context-free grammars, since there are parsing algorithms (for example CYK, page 80), the membership problem is, in that case, decidable.

The second problem is the **equivalence problem**:

Given G and G' in $\mathcal{G}$, do we have $G \equiv G'$, i.e. $\mathbb{L}(G) = \mathbb{L}(G')$?

Continuing with the example, it is known that the equivalence problem is undecidable for context-free grammars, whereas it is decidable in the case of finite automata, with an important difference in complexity depending on whether the automaton is deterministic or not.

When we use the term 'decision problems' we should add that we consider these under the point of view of complexity theory, not just of computability. In the above we can think of classes of languages as being the regular languages or the context-free languages, and classes of representations, in turn, as being those of context-free grammars, deterministic finite state automata or regular expressions. The situation of these two decision problems with respect to the main classes of language representations is summarised in Table 7.1. In each case we denote by n the size of the grammar, measured by the number of states in the case of automata, and measured as the length of the pattern or of the regular expression,

Table 7.1. *Status of problems for different classes of language representations. Here, m is the length of the input string and n is the size of the grammar.*

	Membership problem	Equivalence problem
$\mathcal{DFA}(\Sigma)$	$\mathcal{O}(m)$	$\mathcal{O}(n \log n)$
$\mathcal{NFA}(\Sigma)$	$\mathcal{O}(n \cdot m)$	$\mathcal{P}$-space complete
$\mathcal{REGEXP}(\Sigma)$	$\mathcal{O}(n \cdot m)$	$\mathcal{P}$-space complete
$\mathcal{PATTERNS}(\Sigma)$	$\mathcal{NP}$-hard	decidable (if non-erasing)
$\mathcal{CFG}(\Sigma)$	$\mathcal{O}(m^3)$	undecidable
$\mathcal{LIN}(\Sigma)$	$\mathcal{O}(m^2)$	undecidable

or the number of rules in a context-free grammar, and by m the length of the input string for the membership problem. We discuss these sizes in Section 7.3.1 (page 152). The results hold for an alphabet Σ of size at least two and note that the complexities we indicate can be optimised in certain cases. Recall that most classes have been defined in Chapter 4; patterns and pattern languages denoted by $\mathcal{PATTERNS}$ (Σ) are studied in Section 11.3 (page 230).

7.1.3 Why we need targets

An attractive notion is that of considering machine learning as a field where *good ideas* can be tested and lead to algorithms, and where learning is something natural, like for human beings, where only experience will tell if what we have learnt is correct or not. Yet this idea conveys a number of problems:

(i) What about reproducibility? How do we know that the method or algorithm we have used is in some sense going to reproduce the results in another setting or on another set of data? What conditions does the new setting need for our algorithm to work?
(ii) If we do use the method/algorithm, it will be on a given set of data, or using some form of knowledge. Have we got enough data/knowledge for us to believe that what we have learnt is meaningful? Suppose, for instance, that the learning algorithm returns a DFA with 1,000,000 states when given just a bunch of short strings to learn from. One might consider that something has gone wrong.
(iii) If someone else comes up with a new method/algorithm, and gets better results over the data we had experimented on, does this mean that his or her method is better? Or does it only mean that it is better on this particular set of data?
(iv) If we have tried for six months to learn in a particular setting and after six months we have not found anything convincing, should we consider that the task is impossible or that we are the one person incapable of producing a learning algorithm for this task?

None of these questions can reach an entirely convincing answer through benchmarking, if what we are looking for is a generalisable, characterisable method. Thus some form of *convergence towards a correct answer* needs to be defined and studied. In which case we could hope to answer the previous questions as follows:

(i) If we have a learning algorithm which, given data complying with conditions C, converges towards a correct answer for any language in $\mathcal{L}$, then we can use C as a *bias* and state that our algorithm can learn a *biased* language.

(ii) We can study the convergence rate of our algorithm and then hope to put some upper or lower bound on the amount of data needed for convergence towards a correct answer to hold.

(iii) A formal framework allows us to compare methods: which one has the stronger bias? Which one needs the most data to converge?

(iv) A formal framework allows us to state that some problems are just impossible to solve, or that the quantity and quality of data in a particular setting are insufficient.

The trick we will use is to consider that the problem we are really interested in is not about discovering some rules (that are not necessarily present) that would somehow explain the data, but about *identifying* some **hidden target rules**. The first problem then is that there are many learning situations where there is no rule nor language belonging to the intended class (the one we are voluntarily restricting ourselves to) to be found. A typical example is that of natural language: over the years there have been lengthy discussions concerning the question of deciding whether natural language is context-free or not. Therefore, studying the learnability of context-free grammars in an identification situation may seem to be the wrong approach. Another example corresponds to the case where, due to experimental circumstances, the data are corrupted and therefore the target cannot be found, because it would actually be inconsistent with the data. So we have to admit that the hypothesis of a hidden target does have drawbacks.

These criticisms have motivated the introduction of an alternative formalism for convergence where there seems to be no target: the idea is just to induce a grammar from the data in such a way as to minimise some statistical criterion (a measure of the errors to be committed over unseen data, for instance). But again, whether explicitly or implicitly, there is somewhere, hidden, an ideal solution that we can call a *target*.

We will not debate further here, and we propose a simple idea:

- The issue is not about if there is a hidden grammar to be found or not. The issue is about believing there is: deciding that there is a context-free grammar consists of adding the bias and searching for a biased solution.
- If you have decided to learn some particular form of functions or grammars from the data, it is reasonable and necessary to believe there is such a grammar or function to be found.

7.2 Identification in the limit and variants

We follow lines close to complexity theory, which is a field where *reductions* are a powerful tool allowing us to reduce one problem to another, which means that solutions for one problem can be used to solve another one in a systematic way.

In most works concerned with identification in the limit, definitions are irksome and comparisons are hard due to a general lack of common notions and notations. We propose a general definition and use it to derive reduction results with a categorical flavour.

7.2.1 The learning setting

There are four questions that have to be settled in order to describe a language learning task:

(i) The first concerns the sort of languages that we are expecting to have to learn. The task will clearly not be the same if we are thinking about finite languages or about context-free languages.

(ii) The second is about the way we intend to represent the languages. The issue will not be so much about learning, but about sizes of what we want to learn and what we are manipulating. It is well known that a regular language like $L_n = \{x \in \Sigma^\star : \exists m \in \mathbb{N} \wedge x = (a+b)^m a(a+b)^n\}$ can be represented by a regular expression or an NFA of size polynomial in n, whereas the same language when represented by a DFA requires an exponential (in n) number of states. This has two implications:
   - The first is that there is no generally accepted definition of a *natural* representation. Even using Kolmogorov complexity we would only reach relative definitions.
   - Using a class where representatives are larger will mean that as the size of the representation is usually linked with the inherent complexity of the language, we will be allowed more resources (examples, computation time) to learn grammars from these classes. There is therefore an advantage in using the less condensed classes in that more resources will be allowed in order to learn the same language! But this advantage may be unfair. We discuss these issues in Section 7.3 (page 152).

(iii) The third question is about what information will be available to us. This can be strings from the language, labelled strings indicating if they belong to the language (and are called examples) or not (and then they are called counter-examples), but also other pieces of information about the language: noisy strings, prefixes, substrings...

(iv) The fourth item is the presentation protocol: how are the examples presented? Do they follow some particular order? Are they all there? How much time are we allowed to learn with and how should convergence be reached? Questions of sampling are also important, in a probabilistic setting (we will discuss this in Chapter 10).

We formalise these notions now:

Definition 7.2.1 Let $\mathcal{L}$ be a class of languages. A **presentation** is a function $\phi : \mathbb{N} \to \boldsymbol{X}$ where $\boldsymbol{X}$ is some set. For a given task the set of all admitted presentations for $\mathcal{L}$ is denoted by $\mathbf{Pres}(\mathcal{L})$, which is a subset of $[\mathbb{N} \to \boldsymbol{X}]$.

In some way these presentations *denote* languages from $\mathcal{L}$, *i.e.* there exists a function YIELDS: $\mathbf{Pres}(\mathcal{L}) \to \mathcal{L}$.

If $L = \text{YIELDS}(\phi)$ then we will say that ϕ is a presentation of L.

We also denote by $\mathbf{Pres}(L)$ the set $\{\phi \in \mathbf{Pres}(\mathcal{L}) : \text{YIELDS}(\phi) = L\}$.

A **presentation mode** refers to the way in which set $\boldsymbol{X}$ is built and to what the valid presentations are.

Let us comment upon the above definition:

- A presentation is an enumeration of elements in some set. This set can be $\Sigma^\star$ if we are learning from text, but can be something entirely different: basically, set X is the set in which the elements of information are chosen. With this definition one should not think of presentations as ***text*** or ***informant*** (which we will define now), but in a broader sense as a sequence of information of some type that hopefully informs us about the language we are to learn.
- **Pres**($\mathcal{L}$) is the set of all admissible presentations for a given class of languages and a given task. It should comprise, for example, all possible enumerations of each language L in $\mathcal{L}$ in a text learning task.
- YIELDS is an implicit and non-constructive function. It depends on the class of presentations and the class of languages. For example, we would use YIELDS(ϕ)=$\{\phi(n) : n \in \mathbb{N}\}$ in the case of learning a language from text.

Here are some examples of possible presentations of a language L:

Example 7.2.1

- TEXT(L)=$\{\phi : \mathbb{N} \to \Sigma^\star : \phi(\mathbb{N}) = L\}$. Some care is needed to deal with the empty language: as the presentation is a total function, one possibility is to introduce a special symbol # which does not belong to L and which, when presented, has the intended meaning 'nothing'.
- ORDERED_TEXT(L) is the set of ordered presentations of a language L: $\{\phi : \mathbb{N} \to \Sigma^\star : \phi(\mathbb{N}) = L \text{ and } (i < j \implies \phi(i) \leq_{\Sigma^\star} \phi(j))\}$. Note that the order $\leq_{\Sigma^\star}$ we choose to use has to be specified.
- INFORMANT(L)=$\{\phi : \mathbb{N} \to \Sigma^* \times \{0, 1\} : \phi(\mathbb{N}) = L \times \{1\} \cup \overline{L} \times \{0\}\}$.
- PREFIXES(L) is the set of all presentations by prefixes of a language L: $\{\phi : \mathbb{N} \to \Sigma^* \times \{0, 1\} : \phi(\mathbb{N}) = \textsc{Pref}(L) \times \{1\} \cup \textsc{Pref}(\overline{L}) \times \{0\}\}$ where $L \subseteq \Sigma^\infty$.

Definition 7.2.2 Let $\mathcal{L}$ be a class of languages and **Pres**($\mathcal{L}$) be the set of presentations for $\mathcal{L}$, with associated function YIELDS. The setting is said to be **valid** when given two presentations ϕ and ψ, whenever their range is equal (*i.e.* if $\phi(\mathbb{N}) = \psi(\mathbb{N})$) then YIELDS($\phi$) = YIELDS($\psi$).

If a setting is not valid, $\mathcal{L}$ is not going to be identifiable from **Pres**($\mathcal{L}$), (see Exercise 7.4).

For learnability to be studied we need to choose some representation scheme for $\mathcal{L}$. We consider a class of grammars $\mathcal{G}$ with a surjective (or onto) naming function $\mathbb{L} : \mathcal{G} \to \mathcal{L}$.

We will use the generic term ***grammar*** for a representation of a language.

In the same way as above we may, with slight abuse of notation, consider presentations of a grammar G as **Pres**(G) = **Pres**($\mathbb{L}(G)$).

Definition 7.2.3 Given a presentation ϕ we denote the set $\{\phi(j) : j \leq n\}$ by ϕ_n. Given a presentation ϕ we denote by $G \models \phi(n)$ (conversely $G \not\models \phi(n)$) when $\phi(n)$ is **consistent** with $\mathbb{L}(G)$.

Consistency holds when the current hypothesis may still be the target, i.e. the new information ($\phi(n)$) does not contradict the current hypothesis.

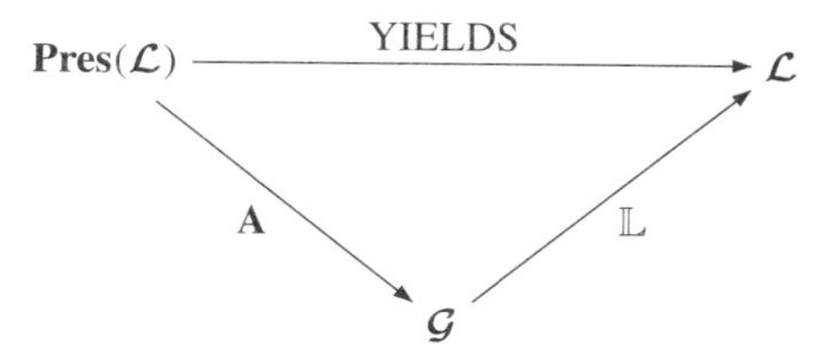

Fig. 7.1. The learning setting.

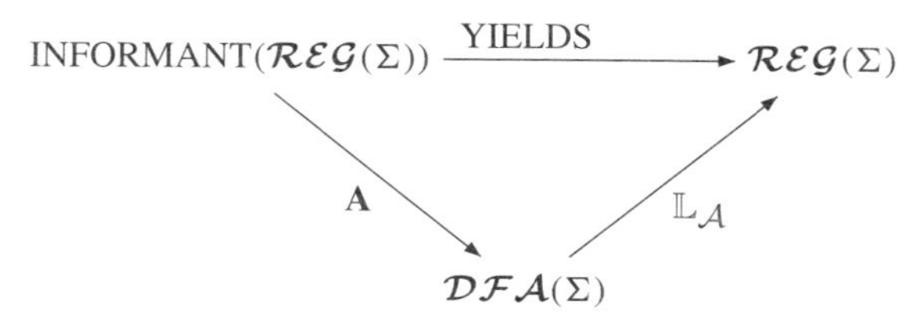

Fig. 7.2. The learning setting for DFAs from an informant.

Definition 7.2.4 A **learning algorithm A** is a program that takes the first n elements of a presentation and returns a grammar as output: $\mathbf{A} : \{\phi_i : i \in \mathbb{N}, \phi \in \mathbf{Pres}(\mathcal{L})\} \to \mathcal{G}$.

We summarise these notions in Figure 7.1.

A particular instance of the triangular representation of the learning setting is represented in Figure 7.2. The function YIELDS is as follows: YIELDS$(\phi) = \{w \in \Sigma^\star : \exists i \in \mathbb{N}\ (w, 1) = \phi(i)\} = \Sigma^\star \setminus \{w \in \Sigma^\star : \nexists i \in \mathbb{N}\ (w, 0) = \phi(i)\}$.

Even if this setting has a clear flavour of online or incremental learning, the idea is not to limit the definition to that setting. Indeed, with given data, algorithm **A** will learn a grammar. The incremental description of the above process just tells us how the algorithm performs with more data and time.

A more important question is raised by the possibility that the learning problem may be randomised, in which case, given the initial elements of a presentation, it may not return the same grammar each time it is run. In Chapter 16 this will not be a problem since we will be working in a probabilistic environment, but here we are still requiring deterministic results. This means that the type of randomisation the learner is allowed is only limited. We will study these questions in Exercise 7.8 and discuss these issues only when needed, but we will suppose that unless otherwise mentioned the learning algorithms under consideration are deterministic.

Example 7.2.2 Let $\mathcal{L}$ be the class of singleton languages over alphabet $\Sigma = \{\mathtt{a}, \mathtt{b}\}$ and $\mathbf{Pres}(\mathcal{L})$ be the class of presentations by counter-examples of this class:

$$\phi \in \mathbf{Pres}(\mathcal{L}) \iff \exists w \in \Sigma^\star : \phi(\mathbb{N}) = \Sigma^\star \setminus \{w\}.$$

Obviously, if $\phi(\mathbb{N}) = \Sigma^\star \setminus \{w\}$ then we will have: YIELDS$(\phi) = \{w\}$.

7.2.2 The definition of identification

Even if the information the learner receives concerns the languages, the learning algorithm returns a grammar. We therefore present our definitions and results, as much as possible, in terms of grammars. This requires a first simplification: we write **Pres**($\mathcal{G}$) for **Pres**($\mathcal{L}$), where $\mathcal{G}$ and $\mathcal{L}$ are as in Definition 7.2.4.

Definition 7.2.5 $\mathcal{G}$ is identifiable in the limit from **Pres**($\mathcal{G}$) if_{def} there exists a learning algorithm **A** such that for all $G \in \mathcal{G}$ and for any presentation ϕ of $\mathbb{L}(G)$ (belonging to **Pres**($\mathcal{G}$)), there exists a rank n such that for all $m \geq n$, $\mathbb{L}(\mathbf{A}(\phi_m)) = \mathbb{L}(G)$, and $\mathbf{A}(\phi_m) = \mathbf{A}(\phi_n)$.

Notice that the above definition does not force us to learn the target grammar but only to learn a grammar equivalent to the target. Furthermore there is a point from which the learner does not change its mind. This is different from what is known as ***behaviourally correct identification***, where only semantic identification is required, i.e. the last condition($\forall m \geq n$, $\mathbf{A}(\phi_m) = \mathbf{A}(\phi_m)$) is not needed.

Example 7.2.3 Continuing with the previous example, it is easy to see that the algorithm **A** such that $\mathbf{A}(\phi_n) = \min_{\leq_{lex\text{-}length}}\{u \in \Sigma^\star : u \notin \phi_n\}$ will identify $\mathcal{G}$ in the limit. To prove this, consider some particular singleton language $\{w\}$ and any presentation by counter-examples of $\{w\}$ (as in Example 7.2.2). This means that every u (with $u \leq_{lex\text{-}length} w$) appears in the presentation a first time at some rank $n_u = \min\{k \in \mathbb{N} : f(k) = u\}$. Since $\leq_{lex\text{-}length}$ is a good order, there is only a finite number of such u smaller than w and therefore the following quantity is well defined: $N_w = \max\{n_u \in \mathbb{N} : u <_{lex\text{-}length} w\}$.

Notice that in the definition above (and in the rest of this book) we have chosen to identify grammars, not languages. In the case where just identification in the limit is wanted, without taking into account complexity issues, it usually does not matter if we learn languages instead of grammars. But as soon as one is interested in the actual cost of the identification, the way the target and the hypotheses are encoded is essential. Definition 7.2.5 corresponds to what is known as *behaviourally correct identification*: notice that the learning algorithm may in fact switch grammars indefinitely often, provided the corresponding language does not change.

When needed we will say that the associated class of languages is identifiable in the limit.

Let us call the ***convergence point*** of a presentation for a given algorithm the moment the algorithm has converged on that presentation:

Definition 7.2.6 The value CONV(**A**, ϕ) is the smallest n such that $\forall j \geq n, \mathbb{L}(\mathbf{A}(\phi_j)) =$ YIELDS(ϕ). This is the ***convergence point*** of the presentation, for this learner.

Example 7.2.4 In the previous example (7.2.3), we have CONV(**A**, ϕ) $= N_w$ where YIELDS(ϕ) $= \{w\}$.

The following lemma states simply that if two presentations supposed to differ have not done so (yet) then one has not reached its convergence point.

Lemma 7.2.1 *Given any deterministic learner* $\mathbf{A}$ *for a class* $\mathcal{L}$*, and two presentations* ϕ^i *and* ϕ^j *of two languages* L_i *and* L_j *(in* $\mathcal{L}$*),* $\phi^i_n = \phi^j_n \implies \max\big(\mathrm{CONV}(\mathbf{A}, \phi^i), \mathrm{CONV}(\mathbf{A}, \phi^j)\big) > n$.

Proof Since the algorithm is deterministic, while it has the same data to learn from, it will return the same grammar. □

Notice that if we accepted randomised learners, this key lemma would also hold: one of the two runs cannot have converged 'for sure' after having seen the same elements.

From this definition E. Mark Gold's well-known results can be derived:

Theorem 7.2.2 *Any recursively enumerable class of recursive languages is identifiable in the limit from an informant.*

Proof Let $\mathcal{L}$ be a recursively enumerable class of recursive languages. As the class is recursively enumerable we can effectively generate grammars $G_0, \ldots, G_n, \ldots$ Remember also that since the languages are recursive, the question $w \in \mathbb{L}(G)$ is decidable. Then Algorithm 7.1 clearly identifies in the limit. □

Definition 7.2.7 A **super-finite** language class is a class that contains all finite languages and at least one infinite language.

This is the case, for example, of the class of regular languages or of the context-free languages.

Theorem 7.2.3 *No super-finite class of languages is identifiable in the limit from text.*

Proof Now take a class of languages that is super-finite and consider first the (there is at least one) infinite language L_∞, and then the sequence of languages $L_0, \ldots, L_n, \ldots$ with $L_0 = \{x_0\}$, $L_{i+1} = L_i \cup \{x_i\}$, $L_i \subset L_\infty$ and $L_\infty = \bigcup_{i \in \mathbb{N}} L_i$.

Now suppose for the sake of contradiction that we have an identifying algorithm **A**, for a correct class of grammars $\mathcal{G}$.

We now construct a family of valid text presentations as follows: given a presentation ϕ^0 of L_0, let $i_0 = \mathrm{CONV}(\mathbf{A}, \phi^0)$. We build a presentation ϕ^1 of L_1 where $\forall j \le i_0$, $\phi^1(j) = \phi^0(j)$ and $\phi^1(i_0 + 1) = x_1$. Again denote $i_1 = \mathrm{CONV}(\mathbf{A}, \phi^1)$.

The same process can be repeated indefinitely: we build a presentation ϕ^k of L_k where $\forall j \le i_{k-1}$, $\phi^k(j) = \phi^{k-1}(j)$ and $\phi^k(i_{k-1} + 1) = x_k$.

Now consider the presentation ϕ^∞ such that $\phi^\infty(i) = \phi^k(i)$ with $k = \min\{j \in \mathbb{N} : j \le i_k\}$.

A does not converge from this presentation, which is a presentation of L_∞, because by construction $\mathbf{A}(\phi^\infty(i)) = \mathbf{A}(\phi^k(i)) = L_k$. □

Algorithm 7.1: Learning a recursively enumerable class.

Data: a presentation ϕ
Result: –
while TRUE **do**
  print the smallest G_m consistent with ϕ_n;
  $n \leftarrow n + 1$
end

7.3 Complexity aspects of identification in the limit

If identification in the limit tells us that we will eventually find what we were looking for, it neither tells us how we know when we have found it nor how long it is going to take.

7.3.1 About the sizes of representations

We first need to define the sizes of the objects we are manipulating. These ought to be defined in a way consistent with what is usually done in complexity theory.

Intuitively, the size of an object from class $\mathcal{G}$ is (polynomially linked to) the number of bits in a *reasonable encoding* of the object.

Let us examine what is an encoding and furthermore a reasonable one: an encoding for $\mathcal{G}$ is an injective function ξ: $\mathcal{G} \rightarrow \{0, 1\}^*$. It is usually a good idea to be able to decode, i.e. to have another function $\gamma : \{0, 1\}^* \rightarrow \mathcal{G}$ such that $\gamma \circ \xi$ is the identity function. An encoding ξ is ***reasonable*** if there is a polynomial relationship between the number of encodings and the number of objects encoded. In other words, even if some strings do not encode anything, and if others may encode redundantly, we want a sufficient fraction (a polynomial fraction) of the encodings to correspond to different objects, here languages. More precisely there should exist a polynomial $p()$ such that, for all but a finite number of values of n:

$$\left|\{G \in \mathcal{G} : |\xi(G)| \leq n\}\right| \geq p\left(\frac{1}{2^n}\right).$$

This means, in other words, that with n bits at least a strictly increasing polynomial number of grammars can be encoded.

On the other hand, the above is not entirely satisfactory: all these grammars could in fact be equivalent to each other, and therefore only a logarithmic (in 2^n) number of languages would be encoded with at most n bits. Therefore, more correctly we will say that an encoding ξ is ***reasonable*** *if*$_{def}$

$$\left|\{L \in \mathcal{L} : \exists G \in \mathcal{G} : \mathbb{L}(G) = L \wedge |\xi(G)| \leq n\}\right| \geq p\left(\frac{1}{2^n}\right).$$

In the equation above it is important to understand the idea: an encoding is reasonable if there is a non-trivial polynomial link between the number of encodings of size n and the number of different languages represented by encodings of size n. Following this, in order to measure sizes we define the following quantities:

- $\|G\|$ is the size of a grammar (related polynonomially to $|\xi(G)|$),
- $|\phi_n|$ is the number of items in the first $(\phi(0), \ldots, \phi(n))$ elements of a presentation (therefore $n+1$)
- $\|\phi_n\|$ is the number of symbols in the first $n+1$ elements of a presentation (number of bits, in the case of a binary alphabet)
- $|L| = \min\{\|G\| : \mathbb{L}(G) = L, G \in \mathcal{G}\}$. The 'size' of a language is the size of the smallest grammar of the considered class that can generate it.

Practically, this justifies the sizes proposed in Table 7.1 (page 145). The size of an automaton, whether deterministic or not, can be the number of states, the size of a pattern or of a rational expression can be its length, and the size of a context-free grammar can be the number of non-terminal symbols if the grammar is in quadratic normal form, or this number multiplied by the length of the longest right-hand side of a rule if not.

7.3.2 *Counting only time*

The question now is to define what *efficient* identification should mean. In computer science, 'efficient' is usually linked with the polynomial complexity classes. But in the case of identification in the limit, there are several ways to count.

The first definition is by far too optimistic and can only be used on very limited classes of language with extremely precise presentations since the idea is to want to identify as soon as a polynomial quantity of information has been presented to the learner. It implies that all the necessary information is given early: this is a far too strong constraint. Indeed with no additional constraint if there are two languages which cannot be differentiated through one element of presentation, then it suffices to present that element an exponential (in the size of the largest language) number of times for no identification to take place. Again, we are considering the problem for a language class $\mathcal{L}$, a grammar class $\mathcal{G}$ and presentations which are functions in $[\mathbb{N} \to \mathbf{X}]$.

Definition 7.3.1 (Overall polynomial time) An algorithm $\mathbf{A}$ is said to have **overall polynomial time** if_{def} there exists a polynomial $p()$ such that $\forall G \in \mathcal{G}$, $\forall n \geq p(\|G\|)$, $\forall \phi \in$ $\mathbf{Pres}(\mathbb{L}(G))$, $\mathbb{L}(\mathbf{A}(\phi_n)) = \mathbb{L}(G)$.

The next definition just states that to produce its next hypothesis the algorithm only requires polynomial time. In this case what can happen is that any identifying algorithm can be transformed into a ***polynomial update time*** learner through the following trick. Let the learner not only compute its solution but also count the time it is using; if at step n the learner notices that the computation of the hypothesis takes too long (longer than the time allowed at that point), the learner returns the previous hypothesis. It will then continue its computation as soon as a new item of data is provided and it gives the learner an extra time allowance. In this case the moment at which convergence is reached will certainly be postponed, but identification in the limit is no problem.

Definition 7.3.2 (Polynomial update time) An algorithm **A** is said to have **polynomial update time** if_{def} there is a polynomial $p()$ such that, for every presentation ϕ and every integer n, constructing $H_n = \mathbf{A}(\phi_n)$ requires $\mathcal{O}(p(\|\phi_n\|))$ time.

A curious alternative (discussed in Exercise 7.15) is to accept that the time can be polynomial in the size of the previous hypothesis ($\|\mathbf{A}(f_{n-1})\|$) and the size of the new example ($|f(n)|$).

7.3.3 Counting the number of examples

An alternative to bounding the number of examples needed to identify is to bound the number of good examples a learner needs. A 'good example' is one a teacher would want to show the learner:

Definition 7.3.3 (Polynomial characteristic sample) A grammar class $\mathcal{G}$ admits **polynomial characteristic samples** if_{def} there exist an algorithm **A** and a polynomial $p()$ such that $\forall G \in \mathcal{G},\ \exists \text{CS} \subseteq \mathbf{X}$,

$$\|\text{CS}\| \leq p(\|G\|) \wedge$$
$$\forall \phi \in \mathbf{Pres}(\mathbb{L}(G)),\ \forall n \in \mathbb{N} : \text{CS} \subseteq \phi_n \implies \mathbb{L}(\mathbf{A}(\phi_n)) = \mathbb{L}(G).$$

Such a set CS is called a ***characteristic sample*** of G for **A**. If such an algorithm **A** exists, we say that ***A identifies $\mathcal{G}$ in the limit in CS polynomial time***.

What is implied is that there is a sufficiently small set (of polynomial size), called the characteristic sample, or sometimes the teaching sample, which, when included in a presentation, makes the learner identify. We make a few additional points here:

- The first point is that the sample is only characteristic for a specific learner. The existence of a general characteristic sample (usually called a *tell tale set*) is also a question worth considering (see Exercise 7.6, page 169).
- The second point is that the size of the characteristic sample should be counted as a polynomial function of the number of bits needed to encode it. This is due to the possibility of being presented with very long strings. Therefore just counting the number of strings is not enough (see Exercise 7.7, page 169).
- If from an identification point of view having a characteristic sample of unmanageable size is clearly a problem, this is not so from a learning perspective: what is wrong with learning a logarithmically small grammar from a sample, one that would compress the data enormously?

7.3.4 Counting the number of implicit errors

In the above definitions, what causes problems and delay in identifying seems to be the presence of useless non-informative examples. An example that comes to reinforce the current hypothesis should perhaps not count negatively towards identification.

Definition 7.3.4 (Implicit prediction errors) Given a learning algorithm **A** and a presentation ϕ, **A makes an implicit prediction error** (IPE) at time n if_{def} $\mathbf{A}(\phi_{n-1}) \not\models \phi(n)$.

An algorithm that changes its mind when the current hypothesis is seen as an error with the new presented element is said to be **consistent**.

Algorithm **A** makes a **polynomial number of implicit prediction errors** if_{def} there is a polynomial $p()$ such that, for each grammar G and each presentation ϕ of $L = \mathbb{L}(G)$, $\left|\{k \in \mathbb{N} : \mathbf{A}(\phi_k) \not\models \phi(k+1)\}\right| \le p(\|G\|)$.

Combining ideas, one gets:

Definition 7.3.5 An algorithm **A** identifies a class $\mathcal{G}$ in the limit in **IPE-polynomial time** if_{def}

- **A** identifies $\mathcal{G}$ in the limit,
- **A** has polynomial update time,
- **A** makes a polynomial number of implicit prediction errors.

Note that neither condition is implied by the other two.

7.3.5 Counting the number of mind changes

A nice alternative to counting the number of errors is that of counting the number of changes of hypothesis one makes. On its own, this is meaningless (why change?), but if combined with identification in the limit the definition makes sense:

Definition 7.3.6 (Polynomial number of mind changes) Given a learning algorithm **A** and a presentation ϕ, we say that **A changes its mind at time** n if_{def} $\mathbf{A}(\phi_n) \neq \mathbf{A}(\phi_{n-1})$.

An algorithm that never changes its mind when the current hypothesis is consistent with the new presented element is said to be **conservative**.

Algorithm **A** makes a **polynomial number of mind changes** (MCS) if_{def} there is a polynomial $p()$ such that, for each grammar G and each presentation ϕ of $L = \mathbb{L}(G)$, $\left|\{k \in \mathbb{N} : \mathbf{A}(\phi_k) \neq \mathbf{A}(\phi_{k+1})\}\right| \le p(\|G\|)$.

Combining ideas, one gets:

Definition 7.3.7 An algorithm **A** identifies a class $\mathcal{G}$ in the limit in MC-polynomial time if_{def}

- **A** identifies $\mathcal{G}$ in the limit,
- **A** has polynomial update time,
- **A** makes a polynomial number of mind changes.

Here also, neither condition is implied by the two others.

Finally, it is easy to see that a consistent algorithm makes at least as many mind changes than implicit prediction errors, whereas a conservative algorithm makes at least as many implicit prediction errors as mind changes. So we deduce the following theorem:

Theorem 7.3.1 *If* **A** *identifies the class* $\mathcal{G}$ *in the limit in MC-polynomial time and is consistent, then* **A** *identifies* $\mathcal{G}$ *in the limit in IPE-polynomial time.*
Conversely, if **A** *identifies* $\mathcal{G}$ *in the limit in IPE-polynomial time and is conservative, then* **A** *identifies* $\mathcal{G}$ *in the limit in MC-polynomial time.*

7.4 Commuting diagrams

As in many fields, it is of interest to provide a technique enabling us to derive a learning algorithm for a new class or a new type of presentation from a known learning algorithm.

A learning/identifying situation can be understood by stating the class of languages under study, the representations one is interested in and the sort of presentations one admits.

Let $\mathcal{L}_{\mathcal{A}}$ and $\mathcal{L}_{\mathcal{B}}$ be two classes of languages represented by grammars from (respectively) the grammar classes $\mathcal{G}_{\mathcal{A}}$ and $\mathcal{G}_{\mathcal{B}}$.

We denote by $\mathbb{L}_{\mathcal{A}}$ (respectively $\mathbb{L}_{\mathcal{B}}$) the surjective mapping $\mathcal{G}_{\mathcal{A}} \rightarrow \mathcal{L}_{\mathcal{A}}$ (respectively $\mathbb{L}_{\mathcal{B}} : \mathcal{G}_{\mathcal{B}} \rightarrow \mathcal{L}_{\mathcal{B}}$).

Given a surjective mapping $\zeta_{\mathbf{G}} : \mathcal{G}_{\mathcal{A}} \rightarrow \mathcal{G}_{\mathcal{B}}$, we denote by $\zeta_{\mathbf{L}}$ a (surjective) mapping $\mathcal{L}_{\mathcal{A}} \rightarrow \mathcal{L}_{\mathcal{B}}$ for which Diagram 7.1 commutes.

$$\begin{array}{ccc} \mathcal{G}_{\mathcal{A}} & \xrightarrow{\zeta_{\mathbf{G}}} & \mathcal{G}_{\mathcal{B}} \\ {\scriptstyle \mathbb{L}_{\mathcal{A}}}\downarrow & & \downarrow{\scriptstyle \mathbb{L}_{\mathcal{B}}} \\ \mathcal{L}_{\mathcal{A}} & \xrightarrow{\zeta_{\mathbf{L}}} & \mathcal{L}_{\mathcal{B}} \end{array} \tag{7.1}$$

Hence $\zeta_{\mathbf{L}}$ is the only mapping such that:

$$\zeta_{\mathbf{L}} \circ \mathbb{L}_{\mathcal{A}} = \mathbb{L}_{\mathcal{B}} \circ \zeta_{\mathbf{G}}$$

Example 7.4.1 There is clearly the possibility of transforming a DFA into a context-free grammar, whose language is obtained also as the image by the identity of the regular language. Such a grammar is usually called regular.

$$\begin{array}{ccc} \mathcal{DFA}(\Sigma) & \xrightarrow{\zeta_{\mathbf{G}}} & \mathcal{CFG}(\Sigma) \\ {\scriptstyle \mathbb{L}_{\mathcal{A}}}\downarrow & & \downarrow{\scriptstyle \mathbb{L}_{\mathcal{B}}} \\ \mathcal{REG}(\Sigma) & \xrightarrow{\zeta_{\mathbf{L}}} & \mathcal{CFL}(\Sigma) \end{array}$$

We now concentrate on presentations. Suppose the presentations of $\mathcal{L}_{\mathcal{A}}$ are functions in $[\mathbb{N} \rightarrow \mathbf{X}]$, whereas those of $\mathcal{L}_{\mathcal{B}}$ are functions in $[\mathbb{N} \rightarrow \mathbf{Y}]$,

$$\begin{array}{ccc} \mathcal{L}_{\mathcal{A}} & \xrightarrow{\zeta_{\mathbf{L}}} & \mathcal{L}_{\mathcal{B}} \\ \text{YIELDS}\uparrow & & \uparrow\text{YIELDS} \\ \mathbf{Pres}(\mathcal{L}_{\mathcal{A}}) & \xrightarrow{\zeta_{\mathbf{P}}} & \mathbf{Pres}(\mathcal{L}_{\mathcal{B}}) \end{array} \tag{7.2}$$

Again we need diagram (7.2) to commute. Given a surjective mapping $\zeta_{\mathbf{L}} : \mathcal{L}_{\mathcal{A}} \to \mathcal{L}_{\mathcal{B}}$, we denote by $\zeta_{\mathbf{P}}$ a (surjective) mapping $\mathbf{Pres}(\mathcal{L}_{\mathcal{A}}) \to \mathbf{Pres}(\mathcal{L}_{\mathcal{B}})$ for which the diagram commutes. So we have:

$$\zeta_{\mathbf{L}} \circ \text{YIELDS} = \text{YIELDS} \circ \zeta_{\mathbf{P}}.$$

Function $\zeta_{\mathbf{P}}$ transforms a presentation of one language into a presentation of another. Note that this does not mean a point-by-point transformation: one element of the first transformation can be transformed by $\zeta_{\mathbf{P}}$ into a finite sequence in the second.

Example 7.4.2 One can transform the learning setting of finite languages from text into a setting where one wants to learn co-finite languages (the complement being finite) from counter-examples.

$$\begin{array}{ccc}
\mathcal{FIN}(\Sigma) & \xrightarrow{\zeta_{\mathbf{L}}} & \mathcal{CO}\text{-}\mathcal{FIN}(\Sigma) \\
\text{YIELDS}\uparrow & & \uparrow\text{YIELDS} \\
\text{TEXT}(\mathcal{FIN}(\Sigma)) & \xrightarrow{\zeta_{\mathbf{P}}} & \text{CO-TEXT}(\mathcal{FIN}(\Sigma))
\end{array}$$

As a presentation may not be a computable function, describing the computation aspects of function $\zeta_{\mathbf{P}}$ is as follows:

Definition 7.4.1 A **reduction** of presentations is a function $\zeta_{\mathbf{P}} : \boldsymbol{Pres}(\mathcal{L}_{\mathcal{A}}) \to \boldsymbol{Pres}(\mathcal{L}_{\mathcal{B}})$. The reduction is **computable** if_{def} there exists a computable function $\overline{\zeta_{\mathbf{P}}} : \mathbf{X} \to 2^{\mathbf{Y}}$ such that $\bigcup_{i \in \mathbb{N}} \overline{\zeta_{\mathbf{P}}}(\phi(i)) = \psi(\mathbb{N})$ where $\zeta_{\mathbf{P}}(\phi) = \psi$.

Note that $\zeta_{\mathbf{P}}$ is a mathematical function that takes a presentation (which is also a function) and transforms it into a function. But in order to talk about its computational properties we have to reduce or transform the first presentation into the second one point-to-point. So $\forall\, i \in \mathbb{N},\ \overline{\zeta_{\mathbf{P}}}(\phi(i))$ is a finite set. $\overline{\zeta_{\mathbf{P}}}$ is therefore the description at each point of function $\zeta_{\mathbf{P}}$. We will say that $\zeta_{\mathbf{P}}$ is ***polynomial*** if_{def} there exists a polynomial $p()$ for which given any presentation ϕ, we have $\forall\, i \in \mathbb{N}, \|\overline{\zeta_{\mathbf{P}}}(\phi(i))\| \leq p(\|\phi(i)\|)$, and furthermore there is a polynomial time algorithm allowing us to build $\overline{\zeta_{\mathbf{P}}}(\phi(i))$ given $\phi(i)$.

Example 7.4.3 Suppose as above that ϕ is a presentation of a language by ordered text. If ψ is a presentation by informant of the same language, the transformation consists, with every new string that is presented, of adding this string to the positive examples and adding all intermediate strings between the last presented string and this new string to the set of negative examples. It can be checked that the result is a complete presentation by informant of the language. In this case $\overline{\zeta_{\mathbf{P}}}$ is computable but not polynomial (unless the alphabet is only of size one).

Definition 7.4.2 A reduction of presentations $\zeta_{\mathbf{P}} : \mathbf{Pres}(\mathcal{L}_{\mathcal{A}}) \to \mathbf{Pres}(\mathcal{L}_{\mathcal{B}})$ such that $\zeta_{\mathbf{P}}(\phi) = \psi$ is ***polynomial*** if_{def} there exists a polynomial function $\overline{\zeta_{\mathbf{P}}} : \mathbf{X} \to 2^{\mathbf{Y}}$ such that $\bigcup_{i \in \mathbb{N}} \overline{\zeta_{\mathbf{P}}}(\phi(i)) = \psi(\mathbb{N})$.

Now combining both diagrams we have:

$$\begin{array}{ccc} \mathcal{G}_{\mathcal{A}} & \xrightarrow{\zeta_{\mathbf{G}}} & \mathcal{G}_{\mathcal{B}} \\ \mathbb{L}_{\mathcal{A}}\big\downarrow & & \big\downarrow\mathbb{L}_{\mathcal{B}} \\ \mathcal{L}_{\mathcal{A}} & \xrightarrow{\zeta_{\mathbf{L}}} & \mathcal{L}_{\mathcal{B}} \\ \text{YIELDS}\big\uparrow & & \big\uparrow\text{YIELDS} \\ \mathbf{Pres}(\mathcal{L}_{\mathcal{A}}) & \xrightarrow{\zeta_{\mathbf{P}},\overline{\zeta_{\mathbf{P}}}} & \mathbf{Pres}(\mathcal{L}_{\mathcal{B}}) \end{array} \tag{7.3}$$

Theorem 7.4.1 *Let $\mathcal{L}_{\mathcal{A}}$, $\mathcal{L}_{\mathcal{B}}$, $\mathcal{G}_{\mathcal{A}}$ $\mathcal{G}_{\mathcal{B}}$, $\zeta_{\mathbf{L}}$ and $\zeta_{\mathbf{G}}$ be as in Definition 7.4.2 and Diagram 7.3.*

If $\mathcal{G}_{\mathcal{B}}$ is identifiable in the limit from **Pres**($\mathcal{L}_{\mathcal{B}}$)*, and there exists a computable function $\zeta_{\mathbf{M}} : \mathcal{G}_{\mathcal{B}} \to \mathcal{G}_{\mathcal{A}}$ such that $\zeta_{\mathbf{G}} \circ \zeta_{\mathbf{M}} = Id$, and $\zeta_{\mathbf{P}}$ is a computable reduction, then $\mathcal{G}_{\mathcal{A}}$ is identifiable in the limit from* **Pres**($\mathcal{L}_{\mathcal{A}}$)*.*

Before proving the theorem it will be more useful to represent the situation as follows:

$$\begin{array}{ccc} \mathcal{G}_{\mathcal{A}} & \xleftarrow{\zeta_{\mathbf{M}}} & \mathcal{G}_{\mathcal{B}} \\ \mathbb{L}_{\mathcal{A}}\big\downarrow & & \big\downarrow\mathbb{L}_{\mathcal{B}} \\ \mathcal{L}_{\mathcal{A}} & \xrightarrow{\zeta_{\mathbf{L}}} & \mathcal{L}_{\mathcal{B}} \\ \text{YIELDS}\big\uparrow & & \big\uparrow\text{YIELDS} \\ \mathbf{Pres}(\mathcal{L}_{\mathcal{A}}) & \xrightarrow{\zeta_{\mathbf{P}},\overline{\zeta_{\mathbf{P}}}} & \mathbf{Pres}(\mathcal{L}_{\mathcal{B}}) \end{array}$$

Proof Let $\mathbf{A_2}$ be a learning algorithm that identifies $\mathcal{G}_{\mathcal{B}}$ from a presentation in **Pres**($\mathcal{L}_{\mathcal{B}}$). Consider algorithm $\mathbf{A_1}$ below (Algorithm 7.2), which takes the n first items (ϕ_n) and then executes:

Algorithm 7.2: Reduction between presentations.

Data: a presentation ϕ_n
Result: a grammar
$\psi_m \longleftarrow \bigcup_{i \leq n} \overline{\zeta_{\mathbf{P}}}(\phi(i))$;
$G_{\mathcal{B}} \longleftarrow \mathbf{A_2}(\psi_m)$;
$G_{\mathcal{A}} \longleftarrow \zeta_{\mathbf{M}}(G_{\mathcal{B}})$;
return $G_{\mathcal{A}}$

As $\zeta_{\mathbf{P}}$ is computable, the set ψ_m can be constructed. □

Also if $\zeta_{\mathbf{P}}$ and $\zeta_{\mathbf{M}}$ can be computed by polynomial time algorithms, then $\mathbf{A_1}$ is polynomial time if $\mathbf{A_2}$ is. This can be extended with care to definitions of polynomial learning such as those from Section 7.3.

We visit in the following sections different results obtained by using the reduction technique.

7.4.1 *With more, get more*

Our first example is very simple. The idea is just that any class we can learn from text is also learnable from an informant. Surprisingly, this may not be true if complexity considerations are taken into account.

$$
\begin{array}{ccc}
\mathcal{G}_{\mathcal{A}} & \xleftarrow{Id} & \mathcal{G}_{\mathcal{A}} \\
\mathbb{L}_{\mathcal{A}}\big\downarrow & & \big\downarrow\mathbb{L}_{\mathcal{A}} \\
\mathcal{L}_{\mathcal{A}} & \xrightarrow{Id} & \mathcal{L}_{\mathcal{A}} \\
\text{YIELDS}\big\uparrow & & \big\uparrow\text{YIELDS} \\
\text{INFORMANT}(\mathcal{L}_{\mathcal{A}}) & \xrightarrow{\zeta_{\mathbf{P}},\overline{\zeta_{\mathbf{P}}}} & \text{TEXT}(\mathcal{L}_{\mathcal{A}})
\end{array}
$$

$\overline{\zeta_{\mathbf{P}}} : \Sigma^* \cup \{0, 1\} \to 2^{\Sigma^*}$, $\overline{\zeta_{\mathbf{P}}}(w, 0) = \emptyset$, and $\overline{\zeta_{\mathbf{P}}}(w, 1) = \{w\}$. Hence any class identifiable from text is also identifiable from an informant.

7.4.2 *About safe languages*

Safe languages are languages of infinitary strings. In substance, a language is safe when the limits of all the prefixes are themselves in the language. As an example suppose strings $\mathtt{a}^n\mathtt{b}^\omega$ belong to L ($\forall n \in \mathbb{N}$), then, for the language to be safe, string $\mathtt{a}^\omega$ should belong too.

Definition 7.4.3 An ω-language L is **safe** if_{def}

$$\forall w \in \Sigma^\omega, (\forall u \in \text{PREF}(w), \exists v \in \Sigma^\omega \text{ such that } u.v \in L) \Longrightarrow w \in L.$$

It can be easily shown that this definition is equivalent to:

$$\forall w \in \Sigma^\omega, \text{PREF}(w) \subseteq \text{PREF}(L) \Longrightarrow w \in L$$

or even by taking the negation of the last line:

$$\forall w \in \Sigma^\omega, w \notin L \Longrightarrow (\exists u \in \text{PREF}(w) \text{ such that } \forall v \in \Sigma^\omega, u.v \notin L).$$

Definition 7.4.4 A **Büchi automaton** is a quintuple $\mathcal{A} = \langle Q, \Sigma, \delta, F, q_1\rangle$ where Σ is an alphabet, Q is a finite set of states, $q_1 \in Q$ is an initial state, $\delta : Q \times \Sigma \to 2^Q$ is a transition function and $F \subseteq Q$ is a set of marked states. These states are distinguished from the others but not 'final', as the strings are infinite.

A run of $\mathcal{A}$ on an ω-string u is a mapping $\rho_u : \mathbb{N} \to Q$ such that:

- $\rho_u(0) = q_1$ and
- $\rho_u(i+1) \in \delta(\rho_u(i), u_i)$, for all $i \in \mathbb{N}$.

Note that ρ_u is undefined if at some point, $\rho_u(i)$ is undefined.

Definition 7.4.5 An ω-string u is **accepted** by $\mathcal{A}$ if_{def} there exists a state of F which appears infinitely often in a run of $\mathcal{A}$ on u. Let $\mathbb{L}(\mathcal{A})$ be the set of all accepted ω-strings by $\mathcal{A}$. We can show that an ω-language L is ω-regular if_{def} $L = \mathbb{L}(\mathcal{A})$ for some Büchi automaton $\mathcal{A}$.

A Büchi automaton is **deterministic** if_{def} $|\delta(q, a)| \leq 1$ for all states q and letters a.

A **DB**-machine is a deterministic Büchi automaton where $F = Q$. As the automaton need not be complete, the strings that are not accepted are those that simply cannot be parsed through the automaton.

Theorem 7.4.2 *L is a safe ω-regular language* if and only if *L is recognised by a **DB**-machine.*

Let $\mathcal{SAFE}(\Sigma)$ denote the class of all safe ω-regular languages, and $\mathcal{DB}(\Sigma)$ the class of ***DB***-machines over the alphabet Σ. The learning problem we are interested in is that of learning ***DB***-machines from positive and negative examples. If we are to consider that asking for infinite examples is too much, let us suppose that we want to learn from finite prefixes: a positive prefix is one that has at least one good continuation, and a negative one is a string which, whatever comes next, will not be accepted.

One can prove by building a new algorithm that safe languages are identifiable from such prefixes. But the result can be reached in a much simpler way through Theorem 7.4.1:

$$\begin{array}{ccc}
\mathcal{DB}(\Sigma) & \xleftarrow{\chi} & \mathcal{DFA}(\Sigma) \\
\mathbb{L}_{\mathcal{A}}\downarrow & & \downarrow\mathbb{L}_{\mathcal{B}} \\
\mathcal{SAFE}(\Sigma) & \xrightarrow{\zeta_{\mathbf{L}}} & \mathcal{REG}(\Sigma) \\
\text{YIELDS}\uparrow & & \uparrow\text{YIELDS} \\
\text{PREF-INF}(\mathcal{SAFE}(\Sigma)) & \xrightarrow{\zeta_{\mathbf{P}},\overline{\zeta_{\mathbf{P}}}} & \text{INFORMANT}(\mathcal{REG}(\Sigma))
\end{array}$$

$\zeta_{\mathbf{P}}(u, l) = \{(w, l) \in \Sigma^\star : w \in \text{PREF}(u)\}$, and $\zeta_{\mathbf{L}}$ and χ are the natural transformations. Details of the transformations are not given, but these can be reconstructed.

7.4.3 *Even linear grammars*

Even linear languages and grammars have been analysed in the context of grammatical inference in several papers.

Definition 7.4.6 (Linear context-free grammars) A context-free grammar $G = \langle \Sigma, V, P, S\rangle$ is a **linear** grammar if_{def} $P \subset V \times (\Sigma^* V \Sigma^\star \cup \Sigma^\star)$.

Definition 7.4.7 (Even linear context-free grammars) A context-free grammar $G = \langle \Sigma, V, P, S\rangle$ is an **even linear** grammar if_{def} $P \subset V \times (\Sigma V \Sigma \cup \Sigma \cup \{\lambda\})$.

Thus languages like the set of all palindromes, or the language $\{a^n b^n : n \in \mathbb{N}\}$, are even linear without being regular.

We denote by $\mathcal{ELG}(\Sigma)$ and $\mathcal{ELL}(\Sigma)$ the classes of even linear grammars and even linear languages for an alphabet Σ. Given an alphabet Σ, we denote by $d(\Sigma)$ the alphabet composed of symbols $\binom{a_1}{a_2}$ where $a_1, a_2 \in \Sigma \times (\Sigma \cup \{\#\})$, with # a symbol not in Σ.

$$
\begin{array}{ccc}
\mathcal{ELG}(\Sigma) & \xleftarrow{\zeta_{\mathbf{M}}} & \mathcal{DFA}(d(\Sigma)) \\
\mathbb{L}\downarrow & & \downarrow\mathbb{L} \\
\mathcal{ELL}(\Sigma) & \xrightarrow{Id} & \mathcal{REG}(d(\Sigma)) \\
\text{YIELDS}\uparrow & & \uparrow\text{YIELDS} \\
\text{INFORMANT}(\mathcal{ELL}(\Sigma)) & \xrightarrow{\zeta_{\mathbf{P}}, \overline{\zeta_{\mathbf{P}}}} & \text{INFORMANT}(\mathcal{REG}(d(\Sigma)))
\end{array}
$$

$\overline{\zeta_{\mathbf{P}}}$ takes a string of even length $a_1 a_2 \cdots a_{2n}$ and returns $\binom{a_1}{a_{2n}}\binom{a_2}{a_{2n-1}} \cdots \binom{a_{n-1}}{a_n}$ or a string of odd length $a_1 a_2 \cdots a_{2n+1}$ and returns $\binom{a_1}{a_{2n+1}} \cdots \binom{a_{n-1}}{a_{n+1}}\binom{a_n}{\#}$.

$\zeta_{\mathbf{M}}$ transforms a DFA over alphabet $\Sigma \times (\Sigma \cup \{\#\})$ into an even linear grammar over Σ.

Example 7.4.4 Let $G = \langle \Sigma, V, P, N_1\rangle$ be an *even linear* grammar with

- $\Sigma = \{\mathtt{a}, \mathtt{b}\}$,
- $V = \{N_1, N_2\}$,
- $P = \{N_1 \to \mathtt{a}T\mathtt{b},\ N_1 \to \mathtt{b}N_2\mathtt{a},\ N_2 \to \mathtt{a}N_1\mathtt{a},\ N_2 \to \mathtt{a},\ N_2 \to \mathtt{b}\}$.

Then $\overline{\zeta_{\mathbf{P}}}$ takes string abbababbaab and returns string $\binom{\mathtt{a}}{\mathtt{b}}\binom{\mathtt{b}}{\mathtt{a}}\binom{\mathtt{b}}{\mathtt{a}}\binom{\mathtt{a}}{\mathtt{b}}\binom{\mathtt{b}}{\#}$.

The learning algorithm we might use may then return from an informed sample the automaton depicted in Figure 7.3, which can be transformed into a grammar equivalent to the above.

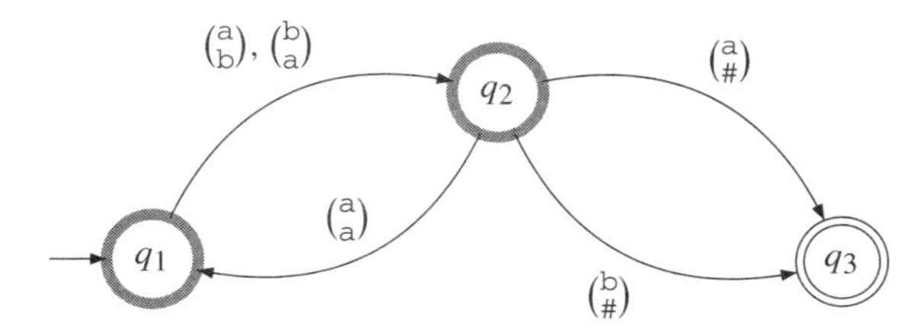

Fig. 7.3. A DFA learnt from data transformed by $\overline{\zeta_{\mathbf{P}}}$.

7.4.4 Trees

As the theory for trees has very often developed in a parallel way to that on strings, it is very tempting to present a reduction like that of Diagram 7.4, which would allow us to learn a grammar for trees by transforming the tree data into strings, then using our string grammar algorithm, and finally transforming the string grammar into a tree grammar.

$$
\begin{array}{ccc}
\mathcal{TG}(\Sigma) & \xleftarrow{\zeta_{\mathbf{M}}} & \mathcal{DFA}(\Sigma) \\
\mathbb{L}\big\downarrow & & \big\downarrow\mathbb{L} \\
\mathcal{TL}(\Sigma) & \xrightarrow{Id} & \mathcal{REG}(\Sigma) \\
\text{YIELDS}\big\uparrow & & \big\uparrow\text{YIELDS} \\
\mathbf{Pres}_1 & \xrightarrow{\zeta_{\mathbf{P}},\overline{\zeta_{\mathbf{P}}}} & \mathbf{Pres}_2
\end{array}
\qquad (7.4)
$$

But the above construction only works for very simple classes of language: The expressive power of trees is much stronger than that of strings.

7.5 Active learning

In the above formalism the presentations were uncontrolled by the learner. The most we could hope for was a fair deal: if the data were to be labelled then the labelling should be correct (at least in a non-noisy setting) and if specified, the presentations had to comply with some *completeness* condition. For instance, imposing that $\phi(\mathbb{N}) = L$ ensured that in the limit no essential piece of data was missing. There is nevertheless a case for wanting to be able to control the data we receive further, and this for at least two reasons:

- The first reason is that if, when placing ourselves in the most favourable conditions, we still cannot learn, we should be able to derive negative results for the general case.
- The second reason is that there may be realistic situations where we might be able to choose the data. This is the case in testing, in situations where robots may want to explore or interact with their environment. Another related setting appears when there may be too much data around, and then learning is going to depend on interactively choosing the data. This is clearly the case when wanting to learn from web data.

7.5.1 The Oracle

An Oracle is just some abstract machine that knows the target and answers some queries. An Oracle (*she*) is generally supposed to be perfect: she can answer any specific query (provided the learning algorithm is allowed to ask it). She can even answer queries that a concrete machine would not be able to cope with and therefore solve undecidable problems. The ability of the Oracle is determined by the learning setting.

In some cases the Oracle may have various possible answers. In this case she should be allowed to give any admissible answer. As our goal when studying Oracle learning is to consider worst case scenarios, we will always have to suppose the Oracle is giving us the least informative of all possible answers.

In this sense, the analysis will be similar to that made in complexity theory. The best case is of hardly any interest, mainly because the Oracle could be tempted to *collude* and give us the answer through a very specially encoded example. The second possibility would be to consider an *average* response by the Oracle. This requires a distribution over the possible responses and brings us into an altogether different setting. Remarkably, the key problem is that of having access to an unknown distribution: one can for instance wonder what a 'typical' distribution of web pages can look like. The worst case analysis may seem to correspond to a very unlikely situation, but it does give us an upper bound on the effort needed.

7.5.2 Some queries

There are many types of query one can make to an Oracle. Some are natural in the sense that there is at least some application with a biological, mechanical or cognitive instantiation of these queries; others are defined to build negative results only.

- ***Membership queries*** (MQs) are made by asking the Oracle if a given string is in the target language or not. The Oracle answers YES or NO. Extensions of these membership queries are ***correction queries***, where the Oracle answers YES or returns a close string in the language, and ***extended membership queries*** where for a given string, the probability of this string being in the language is returned. An ***extended prefix language query*** is made by submitting to the Oracle a string w. The Oracle then returns the probability $p(w\Sigma^*)$. It can be noticed that an extended membership query for a string w can be simulated through $|w| + 1$ extended prefix language queries.
- ***Equivalence queries*** can be strong (EQ) or weak (WEQ). They are in both cases made by proposing some hypothesis to the Oracle. The hypothesis is a grammar representing the unknown language. When the query is weak, the Oracle just answers YES or NO. When it is strong the Oracle has, in the negative case, to return a counter-example. A counter-example is a string in the symmetric difference between the target language and the submitted hypothesis.
- ***Subset queries*** (SSQ) are usually strong: a hypothesis language is submitted (by proposing a grammar). The Oracle answers YES if the hypothesis language is included in the target language, and returns a counter-example from the hypothesis and not in the target language if not. ***Superset queries*** can be defined in a similar way.
- ***Sampling queries*** (EX) exist in various forms. In all cases we have to suppose the existence of a distribution, known or unknown but permanent, over the strings. Then we can sample from the positive examples only, from the negative examples only, from the set of all strings, or even from a subset of strings having some particularity. A ***specific sampling query*** is made by submitting a pattern: the Oracle draws a string that matches some chosen pattern sampled according to the distribution $\mathcal{D}$. Specific sampling queries are intended to fit the idea that the user can ask for examples matching some pattern he is interested in.

For example, a specific sampling query for aΣ^*b requires an example starting with a and ending with b, sampled following the distribution induced by $\mathcal{D}$ over a $\Sigma^\star$ b.

For a given task we will only be allowed to ask some sort of queries. We will denote this set by QUER. For instance:

- QUER = {MQ}. In this situation the learner is to identify with the help of membership queries only.
- QUER = {MQ, EQ}. The learner can ask both membership and strong equivalence queries. This query combination is known as an MAT (minimally adequate teacher).

7.5.3 Polynomial learning with queries

When learning with an Oracle, the goal is to bound the number of queries needed to identify. This number should be:

- Polynomial in the size of the target. Obviously, complex languages (or better said, languages whose description is complex) will need more questions than simpler ones.
- Polynomial in the amount of information received. If in the case of membership queries the Oracle only requires one bit to give her answer, things are more complex when the Oracle has to return a string (for example in the case of a counter-example to an equivalence query). This is a tricky matter: as the Oracle can return a counter-example that is worse to us, without this, an exponentially long counter-example would not allow us to learn in polynomial time. Just reading the string is itself too long! It should be noted that the amount of information received varies with time, so we should measure this at every step for the algorithm. Indeed we can imagine (in certain settings) a learning algorithm that would use too many resources to learn correctly, but then defer the moment of returning the correct solution and 'gain' extra resources by querying in such a way that a very long counter-example is returned, thus justifying *a posteriori* the resources spent.

Formally, let QUER be a fixed set of queries, G be a grammar class and $\mathbf{A}$ be a learning algorithm. Suppose that to identify G, algorithm $\mathbf{A}$ makes on a particular run ρ queries $Q_1, \ldots, Q_n$. We denote by $LQ(\rho, i)$ the length of the longest counter-example (or information) returned by the queries $Q_1, \ldots, Q_i$ in ρ, and $RT(\rho, i)$ the running time of $\mathbf{A}$ before interrogating the Oracle with query Q_i. $LQ(\rho, n+1)$ and $RT(\rho, n+1)$ refer to the values of LQ and RT at the end of the entire run ρ.

Definition 7.5.1 A class of grammars is **polynomially identifiable** with queries from QUER if_{def} there exists an algorithm $\mathbf{A}$ and a polynomial $p()$ for which $\forall G \in \mathcal{G}$, $\mathbf{A}$ identifies G in the limit, and, for any valid run ρ with n queries, we have

$$\forall i \leq n+1,\ RT(\rho, i) \leq p(\|G\|,\ LQ(\rho, i)).$$

In the above definition, we first consider that the learning algorithm $\mathbf{A}$ identifies in the limit any target from correct answers by the Oracle. The complexity limitations are that at any moment, the time needed by the algorithm to build its next hypothesis is polynomially bounded by the size of the longest piece of information it has received so far, and the size of the target.

There are several other definitions that seem possible, but one should steer clear of at least the following problems:

- The definition should allow that the Oracle returns unbounded information
- The definition should take into account the complexity during the run: it should not allow a learner to spend a lot, then make queries at the end that it can solve easily in order to have a global runtime polynomial with the information received.

7.5.4 Approximate fingerprints

The typical technique to prove negative learning results is by *approximate fingerprints*. The basic proof method goes as follows. Let $\mathcal{H}_n$ be a set of hypotheses of size 2^n. Suppose that given any query there is an answer the Oracle can make (of size at most polynomial in n) such that the amount of hypotheses eliminated due to that answer is only a polynomial in n. Then it is clear that with only a polynomial number of queries there will always remain more than one hypothesis (actually most of them!) consistent with all the answers seen so far. Therefore, the class is not identifiable in the limit. We will again write '$L \not\models a$' to indicate that the information a is inconsistent with the language L.

Theorem 7.5.1 *Let $\mathcal{L}$ be a class containing at least 2^n different languages. Now suppose that for every query q from* QUER *the Oracle gives an answer a such that $\left|\{L \in \mathcal{L} : L \not\models a\}\right| < p(n)$. Then the class $\mathcal{L}$ is not polynomially identifiable with queries from* QUER.

Proof The number of hypotheses that are consistent with all the information received after having asked a polynomial ($r(n)$) number of queries is at least $2^n - r(n) \cdot p(n)$ which remains exponential in n. □

It should be noted that the technique says nothing about the way the class is encoded. It relies only on the fact that there are simply too many grammars consistent with the answers to the queries. In this sense the technique is *information-theoretic*.

Example 7.5.1 Let $\mathcal{L}$ be the class of all finite languages over $\{a, b\}$. We consider $\mathcal{H}_n$ the class of all singletons and prove that we cannot learn $\mathcal{H}_n$ by membership queries. Indeed the Oracle answers MQ(w) with **no** and eliminates just one language ($\{w\}$).

7.5.5 The halving algorithm

But information-theoretic arguments should be handled with care. The following puzzling algorithm (Algorithm HALVING, 7.3) seems to imply that learning with equivalence queries is always possible. Let Σ^n be the set of all strings of size n and let $\mathcal{H}$ be the set of all possible candidate languages.

Algorithm 7.3: HALVING.

Data:
Result: a grammar G
$H \leftarrow \emptyset$;
while ***not*** EQ(H) **do**
  Let w be the counter-example for EQ(H);
  if $w \in \mathbb{L}(H)$ **then**
    $l \leftarrow 0$
  else
    $l \leftarrow 1$
  end
  Eliminate from $\mathcal{H}$ the languages that disagree with (w, l) ;
  $H \leftarrow \left\{x \in \Sigma^\star : \ |L \in \mathcal{H} : x \in L| \geq \frac{|\mathcal{H}|}{2}\right\}$
end
return H

Algorithm 7.3 makes an equivalence query based on a majority vote: the hypothesis language H contains exactly those strings that belong to at least half of the possible hypothesis languages. Therefore whatever counter-example we get is going to allow us to eliminate at least half of the hypotheses: if the counter-example is a string w labelled '1', this string doesn't belong to H so therefore belongs to less than half of the languages. Conversely, if the string is a w with label '0', it belongs to H and therefore to at least half of the languages left. Obviously, in general, these unlimited equivalence queries do not exist. It is usually required that equivalence queries are *proper* (i.e. are made inside the solution class) in order to avoid problems of this type.

7.5.6 How to obtain positive results

There is no general technique to learn with queries, because of the variety of queries one may be allowed to use, but there is one key algorithm. The most important positive result is that DFAs are learnable with membership and strong equivalence queries. The algorithm (LSTAR) is discussed in Chapter 9. There are also practical reasons for studying this setting, as it corresponds to problems that appear in many interactive applications.

7.6 Learning with noise

As a consequence of the negative results described in the previous sections, learning grammars in a noisy setting is going to be a very hard task, since we must add the difficulty of learning in a reasonable amount of time to the fact that the noisy conditions are even more challenging. Whereas in other areas of machine learning and pattern recognition there are many methods and results allowing us to work in cases where the data may be imperfect, this is not (yet?) the case in grammatical inference. One of the few hopes to do anything of use in this case is to learn probabilistic automata instead, even if this corresponds to tackling an altogether different problem (see Chapter 16).

7.6.1 Systematic noise

As a first approach to learning with noisy data we introduce a model of noise called *systematic*: each item of data will appear in the presentation with all its noisy variants, a noisy variant being the string to which up to n edit operations have been added. Intuitively the intended type of noise can be described by a spot of paint on which a heavy object is pressed. The spot of paint will then become a blur. In an ideal world this blur is a disk. Note that the edit distance is used because in most applications this is what best models noise.

It is reasonable to study this kind of noise in the paradigm of identification in the limit. A first straightforward result is that if once noise is added two languages are not distinguishable from each other, then the class of languages is not resistant to systematic noise.

It is easy to notice that this is the case for the class of the regular languages, and in a broader way for all usual classes of languages in the Chomsky hierarchy.

The fact that parity functions can be represented as DFAs can convince us easily that they are ill-suited to deal with noise. Parity functions are formally defined in Chapter 6. They are functions $\{\mathtt{0}, \mathtt{1}\}^n \rightarrow \{\mathtt{0}, \mathtt{1}\}$ defined by a string u in $\{\mathtt{0}, \mathtt{1}\}^n$ with $f_u(w) = \sum_{i \leq n}(u_i \wedge w_i)$ mod 2. Interestingly each function can be described by a DFA with at most $2n + 1$ states (the construction is given in Section 6.5.1, page 132).

These difficulties are also an argument in favour of considering classes of languages defined outside the Chomsky hierarchy.

Definition 7.6.1 (Noisy presentation) A **noisy presentation** of a language is a presentation $\phi : \mathbb{N} \rightarrow \mathbf{X}$ with which is associated a function **isnoise** : $\mathbf{X} \rightarrow \{0, 1\}$ indicating if a particular element of the presentation is noise or not.

For example a noisy text is just a text presentation of the language to which its noise is added. The noise function N is then a function $\Sigma^\star \rightarrow 2^{\Sigma^\star}$. An N-noisy text presentation of a language L is therefore a $\phi : \mathbb{N} \rightarrow \Sigma^\star$ such that $\phi(\mathbb{N}) = N(L)$. We can then give a simple preliminary result:

Proposition 7.6.1 *Let $\mathcal{L}$ be such that there exist L_1 and L_2 in $\mathcal{L}$ with $L_1 \neq L_2$ but $N(L_1) = N(L_2)$. Then $\mathcal{L}$ is not identifiable in the limit from N-noisy text.*

The proof is trivial since in this case any presentation valid for L_1 is also valid for L_2.

7.6.2 Statistical noise

In a classification problem, the noise can be on the labels: some positive examples are wrongfully labelled as negative, whereas on the contrary some strings that should be in S_- are labelled as if they belonged to S_+.

There are very few results in this setting. On the negative side it is known that even if only a very small fraction of strings of length at most $2n + 1$ are missing, the problem of finding the correct DFA with n states is $\mathcal{NP}$-hard. So obviously this is a strong indication of the hardness of working with wrongly labelled strings.

On the positive side there have been attempts, with techniques from artificial intelligence, to solve the problem with varying quantities of noise. One alternative then is, in the state-merging algorithms, to allow a merge when the quantity of errors made is under a certain threshold.

Statistical noise corresponds to the case where either the strings can appear randomly or the labels can be wrong, depending on some distribution. In the case of noise over the labels it is essential to deal with permanent noise. Otherwise, re-sampling tends to solve the problem. Again, one of the difficulties concerns the classes of language: if the languages offer no robustness to noise, then learning is going to be hard. The better chances are with classes that are topologically sound, like balls of strings.

7.6.3 Learning the noise model

Noise over strings is very often modelled by using the edit distance, the parameters of which are supposed to reflect how close one symbol is to another. Indeed there are many reasons to suggest that the weights between the symbols are not equal, but then the question of obtaining the cost matrix is posed.

If we are given this cost matrix, the noise model can also be used for parsing. The different parsers proposed in Chapters 4 and 5 can be adapted in order to parse taking the distance into account. This allows us to define alternative languages: the sets of all strings at distance at most k, called balls of strings. In the probabilistic case, the distance allows us to smooth the language model.

But how do we get hold of this cost matrix? There may be some hope with heuristics, but there is another way around the problem: instead of learning the weights, learn probabilities! The idea is to suppose that the noise model is given by a probabilistic transducer (transducers are studied in Chapter 18). The transducer is then used by feeding a first string as input and parsing it in a probabilistic way, with possibly different outputs. The weights can be learnt in such a way as to define the probability that string w is rewritten into string w'.

The setting is important, with a certain number of interesting questions that deserve attention:

- How well suited is the best transducer to model non-probabilistic edit noise?
- How do we get hold of learning pairs? Can we simulate these from data that are noisy in all cases?

- A transducer can also handle the fact that the edit weights can have different values at the beginning or at the end of a string. Therefore, learning transducers (see Chapter 18 for some ideas) is an alternative approach.

7.7 Exercises

7.1 What conditions should the presentation function ϕ meet when learning from counter-examples only?

7.2 What conditions should the presentation function ϕ meet when learning from substrings only?

7.3 Consider learning by prefixes as proposed in Example 7.2.1 (page 148). Show that the same string can appear both as positive and as negative. Propose a limitation over the class of languages in order to avoid this problem.

7.4 Prove that if there can be two presentations that in the limit give the same language but do not yield a unique language, identification in the limit is impossible.

7.5 Prove that identification in the limit from text is always possible if $\mathcal{L}$ is a finite set.

7.6 Let us say that a learning algorithm **A** is an identification algorithm for a class $\mathcal{G}(\Sigma)$ *if_{def}* **A** identifies $\mathcal{G}(\Sigma)$ in the limit. Let us now consider the set $\mathcal{A}(\mathcal{G}(\Sigma))$ of all identification algorithms for a class $\mathcal{G}(\Sigma)$. A ***universal characteristic sample*** for $\mathcal{A}(\mathcal{G}(\Sigma))$ is a sample U_{CS} such that $\forall \mathbf{A} \in \mathcal{A}(\mathcal{G}(\Sigma)),\ \forall G \in \mathcal{G},\ \forall \phi \in \mathbf{Pres}(\mathbb{L}(G)),\ \forall n \in \mathbb{N} :$ $U_{CS} \subseteq \phi_n \implies \mathbb{L}(\mathbf{A}(\phi_n)) = \mathbb{L}(G)$.

Prove that in the case where $\mathcal{G}$ is the class of DFAs there is no finite universal characteristic sample.

7.7 Suppose we used the following definition of characteristic samples for learning from text:

Definition 7.7.1 (Polynomial characteristic sample (2)) A class $\mathcal{G}$ admits **polynomial characteristic samples** *if$_{def}$* there exists an algorithm **A** and a polynomial $p()$ such that $\forall G \in \mathcal{G},\ \exists \mathrm{CS} \subseteq \mathbf{X}\ \forall \phi \in \mathbf{Pres}(\mathbb{L}(G)),\ \forall n \in \mathbb{N} : \big(|\mathrm{CS}| \le p(\|G\|) \wedge \mathrm{CS} \subseteq \phi_n\big) \implies \mathbb{L}(\mathbf{A}(\phi_n)) = \mathbb{L}(G)$.

The difference between this definition and Definition 7.3.3 (page 154) relies on the fact that the size of the sample is now only counted as the number of strings it contains: the length of these strings is no longer an issue.

Prove (by producing a convenient, albeit artificial) class, that this definition is weaker than Definition 7.3.3.

7.8 Propose a randomised algorithm to learn $\mathcal{CONE}(\Sigma)$. Prove that this algorithm, given any presentation, has a convergence point.

7.9 Prove that if an algorithm does not have a convergence point for some presentation, it does not identify in the limit. What do you think of this definition?

7.10 Prove that $\mathcal{BALL}(\Sigma)$ (Section 3.6.1, page 65) is not identifiable in the IPE, MC and CS settings.

7.11 Prove that $\mathcal{GB}(\Sigma)$ (Section 3.6.1, page 65) is identifiable in the IPE, MC and CS settings.

7.12 Prove that $\mathcal{CONE}(\Sigma)$ (Section 3.6.2, page 65) is identifiable in the IPE, MC and CS settings.

7.13 Prove that $\mathcal{COCONE}(\Sigma)$ (Section 3.6.3, page 65) is identifiable in the IPE, MC and CS settings.

7.14 Prove that if we restrict only to conservative and consistent learning algorithms, then the IPE and MC classes coincide.

7.15 There are various definitions of incremental identification. This is one:

Definition 7.7.2 An algorithm **A incrementally** identifies grammar class $\mathcal{G}$ in the limit if_{def} given any T in $\mathcal{G}$, and any presentation ϕ of $\mathbb{L}(T)$, there is a rank n such that if $i \geq n$, $\mathbf{A}(\phi(i), G_i) \equiv T$.

Prove that $\mathcal{DFA}(\Sigma)$ are not incrementally identifiable in the limit.

7.16 Using Definition 7.7.2, find a (simple) class that is incrementally identifiable in the limit.

7.17 Let us call a presentation ***infinitely redundant*** if_{def} every element appears infinitely often. For example an infinitely redundant text presentation is a ϕ for L, such that given any w in L, the number of $i \in \mathbb{N}$ for which $\phi(i) = w$ is infinite. How can incremental learning of balls of strings now become possible?

7.18 Adapt Definition 7.7.2 to the case of learning with a fixed (instead of one) number of examples in memory.

7.8 Conclusions of the chapter and further reading

7.8.1 Bibliographical background

The introduction of this chapter (Section 7.1) is built from different surveys of the literature and most specifically those by Yasubumi Sakakibara (Sakakibara, 1997) and Colin de la Higuera (de la Higuera, 2005). Alternatively a presentation of the field for theoreticians can be found in (Fernau & de la Higuera, 2004) whereas a presentation for linguists is (Adriaans & van Zaanen, 2004). Satoshi Kobayashi also discusses these questions in (Kobayashi, 2003).

Between the elements we used to back up the point about why theory should be preferred, we mentioned the fact that grammatical inference competitions had provided evidence to this respect. The best known competitions are: the ABBADINGO competition (Lang & Pearlmutter, 1997), the GOWACHIN challenge (Lang, Pearlmutter & Coste, 1998), the OMPHALOS competition (Starkie, Coste & van Zaanen, 2004b) and the TENJINNO competition (Clark, 2006, Starkie, van Zaanen & Estival, 2006). A competition involving noisy data also was organised inside the community working on genetic algorithms (Lucas, 2004).

Section 7.2 is an attempt to cover the main results in identification in the limit, without entering in the algorithmic details. There have been two main lines of direction to study convergence in a formal manner: Gold's model of learning called *identification in the limit*, with two main variants: learning *from text* and learning *from an informant* (Gold, 1967, 1978). From these pioneer works alternative models have been proposed adding complexity constraints (de la Higuera, 1997), the possibility of interrogating an Oracle (Angluin, 1987a) or probabilistic issues. We may notice that the exposition here is through four criteria to define the problem, whereas more classically (Angluin & Smith, 1983, Gold, 1978) five have been proposed. The fifth (which we do not use) is about the type of learner we have, which could be machine or human.

We concentrate in Section 7.3 on the first model and follow lines close to complexity theory where reductions can be found allowing us to reduce one problem to another. By doing this we follow the work by Lenny Pitt and Manfred Warmuth (Pitt & Warmuth, 1988). The question of polynomial learning is also of interest. A first discussion has taken place in (Pitt, 1989), with discussions and further ideas in (de la Higuera, 1997) and (Yokomori, 2003). Many results relating the different classes of polynomial learning can be found in (de la Higuera, Janodet & Tantini, 2008).

Definition 7.2.5 is by E. Mark Gold (Gold, 1967). The insufficiencies of this definition alone were shown in (Pitt, 1989). Another definition which is shown to be insufficient by Lenny Pitt (Pitt, 1989) is Definition 7.3.2.

Definition 7.3.3 can be tracked back to several sources. It is *de facto* used in (Gold, 1978), formalised in the teaching setting by Sally Goldman and David Mathias (Goldman & Mathias, 1996), and studied in a systematic way in (de la Higuera, 1997).

Definitions 7.3.4 and 7.3.7 are also variations of definitions by Nick Littlestone (Littlestone, 1987), or by Lenny Pitt (Pitt, 1989) or Takashi Yokomori (Yokomori, 1995, 2003).

The commuting diagrams presented in Section 7.4 were introduced in (de la Higuera, 2005, Tantini, de la Higuera & Janodet, 2006). The safe languages are defined in (Alpern, Demers & Schneider, 1985) and their learnability (without reductions) is proved in (de la Higuera & Janodet, 2004). Even linear languages and grammars have been analysed in the context of grammatical inference in many papers, for instance (Koshiba, Mäkinen & Takada, 1997, Mäkinen, 1996, Sempere & García, 1994, Takada, 1988). There have been many results over learning from strings that have been adapted to the case where the data are trees. A very limited list can be found in (Fernau, 2002, Knuutila & Steinby, 1994, Sakakibara, 1987, 1990).

We shall study active learning in detail in Chapter 9. The main landmarks are the papers by Dana Angluin (Angluin, 1981, 1987a, 1987b, 1990). Most of the different results presented in Section 7.5 are all due to her work even if the formalism we propose here to deal with the complexity aspects is new. Between the special sorts of query we mention, extended membership queries were introduced in (Bergadano & Varricchio, 1996) and extended prefix language queries in (de la Higuera & Oncina, 2004). The relationships between PAC learning, equivalence queries and active learning have been studied in

(Angluin & Kharitonov, 1991, Gavaldà, 1993). Jose Balcázar *et al.* have studied the query complexities for different combinations of queries (Balcázar *et al.*, 1994a, 1994b).

Learning in noisy settings (Section 7.6) corresponds to work by Frédéric Tantini (Tantini, de la Higuera & Janodet, 2006) for the systematic noise, and by Marc Sebban and Jean-Christophe Janodet for the classification noise (Sebban & Janodet, 2003). There have also been attempts to learn a transducer to model the probabilistic edit distance (Bernard, Janodet & Sebban, 2006, Oncina & Sebban, 2006).

7.8.2 Some alternative lines of research

There are several alternative lines of research. The main one is followed by researchers in inductive inference, and specifically in algorithmic learning theory. Yet concrete complexity issues have not been explored.

7.8.3 Open problems and possible new lines of research

The paradigms described in this chapter are still not fully understood. Between the many possible open research lines, let us mention:

- The question of reducing from strings to trees is puzzling. There are many papers with complex proofs to export a result on learning classes of strings to an equivalent one on trees. Yet there is no general transformation. Is there one? Can we show an example where something is feasible with strings but not with trees?
- Relating the models to each other is important. Where are the implications?
- Several researchers (Castro, 2001, Watanabe, 1994) have noticed that expecting exponentially long counter-examples in query learning was a problem. In the same sense, one can argue that some grammars are too small for strings: if the grammar is of logarithmic size in the size of the basic strings it generates, we will say that this grammar does not have characteristic samples. Developing a nice model for this is of real interest.
- Noise is a topic where little has been studied theoretically. There is room for new classes of languages, motivated by topological questions, by algorithmic considerations and, if possible, based on the knowledge we have of formal language theory.
- Correction queries were introduced as a nice alternative to equivalence queries (Becerra-Bonache, 2006, Becerra-Bonache, Horia Dediu & Tîrnauca, 2006, Becerra-Bonache & Yokomori, 2004). These are a compromise between equivalence and membership queries. Along the same lines, studying alternative types of query is of interest.

8
Learning from text

No quería componer otro Quijote -lo cual es fácil- sino el Quijote. Inútil agregar que no encaró nunca una transcripción mecánica del original; no se proponía copiarlo. Su admirable ambición era producir unas páginas que coincidieran palabra por palabra y línea por línea con las de Miguel de Cervantes. Mi empresa no es difícil, esencialmente leo en otro lugar de la carta. Me bastaría ser inmortal para llevarla a cabo.

***Jorge Luis Borges**, Pierre Menard: autor del Quijote, Ficciones © 1995. Maria Kodama. All rights reserved.*

Apart from the fascinating (and phoney) linguistic challenge (could a computer, like the young Tarzan of the Apes, learn a language by simply reading books written in it?), it has an interesting position in syntactic pattern recognition.

***Laurent Miclet** (on grammatical inference) on (Miclet, 1990)*

Learning from text consists of inferring from a presentation of examples that all come from the target language. The learner is asked to somehow generalise from the data it sees while not having counter-examples that would help it refrain from over-generalising.

8.1 Identification in the limit from text

Learning from text is considered by many to be the essence of language learning. It is in a sense the *initial* problem, the one with least constraints, and the one that, once we show it cannot be solved, allows us to consider making the problem easier by adding some helpful information like negative examples, knowledge about the structure or the possibility to interrogate an Oracle.

We survey the problem, give alternative ways of seeing it and give conditions that have to be met for learning to be possible. We also discuss the issue of *polynomial learning from text*.

8.1.1 Why is this a hard problem?

Given only *text*, that is, strings from a language, guess this language.

This can be considered as the most pure problem of grammatical inference or at the very least as the basic problem from which the others are derived. In many cases the other questions are specialisations of this one; they could be called *learning with extra help*.

Furthermore there are arguments, in many applications, that negative information or added bias are not always available. Take the very much discussed question of language acquisition by children: while it is clear that text is presented to the child (by the mother), it is quite unreasonable to hope for labelled negative examples, especially if we require a presentation of all possible counter-examples. Even if this point deserves a longer discussion (one alternative being to describe an interactive learning setting where some sort of *corrections* are given, which should be returned in an exploratory dialogue), the general agreement is that learning from text is the paradigm in which we should study language acquisition.

Another argument is that even in those cases where negative data may exist, there are many reasons to believe that these negative strings arrive in a biased way: it is obviously not the same to learn from near-misses as from random strings that are not in the language. Taking again the case of language acquisition, a negative example could be a string that is slightly wrong, 'this is me car', a string in a different language, 'ceci est ma voiture', a random set of words that cannot be parsed, 'car when cooking whereas', or even any sequence of symbols, 'gyilgfcliq saauas cfjaeafea'.

Another point worth discussing concerns the type of convergence that we may want or hope for. Identification in the limit is in this case the better candidate, but trying to add complexity criteria is again going to be an issue. PAC learning is possible but not very exciting as noted in Section 10.4: only positive data are given for the inference yet positive and negative data could be used in testing. As soon as there are instances of data for which there are two different consistent minimal (for inclusion) languages, this proves to be impossible.

8.1.2 About languages and grammars

Most of the results in this section have been obtained inside the algorithmic learning theory community and by inductive inference specialists. In many cases the words 'grammar' and 'language' are interchangeable. We will try to avoid following this trend as we are not only concerned by learnability but also by 'efficient' learnability, and in this case we will consider with care the question of the representations of the languages.

Adapting from Section 7.2:

Definition 8.1.1 Let $\mathcal{L}$ be a class of languages over an alphabet Σ. A **TEXT presentation** of a language L is a function $\phi : \mathbb{N} \to \Sigma^\star$ where $\phi(\mathbb{N}) \in \mathcal{L}$. If $L = \phi(\mathbb{N})$ then we will say that ϕ is a presentation of L.

$\text{TEXT}(L) = \{\phi : \mathbb{N} \to \Sigma^\star : \phi(\mathbb{N}) = L\}$.

Note that in the above definition we have not included the special case of the empty language, for which we either have to allow partial functions or the presentation of a special symbol not in Σ.

Remember that $\mathbb{L}$ is the naming function; it takes a grammar G and returns a language $\mathbb{L}(G)$ in the class $\mathcal{L}$.

Definition 8.1.2 A class of languages $\mathcal{L}$ is **identifiable in the limit from text** if_{def} there is an algorithm $\mathbf{A}$ for which, given any language L and any presentation ϕ in TEXT(L), there exists a rank n such that $\forall i \geq n,\ \mathbb{L}(\mathbf{A}(\phi_i)) = L$ and $\mathbf{A}(\phi_i) = \mathbf{A}(\phi_n)$.

In the above definition the class of grammars concerned is implicit. If explicit, we can rewrite this as follows:

Definition 8.1.3 A class of grammars $\mathcal{G}$ is **identifiable in the limit from text** if_{def} there is an algorithm $\mathbf{A}$ for which, given any grammar G in $\mathcal{G}$ and any presentation ϕ in TEXT$(\mathbb{L}(G))$, there exists a rank n such that $\forall i \geq n$, $\mathbb{L}(\mathbf{A}(\phi_i)) = \mathbb{L}(G)$ and $\mathbf{A}(\phi_i) = \mathbf{A}(\phi_n)$.

We will use both definitions, but will prefer the second one when dealing with complexity issues that depend usually on the type of grammars under scrutiny. In both definitions, not using the restriction that $\forall i \geq n,\ \mathbf{A}(\phi_i) = \mathbf{A}(\phi_n)$ leads to the definition of ***behaviourally correct identification***: the algorithm may switch indefinitely often between various equivalent descriptions of the language.

8.1.3 The main results, grammar-independent

We proved in Section 7.2 (Theorem 7.2.3, page 151) that no super-finite class of languages is identifiable in the limit from text. A class is super-finite if it contains all the finite languages and at least one infinite one. The goal of this section is to generalise this result, and, more importantly, to try to understand the reasons for which it holds, mainly in order to get around it.

We first provide negative results which show that learning from text is going to be an arduous task. These do not depend on a specific representation of the languages. The first two theorems indicate that constructive proofs may not be available:

Theorem 8.1.1 (Non-union theorem) *Let $\mathcal{L}_1$ and $\mathcal{L}_2$ be two classes of languages, each identifiable from text. Then $\mathcal{L}_1 \cup \mathcal{L}_2$ may not be identifiable from text.*

Proof Let $\mathcal{L}_1 = \mathcal{FIN}(\Sigma)$ be the set containing all finite languages over some alphabet Σ and $\mathcal{L}_2 = \{\Sigma^\star\}$. $\mathcal{L}_1$ and $\mathcal{L}_2$ are identifiable from text but not $\mathcal{L}_1 \cup \mathcal{L}_2$. □

Theorem 8.1.2 (Non-concatenation theorem) *Let $\mathcal{L}_1$ and $\mathcal{L}_2$ be two classes of languages, each identifiable from text. $\mathcal{L}_1 \cdot \mathcal{L}_2$ may not be identifiable from text.*

Proof Let $\mathcal{L}_1 = \mathcal{FIN}(\Sigma)$ as above and $\mathcal{L}_2 = \{\Sigma^\star, \{\lambda\}\}$. Then both $\mathcal{L}_1$ and $\mathcal{L}_2$ are identifiable from text but $\mathcal{L}_1 \cdot \mathcal{L}_2$ is not. □

We turn now to finding necessary conditions for identification in the limit from text.

Definition 8.1.4 (Limit point) A class $\mathcal{L}$ of languages has a **limit point** if_{def} there exists an infinite sequence $L_n, n \in \mathbb{N}$ of languages in $\mathcal{L}$ such that $L_0 \subsetneq L_1 \subsetneq \ldots \subsetneq L_n \subsetneq \ldots$, and there exists another language L in $\mathcal{L}$ such that $L = \bigcup_{n\in\mathbb{N}} L_n$.

L is called a limit point of $\mathcal{L}$.

Clearly the class of all regular languages has a limit point since one can consider an infinite sequence of finite languages, and this is the case for any super-finite class.

Theorem 8.1.3 *If $\mathcal{L}$ admits a limit point, then $\mathcal{L}$ is not identifiable from text.*

Proof In the proof of Theorem 7.2.3 (page 151) L_∞ is the limit point. □

Note that the absence of a limit point is a necessary condition for learnability; for sufficient conditions, strong computability constraints are required, among others (see Exercise 8.4).

Definition 8.1.5 (Accumulation point) A class $\mathcal{L}$ of languages has an **accumulation point** L if_{def} $L = \bigcup_{n\in\mathbb{N}} S_n$ where $S_0 \subseteq S_1 \subseteq \ldots \subseteq S_n \subseteq \ldots$, and given any $n \in \mathbb{N}$ there exists a language L' in $\mathcal{L}$ such that $S_n \subseteq L' \subsetneq L$.

The language L is called an *accumulation point* of $\mathcal{L}$.

Clearly, if L is a limit point then it is also an accumulation point. In both cases we will also say that the associated class of grammars $\mathcal{G}$ has a limit or an accumulation point. Not having an accumulation point means that given any increasing sequence of examples there is always a moment where the target is one of the smallest consistent languages.

But this can also be a sufficient condition. To prove this we need to rely on a grammar formalism. We leave this question to the next section.

In the above theorems and definitions the goal, to prove non-identifiability, is to build an infinite sequence of languages such that identification of at least one of these languages is impossible. If a learning algorithm takes the risk of naming a language, that means that no inductive bias is possible: why should the learner want to 'skip' all the smaller languages and reach directly for the larger one?

Definition 8.1.6 (Infinite elasticity) A class of languages $\mathcal{L}$ has **infinite elasticity** if_{def} there exists an infinite sequence of languages $L_1, \ldots, L_i, \ldots$ in $\mathcal{L}$, and an infinite sequence of strings $x_0, x_1, \ldots, x_i, \ldots$ such that $\forall\, i \in \mathbb{N}, \{x_0, \ldots, x_i\} \subseteq L_i$ and $x_{i+1} \notin L_i$.

Theorem 8.1.4 *If $\mathcal{L}$ admits a limit point, it has infinite elasticity.*

Proof The proof is straightforward. □

Definition 8.1.7 (Finite elasticity) $\mathcal{L}$ has **finite elasticity** if_{def} it does not have infinite elasticity.

Again, this will be used to state that a class is identifiable, but will still depend on a grammar representation for the class of languages. And an even stricter restriction allows us to obtain positive results even more nicely:

Definition 8.1.8 (Finite thickness) $\mathcal{L}$ has **finite thickness** if_{def} given any non-empty set X, the number of languages in $\mathcal{L}$ that contain X is finite.

If a class has finite thickness, identification from text will be possible as soon as (1) $\mathcal{L}$ is recursively enumerable, and (2) we have grammars for each language in $\mathcal{L}$.

Definition 8.1.9 Let $\mathcal{L}$ be a class of languages. A set $T_L \subseteq L$ is a **tell-tale set** for language L if_{def} $\forall L' \in \mathcal{L}, T_L \subseteq L' \implies L' \not\subset L$.

Notice that a tell-tale set is a set defined by a language, whereas a characteristic sample depends also on the learning algorithm. Notice also that a tell-tale set is not the set for which the minimum language containing it is the target, but only for which the target is a minimal language in the class of all languages containing this set. Furthermore, we do not (contrarily to the original definition) require the tell-tale set to be computable.

Proposition 8.1.5 *A language L in $\mathcal{L}$ has a tell-tale set* if and only if *L is not an accumulation point of $\mathcal{L}$.*

Proof If a language L is an accumulation point, then given any finite set S of strings in L, there is another language L' in $\mathcal{L}$ with $S \subseteq L' \subsetneq L$. Therefore there is no tell-tale set.

Conversely let L be a language without a tell-tale set. Then it means that for any finite subset S of L, there exists a language L_S with $S \subseteq L_S \subsetneq L$, i.e. we can always build a smaller language containing any set, so no set can be a tell-tale. This enables the construction of an accumulation point. □

8.1.4 Grammar-dependent results

For sufficient conditions for polynomial identification to be expressed we will have to use grammars, since computation issues are at stake. So let us suppose that we have associated with the class of languages $\mathcal{L}$ a class of grammars $\mathcal{G}$, for which the membership problem ($w \in \mathbb{L}(G)$?) is decidable.

Suppose we have an algorithm that given a positive sample can always return the smallest consistent solution for that sample, in the following sense:

Definition 8.1.10 MINCONS is an algorithm which, given a sample S, returns the minimal grammar G in M for a good total order $<_{\mathcal{G}}$, where M is the set of smallest (for language inclusion) consistent grammars, i.e. those such that if $S \subseteq \mathbb{L}(G') \subseteq \mathbb{L}(G)$ then $\mathbb{L}(G') = \mathbb{L}(G)$.

Notice that we require that the algorithm MINCONS returns a grammar corresponding to a language such that there is no smaller one that contains the sample, and furthermore which is the smallest grammar for some well-defined (Noetherian) order.

If $\mathcal{G}$ (with corresponding class of languages $\mathcal{L}$), which admits a MINCONS algorithm, is *algorithmically minimisable*, the following two theorems hold:

Theorem 8.1.6 *If $\mathcal{G}$ is algorithmically minimisable and has finite elasticity, then $\mathcal{G}$ is identifiable in the limit from text.*

Proof If $\mathcal{G}$ has finite elasticity, then MINCONS identifies $\mathcal{G}$ in the limit. □

It should be added that without the condition that the class of grammars is algorithmically minimisable, identification remains possible.

Theorem 8.1.7 *Let $\mathcal{G}$ be algorithmically minimisable. $\mathcal{G}$ admits an accumulation point* if and only if *$\mathcal{G}$ is not identifiable from text.*

Proof We can use the proof of Theorem 7.2.3 (page 151), to show that $\mathcal{G}$ is not identifiable from text.

Conversely suppose the class has no accumulation point. Then every sequence $S_0 \subsetneq S_1 \subsetneq \ldots \subsetneq S_n \subsetneq \ldots$ is either finite or, if it is infinite, then there is a point n where there is only one language L such that $S_n \subseteq L$. If we have an algorithm that can find a minimally consistent language, we are done. □

One way to learn is to associate with each language some typical examples, in such a way that we have some enumeration of the languages and the algorithm returns the first (for the enumeration) language for which all the typical examples have been seen. This leads to the following definition:

Definition 8.1.11 (Characteristic sets) Let $\mathcal{L}$ be a class of languages and $L \in \mathcal{L}$. Let $\mathbf{A}$ be a learning function that identifies $\mathcal{L}$ from text. Let $L \in \mathcal{L}$.

Then CS $\subset \Sigma^\star$ is a **characteristic set** for $\langle L, \mathbf{A} \rangle$ if_{def} for any presentation $\phi \in \text{TEXT}(L), \forall n \in \mathbb{N}, \text{CS} \subseteq \phi_n \implies \mathbb{L}(\mathbf{A}(\phi_n)) = L$.

The same definition, if centred on grammars, becomes:

Definition 8.1.12 (Characteristic sets (grammars)) Let $\mathcal{G}$ be a class of grammars and $G \in \mathcal{G}$. Let $\mathbf{A}$ be a learning function that identifies $\mathcal{G}$ from text.

Then CS $\subset \Sigma^\star$ is a **characteristic set** for $\langle G, \mathbf{A} \rangle$ if_{def} for any presentation $\phi \in \text{TEXT}(G), \forall n \in \mathbb{N}, \text{CS} \subseteq \phi_n \implies \mathbb{L}(\mathbf{A}(\phi_n)) = \mathbb{L}(G)$.

Note that in the case of learning from text, we can use *characteristic set* for *characteristic sample*.

The above definition deserves to be discussed. No condition is put on the way the examples should be chosen or even on some inclusion relation between the languages and the characteristic sets. Furthermore the characteristic set obviously depends on the algorithm we are using: take an algorithm that decides to reject every odd example and learn only from the even ones. This algorithm would not be able to use the same characteristic set as one that uses all the data. We leave as an exercise (8.3) to prove that there is no universal characteristic set.

Example 8.1.1 Let $\Sigma = \{\mathtt{a}, \mathtt{b}\}$ and consider the class $\mathcal{FIN}(\Sigma)$ represented by finite sets of strings. Consider the algorithm RETURNSAMPLE which returns as a grammar the exact set of strings seen so far. Then the characteristic sample is the exact set of strings composing the language.

We can relate characteristic samples to identification in the limit:

Theorem 8.1.8 *A language class $\mathcal{L}$ is identifiable in the limit by an algorithm* **A** *if and only if each language in the class has a characteristic sample for $\langle L, \mathbf{A}\rangle$.*

To prove Theorem 8.1.8, we need Lemma 8.1.9 which tells us that the order in which the presentation is given does not matter for identification.

Definition 8.1.13 An **order-independent learner A** is a learning algorithm such that $\forall n \in \mathbb{N}, \forall \phi,\ \psi \in \text{TEXT}(L)$, if $\phi_n = \psi_n$, $\mathbf{A}(\phi_n) = \mathbf{A}(\psi_n)$.

Remember that ϕ_n is the set of the first n elements from presentation ϕ, so $\phi_n = \psi_n$ corresponds to set (and therefore order-independent) equality.

Lemma 8.1.9 *If a class of languages $\mathcal{L}$ is identifiable from text, then it is identifiable by an order-independent learner.*

Proof Let $\mathbf{A_1}$ be eventually an order-dependent learner. We construct a new learner $\mathbf{A_2}$ that reads its input, reorders it in some total order, eliminates repetitions, runs $\mathbf{A_1}$ on this data and returns the result. It is easy to see that identification will be reached by $\mathbf{A_2}$, which is no longer order-dependent. □

Proof [of Theorem 8.1.8] If $\mathcal{L}$ admits characteristic sets, the associated algorithm $\mathbf{A_1}$ identifies in the limit. Conversely, if $\mathcal{L}$ is identifiable from text, it is also (by Lemma 8.1.9) identifiable by some order-independent algorithm $\mathbf{A_2}$. And consider any set Y such that $\mathbf{A_2}$ does not make any further mind changes. Then this set is characteristic. □

Theorem 8.1.10 *If $\mathcal{L}$ is identifiable in the limit then every language in $\mathcal{L}$ admits a tell-tale set.*

Proof If a language does not admit a tell-tale set, it is an accumulation point by Proposition 8.1.5, and we can conclude by using Theorem 8.1.7. □

Example 8.1.2 Let us consider five simple classes of languages over a fixed alphabet Σ of size at least 2. In each case the associated grammars are implicit:

- $\mathcal{SINGLE}(\Sigma)$ is the set of all singleton languages, i.e. $L_u = \{u\}$.
- $\mathcal{ABO}(\Sigma)$ is the set of the *'all but one'* languages $L_{\overline{u}}$, where for each u in Σ, $L_{\overline{u}} = \Sigma^\star \setminus \{u\}$.
- $\mathcal{FIN}(\Sigma)$ is the set containing all finite languages over Σ.
- Let $K(u) = \{x \in \Sigma^\star : x \leq_{sub\text{-}seq} u\}$. $\mathcal{CONE}(\Sigma) = \{K(u) : u \in \Sigma^\star\}$.
- Similarly let $KK(u) = \{x \in \Sigma^\star : u \leq_{sub\text{-}seq} x\}$. $\mathcal{COCONE}(\Sigma) = \{KK(u) : u \in \Sigma^\star\}$.

Some of the above classes were discussed in Section 3.6.

Table 8.1. *Some classes and their status for learning from text.*

Class	Accumulation point	Finite elasticity	Finite thickness	Tell-tale sets	Characteristic sets
$\mathcal{FIN}(\Sigma)$	No	No	No	Yes	Yes
$\mathcal{SINGLE}(\Sigma)$	No	Yes	Yes	Yes	Yes
$\mathcal{ABO}(\Sigma)$	No	Yes	No	Yes	Yes
$\mathcal{CONE}(\Sigma)$	No	No	No	Yes	Yes
$\mathcal{COCONE}(\Sigma)$	No	Yes	Yes	Yes	Yes

It can be easily checked that if we consider the definitions of finite thickness, finite elasticity, characteristic sets, tell-tale sets and accumulation points, one obtains the results represented in Table 8.1.

8.1.5 Polynomial aspects

There are a number of ways of bounding the resources: one can ask for only a polynomial (in the size of the target) number of examples before identification. We can want to only change our mind a small number of times or even only make a number of errors that is reasonable before converging.

More precisely, suppose we are given a learning algorithm and we want to measure the quality of the algorithm. Then we should be able to limit:

- the complete runtime of the algorithm, over a typical example or over a worst case,
- the runtime of the algorithm when needing to update the hypothesis, i.e. to get from $\mathbf{A}(f_n)$ to $\mathbf{A}(f_{n+1})$,
- the size of the characteristic set,
- the number of implicit prediction errors,
- the number of mind changes.

This leads to a number of definitions, for instance the one already defined in the general case (as Definition 7.3.7, page 155):

Definition 8.1.14 An algorithm **A** identifies from text a class $\mathcal{G}$ of grammars in the limit in POLY-IPE time if_{def}

- **A** identifies in the limit $\mathcal{G}$ from text,
- **A** has polynomial update time,
- **A** makes a polynomial number of implicit prediction errors.

Note that the process of counting implicit prediction errors, sizes of characteristic sets, or mind changes depends on the intended application. One should naturally count whatever resource is going to be expensive.

8.2 Exercises

8.1 Propose a learning algorithm that identifies in the limit from text a specific class, but that does not admit characteristic samples. Hint: consider a learner that is order-dependent, i.e. that will be influenced by the order in which it sees the examples.

8.2 The following version of a definition of infinite elasticity (to be compared with Definition 8.1.6, page 176) was proposed in the literature:

Definition 8.2.1 (Infinite elasticity 2) A class of languages $\mathcal{L}$ has **infinite elasticity** *if*$_{def}$ there exists an infinite ascending chain of languages $L_0 \subsetneq L_1 \subsetneq \cdots \subsetneq L_n \subsetneq \ldots$.

A class would again have finite elasticity if it did not have infinite elasticity. Let us call this form of finite elasticity *restricted finite elasticity*. Find a class that has restricted finite elasticity but that cannot be identified in the limit from text. Hint: one may want to use as a starting point the set of the 'all but one' languages $\mathcal{ABO}(\Sigma)$ introduced in Example 8.1.2, page 179.

8.3 Prove that as soon as a class of languages is non-trivial, there is no universal characteristic set, i.e. one from which any identifying algorithm will identify each language **as soon as** the corresponding set has been presented.

8.4 Find a class that admits accumulation points but not limit points.

8.5 Let $\mathcal{BLC}(\Sigma)$ be the class $L_n = \{w \in \Sigma^\star : |w| \leq n\}$. Find an algorithm that identifies $\mathcal{BLC}(\Sigma)$ in the limit from text.

8.6 Prove that the algorithm from Exercise 8.5 works in polynomial update time.

8.7 Prove that the algorithm from Exercise 8.5 does not admit a polynomial characteristic sample.

8.8 Prove that the algorithm from Exercise 8.5 makes an exponential number of implicit prediction errors, and also an exponential number of mind changes.

8.9 Prove that $\mathcal{BALL}(\Sigma)$ and $\mathcal{GB}(\Sigma)$ (see page 65) are both identifiable in the limit from text.

8.10 Prove that $\mathcal{BALL}(\Sigma)$ cannot be identified by a polynomial number of mind changes, nor of implicit prediction errors.

8.3 Conclusions of the chapter and further reading

8.3.1 Bibliographical background

The approach followed here has concentrated on the algorithmic (and complexity-related) aspects of learning from text. The alternative approach, closer to the computability issues, uses the notation TXTEX for the class of language families that are identifiable in the limit from text.

Identification from text (Section 8.1) was introduced in E. Mark Gold's seminal paper (Gold, 1967). A lot of work in the field has been done by researchers in the field of inductive inference, and more specifically in that of algorithmic learning theory. There have been a lot of variants. We have chosen here to present things through the questions of complexity theory.

The general picture of identification in the limit from text we have given here is chiefly based on Christophe Costa Florêncio's PhD (Costa Florêncio, 2003) and on Satoshi Kobayashi's ideas as expressed in (Kobayashi, 2003). Other sources for this brief presentation are Takashi Yokomori's analysis (Yokomori, 2004), and Dana Angluin and Carl Smith's survey (Angluin & Smith, 1983). Theorem 8.1.6 (page 178) is usually attributed to Keith Wright (Wright, 1989) who also introduced the definition of infinite elasticity, later corrected by Tatsuya Motoki *et al.* (Motoki, Shinohara & Wright, 1991). Theorem 8.1.7 (page 178) is by Shyam Kapur (Kapur, 1991).

Henning Fernau generalised a number of such results by introducing the elegant notion of function distinguishability (Fernau, 2000, 2002): if the language admits a (special sort of) function to sort out the cases of non-determinism, then learning from text is easier.

The results on characteristic sets are to be compared with Dana Angluin's definition of tell-tale sets (Angluin, 1980) or Lenore and Manuel Blum's definition of locking sequences (Blum & Blum, 1975). The flavour here is more that of grammatical inference, along the lines of (de la Higuera, 1997).

Theorem 8.1.10 is by Dana Angluin, who introduced the tell-tale sets (page 179), even if her definition also takes into account computability aspects (Angluin, 1980).

Section 8.1.5 describes work on polynomial identification from text: actually too little has been studied in this setting, two notable exceptions being papers by Takashi Yokomori (Yokomori, 2003, 2005).

8.3.2 Some alternative lines of research

The inductive inference community has obtained a variety of results in the setting of identification from text. The classes of grammars under scrutiny are sometimes more artificial, but one can find ideas concerning how to do one-shot learning (the learner has to say 'halt' when it knows it has learned), or how to refute (the learner has to declare, when necessary, that no concept in the class is going to be acceptable) (Angluin & Smith, 1983, Jain *et al.*, 1999, Osherson *et al.*, 1997).

A particular issue arises when probabilities are introduced. One can either draw a distribution over the presentations or over the examples themselves. Typically, the definition of identification in the limit will include the addendum 'with probability one'. We will explore some of these ideas in Chapter 10. Other lines of research have been followed in (Zeugmann, 2006) and in the inductive inference papers.

8.3.3 Open problems and possible new lines of research

Identification in the limit and the polynomial variants may be of use when trying to study complexity for online and incremental problems. The key question of being able to study complexity issues in problems that do not fit the traditional model ('here is the data, this is the decision or optimisation problem you have to solve, solve it') may be able to receive at least a partial answer through more effort on the question of polynomial identification in the limit. Indeed, there is a mixture of short-term goals (do as well as you can with the data available at that point) and long-term goals.

9
Active learning

No one can tell you how to do it. The technique must be learned the way I did it, by failures.

***John Steinbeck**, Travels with Charlie*

Similarly, a responsive informant could answer questions involving non-terminals, or instead of responding 'No' could give the closest valid string.

***Jim Horning** (Horning, 1969)*

There are several situations where the learning algorithm can actively interact with its environment. Instead of using given data, the algorithm may be able to perform tests, create new strings, and find out how far he may be from the solution. The mathematical setting to do this is called *active learning*, where queries are made to an *Oracle*.

In this chapter we cover positive and negative aspects of this important paradigm in grammatical inference, but also in machine learning, with again a special focus on the case of learning deterministic finite automata.

9.1 About learning with queries

In Section 7.5 we introduced the model of learning from queries (or active learning) in order to produce negative results (which could then also apply to situations where we have less control over the examples) and also to find new inference algorithms in a more helpful but credible learning setting.

9.1.1 Why learn with queries?

Active learning is a paradigm first introduced with theoretical motivations but that for a number of reasons can today be considered also as a pragmatic approach.

Some of the theoretical reasons for introducing the model are:

- To propose a model sufficiently general to cover most cases, not only situations where precise questions are asked to the environment (like 'does this string belong?') but also where asking for a new example is an action. Indeed, even if this is not the chosen way, sampling can be seen as a dialogue with an Oracle.

- To make use of additional information that can be measured; if the initial setting is that of imposed data, here we consider that asking some specific questions is allowed, but the type of question and the size of the answers should be measured.
- To explain thus the difficulty of learning certain classes: if you can prove that the number of queries to be asked is too large, then learning with imposed examples is going to be at least as hard.
- To discover new algorithms making use of this new information.

On the other hand, there are a number of reasons for believing that this setting is not just theoretical:

- In specific applications, testing is possible. For instance, this is the case of circuit testing, where the automaton may represent a circuit and the testing data would be the input sequences to the circuit, chosen with the idea to do as few (expensive) tests as possible. This is also the case in robotics where the robot can perform some experiment in order to interrogate its environment/Oracle before taking decisions or building some map of its situation.
- In World Wide Web applications, we may want to consider the Web as a formidable Oracle, and search engines as the means to use queries, even very elaborate ones.
- There are more and more situations where there is too much data, or where the labelling of the data is too expensive. A typical situation is that of automatic translation: the task of translating an individual sentence may not require the construction and use of a huge bilingual corpus. In an approach closer to transductive learning, it may be easier to obtain the translation through the labelling of just some sentences.
- In modelling language acquisition situations where children learn from their parents, the mother can be described in the paradigm of learning from an Oracle or active learning. This can be extended to other situations where interaction with a partner or a teacher is part of the learning process.

To ensure better readability, and also to pinpoint the mother-child learning setting, we will call the Oracle a ***she*** and the learner a ***he***.

9.1.2 Some types of queries

The active learning paradigm is based on the existence of an *Oracle* which can be seen in principle as a device that:

- knows the language and has to answer correctly,
- can only answer queries from a given set of queries.

When the Oracle finds herself in a situation where various legal answers are possible, if she is asked to sample following a given distribution, she does so. In all other cases, she does not handle probabilities: there is no distribution over the possible answers she may give. But then, if we want to analyse if convergence of the learning algorithm is possible or not, we should rely on a worst-case policy: the Oracle does not 'want' to help.

Different types of queries (corresponding to Oracles that will be more or less powerful) have been defined in the literature. The main ones in the non-probabilistic setting are:

- Membership queries: MQ. A membership query is made by proposing a string to the Oracle, who answers YES if the string belongs to the language and NO if not. We will denote this formally by:

$$\text{MQ} : \Sigma^{\star} \rightarrow \{\text{YES}, \text{NO}\}$$

- Equivalence queries (weak): WEQ. A weak equivalence query is made by proposing a grammar G to the Oracle. The Oracle answers YES if the grammar is equivalent to the target and NO if not:

$$\text{WEQ} : \mathcal{G} \rightarrow \{\text{YES}, \text{NO}\}$$

- Equivalence queries (strong): EQ. A strong equivalence query is made by proposing a grammar G to the Oracle. The Oracle answers YES if the grammar is equivalent to the target and returns a string in the symmetrical difference between the target and $\mathbb{L}(G)$ if not:

$$\text{EQ} : \mathcal{G} \rightarrow \{\text{YES}\} \cup \Sigma^{\star}$$

- Subset queries: SSQ. A subset query is made by proposing a grammar G to the Oracle. The Oracle answers YES if $\mathbb{L}(G)$ is a subset of the target language and returns a string from $\mathbb{L}(G)$ that is not in the target language if not:

$$\text{SSQ} : \mathcal{G} \rightarrow \{\text{YES}\} \cup \Sigma^{\star}$$

9.1.3 Correct learning with queries

Describing a problem of learning with queries requires defining precisely how the languages are represented and what the admissible queries are. It will also be necessary, when using queries where the Oracle might return some counter-example, to count the length of this counter-example as the size of the data and not as the result. This can be argued and may seem strange, but is required if one wants to have an independent Oracle.

This point should be taken seriously (and has been by the different researchers working in the field). Some alternative ideas that might be worth considering are:

- One may want to only consider strings of bounded length as counter-examples: the Oracle should try to return one such string if possible.
- One should also be careful with not wanting to allow some tricky strategy where an equivalence query can be made with the sole objective of gaining time, not information!

We will therefore present our results in terms of classes of grammars (or representations) with the intended meaning that:

- The language identified or learnt is the one corresponding ($\mathbb{L}(G)$) to the learnt grammar G.
- If a grammar G is the target then learning an equivalent grammar G' corresponds to completing the identification task successfully.

We first describe the interactive active learning process. We define a class of grammars $\mathcal{G}$ and the sort of queries we are allowed to make (and that therefore the Oracle will have to answer). We call this class of queries QUER. Typically if the learner is only allowed to make membership queries, we will have QUER $= \{\text{MQ}\}$. Given a run of our learning algorithm, we can break this run into the different phases leading from one query to the next. Let us call ***query point*** the moment (defined by an integer corresponding to the number of queries made up to that point) before a new query is made. We can count:

- the overall runtime,
- the length of the longest counter-example seen at some given query point,
- the update runtime, consisting of the time spent from one query point to the next.

Correct learning can be defined as follows:

Definition 9.1.1 A class $\mathcal{G}$ is **identifiable in the limit with queries** from QUER if_{def} there exists an algorithm **A** such that given any grammar G in $\mathcal{G}$, **A** identifies G in the limit, i.e. returns a grammar G' equivalent to G and halts.

Definition 9.1.2 Let ρ be a run of a learner **A**. Let $\langle r_1, r_2, \dots r_m\rangle$ be the sequence of **replies** to queries $\langle q_1, q_2, \dots q_m\rangle$ the Oracle makes during run ρ. We will say that **A** is **polynomially bounded** if_{def} there exists a polynomial $p(,)$ which given any target grammar G and any run ρ, at any query point k of the run, denoting the runtime before that point by t_k, we have:

- $k \leq p\big(\|G\|, \max\{|r_i| : i < k\}\big)$
- $|q_k| \leq p\big(\|G\|, \max\{|r_i| : i < k\}\big)$
- $t_k \in \mathcal{O}\big(p(\|G\|, \max\{|r_i| : i < k\}\big)$

In other words at any query point k of the run, indicating the moment before query q_k is made, the number of queries made, the computation time spent up to then and the size of the next query are all bounded by the size of the target and the length of longest information returned by the Oracle up to that point.

Definition 9.1.3 A class $\mathcal{G}$ is **polynomially** identifiable in the limit with queries from QUER if_{def} there exists a **polynomially bounded** learner **A** which given any grammar G in $\mathcal{G}$, identifies G.

In the above definition a couple of elements need to be discussed:

- The length of the information (usually a counter-example) returned by the Oracle is a parameter (not a result) for the complexity of a learner. Indeed, the Oracle can choose to return any consistent result. But if this result is exponentially longer than the size of the encoding of the target, we should still be allowed to read it and deal with it.
- There is the possibility of building a degenerate learning algorithm if the complexity is only measured at the end of the run of the algorithm: one could choose to spend time freely, identify the correct target and make a query where the only possible answer is exponentially long, thus allowing *a posteriori* polynomial time.

9.2 Learning from membership queries alone

In the previous section we used algorithm LSTAR to prove that DFAs were learnable when we could access an Oracle that answered membership and strong equivalence queries. A reasonable question is to try to learn with less powerful queries. We prove here that membership queries alone are insufficient to learn DFAs. In fact, we prove the stronger result that we can also use subset queries and weak equivalence queries, without doing much better.

The intuition is to keep in mind ***lock automata***: these automata recognise just one string of length n. With a two-letter alphabet there are 2^n such automata, each of size n. It is easy to see that to discard the $2^n - 1$ other automata, one will need $2^n - 1$ queries...

Lemma 9.2.1 *If a class $\mathcal{L}$ contains a non-empty set $L_\cap$ and n sets $L_1, \ldots, L_n$ such that $\forall i, j \in [n]\ L_i \cap L_j = L_\cap$, any algorithm using membership, weak equivalence and subset queries needs in the worst case to make n–1 queries.*

Proof The Oracle will answer each query as follows:

- To MQ(x) with x in $L_\cap$, YES, and the learner cannot discard any language.
- To MQ(x) with x in $L_i \setminus L_\cap$, NO, and only language L_i can be discarded by the learner.
- To WEQ($L_\cap$), NO, and only language $L_\cap$ can be discarded by the learner.
- To WEQ(L_i), NO, and only language L_i can be discarded.
- To SSQ($L_\cap$), YES, and no language can be discarded.
- To SSQ(L_i), NO, with counter-example x_i from $L_i \setminus L_\cap$ and only language L_i can be discarded. □

Corollary 9.2.2 *$\mathcal{DFA}(\Sigma)$ cannot be identified by a polynomial number of membership, weak equivalence and subset queries.*

Proof Let us consider $L_\cap = \emptyset$, and $L_i = \{m_i\}$ where m_i is the number i written in base 2 over n bits using alphabet $\{0, 1\}$. Therefore, as this class is of size $2^n + 1$ and corresponds exactly to the situation covered by Lemma 9.2.1, a total of 2^n membership queries, subset queries and weak equivalence queries will be needed to learn. □

9.3 Learning from equivalence queries alone

The goal of this section is to give the ***approximate fingerprints*** proof (see Section 7.5.4) that equivalence queries alone are insufficient for tractable learning of DFAs. The result, as a corollary, applies to more complex classes. The proof is due to Dana Angluin, and is as instructive as the result itself.

Theorem 9.3.1 *$\mathcal{DFA}(\Sigma)$ cannot be identified by a polynomial number of strong equivalence queries.*

The rest of the section is devoted to the proof of Theorem 9.3.1. Let us first describe the idea.

The goal is to make use of ***approximate fingerprints***: let us consider for each value of n, a special class of languages (represented by automata of size n) $\mathcal{H}_n$ over a fixed alphabet $\Sigma = \{\mathtt{a}, \mathtt{b}\}$.

We aim to build two languages I_n and E_n (indexed by an integer n) where I_n contains strings that belong to all languages in $\mathcal{H}_n$ and E_n contains strings that belong to only very few languages in $\mathcal{H}_n$.

Then given any candidate language C (perhaps not in $\mathcal{H}_n$), and given any polynomial $r()$, with an integer n sufficiently large, we can always find a string w of length less than $r(n)$ such that:

$$\left|\{L \in \mathcal{H}_n : [w \in I_n \iff w \in L]\}\right| < \frac{|\mathcal{H}_n|}{r(n)}$$

Answering NO to EQ(C) and returning w as a counter-example therefore will result in only eliminating a sub-polynomial fraction of $\mathcal{H}_n$, i.e. those automata that are consistent with C on this particular string.

By ***sub-polynomial fraction*** we mean a fraction of $\mathcal{H}_n$: we intend that by repeating this process a polynomial number of times, there will always be too many candidates left for identification to take place.

We now start the technical proof of Theorem 9.3.1.

Proof We build $\mathcal{H}_n$ as follows:

$$L(i, n) = \{ucvcw : u \in \Sigma^{i-1}, c \in \Sigma, v \in \Sigma^{n-1}, w \in \Sigma^{n-i}\},\ i \in [n]$$

Note: $L(i, n)$ is accepted by a DFA with $3n + 1$ states (see Figure 9.1). We also notice that there are only n languages $L(i, n)$, for a given value of n. We continue the construction as follows. By concatenating n of these $L(i, n)$ we get languages that only contain strings of length $2n^2$:

$$\mathcal{H}_n = \{L(i_1, n)L(i_2, n) \cdots L(i_n, n)\},\ \forall j \in [n], i_j \in [n]$$

Each language $L(i_1, n)L(i_2, n) \cdots L(i_n, n)$ is accepted by a DFA which has $3n^2 - 1$ states. The number $|\mathcal{H}_n|$ of such languages is n^n.

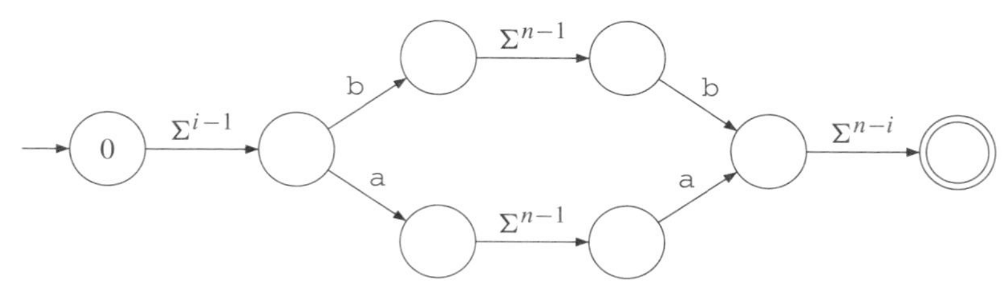

Fig. 9.1. The automaton for $L(i, n)$.

We now consider I_n, the intersection of all the languages in $\mathcal{H}_n$. The strings in this language have the property of coinciding, in each of the n substrings, in each of the n positions. Thus:

$$I_n = \{x_1x_1x_2x_2\cdots x_nx_n : \forall i,\ x_i \in \Sigma^n\}$$

Now notice that if querying with a language C whose intersection with I_n is not I_n, all the Oracle has to do is answer with a string from $I_n \setminus C$ and no language in $\mathcal{H}_n$ is eliminated.

Notice also that I_n can be recognised by a DFA (it is a finite language). But it can be easily proved that this DFA has to have an exponential number of states.

We now turn to defining E_n.

Let $d_{Hamming}$ be the Hamming distance (used in this case because all strings have identical length, see Definition 3.4.1, page 55). We build E_n as a set containing strings that are 'far away' from strings in I_n.

$$E_n = \left\{x_1y_1x_2y_2\cdots x_ny_n : \forall i \in [n],\ x_i \in \Sigma^n\ y_i \in \Sigma^n \wedge d_{Hamming}(x_i, y_i) > \frac{n}{4}\right\}$$

E_n contains strings that are far (for the Hamming distance) from strings in I_n. Furthermore since a string xy belongs to $L(i, n)$ only if $x(i) = y(i)$ we have:

$$\forall w \in E_n,\ \left|\{L \in \mathcal{H}_n : w \in L\}\right| \leq \left(\frac{3n}{4}\right)^n$$

Therefore since the size of $\mathcal{H}_n$ is exponentially larger than the above quantity†, E_n is defined as required, and if an equivalence query EQ($\mathcal{A}$) is made for $\mathbb{L}(\mathcal{A}) \cap E_n \neq \emptyset$ the Oracle can return a string from E_n and only a sub-polynomial fraction of languages in $\mathcal{H}_n$ is eliminated.

The situation is summarised in Figure 9.2: a DFA that would recognise all strings from I_n and none from E_n would be of exponential size.

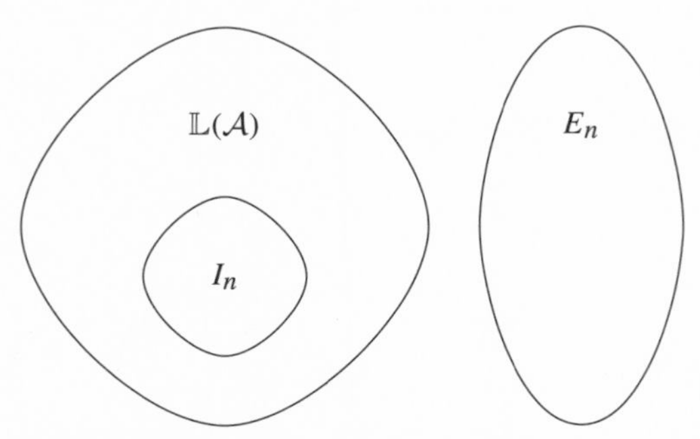

Fig. 9.2. The impossible situation: no automaton with the behaviour of $\mathcal{A}$ can be of polynomial size.

† n^n is exponentially larger than $\left(\frac{3n}{4}\right)^n$ because $\frac{\left(\frac{3n}{4}\right)^n}{n^n} = \left(\frac{3}{4}\right)^n$.

Formally:

Let $r()$ be any polynomial. There exists n such that given any DFA $\mathcal{A}$ with $r(n)$ states, • either $\exists w \in E_n : w \in \mathbb{L}(\mathcal{A})$, • or $\exists w \in I_n : w \notin \mathbb{L}(\mathcal{A})$.

To prove this we make three claims:

Claim 1. $\forall w \in \Sigma^n, \exists c < 1 : |\{x \in \Sigma^n : d_{Hamming}(x, w) \leq \frac{n}{4}\}| < 2^{cn}$.
From this it follows that there are fewer than 2^{cn} strings ***close*** to w.

Claim 2. Given any polynomial $r()$, there exists n such that if $\mathcal{A}$ is a complete DFA of size less than $r(n)$ with $I_n \subseteq L_{\mathcal{A}}$ and q_i is any state of $\mathcal{A}$, then $\exists x, y \in \Sigma^n, d_{Hamming}(x, y) > \frac{n}{4}$ such that $\delta(q_i, x) = \delta(q_i, y)$.

The reason for this is that there are 2^n strings of size n and since the DFA is complete, given any state q_i there is another state q_j reached by at least $\frac{2^n}{r(n)}$ strings. And since $\frac{2^n}{r(n)} > 2^{cn}$, then necessarily two of these strings are at distance at least $\frac{n}{4}$ from each other.

Claim 3. Let $\mathcal{A}$ be a DFA of size less than $r(n)$ such that $I_n \subseteq L_{\mathcal{A}}$. Then $L_{\mathcal{A}} \cap E_n \neq \emptyset$.

Indeed from Claim 2, $\exists x_1, y_1 \in \Sigma^n, d(x_1, y_1) > \frac{n}{4}$ and $\delta(q_\lambda, x_1) = \delta(q_\lambda, y_1)$.

Then $\delta(q_\lambda, x_1x_1) = \delta(q_\lambda, x_1y_1) = q_1$. By induction, $\forall i, \exists x_i, y_i \in \Sigma^n$, $d_{Hamming}(x_i, y_i) > \frac{n}{4}$ such that $\delta(q_{i-1}, x_i) = \delta(q_{i-1}, y_i)$.

Therefore $\delta(q_{i-1}, x_ix_i) = \delta(q_{i-1}, x_iy_i) = q_i$

Concluding: $\exists x_1, .., x_n, y_1, .., y_n \in \Sigma^n$, with $\forall i \leq n\ d_{Hamming}(x_i, y_i) > \frac{n}{4}$ such that $\delta(q_\lambda, x_1x_1x_2x_2 \cdots x_nx_n) = \delta(q_\lambda, x_1y_1x_2y_2 \cdots x_ny_n)$.

And since $x_1x_1x_2x_2 \cdots x_nx_n \in I_n \subseteq L_{\mathcal{A}}$ then $x_1y_1x_2y_2 \cdots x_ny_n \in L_{\mathcal{A}}$. □

Corollary 9.3.2 *$\mathcal{DFA}(\Sigma)$ cannot be polynomially identified in the limit with a polynomial number of prediction errors when learning from an informant.*

Proof [sketch] A prediction error (see Definition 7.3.4, page 155) is made when the new string is inconsistent with the current hypothesis. If we could make only a polynomial number of prediction errors we could deduce an algorithm that identifies with a polynomial number of strong equivalence queries. □

9.4 PAC active learning results

There is a general agreement that equivalence queries play the same part as a structurally complete sample (all rules of the grammar are exercised), but also as the error parameter ϵ in PAC learning. The idea therefore is to use the learning algorithm LSTAR but replace the equivalence queries by sampling: if on a random sample the target automaton and the hypothesis automaton coincide, then (with high confidence) the error, if any, must be very low.

We can trade off equivalence queries for sampling queries in exchange for a small mistake. We will also need to discuss the issue that sampling queries may be just as difficult to get hold of.

It can be noted that equivalence queries are used in the algorithm LSTAR in order to check at each moment where the algorithm has come up with a closed and consistent solution whether this solution is the correct one. An attractive alternative is to use sampling in order to have some sort of a statistical equivalence query. The algorithm would work as follows: at each moment an equivalence query is required, draw instead m labelled examples $x_1, \ldots, x_m$ and check if the current hypothesis labels them in the same way as the target does (the true labelling). If $\forall i \in [m]\ x_i \in \mathbb{L}(G) \iff x_i \in \mathbb{L}(G_T)$ then the error is most likely small... but just how small? We want this error to be at most ϵ. Let us suppose for contradiction that the true error is more than ϵ. Then the probability of selecting randomly one example where G and G_T coincide is less than $1 - \epsilon$ and the probability of selecting randomly m examples where G and G_T coincide (all the time) is now less than $(1 - \epsilon)^m$.

But now we have the following bound: $(1 - \epsilon)^m \leq e^{-\epsilon m}$. So by bounding this value by δ we have:

$$\begin{aligned} \delta \geq e^{-\epsilon m} &\iff \ln \delta \geq -\epsilon m \\ &\iff \ln\left(\frac{1}{\delta}\right) \leq \epsilon m \\ &\iff m \geq \frac{1}{\epsilon} \ln\left(\frac{1}{\delta}\right) \end{aligned}$$

Therefore, by sampling at least $\frac{1}{\epsilon}\ln(\frac{1}{\delta})$ labelled examples and testing them against the current hypothesis, the algorithm can make a ***stochastic*** equivalence query.

But the number of equivalence queries is unknown in practice when the learner starts, and therefore the algorithm has no control over the δ parameter. In order to make sure the total confidence is at least $1 - \delta$, the size of each sample (or of the calls to EX()) should be at least $m_i = \frac{1}{\epsilon}\left(\ln\left(\frac{1}{\delta}\right) + i \ln 2\right)$. By doing this, one ensures that the confidence parameter δ is at most $\sum_{i>0}(1 - \epsilon)^{m_i}$.

9.5 Exercises

9.1 An equivalence query is called *proper* if it has to be made with a function from the class under inspection. Does the proof from Section 9.3 hold for improper equivalence queries?

9.2 What happens if the Oracle returns the smallest counter-example? Write an algorithm for the case of DFAs.

9.3 Learn balls (see Section 3.6.1) of strings with equivalence queries and/or membership queries.

9.4 Definition 9.1.2 (page 187) still seems to leave a loophole as the total number of queries is not bounded *a priori*. Is it possible to build an algorithm that uses this to make many more queries than necessary and nevertheless learn?

9.5 A ***correction query*** (CQ) is made by presenting a string to the Oracle. She will answer YES if the string is in the target language, and return a correction of the string if not. A ***suffix correction*** is the shortest string WV (for the lex-length order) in L that admits w as a prefix. For example, given the DFA from Figure 13.9, the correction of string ba is itself whereas the correction of b is ba. If no correction is possible (i.e. if $w\ \Sigma^\star \cap L = \emptyset$) the Oracle answers NO. Prove that 0-reversible languages (see Section 11.2) are learnable from suffix correction queries.

9.6 Build an example where correction queries don't help over membership queries: prove that $\mathcal{DFA}(\Sigma)$ may need an exponential number of suffix correction queries.

9.6 Conclusions of the chapter and further reading

Active learning is becoming an increasingly important topic, partly because researchers have shifted from considering this as a purely theoretical question to studying it as a very pragmatic one!

9.6.1 Bibliographical background

The model of learning from queries is due to Dana Angluin, who introduced it (Angluin, 1987b) and presented the first negative proofs (Angluin, 1987a), and also the algorithm LSTAR (see Chapter13) (Angluin, 1990).

The different hardness proofs of Sections 9.2 and 9.3 are due to Dana Angluin. For the case of the membership queries only, this can be found in (Angluin, 1987b). The proof of the non-learnability of the DFA with equivalence queries only can be found in (Angluin, 1990): we have followed it closely.

In that paper Dana Angluin introduced the combinatorial notion of *approximate fingerprints*, of independent interest: these correspond to a subset of hypotheses out of which only a small fraction can be excluded, given any counter-example, resulting in the necessity of using an exponential number of equivalence queries to isolate a single hypothesis. Ricard Gavaldà furthered this study in (Gavaldà, 1993).

Lenny Pitt (Pitt, 1989) used this result to prove the intractability of the task of identifying DFAs with only a polynomial number of prediction errors. This result is Corollary 9.3.2.

The links between learning with queries and PAC learning have been studied in a number of papers including. A combinatorial point of view of learning with queries was initiated in (Balcázar *et al.*, 1994a), and an overview of results related with ***dimensions*** can be found in (Angluin, 2004).

Most studies are based on the general idea that the Oracle is some perfect machine. Indeed, Dana Angluin (Angluin, 1987a) actually proposed techniques to implement the

Oracle. *A contrario*, one can thing of an Oracle as something quite different from a deterministic machine. Some examples of this approach are as follows:

- Learning neural networks (Giles, Lawrence & Tsoi, 2001): An alternative approach to learning from sequences is to train recurrent neural networks (Alquézar & Sanfeliu, 1994). But once these are learnt, they need to be interpreted or else we are left with black boxes. One way to do this is to use the black box/neural network as an Oracle.
- Learning test sets (Bréhélin, Gascuel & Caraux, 2001) and testing hardware (Hagerer *et al.*, 2002) are other activities where there is an unsuspected Oracle: the actual electronic device we are testing can be physically tested, by entering a sequence. The device will then be able to answer a membership query. Note that in that setting equivalence queries will usually be simulated by sampling.
- System SQUIRREL (Carme *et al.*, 2005) is used for wrapper induction. The system will interrogate the (human) user who will mark web pages. The marking will be used by SQUIRREL to learn a (sort of) tree automaton. The markings correspond to equivalence queries.
- The World Wide Web can also be seen as an Oracle. The knowledge is there, you cannot expect it to be sampled for you, nor to use it all. Interrogating the web in order to pick the useful information for learning is becoming an important task.
- In the field of robotics, a generally important task is that of building a map. For that, the robot is allowed to explore and therefore to interact with the environment (Rieger, 1995, Rivest & Schapire, 1993).

The real problem is with the equivalence queries which cannot be simulated. A common mistake is to take the argument, 'you can trade off the equivalence queries for a (δ, ϵ) pair in the PAC framework'. But this is not exactly true. You can only do this if you have access to another Oracle that can sample the set of all strings according to the distribution. If one takes the above applications as examples, doing this sampling is actually going to be a really hard problem. There is therefore the need to suppress equivalence queries and find some alternative to learn from membership queries only, or from some form of membership queries.

9.6.2 Some alternative lines of research

Active learning, in a broad sense, corresponds to the capacity of learning by choosing one's examples. In the machine learning literature, it is more often related with the fact that the learner can change in some way the distribution it is working with. There is therefore still a gap to be bridged between the two conceptions of active learning, and so there is a lot of room for further research.

9.6.3 Open problems and possible new lines of research

There are a number of lines of research that might be investigated. And the interest in many applications towards active learning means that this is to be a very active field of research indeed. To name just a few, we may want to consider some of these with high priority:

(i) Prove that neither context-free grammars nor non-deterministic finite automata can be learnt by a minimally adequate teacher (MAT). A related result is that it can be proved that these are not learnable with membership queries in a PAC setting (Angluin, 2004).
(ii) Provide a reasonable definition of *noisy* membership queries, or *fuzzy* membership queries that would somehow tell us (in a consistent way) that a string probably/possibly belongs to the target language, and devise a learning algorithm for an interesting class of languages.
(iii) In the case of robotics, the customary approach is to learn a map (which can, in a simplified way, be a graph) of an area. What happens when there are several robots (a swarm) who intend to learn this map in a cooperative way? Work on distributed versions of active learning algorithms can be of help for this (Balcázar *et al.*, 1994a).
(iv) Learn negotiation protocols. This problem was formally proposed in (de la Higuera, 2006b):

Consider the situation where two adversaries have to negotiate something. The goal of each is to learn the model of the opponent while giving away as little information as possible. The situation can be modelled as follows:

Let L_1 be the language of adversary 1 and L_2 be the language of adversary 2. We suppose here that the languages are regular and can be represented by deterministic finite automata with respectively n_1 and n_2 states. The goal for each is to learn the common language, i.e. language $L_1 \cap L_2$.

The rule is that each adversary can only query the opponent by asking questions from his own language. This means that when player 1 names string w, then $w \in L_1$. In turn, the adversary will state whether or not string w belongs to language L_2.

The goal of each adversary is to identify language $L_1 \cap L_2$. This means that the protocol goes as follows:

- Player 1 announces some string w from L_1. Player 2 answers YES if this string belongs to L_2, NO otherwise.
- From this answer player 1 may update his hypothesis H_1 of language $L_1 \cap L_2$.
- From the information $w \in L_1$, player 2 may update his hypothesis H_2 of language $L_1 \cap L_2$
- Player 2 announces some string w from L_2.
- And so on...

From this setting there are a number of problems to be solved:

- A good learning algorithm for this task can be defined in alternative ways. One can want to be uniformly better than an adversary, or than all the adversaries... Propose definitions of 'good learning'.
- What happens if both opponents 'agree' on a stalemate position, i.e. are satisfied with an identical language L which in fact is a subset of the target?
- What is a good strategy? Can identification be avoided? Are there any 'no win' situations? Are there strategies that are so close to one another (corresponding to what Dana Angluin called *'lock automata'*) that through membership queries alone, learning is going to take too long?
- Using the definition above, find a winning algorithm in the case where $n_1 = n_2$.

And, along similar lines to (Carmel & Markovitch, 1998a), one may adapt this theoretical analysis to consider learning the strategy of the adversary (see also the introductory example in Chapter 1); if we notice the adversary is daring/cautious, do we have a better strategy?

10

Learning distributions over strings

El azar tiene muy mala leche y muchas ganas de broma.

Arturo Perez Reverte

All knowledge degenerates into probability.

***David Hume**, A Treatise on Human Nature, 1740*

If we suppose that the data have been obtained through sampling, that means we have (or at least believe in) an underlying probability over the strings. In most cases we do not have a description of this distribution, and we describe three plausible learning settings.

The first possibility is that the data are sampled according to an unknown distribution, and that whatever we learn from, the data will be measured with respect to this distribution. This corresponds to the well-known PAC-learning setting (probably approximately correct).

The second possibility is that the data are sampled according to a distribution itself defined by a grammar or an automaton. The goal will now no longer be to classify strings but to learn this distribution. The quality of the learning process can then be measured either while accepting a small error (most of the time, since a particular sampling can have been completely corrupted!), or in the limit, with probability one. One can even hope for a combination of both these criteria.

There are other possible related settings that we only mention briefly here: an important one concerns the case where the distribution in the PAC model is computable, without being generated by a grammar or an automaton. The problem remains a classification question for which we have only restricted the class of admissible distributions. This responds in part to a common attack against the PAC model, namely that wanting to beat any distribution is too hard and the reason why there are so few positive results. This leads to definitions based on Kolmogorov complexity, usually called Simple-PAC or PAC-S: a good distribution is (in a very simplified view) one where the simple strings have higher probability, whereas the complex ones (those that are incompressible) have very low probability.

A common feature in all cases is that the distribution does not change over time.

Another common feature is that the learning algorithm has to learn a grammar given a *confidence* parameter δ and an *error* parameter ϵ. The error parameter ϵ will measure how far from the ideal solution we should accept to be. Obviously the smaller this parameter, the more examples and time the learner should be allowed in order to meet the bound. Typically, these resources will be functions of $\frac{1}{\epsilon}$. On the other hand one can never be sure that the sampling process is significant. Obviously, we will only sample a finite number of times, and therefore, we can expect to be *unlucky* from time to time. This is taken into account by the confidence parameter δ. Inside δ we will put the probability that sampling has gone wrong. Again, the smaller the δ, the more resources the learner will need.

10.1 About sampling

The first problem about sampling is that it may not be that easy to do! There are many situations where there is a distribution over the strings, but just 'picking a string' is not that easy. Let us suppose nevertheless for the moment that this is not the case and that we are in a context where sampling is possible.

From a theoretical point of view let us imagine we are sampling by using a PFA. This defines a distribution over strings so it seems that it would be easy to use the PFA. But, unless the underlying language is finite, the generation process only terminates with probability one! Indeed, since a string is generated character by character, there is a loop in the PFA (if the underlying language is infinite) and therefore the probability of adding another character is not null. Indeed, the probability of generating an infinitely long string is zero, but the problem remains.

We will therefore have to make a distinction between learning using probabilistic queries (such as EX) and learning using exact queries (for instance if we interrogate the Oracle and ask for the probability of a given string).

In the case of learning from strings, when sampling, we have to deal with several different issues: the cost, the distribution and being able to specialise the queries.

Even if in practice, sampling may have a cost, we will consider that sampling is done through using a very simple query EX(), which returns in $\mathcal{O}(1)$ a string, with its label; the length of this string therefore only intervenes when we read it. Again, by doing this we are not addressing the difficult question of the generating process terminating only with probability one. The position we choose is more practical than theoretical.

If we have access to different distributions, we will indicate which one we use when sampling as follows: $\text{EX}_{\mathcal{D}}()$ gets a new example using the distribution $\mathcal{D}$.

We are going to sample examples which will be used to learn from. In the case of strings, there is always the risk (albeit often small) of sampling a string too long to account for in polynomial time. In order to deal with this problem, we can sample from a distribution restricted to strings shorter than a specific value given by the following lemma:

Lemma 10.1.1 *Let $\mathcal{D}$ be a distribution over $\Sigma^\star$. Then given any $\epsilon > 0$ and any $\delta > 0$, with probability at least $1 - \delta$ we have: if we draw, following distribution $\mathcal{D}$, a sample S of size $|S|$ at least $\frac{1}{\epsilon} \ln \frac{1}{\delta}$, the probability that a new string x is longer than any string seen in the sample is less than ϵ. Formally, if we write $\mu_S = \max\{|y| : y \in S\}$, then $Pr_{\mathcal{D}}(|x| > \mu_S) < \epsilon$.*

Proof Remember that the sample can be considered as a random variable. Denote by μ_ϵ the smallest integer such that the probability of a randomly drawn string being longer than μ_ϵ is $Pr_{\mathcal{D}}(\Sigma^{>\mu_\epsilon}) < \epsilon$.

A sufficient condition for $Pr_{\mathcal{D}}(|x| > \mu_S) < \epsilon$ to hold is that we take a sample large enough to be nearly sure (i.e. with probability at least $1 - \delta$) of having at least one string longer than μ_ϵ. On the contrary, the probability of having all n strings in S of length bounded by μ_ϵ is at most $(1-\epsilon)^n$. Using the fact that $(1-\epsilon)^n > \delta$ implies that $n < \frac{1}{\epsilon}\ln\frac{1}{\delta}$, it follows that it is sufficient to take $n > \frac{1}{\epsilon}\ln\frac{1}{\delta}$ to have a convenient value of μ_S. □

The above result should not mislead us to believe that we have gone from distribution-free learning to a setting where the distributions are controlled. We are just indicating that if we are asked to learn for a given pair ϵ, δ then we can test (for $\delta/2$ and $\epsilon/2$) in order to get a bound we can work with. That means that at the end of the learning process the errors will sum up and we will get the desired result over all strings in $\Sigma^\star$.

There are usually two ways of describing a distribution learning algorithm. The first consists of supposing the sample is already present, that it has randomly been drawn according to the distribution. In that case we should learn from there, and the sample being part of the data, it is reasonable to consider tractability issues depending on the size of this sample. In practical settings this is usually what the algorithms do, and we will present them in that way for the task of learning probabilistic automata (in Chapter 16) and for that of estimating the parameters of an automaton or a grammar (in Chapter 17). The second possibility (which we will follow in this chapter) is that the learning algorithm starts from nothing and that sampling is part of the learning process. This allows a better measure of the hardness of the learning problems (which is the goal here). It should be added that both paradigms can reduce to each other, at the price of technical details.

The algorithm can query an Oracle: It may ask for an example randomly drawn according to the distribution $\mathcal{D}$. The query will be denoted EX(). When the Oracle is only being queried for a positive example of a language we will write POS-EX(). Finally, if we pass a value m bounding the length of the admissible strings, we will write EX(m) (or POS-EX(m)) and the Oracle will return a string drawn from $\mathcal{D}(L)$, $\mathcal{D}(\Sigma^{\leq m})$ or from $\mathcal{D}(L \cap \Sigma^{\leq m})$, where we denote by $\mathcal{D}(L)$ the restriction of distribution $\mathcal{D}$ to the strings in L: $Pr_{\mathcal{D}(L)}(x) = \frac{Pr_{\mathcal{D}}(x)}{Pr_{\mathcal{D}}(L)}$ if $x \in L$, 0 if not. $Pr_{\mathcal{D}(L)}(x)$ is undefined if L is the empty set.

This idea is extended to the notion of specific sampling (which we discussed briefly in Chapter 7, page 163), consisting of sampling for a string that has a chosen property.

10.2 Some bounds

A soon as an underlying distribution exists over $\Sigma^\star$, we can consider sampling and the first important question is:

How good is my sample?

Now the question in itself is ill-posed: is a sample of 500 heads and 500 tails *good* or *bad* for an unbiased coin? In other words, once the sample is there, where is the randomness?

The correct question is therefore *a priori*:

How good is my sample going to be?

This means that when measuring statistical properties related to a sample, we will consider the sample as a random variable.

The next issue is to measure this, and we will have to use distances. The key idea is to be able to say that the sample is *with high probability* going to be representative, in which case learning should be able to take place. We recall here some of the main sampling bounds used in machine learning.

Chernov additive bounds measure the probability that in a sample randomly drawn from a distribution $\mathcal{D}$, if we are measuring some event of probability p over this sample, then we have

$$\forall \epsilon > 0,\ \forall n \in \mathbb{N},\ Pr_{\mathcal{D}}\left(\frac{f}{n} - p > \epsilon\right) < e^{-2n\delta^2}$$

where the sample is of size n and the observed frequency is f. Notice that the test is one-sided only.

The Hoeffding bound is derived as follows. For the observed frequency f (over n trials) of a Bernoulli variable of probability p, given some value $\delta > 0$ and $\epsilon = \sqrt{\frac{1}{2n}\log\frac{2}{\delta}}$, then with probability greater than $1 - \delta$,

$$\left|p - \frac{f}{n}\right| < \epsilon \tag{10.1}$$

If we now turn our attention to the size of a sample with respect to δ and ϵ, we can rewrite the above as follows. If we sample at least $n > \frac{1}{2\epsilon^2}\log\frac{2}{\delta}$ elements, then with probability greater than $1 - \delta$:

$$\left|p - \frac{f}{n}\right| < \epsilon$$

Practically, we may be given two samples extracted from the same distribution, and we want to measure if (with high probability) these two samples come from the same distribution or not. We cannot do this without prior knowledge of the parameters, but if we have two samples of respective sizes n_1 and n_2 over which the observed frequencies of the event are respectively f_1 and f_2, we can use:

$$\left|\frac{f_1}{n_1} - \frac{f_2}{n_2}\right| < \epsilon$$

which is true with probability at least $1 - \delta$, with $\epsilon = \sqrt{\frac{1}{n_1+n_2}\log\frac{2}{\delta}}$.

Suppose now we have a distribution $\mathcal{D}$ over $\Sigma^\star$, and suppose we sample n strings following $\mathcal{D}$. We denote such a sample by $X_{\mathcal{D},n}$, which can therefore be used as a random variable. Then the empirical distribution corresponding to this random variable is denoted $\widehat{X_{\mathcal{D},n}}$.

Consider now some testable property ϕ over a set of strings. We want to measure how close $\widehat{X_{\mathcal{D},n}}$ is to $\mathcal{D}$ using ϕ as a measure. This means that in the above equations we use $|X_{\mathcal{D},n}|_\phi$ instead of f and $|X_{\mathcal{D}}|$ instead of n, where the notation $|X_{\mathcal{D},n}|_\phi$ measures the number of times property ϕ is verified in sample $X_{\mathcal{D},n}$ (see Section 5.4.2, page 103).

10.3 PAC-learning languages

The question here is not to learn underlying distributions, but to learn a classifier in a world where the strings belong (or not) to a language and there is a distribution over the strings, a distribution that will be followed both when we sample strings to learn from and when we have to sample strings to check if whatever we have learnt is of any value.

The combinatorial results from Chapter 6 imply that exact polynomial identification is very often too hard to obtain for most classes of grammars; this might still leave room for approximate learning. The PAC paradigm has been widely used in machine learning to provide a theoretical setting for this.

Definition 10.3.1 (ϵ-good hypothesis) Let G be the target grammar and H be a hypothesis grammar over Σ. Let $\mathcal{D}$ be a distribution over $\Sigma^\star$. We say, for $\epsilon > 0$, that H is an **ϵ-good hypothesis** with respect to G *if*$_{def}$ $Pr_{\mathcal{D}}\big(x \in \mathbb{L}(G) \oplus \mathbb{L}(H)\big) \leq \epsilon$.

A learning algorithm is now asked to learn a grammar given a *confidence* parameter δ and an *error* parameter ϵ. The algorithm must also be given an upper bound on the size of the target grammar and on the length of the examples it is going to get (perhaps using an extra sample built thanks to Lemma 10.1.1, page 197).

Definition 10.3.2 (Polynomially PAC-learnable) Let $\mathcal{G}$ be a class of grammars. $\mathcal{G}$ is PAC-learnable *if*$_{def}$ there exists an algorithm **A** with the following property:

For each n in $\mathbb{N}$, for every grammar G in $\mathcal{G}$ of size at most n, for every distribution $\mathcal{D}$ over $\Sigma^\star$, for every $\epsilon > 0$ and $\delta > 0$, if **A** is given access to EX(m), m and n, ϵ and δ then, with probability at least $1 - \delta$, **A** outputs an ϵ-good hypothesis with respect to G.

If **A** runs in time polynomial in $\frac{1}{\epsilon}$, $\frac{1}{\delta}$, $|\Sigma|$, m and n we say that $\mathcal{G}$ is **polynomially PAC-learnable**.

Notice that in order to deal with the unbounded length of the examples we have implicitly used an $\epsilon' = \frac{\epsilon}{2}$ and a fraction of δ to compute m and accepted an error of at most ϵ' over all the strings of length more than m, and then used EX(m) instead of EX().

10.4 PAC-learning from text

In certain cases it may even be possible to PAC-learn from positive examples only. In this setting, during the learning phase, the examples are sampled following POS-EX() whereas during the testing phase the sampling is done following EX(), but in both cases the distribution is identical. Again we can sample using POS-EX(m), where m is obtained by using Lemma 10.1.1 at little additional cost. It is easy to see that if in a class there are two languages L_1 and L_2 such that $L_1 \cap L_2$ is not in the class, then the class is not polynomially PAC-learnable from text.

Lemma 10.4.1 *If $\mathcal{L}$ contains two languages L_1, L_2 such that $L_1 \cap L_2 \neq \emptyset$ and $L_1 \cap L_2 \notin \mathcal{L}$, then $\mathcal{L}$ is not* PAC*-learnable from text.*

Proof Let w_1 be a string in $L_1 - L_2$, w_2 be a string in $L_2 - L_1$ and w_3 be a string in $L_1 \cap L_2$. We now consider the distribution $\mathcal{D}_1$ where $Pr_{\mathcal{D}_1}(w_1) = Pr_{\mathcal{D}_1}(w_3) = \frac{1}{2}$ and the distribution $\mathcal{D}_2$ where $Pr_{\mathcal{D}_2}(w_2) = Pr_{\mathcal{D}_2}(w_3) = \frac{1}{2}$. It is easy to see that learning L_1 from $\mathcal{D}_2$ and learning L_2 from $\mathcal{D}_1$ will necessarily reach an identical result. But when testing, the error will be, in one of the two cases, of at least $\frac{1}{2}$. □

Note that Lemma 10.4.1 is given in terms of languages (and not grammars), because the actual representations are not at stake.

10.5 Identification in the limit with probability one

If we are attempting to learn distributions represented by probabilistic finite automata or grammars, we have the alternatives either to try to measure success through approximation and therefore to learn a distribution close to the intended distribution, or to try to identify the distribution. In this case we are going to follow the distribution itself, and we will hope to obtain results with probability one: the idea is that it is impossible, in the long run, to get empirical distributions too far away from the theoretical one.

Identification in the limit with probability one should be understood as follows: when given an increasing sequence of strings (constituting the sample at each moment), identification can only be avoided for a finite period of time. Just like in the case of identification in the limit, the actual moment at which identification will be achieved is usually not guaranteed. We comment immediately upon some important points:

- Why say 'identification with probability one'? This is better understood if we argue that in the limit, the probability of having an amazingly long sequence of bad luck decreases to zero, i.e. the probability *ad infinitum* of having some bizarre event like only heads is 0. Therefore, the complement, or the probability of not having an amazing sequence of events, is one. This also can be understood as saying that even if some event with probability p can be avoided for a very long time, this cannot be the case forever.
- Again this should primarily be read as a necessary condition. Even if identification in the limit with probability one does not involve the learning algorithm performing well in pratice, on the other hand not having this property always means that some bias is not correctly described.

We introduce some notations first. Suppose we have a learning algorithm **A**, attempting to learn a distribution $\mathcal{D}$. Let ρ be a run of this algorithm. This run consists of iterating:

(i) having a current hypothesis H_{i-1},
(ii) making a query to the Oracle,
(iii) receiving the i-th answer w_i from the Oracle,
(iv) adding this string to the current sample S,
(v) updating the hypothesis and building H_i.

The first hypothesis is H_0 and the first example is w_1.

Definition 10.5.1 (Identification in the limit with probability one) A class of probabilistic grammars $\mathcal{G}$ is **identifiable in the limit with probability one** if_{def} there exists an algorithm **A** that for any grammar G in $\mathcal{G}$, uses $\text{EX}_{\mathcal{D}_G}()$ to build a series of hypotheses $H_0, H_1, \ldots H_n \ldots$ With probability one, for all but a finite number of values of n, $\mathbb{L}(H_n) = \mathbb{L}(G)$.

We will also say that **A** ***identifies*** $\mathcal{G}$ ***in the limit with probability one***.

The key lemma necessary to derive results in the limit and with probability one is the following, based on the iterated logarithm:

Lemma 10.5.1 *Let ϕ be a property over $\Sigma^\star$. Let $\mathcal{D}$ be a distribution over $\Sigma^\star$ and p be the probability that ϕ holds on a string randomly drawn following $\mathcal{D}$. Let c be a constant ($c > 1$) and let $I(n) = \sqrt{\frac{6c(\log n)}{n}}$.*

Then, if we consider a random variable X corresponding to a sample of size n, sampled following $\mathcal{D}$, with $\widehat{X_{\mathcal{D},n}}$ the empirical distribution built from X,

- *with probability at least $1 - n^{-c}$,*
- *with probability one and for all but a finite number of values of n,*

$$\mathbf{L}_\infty\left(\frac{|\widehat{X_{\mathcal{D},n}}|_\phi}{n}, p\right) \leq I(n)$$

The first part of this lemma was introduced as Lemma 5.4.2, page 107.

10.5.1 Identifying probabilities

If one is measuring some event and is counting how often the event is realised, the fraction $\frac{\text{number of successes}}{\text{number of trials}}$ is of course going to vary with time. Therefore, to identify a fraction, one should return some other value than just the direct estimation. From Lemma 10.5.1 above, we can build a test that is false infinitely often for all but the correct fraction.

Thus if we have a way of enumerating all fractions, we can identify any fraction. This is the goal of Algorithm 10.1. We leave the construction of function NEXT_FRACTION

Algorithm 10.1: Enumeration identification algorithm.

Input: a property $\phi : \Sigma^\star \to \{0, 1\}$
Output: $Pr_{\mathcal{D}}(\phi(x))$
$a \leftarrow 0$;
$b \leftarrow 0$;
while true do
  $x \leftarrow \mathrm{EX}_{\mathcal{D}}()$;
  $a \leftarrow a + \phi(x)$;
  $b \leftarrow b + 1$;
  $i \leftarrow 0$;
  $j \leftarrow 1$;
  while $\mathbf{L}_\infty(\frac{a}{b}, \frac{i}{j}) > \sqrt{\frac{12(\log b)}{b}}$ **do**
    $\langle i, j\rangle \leftarrow$ NEXT_FRACTION($\langle i, j\rangle$)
  end
  print $\frac{i}{j}$
end

to Exercise 10.2, page 210: this function has to produce, one after another, all the possible fractions. Note that Algorithm 10.1 never halts and just prints, after each query, the first fraction in the enumeration acceptably close to the recorded relative frequency.

An alternative to enumeration is to explore the set of all probabilities in a different manner. We use Stern-Brocot trees for this in Algorithm 10.2. Stern-Brocot trees are introduced to represent the set of all fractions in normal forms. The number of these fractions is of course infinite, but the spirit of the construction can be understood in Figure 10.1. The idea is that more and more complex fractions can be obtained through summing both numerators and both denominators of two consecutive fractions. This operation can be performed generation after generation. In the first generation the process starts with $\frac{0}{1}$ and $\frac{1}{1}$. The median $\frac{0+1}{1+1}$ is built and the three fractions, once ordered, are $\frac{0}{1}$, $\frac{1}{2}$ and $\frac{1}{1}$. With the same operation in between two consecutive fractions we next obtain $\frac{0}{1}$, $\frac{1}{3}$, $\frac{1}{2}$, $\frac{2}{3}$ and $\frac{1}{1}$. The rest of the process is best described in Figure 10.1. Let us note that it can be proved that any fraction (between 0 and 1) will be generated at some point or another.

Algorithm 10.2 identifies in the limit any fraction in]0, 1[. If we want to add rationals 0 and 1, two lines specific for these cases have to be added before the loop.

Note that the Stern-Brocot encoding allows us to encode fractions as strings. This can be done by means of Algorithm 10.3. For example, the encoding corresponding to fraction $\frac{2}{7}$ is string `llr`. Algorithm 10.3 encodes rational numbers in]0, 1[, but an extension to cope with all fractions is easy to implement.

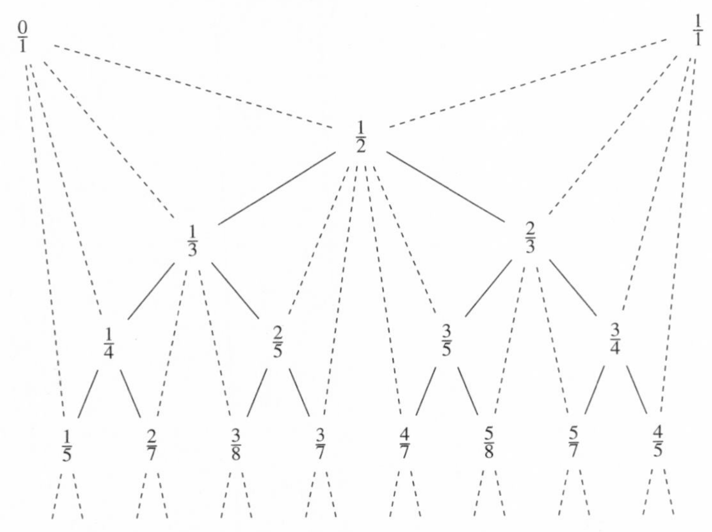

Fig. 10.1. The first fractions using Stern-Brocot trees.

Algorithm 10.2: STERN-BROCOT identification.

Input: z, n
Output: estimation of $\frac{z}{n}$
$a_1 \leftarrow 0$;
$b_1 \leftarrow 1$;
$a_2 \leftarrow 1$;
$b_2 \leftarrow 1$;
while $\left|\frac{z}{n} - \frac{a_1+a_2}{b_1+b_2}\right| > I(n)$ **do** /* Too far from a nice fraction */
  if $\frac{z}{n} > \frac{a_1+a_2}{b_1+b_2}$ **then**
    $a_1 \leftarrow a_1 + a_2$;
    $b_1 \leftarrow b_1 + b_2$
  else
    $a_2 \leftarrow a_1 + a_2$;
    $b_2 \leftarrow b_1 + b_2$
  end
end
return $\frac{a_1+a_2}{b_1+b_2}$

10.5.2 Complexity issues in identification with probability one

The first definition we shall discuss has led to the implicit definition of polynomial identification that has been used in the literature. It basically states that identification has to be obtained by using an algorithm that has polynomial runtime. It therefore does not put any

Algorithm 10.3: Encoding a fraction as a string.

Input: a pair of integers z and n corresponding to the fraction $\frac{z}{n}$
Output: a string over $\{\mathtt{l}, \mathtt{r}\}$

$a_1 \leftarrow 0$;
$b_1 \leftarrow 1$;
$a_2 \leftarrow 1$;
$b_2 \leftarrow 1$;
$s \leftarrow \lambda$;
while $\frac{z}{n} \neq \frac{a_1+a_2}{b_1+b_2}$ **do**
    if $\frac{z}{n} > \frac{a_1+a_2}{b_1+b_2}$ **then**
        $a_1 \leftarrow a_1 + a_2$;
        $b_1 \leftarrow b_1 + b_2$;
        $s \leftarrow s \cdot \mathtt{l}$
    else
        $a_2 \leftarrow a_1 + a_2$;
        $b_2 \leftarrow b_1 + b_2$;
        $s \leftarrow s \cdot \mathtt{r}$
    end
end
return s

requirement on the number of examples needed for the identification to take place. This does not mean that a polynomial number of examples may be sufficient. We will refer to this definition as ***weak*** polynomial identification:

Definition 10.5.2 (Weak polynomial identification in the limit with probability one) A class of probabilistic grammars $\mathcal{G}$ is **weakly** polynomially identifiable in the limit with probability one *if*$_{def}$ there is an algorithm **A** and a polynomial p for which, for any grammar G in $\mathcal{G}$:

1. **A** identifies $\mathcal{G}$ in the limit with probability one,
2. **A** works in time polynomial with the size of the learning data obtained through sampling.

Notice that the quantity of sampling required to learn is not made precise in the above definition. This leads to a serious flaw:

Theorem 10.5.2 *Let $\mathcal{G}$ be a recursively enumerable class of grammars for which the distance $\mathbf{L}_\infty$ is computable* (i.e. *we can compute $\mathbf{L}_\infty(S, G)$ for any sample S). $\mathcal{G}$ is weakly polynomially identifiable in the limit with probability one.*

Proof We sketch the proof as follows. Suppose we have a simple enumerative algorithm, without any polynomial limitations. Then we use this simple enumerative algorithm to find the first grammar consistent with the current sample S in the following sense:

$$\text{CONSISTENT}(S, G) = \mathbf{L}_\infty(S, G) < I(|S|), \text{ with } I(n) = \sqrt{\frac{6a(\log n)}{n}}.$$

It is easy to see that this enumerative algorithm will identify the target in the limit with probability one, since with probability one each of the (finite number of) grammars that appear before the target in the enumeration will be inconsistent with the sample an infinite number of times. And the actual target grammar will (because of Lemma 10.5.1, page 202) be consistent all but a finite number of times.

But then, obviously, this algorithm is not polynomial! To make the algorithm fit with the complexity constraints, we just make the algorithm keep track of the time it is entitled to from the current examples. The algorithm then computes as far as it can go with that time and returns whatever solution it has reached at that point. With the next example the algorithm is given more time and can continue the computation further. There is a point where the algorithm will converge. □

Corollary 10.5.3 *$\mathcal{DPFA}(\Sigma)$ is weakly polynomially identifiable in the limit with probability one.*

An alternative definition which would bound the overall time is as follows:

Definition 10.5.3 (Strong polynomial identification in the limit with probability one) A class of probabilistic grammars $\mathcal{G}$ is **strongly** polynomially identifiable in the limit with probability one *if*$_{def}$ there are an algorithm **A** and two polynomials p and q for which, for any grammar G in $\mathcal{G}$ and any $\delta > 0$:

1 **A** identifies G in the limit with probability one,
2 **A** works in time in $\mathcal{O}(p(\|\text{support}(S)\|, \frac{1}{\delta}))$, where S is a learning sample,
3 if $|S| \geq q(\|G\|, \frac{1}{\delta})$, **A** computes with probability at least $1 - \delta$ a representation H such that $\mathbb{L}(H) = \mathbb{L}(G)$.

The above definition takes into account two aspects. As before, the algorithm is required to identify in the limit and to work in time polynomial in the total lengths of the strings in the sample; furthermore, with high probability, identification is expected from a polynomial number of examples only.

We prove now that even in the case where we have only to choose one probability out of n, identification requires a number of strings that is too large. The argument is independent of the encoding of the grammars and of the priorities. Put simply it says that if you want to distinguish with high certainty between n probabilities, you are going to have to work from a sample of size exponential in n. To do so we do not make any assumption on the way the learning algorithm is to use the information it receives. Let us consider n probabilistic languages $\mathcal{D}_1, \mathcal{D}_2, \ldots, \mathcal{D}_n$ over a set of only two strings, x and y: precisely $Pr_{\mathcal{D}_i}(x) = p_i$ and $Pr_{\mathcal{D}_i}(y) = 1 - p_i$. We have also $\forall i, j \leq n,\ p_i \neq p_j$.

Proposition 10.5.4 *Let $\mathcal{L}$ be a class of distributions that contains distributions $\mathcal{D}_1$, $\mathcal{D}_2,\ldots,\mathcal{D}_n$ as above, let $\mathcal{G}$ be a grammar class for $\mathcal{L}$, let* **A** *be any algorithm that identifies $\mathcal{G}$ with probability one and m be an integer. Given any polynomial $p()$, there is a $\mathcal{D}_i$ in $\mathcal{L}$ such that the probability that $\mathbf{A}(X_{\mathcal{D}_i,m})$ returns a grammar H with $\mathbb{L}(H) = \mathcal{D}_i$ is at most $\frac{1}{p(m)}$.*

Proof [*sketch*] In the above definition, the algorithm may be randomised. One should note that the different distributions are all binomials and an algorithm is certain to make errors between D_i and D_j every time the samples coincide. This corresponds to computing an upper bound on the tails of the binomials. In order for **A** to distinguish each binomial with probability at least $1-\delta$, it needs an exponential number of strings in n. □

The above proposition seems to leave room for polynomial sampling to be sufficient in order to achieve identification in the limit with probability one, and do this with probability at least $1-\delta$, but we should remember that the number of samples here is polynomial with the number of possible probabilities. And if in a reasonable encoding scheme these are encoded in base 2, then we have a problem, since the number of different probabilities that can be encoded with k bits is 2^k.

As this applies to regular languages when represented by DPFA, we have as a corollary:

Theorem 10.5.5 *$\mathcal{DPFA}(\Sigma)$ is not strongly polynomially identifiable in the limit with probability one.*

In Section 10.9.3 we will propose restricting the number of possible probabilities in order to overcome this obstacle.

10.6 PAC-learning distributions

In the previous section, the criterion of identification with probability one suffers from the same problems as identification in the limit in the non-probabilistic setting: from a given sample it is not only impossible to know if identification has taken place, but even how far from being correct we are. In order to do better we turn to the PAC setting, which can be adapted to the problem of learning distributions: errors are no longer counted as misclassifications, but as distances between the expected probabilities and the real ones.

In turn, if we don't access the real (target) probabilities, these can be estimated from a sample.

Definition 10.6.1 (ϵ-good hypothesis) Let G be the target probabilistic grammar and H be a hypothesis grammar. Let $\mathbf{L}_\alpha$ be a distance between distributions. We say, for $\epsilon > 0$, that H is a $\mathbf{L}_\alpha$-ϵ-good hypothesis with respect to G *if*$_{def}$ $\mathbf{L}_\alpha\big(\mathbb{L}(G), \mathbb{L}(H)\big) \leq \epsilon$.

Note the importance of the choice of the distances:

Definition 10.6.2 (Polynomial $\mathbf{L}_\alpha$-PAC-learnable distributions) Let $\mathcal{G}$ be a class of probabilistic grammars. $\mathcal{G}$ is $\mathbf{L}_\alpha$-PAC-learnable *if*$_{def}$ there exists an algorithm **A** with the following property: For every grammar G in $\mathcal{G}$ of size at most n, for every $\epsilon > 0$ and $\delta > 0$, if **A** is given access to EX(m), m and n, ϵ and δ then with probability at least $1 - \delta$, **A** outputs a $\mathbf{L}_\alpha$-ϵ-good hypothesis with respect to G. If **A** runs in time polynomial in $\frac{1}{\epsilon}$, $\frac{1}{\delta}$, $|\Sigma|$, m and n we say that $\mathcal{G}$ is **polynomially $\mathbf{L}_\alpha$-PAC-learnable**.

The above definition's power will very much depend on the distance you use. For instance, using $\mathbf{L}_\infty$:

Proposition 10.6.1 $\mathcal{DPFA}(\Sigma)$ *is polynomially* $\mathbf{L}_\infty$-PAC-*learnable.*

Proof This is a consequence of Lemma 10.5.1. Let us consider the simple algorithm RETURNPTA (see Section 5.4.2, page 103) which, given a learning sample, constructs a DPFA that represents exactly the empirical distribution corresponding to the sample. Obviously, RETURNPTA is no good at identifying DPFAs. Nevertheless RETURNPTA will polynomially $\mathbf{L}_\infty$-PAC-learn DPFAs as, with high probability, the distance according to norm $\mathbf{L}_\infty$ between the empirical distribution and the target distribution converges very fast. □

On the other hand, it can be shown that the above result is no longer true when the distance is taken according to another norm. For instance, for the norms $\mathbf{L}_1$ and $\mathbf{L}_2$, a language which shares the mass of probabilities in a uniform way over an exponential number of strings will not be closely approximated by the empirical distribution drawn from only a polynomial number of strings.

We will not prove the following proposition here:

Proposition 10.6.2 $\mathcal{DPFA}(\Sigma)$ *is not polynomially* $\mathbf{L}_1$-PAC-*learnable.*

10.7 Learning distributions with queries

We first adapt the definitions from Section 7.5 about active learning to the case of learning distributions with probabilistic queries.

Definition 10.7.1 A class $\mathcal{G}$ is identifiable in the limit with probability one with queries from QUER *if*$_{def}$ there exists an algorithm **A** such that $\forall G \in \mathcal{G}$, **A** identifies G in the limit with probability one, i.e. returns with probability one a grammar G' equivalent to G and halts.

And the polynomial aspects are taken into account by the following definition (adapted from Section 7.5.3):

Formally, let QUER be a fixed set of queries, G be a grammar class and **A** be a learning algorithm. Suppose that to identify G, algorithm **A** makes on a particular run ρ queries $Q_1, \ldots, Q_n$. We denote by LQ(ρ, i) the length of the longest counter-example (or information) returned by the queries $Q_1, \ldots, Q_i$ in ρ, and RT(ρ, i) the running time of **A** before interrogating the Oracle with query Q_i. LQ($\rho, n+1$) and RT($\rho, n+1$) refer to the values of LQ and RT at the end of the entire run ρ.

Definition 10.7.2 A class of grammars is **polynomially identifiable** in the limit with probability one with queries from QUER if_{def} there exists an algorithm **A** and a polynomial $p()$ for which $\forall G \in \mathcal{G}$, **A** identifies G in the limit with probability one, and, for any valid run ρ with n queries, we have

$$\forall i \leq n+1,\ \mathrm{RT}(\rho, i) \leq p(\|G\|,\ \mathrm{LQ}(\rho, i)).$$

Let ρ be a run of a learner **A**. Let $\langle r_1, r_2, \ldots, r_m\rangle$ be the sequence of ***replies*** to queries $\langle \mathrm{Q}_1, \mathrm{Q}_2, \ldots, \mathrm{Q}_m\rangle$ the Oracle makes during run ρ. We will say that **A** is ***polynomially bounded*** if_{def} there exists a polynomial $p(,)$, which, with probability one, when given any target grammar G, any run ρ, at any query point k of the run, denoting the runtime up to that point by t_k, we have:

- $|\mathrm{Q}_k| \leq p(\|G\|, \max\{|r_i| : i < k\})$,
- $t_k \in \mathcal{O}(p(\|G\|, \max\{|r_i| : i < k\}))$.

10.7.1 Extended membership queries

An extended membership query is made by providing to the Oracle a string x. The Oracle then has to return the probability of x.

It is easy to show that DPFAs are learnable from extended membership queries only if DFAs are learnable from membership queries, which is known not to be the case as shown in Section 9.2.

Theorem 10.7.1 *$\mathcal{DPFA}(\Sigma)$ is not identifiable in the limit with probability one by a polynomially bounded learner from extended membership queries only.*

Proof If not, $\mathcal{DFA}(\Sigma)$ would be identifiable from membership queries. We transform a completed DFA $\mathcal{A} = (Q, \Sigma, \delta, q_\lambda, F)$ into a DPFA $\mathcal{B} = (Q', \Sigma', \delta_{\mathbb{P}}, q'_\lambda, \mathbb{F}_{\mathbb{P}})$ as follows:

- $Q' = Q \cup \{q_f\}$;
- $\Sigma' = \Sigma \cup \{'+', '-'\}$;
- $\forall q \in Q, \delta_{\mathbb{P}}(q, '+') = q_f, \delta_{\mathbb{P}}(q, '-') = q_f$,
  $\forall q \in Q, \forall a \in \Sigma, \delta_{\mathbb{P}}(q, a) = \delta(q, a)$;
- $q'_\lambda = q_\lambda$;
- $\forall q \in Q, \mathbb{F}_{\mathbb{P}}(q) = 0, \mathbb{F}_{\mathbb{P}}(q_f) = 1$;
- $\forall q \in Q, \forall a \in \Sigma, \mathbb{F}_{\mathbb{P}}(q, a) = \frac{1}{2\cdot|\Sigma|}, \forall a \in \Sigma, \mathbb{F}_{\mathbb{P}}(q_f, a) = 0$,
  if $q \in F$ then $\mathbb{F}_{\mathbb{P}}(q, '+') = \frac{1}{2}, \mathbb{F}_{\mathbb{P}}(q, '-') = 0$,
  if $q \notin F$ then $\mathbb{F}_{\mathbb{P}}(q, '+') = 0, \mathbb{F}_{\mathbb{P}}(q, '-') = \frac{1}{2}$.

The above construction is made from a completed DFA, i.e. a DFA to which eventually an extra non-final state has been added and is reached by all absent transitions. This ensures that through the construction we have $w \in L_A \Leftrightarrow Pr_B(w+) = \frac{(2\cdot|\Sigma|)^{-|w|}}{2}$. An extended membership query therefore gives us the same information on the underlying DPFA as a membership query would. □

10.7.2 Extended prefix language queries

An extended prefix language query is made by submitting to the Oracle a string w. The Oracle then returns the probability $Pr(w\,\Sigma^\star)$. It can be noticed that an extended membership query can easily be simulated through $|\Sigma|$ extended prefix language queries.

Theorem 10.7.2 *$\mathcal{DPFA}(\Sigma)$ is not identifiable in the limit with probability one by a polynomially bounded learner from extended prefix language queries only.*

Proof Let $w \in \Sigma^n$ and consider the following probabilistic language $\mathcal{D}_w$: $\forall x \in \Sigma^n, x = w \Rightarrow p_{\mathcal{D}_w}(x) = 0,\ x \neq w \Rightarrow p_{\mathcal{D}_w}(x) = \frac{1}{2^n}$, $p_{\mathcal{D}_w}(wa) = \frac{1}{2^n}$. For all other strings $p_{\mathcal{D}_w}(x) = 0$.

This language is recognised by a DPFA with at most $2n + 2$ states. Call $\mathcal{L}_n$ the set of all languages $\mathcal{D}_w$ with $w \in \Sigma^n$. Now let the Oracle answer each extended prefix language query 'x' with the quantity $\frac{1}{2^{|x|}}$ if $|x| \leq n$, 0 if not. Then it is straightforward that in the worst case 2^n queries are needed. □

10.8 Exercises

10.1 Compute, for $\epsilon = 0.1$ and $\delta = 0.1$, the values of n derived from the Hoeffding bound and the Chernov additive bound.

10.2 Write the algorithm NEXT_FRACTION that can enumerate all fractions, used in Algorithm 10.1 (page 203).

10.3 Compute how many examples are needed to separate two probabilities. Given $\delta > 0$, how large does a sample have to be in order to tell with probability at least $1 - \delta$ if we are facing one probability or the other? Obviously, if we can choose the two probabilities, the question can be trivial. But what happens in a general case?

10.4 Consider the following game. Player 1 chooses k coins, each with its bias (the probability of 'heads'). Then he gives the coins to Player 2 who chooses one of these coins and gives it back to Player 1. Player 1 must now toss the coin as many times as he needs and when he is sure, with probability at least $1-\delta$ that he knows which coin he has, he must name it. How should Player 1 choose the coins for k=2, 3 and 4? What about the general case?

10.5 Derive from the preceding exercise the minimum number of tosses Player 1 has to make. Note that this gives you the detailed proof of Proposition 10.5.4.

10.6 Give an algorithm that learns DPFAs from both extended membership queries and distribution equivalence queries. A distribution equivalence query would be an equivalence query over distributions, of course.

10.7 Prove that the Stern-Brocot encoding of strings is reasonable in the sense of Definition 7.3.1.

10.9 Conclusions of the chapter and further reading

10.9.1 Bibliographical background

Learning probabilistic grammars was what allowed Ray Solomonoff to introduce his ideas concerning the questions of inductive inference which led to his definitions of intrinsic complexity (Solomonoff, 1997). Jim Horning (Horning, 1969) proved that probabilistic context-free grammars were identifiable with probability one and also proposed an alternative empirical algorithm, relying on finding a (context-free) grammar giving a good compromise between the quality of the probabilities and its simplicity. This approach was followed several times since, with for example algorithm MDI by Franck Thollard *et al.* (Thollard, Dupont & de la Higuera, 2000).

The bounds we give and use in Section 10.2 are standard in computational learning theory (Kearns & Vazirani, 1994). It should be noted that they are the simplest ones, and that further work on these questions would deserve stronger bounds, but also a more detailed analysis of the properties of the distributions one generates with automata. There are many papers and books on the topic of statistical machine learning where this sort of information can be found. PAC-learning is presented in Section 10.4.

An important PAC study concerning the problem of approximating distributions, written by Naoki Abe and Manfred Warmuth (Abe & Warmuth, 1992), shows that even the seemingly simpler problem of estimating the probabilities correctly, given the structure provided by a finite non-deterministic automaton, is hard.

When the structure is unknown, even for the case of DFAs, most results are negative: Michael Kearns and Leslie Valiant (Kearns & Valiant, 1989) linked the difficulty of learning DFAs with that of solving cryptographic problems believed to be intractable (a nice proof is published in Michael Kearns and Umesh Vazirani's book (Kearns & Vazirani, 1994)). Another testimony of the hardness of approximation can be found when examining the different competitions that have been organised for grammatical inference: in each case an error of 1% was allowed, but didn't seem to help.

The model has been adapted by taking into account only certain types of distributions and simple PAC-learning has been considered with more success (Denis, 2001, Denis & Gilleron, 1997). If languages are defined in another way (through equations, these are then sometimes called semi-linear sets), one can obtain some positive PAC-like results (Abe, 1995).

The question of identification in the limit with probability one is studied in (Angluin, 1988). The analysis in this setting of the well-known algorithm ALERGIA is given in (Carrasco & Oncina, 1994b, 1999, de la Higuera, 1998). Adaptation to a special class of context-free grammars is in (de la Higuera & Oncina, 2003), where it is also noted that weak polynomial identification in the limit (Definition 10.5.2, page 205) is insufficient. An alternative analysis can be found in (de la Higuera & Oncina, 2004).

Stern-Brocot trees were used in (de la Higuera & Thollard, 2000) to identify the probabilities of a DPFA. They were introduced in the nineteenth century (Brocot, 1861, Stern, 1858), and then used in computer science (Graham, Knuth & Patashnik, 1994).

Theorem 10.5.2 is based on work by Lenny Pitt and Dana Angluin (Angluin, 1988, Pitt, 1989).

Proposition 10.6.2 is proved in Omri Guttman's thesis (Guttman, 2006), as an extension of a result by Michael Kearns *et al.* (Kearns *et al.*, 1994) in the case of using KL-divergence. In both cases the proof relies on parity functions (presented in the non-probabilistic setting in Section 6.5.1).

Positive results in PAC-learning DPFAs depend on many parameters. The question of dealing with edges in the automaton used by too few strings, and also the question of dealing with the unbounded length of the strings one can get through sampling, are dealt with in different ways (Guttman, Vishwanathan & Williamson, 2005, Palmer & Goldberg, 2005, Thollard & Clark, 2004).

There are few results mixing queries and PAC-learning. Whereas membership queries allow PAC-learning DFAs (Angluin, 1987a), it is proved, under cryptology assumptions, that this is neither the case for NFAs nor context-free grammars (Angluin & Kharitonov, 1991).

We presented in Section 10.7 some results concerning learning distributions with queries. Most results are from (de la Higuera & Oncina, 2004). Alternative sources are (Bergadano & Varricchio, 1996, Guttman, Vishwanathan & Williamson, 2005).

10.9.2 Some alternative lines of research

Simple-PAC and PACS are two lines of research that have been explored, but for different reasons have not prospered. The PACS and simple PAC settings mentioned in the introduction of the chapter rely on work by Ming Li and Paul Vitanyi (Li & Vitanyi, 1991, 1993). One can also read the papers about PACS by Jorge Castro and David Guijarro (Castro & Guijarro, 2000), or François Denis *et al.* (Denis, 2001, Denis & Gilleron, 1997, Denis, d'Halluin & Gilleron, 1996).

One crucial question before using a bias in learning distributions is to see how far off we are going to be, even if we are successful in learning. The question therefore becomes: given a target from one class of distributions, how bad is the closest distribution in a different class? Initial results by Omri Guttman (Guttman, 2006) show that for a regular distribution to approximate an arbitrary distribution over strings, the size of the automaton may have to be exponentially large.

10.9.3 Open problems and possible new lines of research

The importance of using one distance instead of another has only recently started to be understood. Exploratory papers on this topic are (Guttman, Vishwanathan & Williamson, 2005, Palmer & Goldberg, 2005, Thollard & Clark, 2004). These have led the authors to state conditions for restricted classes of DPFAs to be PAC-learnable. But the question concerning the entire class of the DPFAs is open.

Specific sampling queries. Extended membership queries were introduced in (Bergadano & Varricchio, 1996): the learning algorithm may ask for the value $p(x)$ on strings x of its choice.

We may consider refining the concept to the case where the answer to the query can be an approximate answer, obtained through a particular form of sampling. A specific sampling query is made by submitting a pattern π: the Oracle draws a string matching pattern π sampled according to the distribution $\mathcal{D}$. Specific sampling queries are intended to fit the idea that the user can ask for examples matching some pattern he or she is interested in.

For example, the specific sampling query ($\mathtt{a}\,\Sigma^\star\,\mathtt{b}$) requires an example starting with $\mathtt{a}$ and ending with $\mathtt{b}$, sampled following the distribution induced by $\mathcal{D}(\mathtt{a}\,\Sigma^\star\,\mathtt{b})$. We suggest that these queries could be explored in an attempt to help us learn the class of DPFAs.

Fixing the number of probabilities. We suggest limiting the set of probabilities of the automaton in order to be able to obtain positive learning results. The idea here is to learn DPFAs but where the probabilities are chosen inside a fixed set. A typical definition could be:

Definition 10.9.1 (K – DPFA) A K – DPFA $\mathcal{A}$ is a DPFA $(Q, \Sigma, \delta_{\mathbb{P}}, q_\lambda, \mathbb{F}_{\mathbb{P}}, P)$, where all probabilities (transition and final) are from a set $\mathcal{P}_K = \{\frac{i}{K} : 0 \leq i \leq K\}$.

The idea here is that, the variance having diminished, the class of K – DPFAs might now be learnable in a variety of settings. More precisely we conjecture that the class is PAC-learnable for the distances $\mathbf{L}_1$ and $\mathbf{L}_2$.

PART III

Learning Algorithms and Techniques

11
Text learners

> L'homme est à la recherche d'un nouveau langage auquel la grammaire d'aucune langue n'aura rien à dire.
>
> ***Guillaume Apollinaire***

> The method that will be used is equivalent to finding a PSL (context free phrase structure language, Chomsky (1956)) that in some sense best 'fits' the set $[\alpha 1]$. The measure of goodness of fit will be the product of the a-priori probability of the language selected and the probability that the language selected would produce $[\alpha 1]$ as a set of acceptable sentences.
>
> ***Ray Solomonoff (Solomonoff, 1964)***

From the results of the previous section, it appears quite clear that DFAs cannot be identified in the limit from text. At which point, if the chosen task is indeed to learn DFAs from text, we need either some help (perhaps some extra information about the strings or some structure) or else we will have to add an extra bias and reduce the class of automata.

It should be noted that adding more bias consists also of adding some more information (something like *'we know that the automaton has this property'*).

11.1 Window languages

It is well known that regular languages correspond to the class of languages that can be parsed using a bounded memory. But what happens is that this memory is bounded *a priori*. We may consider the subclass of languages for which parsing uses only a memory of size k, meaning that the next letter to be read will depend only on the knowledge of the $k-1$ previous characters.

Formally, this means that the following holds for some k:

$$\forall l, l' \in \Sigma^\star, \forall w \in \Sigma^{\geq k},\ lw \in L \Longleftrightarrow l'w \in L.$$

Better said, the prefixes l and l' can be 'forgotten'.

A language L for which the above equation holds will be called a k-testable language.

Definition 11.1.1 A regular language L is a ***k-testable language*** *if$_{def}$* given every pair of strings lw and $l'w$ in $\Sigma^\star$ such that $l, l', w \in \Sigma^\star$, $|w| \geq k$, the following holds: $lw \in L \iff l'w \in L$.

It should be noted that for strings of length less than k nothing is said. But then for a fixed k there are only a finite number of such strings.

11.1.1 Definition of k-testable machine

To recognise k-testable languages, as introduced in Definition 11.1.1 above, we need special machines:

Definition 11.1.2 Given $k > 0$, a ***k*-testable machine** ($k - \mathcal{TSS}$) is a 5-tuple $Z_k = \langle \Sigma, I, F, T, C \rangle$ with:

- Σ a finite alphabet,
- $I, F \subseteq \Sigma^{k-1}$ (prefixes of length $k-1$ and suffixes (or finals) of length $k-1$),
- $C \subseteq \Sigma^{<k}$ (short strings),
- $T \subseteq \Sigma^k$ (allowed segments).

Definition 11.1.3 Given a k-testable machine, the k-testable language recognised by $Z_k = \langle \Sigma, I, F, T, C \rangle$ is $\mathbb{L}(Z_k) = (I\,\Sigma^\star \cap \Sigma^\star F - \Sigma^\star(\Sigma^k - T)\,\Sigma^\star) \cup C$.

In words, this means that the only admissible strings are those either corresponding exactly to strings in C, or those whose prefix of length $k-1$ is in I, whose suffix of length $k-1$ is in F, and where all substrings of length k belong to T. In other words there are two types of strings: those of length less that k and defined exactly as such in C and those longer than k for which $\Sigma^k - T$ defines the prohibited segments. The key idea is to use as a parser a window of size k to reject the strings that at some moment do not comply with these conditions.

Definition 11.1.4 $k - \mathcal{TSS}_{\mathcal{G}}(\Sigma)$ is the class of all k-testable machines over alphabet Σ. $k - \mathcal{TSS}_{\mathcal{L}}(\Sigma) = \{L \subseteq \Sigma^\star : L \text{ is } k - \mathcal{TSS}\}$.

Given a $k - \mathcal{TSS}$ machine, a DFA recognising the language generated is constructed by Algorithm 11.1. The algorithm just converts the prefixes of the sets I and C into strings and makes final the states corresponding to the elements in F. More interesting is the way set T is used: this set contains strings of length k, whereas the states in the DFA correspond to strings of length (at most) $k-1$. So the algorithm makes two states out of each string u in T: one is $u(1)^{-1} \cdot u$ and the other is $u \cdot u(|u|)^{-1}$. A transition labelled by $u(|u|)^{-1}$ links these two states.

Example 11.1.1 The automaton from Figure 11.1 recognises the language $\mathtt{a} + \mathtt{aa^*b(ab)^*}$. The language is clearly $3 - \mathcal{TSS}$ but is not $2 - \mathcal{TSS}$: with a window of size only two, we would not be able to avoid substring bb. The meaning of state $q_{\mathtt{ab}}$ is no

Algorithm 11.1: Building a DFA from a $k - \mathcal{TSS}$ machine.

Input: a $k - \mathcal{TSS}$ machine $\langle \Sigma, I, F, T, C \rangle$
Output: a DFA $\langle \Sigma, Q, q_\lambda, F_\mathcal{A}, \delta \rangle$
$Q \leftarrow \emptyset$;
$F_\mathcal{A} \leftarrow \emptyset$;
for $pu \in I \cup C,\ p, u \in \Sigma^\star$ **do** $Q \leftarrow Q \cup \{q_p\}$;
for $au \in T,\ a \in \Sigma,\ u \in \Sigma^\star$ **do** $Q \leftarrow Q \cup \{q_u\}$;
for $ua \in T,\ a \in \Sigma,\ u \in \Sigma^\star$ **do** $Q \leftarrow Q \cup \{q_u\}$;
for $pau \in I \cup C,\ a \in \Sigma,\ p, u \in \Sigma^\star$ **do** $\delta(q_p, a) = q_{pa}$;
for $aub \in T,\ a, b \in \Sigma,\ u \in \Sigma^\star$ **do** $\delta(q_{au}, b) = q_{ub}$;
for $u \in F \cup C$ **do** $F_\mathcal{A} \leftarrow F_\mathcal{A} \cup \{q_u\}$;
return $\langle \Sigma, Q, q_\lambda, F_\mathcal{A}, \delta \rangle$

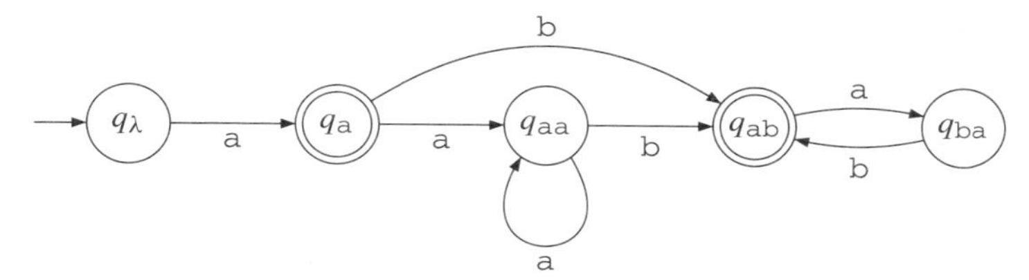

Fig. 11.1. A DFA corresponding to the $3-\mathcal{TSS}$ machine $Z_3 = \langle\{\mathrm{a}, \mathrm{b}\}, I = \{\mathrm{aa}, \mathrm{ab}\}, F = \{\mathrm{ab}\}, C = \{\mathrm{a}\}, T = \{\mathrm{aaa}, \mathrm{aab}, \mathrm{aba}, \mathrm{bab}\rangle\}$.

longer that $\delta(q_\lambda, \mathrm{ab}) = q_{\mathrm{ab}}$ but rather that the only string of length 2 that reaches state q_{ab} is (reached from any state) string ab.

11.1.2 Properties of the k-testable languages

The hierarchy of $k - \mathcal{TSS}$ languages has the following properties:

Proposition 11.1.1

- *If L is finite then $L \in k - \mathcal{TSS}$ where $k = 1 + \max\{|u| : u \in L\}$.*
- $k - \mathcal{TSS}(\Sigma) \subset [k+1] - \mathcal{TSS}(\Sigma)$.
- *There are regular languages that are not $k - \mathcal{TSS}$ (for any k).*

Proof

- All finite languages are in $k - \mathcal{TSS}(\Sigma)$ if k is large enough: let L be a finite language with $k > \max\{|u| : u \in L\}$. Then the machine $Z_k = \langle \Sigma, I, F, T, C \rangle$ with $I = \mathrm{PREF}(L)$, $F = T = \emptyset$ and $C = L$ accepts exactly L.
- $(\mathrm{ba}^k)^* \in [k+1] - \mathcal{TSS}(\Sigma)$ but $(\mathrm{ba}^k)^* \notin k - \mathcal{TSS}(\Sigma)$.
- Consider the language described by the regular expression $\mathrm{a}\,\Sigma^\star\,\mathrm{a} + \mathrm{b}\,\Sigma^\star\,\mathrm{b}$. This language is not $k - \mathcal{TSS}$ (for any k). □

Algorithm 11.2: $\mathbf{A}_{k-\mathcal{TSS}}$.

Input: a sample S
Output: a $k-\mathcal{TSS}$ machine $\langle\Sigma, I(S), F(S), T(S), C(S)\rangle$
Σ is the alphabet used in S;
$I(S) \leftarrow \Sigma^{k-1} \cap \text{PREF}(S)$;
$C(S) \leftarrow \Sigma^{<k} \cap S$;
$F(S) \leftarrow \Sigma^{k-1} \cap \text{SUFF}(S)$;
$T(S) \leftarrow \Sigma^{k} \cap \{v : uvw \in S\}$;
return $\langle\Sigma, I(S), F(S), T(S), C(S)\rangle$

11.1.3 The algorithm

Learning k-testable languages is really only about finding the prefixes, substrings and suffixes that occur in the data.

This can be done through the simple Algorithm $\mathbf{A}_{k-\mathcal{TSS}}$, 11.2.

11.1.4 Running the algorithm

Let us run Algorithm 11.2 on an example:

Example 11.1.2 Let $S = \{\texttt{a}, \texttt{aa}, \texttt{abba}, \texttt{abbbba}\}$ be our learning sample and suppose we choose $k = 3$. We will discuss the issue of how we should choose k later. It follows by construction that:

- $\Sigma = \{\texttt{a}, \texttt{b}\}$
- $I(S) = \{\texttt{aa}, \texttt{ab}\}$
- $F(S) = \{\texttt{a}, \texttt{aa}\}$
- $T(S) = \{\texttt{abb}, \texttt{bbb}, \texttt{bba}\}$
- $C(S) = \{\texttt{a}, \texttt{aa}\}$

Building the corresponding automaton using Algorithm 11.1 once the machine is given is straightforward (see Figure 11.2).

Hence $\mathbf{A}_{3-\mathcal{TSS}}(S) = \texttt{a} + \texttt{aa} + \texttt{abbb}^*\texttt{a}$.

Example 11.1.3 In Figure 11.3 we show the corresponding automata for $k = 1$, $k = 2$ and $k > 6$.

11.1.5 Some properties of the algorithm

Proposition 11.1.2 *$\mathbf{A}_{k-\mathcal{TSS}}(S)$ is the smallest $k - \mathcal{TSS}$ language that contains S.*

Proof If there is a smaller one, some prefix, suffix or substring has to be absent. □

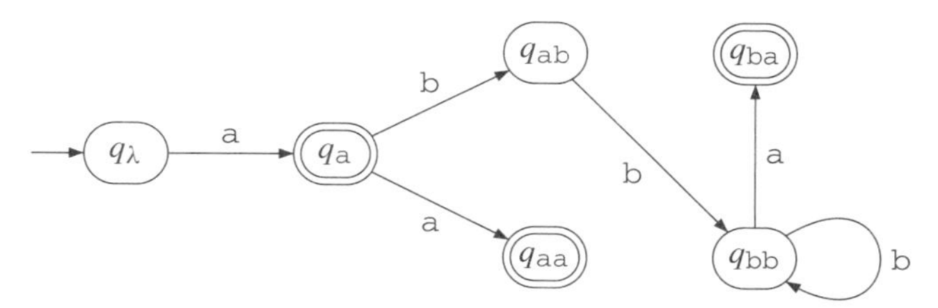

Fig. 11.2. Automaton learnt from sample $S = \{$a, aa, abba, abbbba$\}$.

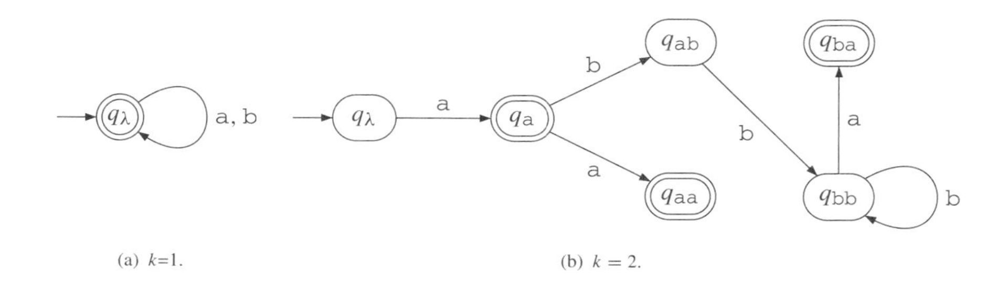

(a) k=1. (b) $k = 2$.

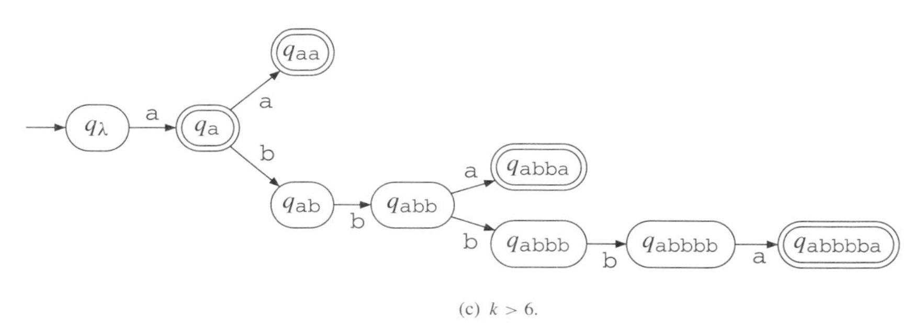

(c) $k > 6$.

Fig. 11.3. Alternative automata learnt from the sample $S = \{$a, aa, abba, abbbba$\}$.

Proposition 11.1.3

- $\mathbf{A}_{[k+1]-\mathcal{TSS}}(S) \subseteq \mathbf{A}_{k-\mathcal{TSS}}(S)$,
- $\forall k > \max\{|x| : x \in S\}$, $\mathbf{A}_{k-\mathcal{TSS}}(S) = S$,
- *If* $Y \subseteq S$, *then* $\mathbf{A}_{k-\mathcal{TSS}}(Y) \subseteq \mathbf{A}_{k-\mathcal{TSS}}(S)$.

Proof Let I_{k+1} (resp. F_{k+1} and T_{k+1}) denote the sets obtained by $\mathbf{A}_{[k+1]-\mathcal{TSS}}(S)$ and I_k (resp. F_k and T_k) those by $\mathbf{A}_{k-\mathcal{TSS}}(S)$. There are fewer allowed prefixes (resp. suffixes or substrings) in I_{k+1} (resp. F_{k+1} and T_{k+1}) than in I_k (resp. F_k and T_k). Therefore, the smaller the value of k, the more the algorithm generalises. The second point is a direct consequence of the construction provided in Proposition 11.1.1. The third item corresponds to the fact that the algorithm simply counts what it sees. Therefore, if during learning, a prefix, a suffix or a substring is found, the algorithm generalises further. □

Proposition 11.1.4 $\mathbf{A}_{k-\mathcal{TSS}}$ *identifies any* $k - \mathcal{TSS}$ *language in the limit from polynomial data, is consistent and conservative and admits* polynomial characteristic samples.

Proof Once all the prefixes, suffixes and substrings have been seen, the correct automaton is returned. Identification in the limit is then an easy consequence of the fact that the language is the smallest. □

Let us denote in a standard way by $\|Z_k\|$ the size of a k-testable machine: This corresponds to the sum of the lengths of the different strings in I, F, C or T. The algorithm is conservative because if the current machine can parse a new string, that means that all the substrings of this string are in the machine, so nothing will be added. On the other hand, if it cannot parse, then some substring is missing which the algorithm will add.

Proposition 11.1.5 (Complexity issues)

- *The runtime of* $\mathbf{A}_{[k]-\mathcal{TSS}}(S)$ *is in* $\mathcal{O}(\|S\|)$.
- *Every* $k - \mathcal{TSS}$ *language admits a characteristic set of size linear with* $\|Z_k\|$.
- *Algorithm* $\mathbf{A}_{[k]-\mathcal{TSS}}$ *makes* $\|Z_k\|$ *implicit prediction errors.*

Proof

- Algorithm $\mathbf{A}_{[k]-\mathcal{TSS}}$ parses the data from left to right.
- The characteristic set must contain strings that exercise all the substrings and the prefixes and suffixes.
- Even if the algorithm is not incremental *per se* it can be easily considered as such. A prediction error is made whenever the current machine can't parse a new example. In this case something is added to either I, F, or T. And the number of such additions is bounded by $\|Z_k\|$. □

It should be noted that defining a k-testable language through the prohibited substrings would lead to complexity problems: remember that the number of strings of length k over an alphabet containing at least two symbols is exponential in k.

11.1.6 Conclusion

Algorithm 11.2 is very simple and straightforward, yet a number of questions remain to be solved:

- There is no room for noise in the scheme: a substring that appears just once and that is completely wrong (with many absent substrings) will introduce many states in the resulting automaton.
- Variants where we count the number of occurrences of each substring lead to statistical models called n-grams; they are also comparable with the spectrum kernels described in Definition 3.5.3 (page 61).
- How to choose k is a crucial issue. Choose too small a k and there is a risk of over-generalisation. Choose too large a k and inversely one may not generalise enough.

- One can use counter-examples to find the smallest k, i.e. to control generalisation (see Exercise 11.8).
- The size of the alphabet is also of great importance: if the size is too small, the resulting automaton (whose number of states is at most $|\Sigma|^k$) may be too small. Having too large an alphabet may result in having to deal with problems of sparseness of data.

11.2 Look-ahead languages

We introduce another subclass of the regular languages for which positive results can also be obtained: the k-reversible languages. The class is composed of regular languages that can be accepted by a DFA such that its reverse is deterministic with a look-ahead of k.

11.2.1 Definition of reversible languages

When reversing the arrows in a DFA we obtain an NFA. This NFA is where the k-reversibility property is best understood. We therefore give the definitions in the context of non-deterministic automata. On the other hand, as we are not working with negative data, we do not need to have rejecting states (see page 71), so we will take $\mathbb{F}_{\mathbb{R}} = \emptyset$. Thus we will not represent the rejecting states ($\mathbb{F}_{\mathbb{R}}$).

Definition 11.2.1 (Reversal automaton) Let $\mathcal{A} = \langle \Sigma, Q, \mathbb{I}, \mathbb{F}_{\mathbb{A}}, \delta_N \rangle$ be an NFA. The **reversal automaton** of $\mathcal{A}$, $\mathcal{A}^T$, is the NFA $\langle \Sigma, Q, \mathbb{F}_{\mathbb{A}}, \mathbb{I}, \delta_N^T \rangle$ with $\delta_N^T(q, a) = \{q' \in Q : q \in \delta_N(q', a)\}$.

We now define as k-successors of a given state the strings of length k that correspond to 'live' paths in the NFA, when parsing from that state.

Definition 11.2.2

- u is a ***k*-successor** of q if_{def} $|u| = k$ and $\delta(q, u) \neq \emptyset$,
- u is a ***k*-predecessor** of q if_{def} $|u| = k$ and $\delta^T(q, u^T) \neq \emptyset$,
- $\text{SUCC}_k(q)$ (respectively $\text{PRED}_k(q)$) is the set of all k-successors of q (respectively all k-predecessors of q).

Note that λ is a 0-successor and 0-predecessor of any state.

Definition 11.2.3 Two states q and q' in an NFA $\mathcal{A} = \langle \Sigma, Q, \mathbb{I}, \mathbb{F}_{\mathbb{A}}, \delta_N \rangle$ are k-**ambiguous** if_{def} $\text{SUCC}_k(q) \cap \text{SUCC}_k(q') \neq \emptyset$.

Two states are therefore k-ambiguous whenever there is a common possible path of length k starting from each. It means that one cannot disambiguate by just looking at the following k symbols. Note that in Figure 11.4, states q_1 and q_3 are k-ambiguous for any value of k.

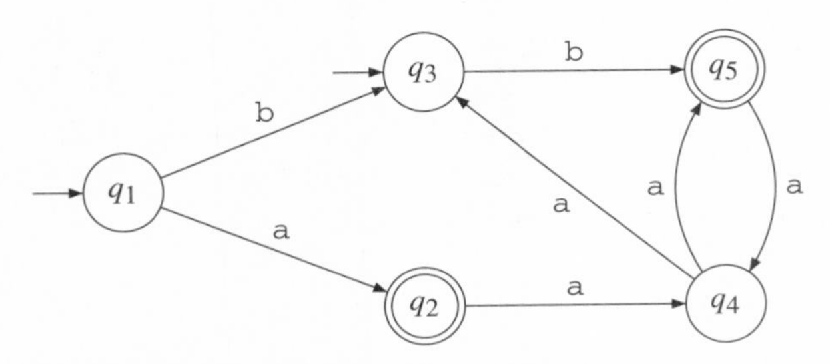

Fig. 11.4. An NFA: bb is a 2-successor of q_1 but not of q_3. a is a 1-successor of q_4. aa is a 2-predecessor of q_4 but not of q_2.

Definition 11.2.4 An NFA $\mathcal{A} = \langle \Sigma, Q, \mathbb{I}, \mathbb{F}_{\mathbb{A}}, \delta_N \rangle$ is **deterministic with look-ahead** k if_{def} given any two different states q and q' in Q, such that either $q, q' \in I$ or $q, q' \in \delta_N(p, a)$ (for some p in Q), q and q' are not k-ambiguous.

In other words, any two states which may result in a conflicting non-deterministic situation should not be k-ambiguous.

Example 11.2.1 The automaton represented in Figure 11.4 is not deterministic with look-ahead 1 because $\texttt{b} \in \text{SUCC}_2(q_1)$ and $\texttt{b} \in \text{SUCC}_2(q_3)$ with $q_1, q_3 \in Q$. But it is deterministic with look-ahead 2. We can compute the SUCC_2 sets: $\text{SUCC}_2(q_1) = \{\texttt{aa}, \texttt{bb}\}$, $\text{SUCC}_2(q_2) = \{\texttt{aa}\}$, $\text{SUCC}_2(q_3) = \{\texttt{ba}\}$, $\text{SUCC}_2(q_4) = \{\texttt{aa}, \texttt{ab}\}$ and $\text{SUCC}_2(q_5) = \{\texttt{aa}\}$. The two ambiguous situations arise between q_1 and q_3 on one hand and between q_3 and q_5 on the other.

Definition 11.2.5 $\mathcal{A}$ is k-**reversible** if_{def} $\mathcal{A}$ is deterministic and $\mathcal{A}^T$ is deterministic with look-ahead k.

An alternative definition can be given using predecessor sets and thus without using the reversal automaton:

Definition 11.2.6 $\mathcal{A}$ is k-reversible if_{def} $\mathcal{A}$ is deterministic and

$$\forall q, q' \in \mathbb{F}_{\mathbb{A}}, q \neq q' \Rightarrow \text{PRED}_k(q) \cap \text{PRED}_k(q') = \emptyset,$$

$$\delta(q, a) = \delta(q', a), q \neq q' \Rightarrow \text{PRED}_k(q) \cap \text{PRED}_k(q') = \emptyset.$$

Example 11.2.2 We show in Figure 11.5 the case of a 3-reversible automaton. Notice that it is not 2-reversible since $\text{PRED}_2(q_2) \cap \text{PRED}_2(q_5) = \{\texttt{aa}\}$ and both q_2 and q_5 are final. On the other hand we can compute the PRED_3 sets: $\text{PRED}_3(q_1) = \text{PRED}_3(q_3) = \emptyset$, $\text{PRED}_3(q_2) = \{\texttt{aaa}\}$, $\text{PRED}_3(q_4) = \{\texttt{aab}, \texttt{abb}, \texttt{bab}, \texttt{bbb}\}$ and $\text{PRED}_3(q_5) = \{\texttt{aaa}, \texttt{baa}\}$. The ambiguous situations arise between q_2 and q_5 on one hand (because they are both final) and between the pairs (q_1, q_2) (because $\delta(q_1, \texttt{a}) = \delta(q_2, \texttt{a})$) and (q_4, q_5) (because $\delta(q_4, \texttt{b}) = \delta(q_5, \texttt{b})$). Each ambiguous pair is tested and an empty intersection of the predecessor sets is found.

What makes a DFA not be k-reversible is the violation of one of the rules.

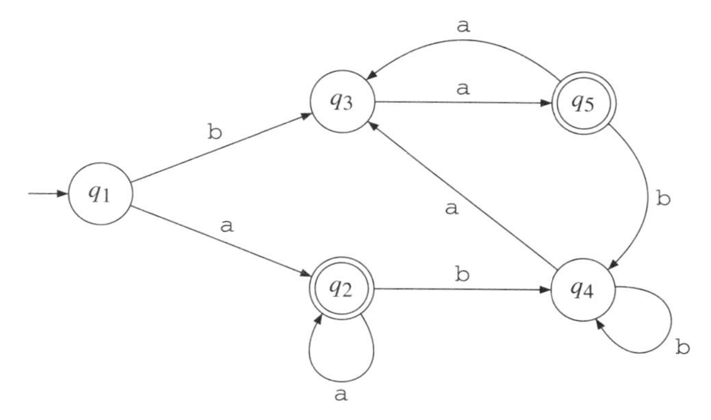

Fig. 11.5. A DFA that is 3-reversible.

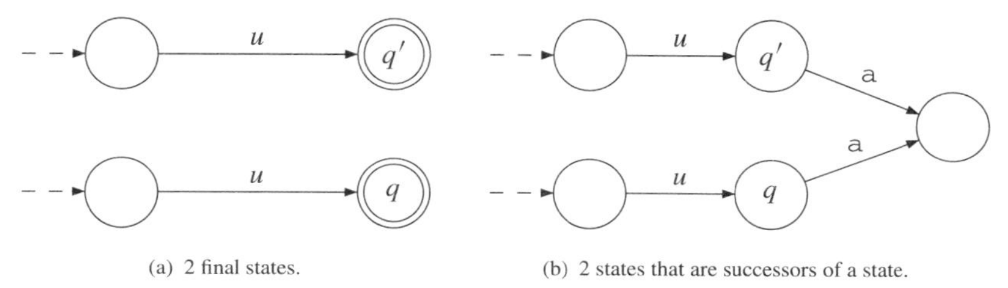

(a) 2 final states. (b) 2 states that are successors of a state.

Fig. 11.6. Prohibited patterns, for $|u| = k$.

Definition 11.2.7 Two states q, q' **violate** the k-reversibility condition if_{def}

- they violate the determinism condition: $\exists q'' \in Q : q, q' \in \delta(q'', a)$,
- or they violate the look-ahead condition:
  - $q, q' \in \mathbb{F}_{\mathbb{A}} \wedge \exists u \in \Sigma^k : u$ is k-predecessor of both,
  - $\exists u \in \Sigma^k, \delta(q, a) = \delta(q', a)$ and u is k-predecessor of both q and q'.

We will call two such states ***violators***.

The idea is to be able, through a look-ahead of length k, to decide what the next state should be. The prohibited situations are depicted in Figures 11.6(a) and 11.6(b). Therefore, for each value of k, there are languages that admit k-reversible automata as recognisers.

Definition 11.2.8
$k - \mathcal{REV}_{\mathcal{G}}(\Sigma)$ is the class of all k-reversible automata over the alphabet Σ.
$k - \mathcal{REV}_{\mathcal{L}}(\Sigma) = \{\mathbb{L}(\mathcal{A}) : \mathcal{A} \in k - \mathcal{REV}_{\mathcal{G}}(\Sigma)\}$.

It should be noted that, similarly to the case of the k-testable languages, there is a hierarchy of $k - \mathcal{REV}_{\mathcal{L}}(\Sigma)$, and there are also languages that are not k-reversible (given any value of k).

Proposition 11.2.1 (Properties of the k-reversible languages)
Given any alphabet Σ,

- $k - \mathcal{REV}_{\mathcal{L}}(\Sigma) \subsetneq (k+1) - \mathcal{REV}_{\mathcal{L}}(\Sigma)$.
- *There is a language* L *over* Σ *such that* $\forall k \in \mathbb{N},\ L \notin k - \mathcal{REV}_{\mathcal{L}}(\Sigma)$.

Proof First, it is clear that using a look-ahead of length $k+1$ is going to allow us to recognise at least the same languages that could be recognised with a look-ahead of size k.

Now, let us suppose without loss of generality that $\mathrm{a} \in \Sigma$. One can check that the finite language $\{\mathrm{a}^k, \mathrm{a}^{k+1}\}$ belongs to $k+1 - \mathcal{REV}_{\mathcal{L}}(\Sigma)$, but $\{\mathrm{a}^k, \mathrm{a}^{k+1}\} \notin k - \mathcal{REV}_{\mathcal{L}}(\Sigma)$.

Finally, given any integer k, $\{\mathrm{a}^n : n \mod 3 \neq 0\} \notin k - \mathcal{REV}_{\mathcal{L}}(\Sigma)$. □

11.2.2 Learning k-reversible automata

One way to learn k-reversible automata from text is to take as a starting point the PTA that corresponds exactly to the sample, and then merge states that are breaking the violation conditions. The key idea of the algorithm we now present is that the order in which the merges are performed does not matter! So the algorithm works by just merging states that do not comply with the conditions for k-reversibility. There is actually a better order in which the states of the PTA should be visited: the breadth-first order. Algorithm $\mathbf{A}_{k-\mathcal{REV}}$ (11.4) tracks the violating pairs and merges them, until there are no more such pairs. At each moment, the algorithm solves the non-determinism violations first. The merging algorithm (MERGE, Algorithm 11.3) is studied in more detail in Section 12.2.2.

Notice that testing the k-reversibility conditions on a DFA consists of updating the sets of k-predecessors of each state. Notice also that each pair of states needs only to be checked once.

Algorithm 11.3: MERGE.

Input: an NFA : $\mathcal{A} = \langle \Sigma, Q, \{q_i\}, \mathbb{F}_{\mathbb{A}}, \delta_N \rangle$, $q_1, q_2 \in Q$, with $q_2 \neq q_i$
Output: NFA : $\mathcal{A} = \langle \Sigma, Q, \{q_i\}, \mathbb{F}_{\mathbb{A}}, \delta_N \rangle$ in which q_1 and q_2 have been merged into q_1
for $q \in Q$ **do**
  for $a \in \Sigma$ **do**
    if $q_2 \in \delta_N(q, a)$ **then** $\delta_N(q, a) \leftarrow \delta_N(q, a) \cup \{q_1\}$;
    if $q \in \delta_N(q_2, a)$ **then** $\delta_N(q_1, a) \leftarrow \delta_N(q_1, a) \cup \{q\}$
  end
end
if $q_2 \in \mathbb{I}$ **then** $\mathbb{I} \leftarrow \mathbb{I} \cup \{q_1\}$;
if $q_2 \in \mathbb{F}_{\mathbb{A}}$ **then** $\mathbb{F}_{\mathbb{A}} \leftarrow \mathbb{F}_{\mathbb{A}} \cup \{q_1\}$;
$Q \leftarrow Q \setminus \{q_2\}$;
return $\mathcal{A}$

Algorithm 11.4: $\mathbf{A}_{k-\mathcal{REV}}$: k-reversible automata learning.

Input: $k \in \mathbb{N}$, S a sample of a k-reversible language L
Output: $\mathcal{A}$, a k-reversible automaton recognising S
$\mathcal{A} \leftarrow \text{PTA}(S)$;
repeat
  changed$\leftarrow$ **false**;
  if $\exists q, q', p \in Q,\ q \neq q',\ \exists a \in \Sigma,\ \{q, q'\} \subset \delta_N(p, a)$ **then**
    $\mathcal{A} \leftarrow \text{MERGE}(\mathcal{A}, q, q')$;
    changed$\leftarrow$ **true**
  end
  if not(changed) **and** $\exists q, q' \in F : \text{PRED}_k(q) \cap \text{PRED}_k(q') \neq \emptyset$ **then**
    $\mathcal{A} \leftarrow \text{MERGE}(\mathcal{A}, q, q')$;
    changed$\leftarrow$ **true**
  end
  if not(changed) **and** $\exists q, q' \in Q,\ q \neq q',\ \exists a \in \Sigma :\ \delta_N(q, a) \cap \delta_N(q', a) \neq \emptyset \wedge \text{PRED}_k(q) \cap \text{PRED}_k(q') \neq \emptyset$ **then**
    $\mathcal{A} \leftarrow \text{MERGE}(\mathcal{A}, q, q')$;
    changed$\leftarrow$ **true**
  end
until not(changed) ;
return $\mathcal{A}$

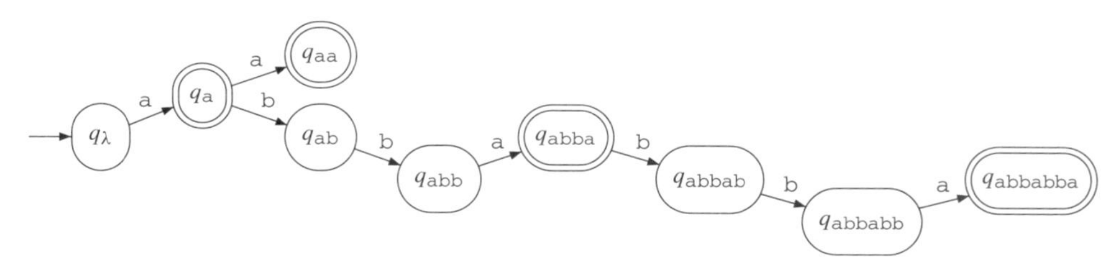

Fig. 11.7. Starting the run.

11.2.3 A run example of Algorithm $\mathbf{A}_{k-\mathcal{REV}}$

Let $S = \{\texttt{a}, \texttt{aa}, \texttt{abba}, \texttt{abbabba}\}$ and $k = 2$. We start by building the prefix tree acceptor (Figure 11.7).

We can then notice that the pair $(q_{\texttt{abba}}, q_{\texttt{abbabba}})$ is violating: indeed both states are final and have ba as common predecessor. Therefore the algorithm merges these two states leading to the automaton represented in Figure 11.8.

States $q_{\texttt{abb}}$ and $q_{\texttt{abbabb}}$ are violating (Figure 11.8); the algorithm merges them, leading to automaton depicted in Figure 11.9.

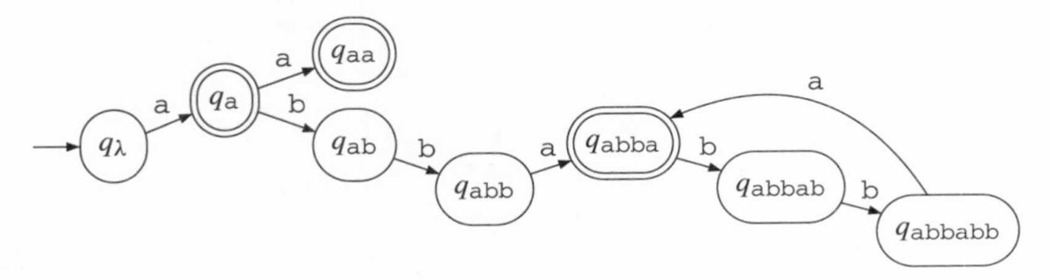

Fig. 11.8. After merging q_{abba} and $q_{abbabba}$.

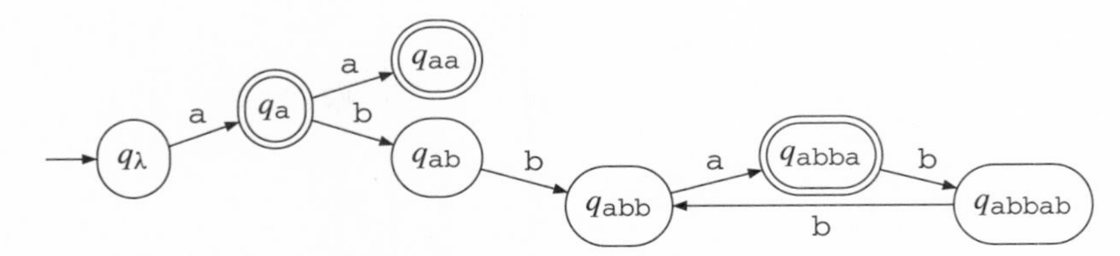

Fig. 11.9. After merging q_{abb} and q_{abbabb}.

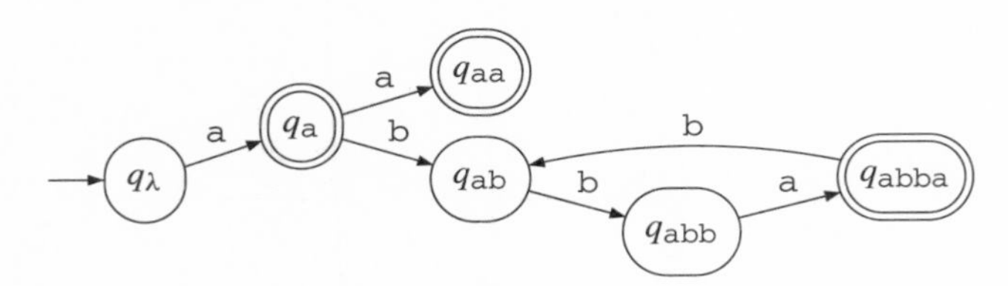

Fig. 11.10. After merging q_{ab} and q_{abbab}.

Now states q_{ab} and q_{abbab} are the violating pair and the algorithm merges them, leading to a new automaton (Figure 11.10).

The algorithm halts and returns the automaton from Figure 11.10 which is 2-reversible. It can be noticed that if k=1, further merges are necessary, such as for instance q_a and q_{aa}.

11.2.4 Properties

We study here the convergence of the algorithm. First we give some more properties concerning the k-reversible languages:

Proposition 11.2.2 *A (regular) language L is k-reversible* if and only if $\forall u_1, u_2, v \in \Sigma^\star$, $(u_1v)^{-1}L \cap (u_2v)^{-1}L \neq \emptyset$ *and* $|v| = k \implies (u_1v)^{-1}L = (u_2v)^{-1}L$.

Proof The proposition states that if two strings are prefixes of a string of length at least k, then the strings are Nerode-equivalent. This corresponds to a rewriting of Definition 11.2.6. □

The key lemma for the proof of the algorithm is that no bad merge can be made:

Lemma 11.2.3 *Let L be a k-reversible language, let S be a positive sample of L, and let u and v be two prefixes of strings in S such that $u \not\equiv_L v$. Then in $\mathbf{A}_{k-\mathcal{REV}}(S)$, $\delta(q_\lambda, u) \neq \delta(q_\lambda, v)$.*

Proof Let us suppose this is not true. We consider the first merge made concerning two non-equivalent (in the target) prefixes u and v. Clearly this merge cannot be due to determinism: if not then $u = u'a$ and $v = v'a$ and both u' and v' end in the same state; yet $u' \not\equiv v'$ (since $u \not\equiv v$). But in that case this would not be the first bad merge.

This merge is therefore made as a consequence of the second violation condition. We can suppose the automaton (before merging) is deterministic. But therefore $\text{PRED}_k(q_u) \cap \text{PRED}_k(q_v) \neq \emptyset$. Let $z \in \text{PRED}_k(q_u) \cap \text{PRED}_k(q_v)$. Consider two strings xz and yz such that $\delta(q_0, xz) = \delta(q_0, u)$ and $\delta(q_0, yz) = \delta(q_0, v)$. Both strings exist because of the definition of PRED_k. Now by Proposition 11.2.2 this means that $xz \equiv_L yz$. And by transitivity either $u = v$ (absurd) or an earlier wrong merge has been made. □

A simple result is that the higher the value of k, the further Algorithm $\mathbf{A}_{k-\mathcal{REV}}$ generalises:

Proposition 11.2.4 $\forall S \subset \Sigma^\star$, $\mathbb{L}(\mathbf{A}_{k-\mathcal{REV}}(S)) \subseteq \mathbb{L}(\mathbf{A}_{[k+1]-\mathcal{REV}}(S))$.

Proof Clearly if you can disambiguate with a look-ahead of size k, then you can also do it with a look-ahead of size $k+1$. Necessarily the successor sets cannot conflict, because they would already have been conflicting for k. □

It is also easy to see that the resulting automaton is always k-reversible.

Proposition 11.2.5 $\forall k \geq 0$, $\forall S \subset \Sigma^\star$, *$\mathbf{A}_{k-\mathcal{REV}}(S)$ is a k-reversible automaton.*

Proof Notice that if the automaton is not k-reversible, then there exists a pair of violators that can be merged. □

And one may conclude:

Theorem 11.2.6 *$\mathbf{A}_{k-\mathcal{REV}}(S)$ is the smallest k-reversible automaton whose language contains S. The class $k - \mathcal{REV}_{\mathcal{G}}(\Sigma)$ is identifiable in the limit from text.*

Proof The second proposition is a direct corollary of the first, by using Theorem 8.1.6, page 178. The first depends on the fact that the order in which the merges are made does not change the final result, which is a consequence of Lemma 11.2.3. In turn, this is due to the fact that the non-violation condition corresponds to an equivalence relation. □

Proposition 11.2.7 (Complexity issues)

- *The time complexity of $\mathbf{A}_{k-\mathcal{REV}}$ is in $\mathcal{O}(k \cdot \|S\|^3)$.*
- *Every $k - \mathcal{REV}$ automaton admits a characteristic set of size linear in the size of the target automaton.*

Proof

- We suppose we have a data structure allowing us to keep track of the sets $\text{PRED}_k(q)$. In order to avoid exponential growth of these sets, one can choose to represent them piecewise: clearly $\text{PRED}_0(q) = \{\lambda\}$ and $\text{PRED}_k(q) = \left[\cup_{q':q\in\delta_N(q',a)} \text{PRED}_{k-1}(q')\right] \cdot a$. Using the Union-Find algorithm allows us to update (each time a merge is done) the $\text{PRED}_k(q)$ sets in near-linear time (in $\mathcal{O}(k \cdot \|S\|)$ and also to see if the intersection of two sets is empty or not (k-ambiguity). Testing if two states are violators is in $\mathcal{O}(n)$. The number of possible merges is in $\mathcal{O}(\|S\|^2)$ but the number of total merges that will be made is in $\mathcal{O}(\|S\|)$.
- A characteristic set can be built by introducing strings for the violation of the k-reversibility conditions to be noticed, on the first states the algorithm encounters. Lemma 11.2.3 tells us that no wrong merge can take place. □

Even if the algorithm $\mathbf{A}_{k-\mathcal{REV}}$ (11.4) is not incremental *per se* it can be considered as such: new strings will result only in more merges being made. Again, key Lemma 11.2.3 shows that no wrong merge takes place.

11.3 Pattern languages

In a very different way from grammars and automata, we can use descriptions of the languages to define pattern languages. For example, patterns are strings containing symbols and variables that allow us, through a matching process, to define complex classes of languages. We introduce here a very simple form of pattern languages, and show how these can be learnt from text.

11.3.1 Definitions

Let Σ be an alphabet and $X = \{x_0, x_1, \ldots x_i, \ldots\}$ be an infinite set of variables.

Definition 11.3.1 A **pattern** π is a string over $\Sigma \cup X$. The size of a pattern is the length of this string.

A matching is a function $\sigma : X \to \Sigma^\star$. Given a pattern π of size n ($\pi = \pi(1)\cdots\pi(n)$) and a matching σ, $\sigma(\pi) = \sigma(\pi(1))\cdots\sigma(\pi(n))$ where σ is extended to Σ by $\sigma(a) = a,\ \forall a \in \Sigma$, a string $w = a_1 \cdots a_n$ ***fits a pattern*** π if there is a matching $\sigma : X \to \Sigma^\star$ such that $\sigma(\pi) = w$.

The language defined by the pattern π, denoted by $\mathbb{L}(\pi)$, is the set of all strings over the alphabet Σ that fit π.

Example 11.3.1 Consider the pattern $\pi = x_3\mathtt{ab}x_{17}x_3\mathtt{b}$. It defines a language over an alphabet that contains at least symbols a and b: notice it is important to know the alphabet as the same pattern does not define the same language if $\Sigma = \{\mathtt{a},\mathtt{b}\}$ or if $\Sigma = \{\mathtt{a},\mathtt{b},\mathtt{c}\}$. Let us suppose here that $\Sigma = \{\mathtt{a},\mathtt{b}\}$. The length of the pattern is 6. The pattern $x_0\mathtt{ab}x_1x_0\mathtt{b}$ is equivalent to π. The following strings fit the pattern π: aabaab, bababb and bbaabaabbab.

It should be noted that when λ is not allowed to be the image by σ of any x_i, the pattern is called ***non-erasing***. In both cases (erasing and non-erasing patterns), checking if a string fits a particular pattern is $\mathcal{NP}$-hard.

The equivalence problem is very simple in the non-erasing case: two patterns are equivalent if and only if they are equal, up to renaming of the variables. Somewhat surprisingly, the difficult question with patterns is not equivalence but inclusion: the question of knowing whether $\mathbb{L}(\pi_1) \subset \mathbb{L}(\pi_2)$ is undecidable for patterns.

Also, most positive learning results hold in the case of non-erasing patterns; this will justify that we restrict our analysis here to the non-erasing setting. We denote by $\mathcal{PATTERNS}(\Sigma)$ the class of non-erasing pattern languages over an alphabet Σ. Note that the class is both of languages and grammars, as there is a one-to-one correspondence.

11.3.2 Learning pattern languages

Theorem 11.3.1 *$\mathcal{PATTERNS}(\Sigma)$ is identifiable in the limit from text.*

Proof Let s be one of the shortest strings in a sample S. Then $|\pi| = |s|$. So we know the length of the pattern. Now if we take other shortest strings in S, we can easily decide which positions correspond to a constant and which to a variable. The last operation corresponds to finding those positions that are occurrences of a same variable. This is done by Algorithm PATTERN_LEARNER (11.5) which identifies patterns in the limit because as soon as a characteristic sample is included, the returned pattern is unique (up to a renaming of the variables). The characteristic set contains strings of minimal length, and for every variable position i there should be in the set at least two strings w and y with $w(i) \neq y(i)$. Also, for each pair of positions i and j corresponding to two different variables, there should be at least one string w in the characteristic set such that $w(i) \neq w(j)$. One should note that if a characteristic sample is not included, the returned solution is only guaranteed to be consistent with strings of minimal length. □

Algorithm PATTERN_LEARNER (11.5) is not at all optimal, but can be used as basis for other pattern language learners.

11.3.3 A run of the algorithm PATTERN_LEARNER

Let $\Sigma = \{\mathtt{a}, \mathtt{b}, \mathtt{c}\}$ and $S = \{\mathtt{abcbb}, \mathtt{aabba}, \mathtt{aacbbac}, \mathtt{aaaba}, \mathtt{acbbbac}\}$.

The algorithm first selects the shortest strings only: $Y = \{\mathtt{abcbb}, \mathtt{aabba}, \mathtt{aaaba}\}$. Therefore the length of the pattern is 5.

The algorithm then selects the constant positions, again by only examining set Y. Since here the first letter of all strings is a and the fourth letter in all strings (of Y) is b, this means that the pattern is something like $\mathtt{a}x_?x_?\mathtt{b}x_?$.

Algorithm 11.5: PATTERN_LEARNER.

```
Input: a sample S
Output: a non-erasing pattern π
m ← min{|s| : s ∈ S};
Y ← ∅;
for s ∈ S do                              /* Use only the shortest strings */
|   if |s| = m then Y ← Y ∪ {s}
end
j ← 1;
for i : 1 ≤ i ≤ m do                      /* Find the constants */
|   if |{y(i) : y ∈ Y}| = 1 then π[i] ← y(i) else
|   |   π[i] ← z_j;
|   |   Pos[j] ← i;
|   |   j++
|   end
end
for k : 1 ≤ k < j do                      /* Find the common variables */
|   for l : 2 ≤ l ≤ j do
|   |   if ∀y ∈ Y, y(Pos[k]) = y(Pos[l]) then π[pos[l]] ← z_k
|   end
end
return π
```

The third step of the algorithm consists of attempting to match the different variables. Here, since in all strings (again of Y), the second symbol is equal to the fifth, the pattern $\mathrm{a}x_0x_1\mathrm{b}x_0$ is built and finally returned.

Note that this pattern could be inconsistent with the rest of the data but since the question of knowing if a string fits a pattern is $\mathcal{NP}$-hard, there is no way around this problem.

11.4 Planar languages

We defined several string kernels in Section 3.5. The idea now is to use them to learn languages. Clearly the classes of languages defined by kernels will not fit perfectly into the Chomsky hierarchy, because through kernels we are essentially counting things, not generating strings.

A language will correspond to a point in the feature space, which is a high (and possibly infinite) dimensional vector space.

If we are given only text, then we may want to consider the set of images of the examples in the feature space, then apply techniques from linear algebra (like extracting the

Eigen-vectors) and computing the smallest (in dimension) hyper-planes containing the images of all the points.

We can then, at least implicitly, take the pre-images of the hyperplanes as languages, and say that these are the inferred languages.

Example 11.4.1 Suppose the learning sample consists of

$$S_+\{\texttt{aabca}, \texttt{abaac}, \texttt{bbacaaa}\}$$

If we use the Parikh kernel, we would like to learn the following equation:

$$x_a - x_b - x_c = 1$$

This in turn corresponds to the language:

$$\{w \in \{\texttt{a}, \texttt{b}, \texttt{c}\}^* : \ |w|_\texttt{a} - |w|_\texttt{b} - |w|_\texttt{c} = 1\}$$

Given that we can do this for any string kernel κ, we call these κ-planar languages. More formally, given a string kernel κ and its associated feature map $\phi : \Sigma^\star \to H$, where H is the feature space (some Hilbert space), we can define:

Definition 11.4.1 (κ-planar language) A language $L \subseteq \Sigma^\star$ is κ-planar *if$_{def}$* there is a finite set of strings $\{u_1, \ldots, u_n\}$ such that $L = \{w \in \Sigma^\star : \exists \alpha_1 \ldots \alpha_n \in \mathbb{R} : \sum_{i \in [n]} \alpha_i = 1 \wedge \phi(w) = \sum_{i \in [n]} \alpha_i \phi(u_i)\}$.

Definition 11.4.2 (Hyper-plane) For any language $L \subseteq \Sigma^*$ the **hyper-plane** defined by (all affine combinations of) L is:

$H(L) = \{h \in H \ : \ \exists n > 0, w_1, \ldots, w_n \in L, \alpha_1, \ldots, \alpha_n \in \mathbb{R}$ such that $\sum_{i \in [n]} \alpha_i = 1$, $\sum_{i \in [n]} \alpha_i \phi(w_i) = h\}$.

$H(L)$ is the smallest hyper-plane that contains the image of L. Notice that L can be finite or infinite.

Definition 11.4.3 (Pre-image of hyper-plane) The language $\hat{L}$ is defined as the pre-image of this hyper-plane:

$$\hat{L} = \{w \in \Sigma^\star : \phi(w) \in H(L)\}$$

Example 11.4.2 Consider the language over $\Sigma = \{\texttt{a}, \texttt{b}\}$:

$$L_{\texttt{ab}} = \{u \in \Sigma^\star : |u|_\texttt{a} = |u|_\texttt{b}\}$$

This language contains all strings such that the number of a is equal to the number of b. aabb, abaababb belong to $L_{\texttt{ab}}$. If we consider the Parikh kernel and the associated feature map, we can see that:

$$\phi^P(L_{\texttt{ab}}) = \{(x, x) : x \geq 0\}$$

We have $\phi^P(\texttt{aabb}) = (2, 2)$ and $\phi^P(\texttt{abaababb}) = (4, 4)$. Clearly, these points lie in a hyper-plane which is a line in the feature space $\mathbb{R}^2$. Moreover, the pre-image of the minimal hyper-plane containing all the points of the language is exactly $L_{\texttt{ab}}$. Therefore, $L_{\texttt{ab}}$ can be represented as a one-dimensional hyper-plane in this feature space. $L_{\texttt{ab}}$ is a κ-planar language for the Parikh kernel.

A way around the hardness of learning context-free languages is to avoid them altogether. In this case an interesting approach is through learning planar languages (Clark *et al.*, 2006), using kernel methods (Shawe-Taylor & Christianini, 2004). Typically the idea is to count something (for example the number of occurrences of some symbol, as in the Parikh map (Parikh, 1966)), then to compute (via solving linear algebra problems in the dual space) the smallest hyper-plane containing the examples. For example the typical context-free language $\{a^n b^n : n \in \mathbb{N}\}$ can be described through just two equations : $|u|_a = |u|_b$ and $|u|_{ba} = 0$. Nevertheless this example may be over-simplifying: in general the number and complexity of the equations we may obtain are much greater. Note that the idea here is to make use of an alternative definition of languages: instead of through generation (grammars) or through recognition (automata), languages are defined by description (Salomaa, 2005). A close approach was followed by Naoki Abe (Abe, 1995): the languages were also defined by equations, but the setting was that of learning from an informant and PAC-learning results were obtained.

11.5 Exercises

11.1 Run $\mathbf{A}_{k-\mathcal{TSS}}$ for $k = 1, 2, 3$ and 15, with $S = \{\texttt{ab}, \texttt{abab}, \texttt{abababab}\}$.

11.2 Run $\mathbf{A}_{k-\mathcal{TSS}}$ for $k = 1, 2, 3$ and 4, with $S = \{\texttt{ab}, \texttt{ababba}, \texttt{abbababab}, \texttt{baababa}, \texttt{aaba}, \texttt{baabaab}\}$.

11.3 Compute the complexity in time and space of $\mathbf{A}_{k-\mathcal{TSS}}$.

11.4 Prove that $(\text{ba}^k)^* \in [k+1] - \mathcal{TSS}(\Sigma)$ but $(\text{ba}^k)^* \notin k - \mathcal{TSS}(\Sigma)$.

11.5 Propose a class that is not $k - \mathcal{TSS}$ whatever the value of k.

11.6 Prove that identification of k-testable languages by $\mathbf{A}_{k-\mathcal{TSS}}$ can be done with only a polynomial number of implicit prediction errors. Prove that algorithm $\mathbf{A}_{k-\mathcal{TSS}}$ has only a polynomial number of mind changes.

11.7 Prove that identifying the entire class of testable languages (i.e. the union for all k) is impossible from positive examples only.

11.8 Let S_+ be a positive example and x be one negative example. Prove that there is a value p for which $k < p \implies x \in \mathbf{A}_{k-\mathcal{TSS}}(S_+)$. Deduce an algorithm that identifies in the limit the entire class of testable languages (i.e. the union of the $k - \mathcal{TSS}(\Sigma)$ for all k) from text and just one negative example.

11.9 Prove that learning the entire class of testable languages, from positive and negative examples, is possible but not with polynomial time (consider implicit prediction errors, mind changes or characteristic samples).

11.10 Construct a language L that is not k-reversible, $\forall k \geq 0$.

11.11 Prove that the class $k - \mathcal{REV}_{\mathcal{L}}$ of all k-reversible languages is not identifiable in the limit from text.

11.12 Run $\mathbf{A}_{k-\mathsf{REV}}$ on $S = \{\mathtt{aa, aba, abb, abaaba, baaba}\}$ for k=0, 1, 2, 3.

11.13 Prove that identification in the limit from text is always possible if $\mathcal{L}$ is a finite set.

11.14 Run $\mathbf{A}_{k-\mathsf{REV}}$ for k = 1, 2, 3, and 15, and $S = \{\mathtt{ab, abab, abababab}\}$

11.15 Compute the complexity of $\mathbf{A}_{k-\mathsf{REV}}$.

11.16 Run algorithm PATTERN_LEARNER (11.5) on $S = \{\mathtt{abaaaa, abaaaaa, bbbabb, abaaba, baabaaa}\}$.

11.17 Build a characteristic sample for the pair $\langle\pi, \mathbf{A}\rangle$ where $\mathbf{A}$ is algorithm PATTERN_LEARNER (11.5) and $\pi = \mathtt{a}x_0\mathtt{a}x_1x_1\mathtt{b}x_2\mathtt{ba}x_0$.

11.6 Conclusions of the chapter and further reading

11.6.1 Bibliographical background

The testable languages we introduce in Section 11.1 are usually called k-testable in the strict sense (García & Vidal, 1990, García, Vidal & Oncina, 1990), k-testable or even local. The differences are technical.

Look-ahead languages as defined in Section 11.2 were introduced by Dana Angluin (Angluin, 1982) where they are called k-reversible, but as shown by Henning Fernau (Fernau, 2003) there is a general pattern here that can be used as soon as you have some way to solve non-determinism when parsing the strings from left to right. We have chosen to use the original class here but Henning Fernau's elegant generalisation should be looked into if variants are needed.

The algorithm from Section 11.2 has been applied, for instance by Boris Chidlovskii (Chidlovskii, 2000) for wrapper generation. Using Henning Fernau's extension (Fernau, 2003), Valter Crescenzi and Giansalvatore Mecca define and learn a subclass of regular languages, called the *prefix mark-up languages*, that abstract the structures usually found in HTML (Crescenzi & Mecca, 2004).

Yasubumi Sakakibara built an extension for context-free grammars whose tree language is k-reversible (Sakakibara, 1992). Adaptations to tree languages (Besombes & Marion, 2004b, Oates, Desai & Bhat, 2002) or even to graph grammars (Oates, Doshi & Huang, 2003) have also been done.

Different authors propose learning these automata and then estimating the probabilities as an alternative to learning stochastic automata. A difficult practical matter is that of choosing k (see Chapter 17 for some techniques). If counter-examples exist, then these can be used to decide (see Exercise 11.12). Alternatively, part of the learning sample can be put aside and used for validation: it is then reasonable to choose the largest k such that the validation set is recognised.

Pattern languages (Section 11.3) were introduced by Dana Angluin, whose original class we have chosen to reproduce here (Angluin, 1979, 1987b). There have been numerous papers on pattern languages since, including, just to cite a few, (Erlebach *et al.*, 1997, Koshiba, 1995, Mitchell *et al.*, 1999, Rossmanith & Zeugmann, 1998).

11.6.2 Some alternative lines of research

As mentioned above, there have been many works on learning pattern languages. The formalism is simple and the classes of languages are rich. Nevertheless, in general the membership problem is $\mathcal{NP}$-hard. One should turn to the abundant literature inside the algorithmic learning theory community for further details. The specific problem of learning pattern languages with more complex types of queries has been studied (Lange & Zilles, 2003), with extensions to correction queries in (Kinber, 2008, Tirnauca, 2008).

One problem with k-testable languages is that for small alphabets and small values of k the languages are not very interesting. An alternative to augmenting the size of the window is to augment the size of the alphabet! The idea can be to consider that an a at the beginning of the string is different from an a at the end of the string. This idea was followed by researchers working in pattern recognition on the *morphic generator* (Castro & Casacuberta, 1996, Vidal *et al.*, 1988).

11.6.3 Open problems and possible new lines of research

One of the key advantages of the k-testable language learning algorithm is that it is fast: computation takes place in linear time. As today the quantity of available data is becoming larger and larger, it would seem that this is a good argument in favour of this algorithm. But the fixed size of the window is one obstacle, and so are the difficulties imposed by a noisy setting.

Work done by Henning Fernau in trying to generalise these ideas (Fernau, 2000, 2002) is another direction to be followed. He characterises the properties necessary to be able to obtain a class identifiable in the limit from text. Combining these ideas with complexity considerations, new classes of automata might be able to be introduced.

12
Informed learners

Understanding is compression, comprehension is compression!

Greg Chaitin *(Chaitin, 2007)*

Comprendo. Habla de un juego donde las reglas no sean la línea
de salida, sino el punto de llegada ¿No?

Arturo Pérez-Reverte*, el pintor de batallas*

'Learning from an informant' is the setting in which the data consists of labelled strings, each label indicating whether or not the string belongs to the target language.

Of all the issues which grammatical inference scientists have worked on, this is probably the one on which most energy has been spent over the years. Algorithms have been proposed, competitions have been launched, theoretical results have been given. On one hand, the problem has been proved to be on a par with mighty theoretical computer science questions arising from combinatorics, number theory and cryptography, and on the other hand cunning heuristics and techniques employing ideas from artificial intelligence and language theory have been devised.

There would be a point in presenting this theme with a special focus on the class of context-free grammars with a hope that the theory for the particular class of the finite automata would follow, but the history and the techniques tell us otherwise. The main focus is therefore going to be on the simpler yet sufficiently rich question of learning deterministic finite automata from positive and negative examples.

We shall justify this as follows:

- On one hand the task is hard enough, and, through showing what doesn't work and why, we will have a precious insight into more complex classes.
- On the other hand anything useful learnt on DFAs can be nicely transferred thanks to reductions (see Chapter 7) to other supposedly richer classes.
- Finally, there are historical reasons: on the specific question of learning DFAs from an informed presentation, some of the most important algorithms in grammatical inference have been invented, and many new ideas have been introduced due to this effort.

The specific question of learning context-free grammars and languages from an informant will be studied as a separate problem, in Chapter 15.

12.1 The prefix tree acceptor (PTA)

We shall be dealing here with *learning samples* composed of labelled strings:

Definition 12.1.1 Let Σ be an alphabet. An **informed learning sample** is made of two sets S_+ and S_- such that $S_+ \cap S_- = \emptyset$. The sample will be denoted as $S = \langle S_+, S_- \rangle$.

We will alternatively denote $(x, 1) \in S$ for $x \in S_+$ and $(x, 0) \in S$ for $x \in S_-$. Let $\mathcal{A} = \langle \Sigma, Q, q_\lambda, \mathbb{F}_\mathbb{A}, \mathbb{F}_\mathbb{R}, \delta \rangle$ be a DFA.

Definition 12.1.2 $\mathcal{A}$ is **weakly consistent** with the sample $S = \langle S_+, S_- \rangle$ *if$_{def}$* $\forall x \in S_+,\ \delta(q_\lambda, x) \in \mathbb{F}_\mathbb{A}$ and $\forall x \in S_-,\ \delta(q_\lambda, x) \notin \mathbb{F}_\mathbb{A}$.

Definition 12.1.3 $\mathcal{A}$ is **strongly consistent** with the sample $S = \langle S_+, S_- \rangle$ *if$_{def}$* $\forall x \in S_+,\ \delta(q_\lambda, x) \in \mathbb{F}_\mathbb{A}$ and $\forall x \in S_-,\ \delta(q_\lambda, x) \in \mathbb{F}_\mathbb{R}$.

Example 12.1.1 The DFA from Figure 12.1 is only weakly consistent with the sample $\{(\texttt{aa}, 1), (\texttt{abb}, 0), (\texttt{b}, 0)\}$ which can also be denoted as:

$$S_+ = \{\texttt{aa}\}$$
$$S_- = \{\texttt{abb}, \texttt{b}\}.$$

String abb ends in state q_λ which is unlabelled (neither accepting nor rejecting), and string b (from S_-) cannot be entirely parsed. Therefore the DFA is not strongly consistent. On the other hand the same automaton can be shown to be strongly consistent with the sample $\{(\texttt{aa}, 1), (\texttt{aba}, 1)(\texttt{abab}, 0)\}$.

A ***prefix tree acceptor*** (PTA) is a tree-like DFA built from the learning sample by taking all the prefixes in the sample as states and constructing the smallest DFA which is a tree $(\forall q \in Q, \mid \{q' : \delta(q', a) = q\}\mid \leq 1\})$, strongly consistent with the sample. A formal algorithm (BUILD-PTA) is given (Algorithm 12.1).

An example of a PTA is shown in Figure 12.2.

Note that we can also build a PTA from a set of positive strings only. This corresponds to building the PTA $(\langle S_+, \emptyset \rangle)$. In that case, for the same sample we would get the PTA represented in Figure 12.3.

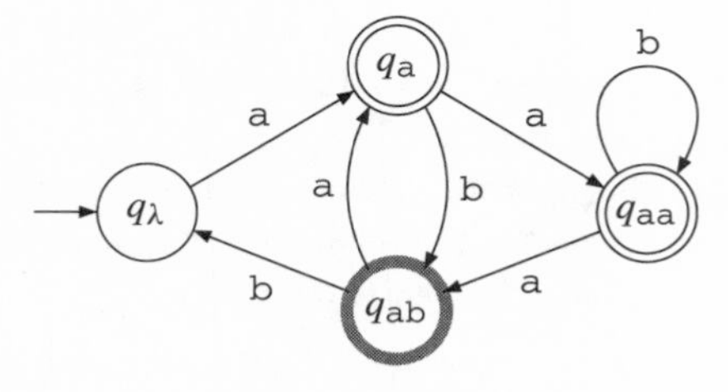

Fig. 12.1. A DFA.

Algorithm 12.1: BUILD-PTA.

Input: a sample $\langle S_+, S_-\rangle$
Output: $\mathcal{A} = \text{PTA}(\langle S_+, S_-\rangle) = \langle \Sigma, Q, q_\lambda, \mathbb{F}_{\mathbb{A}}, \mathbb{F}_{\mathbb{R}}, \delta\rangle$
$\mathbb{F}_{\mathbb{A}} \leftarrow \emptyset$;
$\mathbb{F}_{\mathbb{R}} \leftarrow \emptyset$;
$Q \leftarrow \{q_u : u \in \text{PREF}(S_+ \cup S_-)\}$;
for $q_{u\cdot a} \in Q$ **do** $\delta(q_u, a) \leftarrow q_{ua}$;
for $q_u \in Q$ **do**
  if $u \in S_+$ **then** $\mathbb{F}_{\mathbb{A}} \leftarrow \mathbb{F}_{\mathbb{A}} \cup \{q_u\}$;
  if $u \in S_-$ **then** $\mathbb{F}_{\mathbb{R}} \leftarrow \mathbb{F}_{\mathbb{R}} \cup \{q_u\}$
end
return $\mathcal{A}$

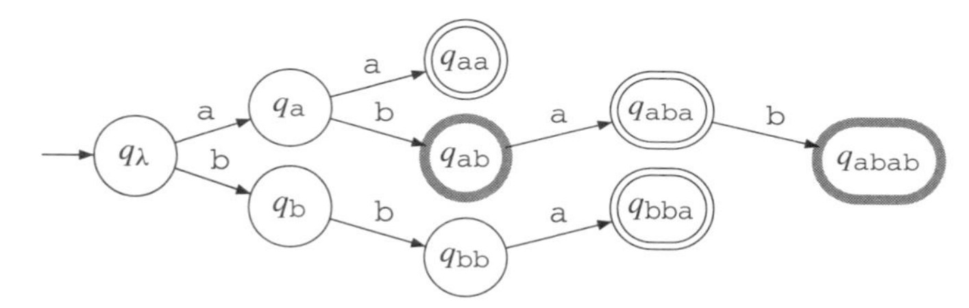

Fig. 12.2. PTA ({(aa, 1), (aba, 1), (bba, 1), (ab, 0), (abab, 0)}).

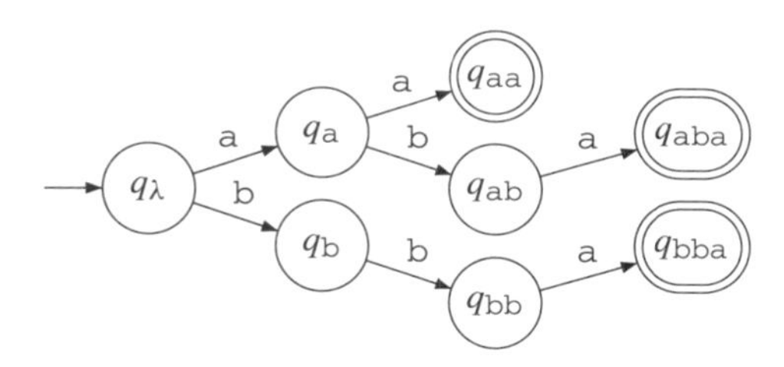

Fig. 12.3. PTA ({(aa, 1), (aba, 1), (bba, 1)}).

In Chapter 6 we will consider the problem of grammatical inference as the one of searching inside a space of admissible biased solutions, in which case we will introduce a non-deterministic version of the PTA.

Most algorithms will take the PTA as a starting point and try to generalise from it by merging states. In order not to get lost in the process (and not undo merges that have been made some time ago) it will be interesting to divide the states into three categories:

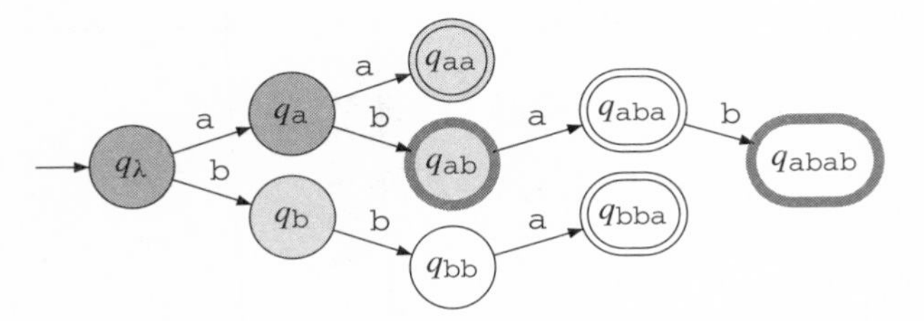

Fig. 12.4. Colouring of states: RED $= \{q_\lambda, q_a\}$, BLUE $= \{q_b, q_{aa}, q_{ab}\}$, and all the other states are WHITE.

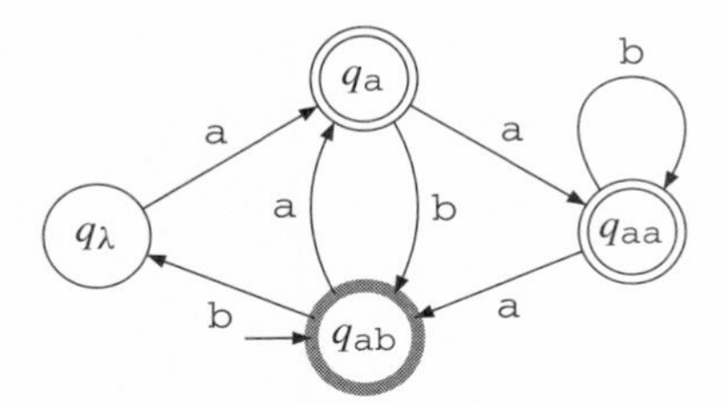

Fig. 12.5. The DFA $\mathcal{A}_{q_{ab}}$.

- The RED states which correspond to states that have been analysed and which will not be revisited; they will be the states of the final automaton.
- The BLUE states which are the ***candidate*** states: they have not been analysed yet and it should be from this set that a state is drawn in order to consider merging it with a RED state.
- The WHITE states, which are all the others. They will in turn become BLUE and then RED.

Example 12.1.2 We conventionally draw the RED states in dark grey and the BLUE ones in light grey as in Figure 12.4, where RED $= \{q_\lambda, q_a\}$ and BLUE $= \{q_b, q_{aa}, q_{ab}\}$.

We will need to describe the suffix language in any state q, consisting of the language recognised by the automaton when taking this state q as initial. We denote this automaton formally by $\mathcal{A}_q$ with $\mathbb{L}(\mathcal{A}_q) = \{w \in \Sigma^\star : \delta(q, w) \in \mathbb{F}_\mathbb{A}\}$. In Figure 12.5 we have used the automaton $\mathcal{A}$ from Figure 12.1 and chosen state q_{ab} as initial.

12.2 The basic operations

We first describe some operations common to many of the ***state merging techniques***. State merging techniques iteratively consider an automaton and two of its states and aim to ***merge*** them. This will be done when these states are ***compatible***. Sometimes, when noticing that a particular state cannot be merged, it gets ***promoted***. Furthermore at any moment all states are either RED, BLUE or WHITE. Let us also suppose that the current automaton is ***consistent*** with the sample. The starting point is the ***prefix tree acceptor*** (PTA). Initially, in the PTA, the unique RED state is q_λ whereas the BLUE states are the immediate successors of q_λ.

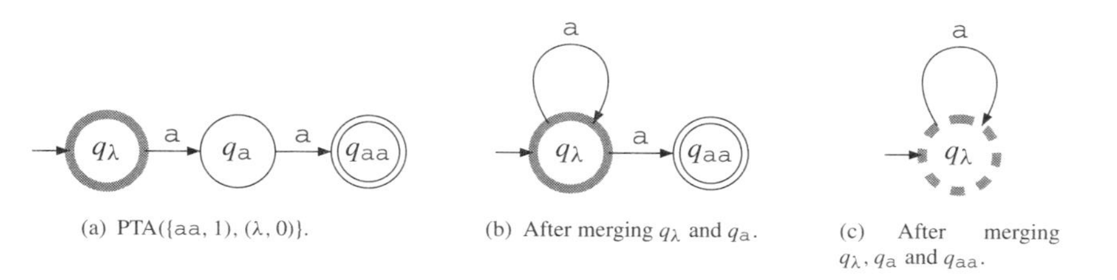

(a) PTA({(aa, 1), (λ, 0)}).

(b) After merging q_λ and q_a.

(c) After merging q_λ, q_a and q_{aa}.

Fig. 12.6. Incompatibility is not a local affair.

There are three basic operations that shall be systematically used and need to be studied independently of the learning algorithms: COMPATIBLE, MERGE and PROMOTE.

12.2.1 COMPATIBLE: deciding equivalence between states

The question here is of deciding if two states are compatible or not. This is the same as deciding equivalence for the Nerode relation, but with only partial knowledge about the language. As obviously we do not have the entire language to help us decide upon this, but only the learning sample, the question is to know if merging these two states will not result in creating confusion between accepting and rejecting states. Typically the compatibility might be tested by:

$$q \simeq_{\mathcal{A}} q' \iff \mathbb{L}_{\mathbb{F}_\mathbb{A}}(\mathcal{A}_q) \cap \mathbb{L}_{\mathbb{F}_\mathbb{R}}(\mathcal{A}_{q'}) = \emptyset \text{ and } \mathbb{L}_{\mathbb{F}_\mathbb{R}}(\mathcal{A}_q) \cap \mathbb{L}_{\mathbb{F}_\mathbb{A}}(\mathcal{A}_{q'}) = \emptyset.$$

But this is usually not enough as the following example (Figure 12.6) shows. Consider the three-state PTA (Figure 12.6(a)) built from the sample $S = \{(\mathtt{aa}, 1), (\lambda, 0)\}$. Deciding equivalence between states q_λ and q_a through the formula above is not sufficient. Indeed languages $\mathbb{L}(\mathcal{A}_{q_\lambda})$ and $\mathbb{L}(\mathcal{A}_{q_\mathtt{a}})$ are weakly consistent, but if q_λ and $q_\mathtt{a}$ are merged together (Figure 12.6(b)), the state $q_\mathtt{aa}$ must also be merged with these (Figure 12.6(c)) to preserve determinism. This results in a problem: is the new unique state accepting or rejecting?

Therefore more complex operations will be needed, involving merging, folding and **then** testing consistency.

12.2.2 MERGE: merging two states

The merging operation takes two states from an automaton and merges them into a single state. It should be noted that the effect of the merge is that a deterministic automaton (see Figure 12.7(a)) will possibly lose the determinism property through this (Figure 12.7(b)). Indeed this is where the algorithms can reject a merge. Consider for instance automaton 12.8(a). If states q_1 and q_2 are merged, then to ensure determinism, states q_3 and q_4 will also have to be merged, resulting in automaton 12.8(b). If we have in our learning sample string aba (in S_-), then the merge should be rejected.

Algorithm 12.2 is given an NFA (with just one initial state, for simplicity), and two states. It updates the automaton.

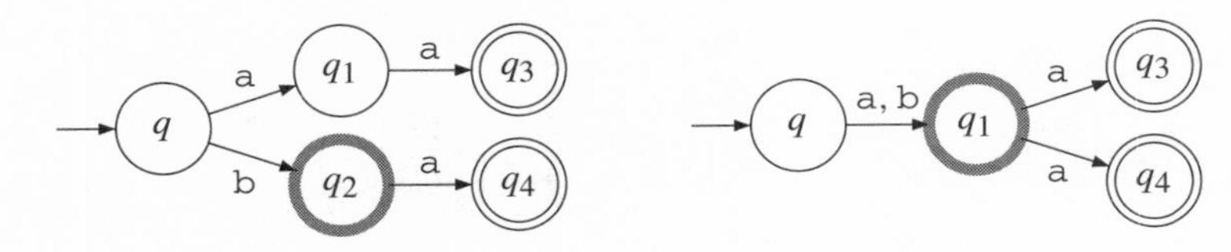

(a) Before merging states q_1 and q_2. (b) After merging the states.

Fig. 12.7. Merging two states may result in the automaton not remaining deterministic.

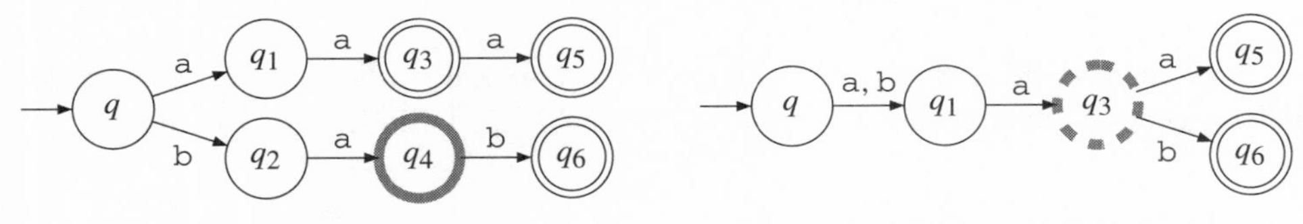

(a) Before merging states 1 and 2. (b) After merging the states recursively (from 1 and 2).

Fig. 12.8. About merging.

Algorithm 12.2: MERGE.

Input: an NFA : $\mathcal{A} = \langle \Sigma, Q, \mathbb{I}, \mathbb{F}_{\mathbb{A}}, \mathbb{F}_{\mathbb{R}}, \delta_N \rangle$, q_1 and q_2 in Q, with $q_2 \notin \mathbb{I}$

Output: an NFA : $\mathcal{A} = \langle \Sigma, Q, \mathbb{I}, \mathbb{F}_{\mathbb{A}}, \mathbb{F}_{\mathbb{R}}, \delta_N \rangle$ in which q_1 and q_2 have been merged into q_1

for $q \in Q$ **do**
  for $a \in \Sigma$ **do**
    if $q_2 \in \delta_N(q, a)$ **then** $\delta_N(q, a) \leftarrow \delta_N(q, a) \cup \{q_1\}$;
    if $q \in \delta_N(q_2, a)$ **then** $\delta_N(q_1, a) \leftarrow \delta_N(q_1, a) \cup \{q\}$
  end
end
if $q_2 \in \mathbb{I}$ **then** $\mathbb{I} \leftarrow \mathbb{I} \cup \{q_1\}$;
if $q_2 \in \mathbb{F}_{\mathbb{A}}$ **then** $\mathbb{F}_{\mathbb{A}} \leftarrow \mathbb{F}_{\mathbb{A}} \cup \{q_1\}$;
if $q_2 \in \mathbb{F}_{\mathbb{R}}$ **then** $\mathbb{F}_{\mathbb{R}} \leftarrow \mathbb{F}_{\mathbb{R}} \cup \{q_1\}$;
$Q \leftarrow Q \setminus \{q_2\}$;
return $\mathcal{A}$

Since non-deterministic automata are in many ways cumbersome, we will attempt to avoid having to use these to define merging when manipulating only deterministic automata.

12.2.3 PROMOTE: promoting a state

Promotion is another deterministic and greedy decision. The idea here is that having decided, at some point, that a BLUE candidate state is different from all the RED states, it

Algorithm 12.3: PROMOTE.

Input: a DFA : $\mathcal{A} = \langle \Sigma, Q, q_\lambda, \mathbb{F}_{\mathbb{A}}, \mathbb{F}_{\mathbb{R}}, \delta \rangle$, a BLUE state q_u, sets RED, BLUE
Output: a DFA : $\mathcal{A} = \langle \Sigma, Q, q_\lambda, \mathbb{F}_{\mathbb{A}}, \mathbb{F}_{\mathbb{R}}, \delta \rangle$, sets RED, BLUE updated
RED $\leftarrow$ RED $\cup \{q_u\}$;
for $a \in \Sigma$: q_{ua} *is not* RED **do** add q_{ua} to BLUE;
return $\mathcal{A}$

should become RED. We call this a ***promotion*** and describe the process in Algorithm 12.3. The notations that are used here apply to the case where the states involved in a promotion are the basis of a tree. Therefore, the successors of node q_u are named q_{ua} with a in Σ.

12.3 Gold's algorithm

The first non-enumerative algorithm designed to build a DFA from informed data is due to E. Mark Gold, which is why we shall simply call this algorithm GOLD. The goal of the algorithm is to find the minimum DFA consistent with the sample. For that, there are two steps. The first is deductive: from the data, find a set of prefixes that have to lead to different states for the reasons given in Section 12.2.2 above, and therefore represent an incompressible set of states. The second step is inductive: alas, after finding the incompressible set of states, we are not done because it is not usually easy or even possible to 'fold in' the rest of the states. Since a direct construction of the DFA from there is usually impossible, (contradictory) decisions have to be taken. This is where artificial intelligence techniques might come in (see Chapter 14) as the problems one has to solve are proved to be intractable (in Chapter 6).

But as more and more strings become available to the learning algorithm (i.e. in the identification in the limit paradigm), the number of choices left will become more and more restricted, with, finally, just one choice. This is what allows convergence.

12.3.1 The key ideas

The main ideas of the algorithm are to represent the data (positive and negative strings) in a table, where each row corresponds to a string, some of which will correspond to the RED states and the others to the BLUE states. The goal is to create through promotion as many RED states as possible. For this to be of real use, the set of strings denoting the states will be *closed* by prefixes, i.e. if q_{uv} is a state so is q_u. Formally:

Definition 12.3.1 (Prefix- and suffix-closed sets) A set of strings S is **prefix-closed** (respectively **suffix-closed**) *if*$_{def}$ $uv \in S \implies u \in S$ (respectively if $uv \in S \implies v \in S$).

No inference is made during this phase of representation of the data: the algorithm is purely deductive.

The table then expresses an inequivalence relation between the strings and we should aim to complete this inequivalence related to the Nerode relation that defines the language:

$$x \equiv y \iff \left[\forall w \in \Sigma^\star \, xw \in L \iff yw \in L\right].$$

Once the RED states are decided, the algorithm 'chooses' to merge the BLUE states that are left with the RED ones and then checks if the result is consistent. If it is not, the algorithm returns the PTA.

In the following we will voluntarily accept a confusion between the states themselves and the names or labels of the states.

The information is organised in a table $\langle \text{STA}, \text{EXP}, \text{OT}\rangle$, called an ***observation table***, used to compare the candidate states by examining the data, where the three components are:

- STA $\subset \Sigma^\star$ is a finite set of (labels of) states. The states will be denoted by the indexes (strings) from a finite prefix-closed set. Because of this labelling, we will often conveniently use string terminology when referring to the states. Set STA will therefore both refer to the set of states and to the set of labels of these states, with context always allowing us to determine which.

  We partition STA as follows: STA $=$ RED $\cup$ BLUE. The BLUE states (or state labels) are those u in STA such that $uv \in \text{STA} \implies v = \lambda$. The RED states are the others. BLUE $= \{ua \notin \text{RED} : u \in \text{RED}\}$ is the set of states successors of RED that are not RED.
- EXP $\subset \Sigma^\star$ is the experiment set. This set is closed by suffixes, i.e. if uv is an experiment, so is v.
- OT : STA $\times$ EXP $\to \{0, 1, *\}$ is a function that indicates if making an experiment in state q_u is going to result into an accepting, a rejecting or an unknown situation: the value OT$[u][e]$ is then respectively 1,0 or $*$. In order to improve readability we will write it as a table indexed by two strings, the first indicating the label of the state from which the experiment is made and the second being the experiment itself:

  $$\text{OT}[u][e] = \begin{cases} 1 & \text{if } ue \in L \\ 0 & \text{if } ue \notin L \\ * & \text{otherwise (not known).} \end{cases}$$

  Obviously, the table should be ***redundant*** in the following sense. Given three strings u, v and w, if OT$[u][vw]$ and OT$[uv][w]$ are defined (i.e. $u, uv \in$ STA and $w, vw \in$ EXP), then OT$[u][vw] =$ OT$[uv][w]$.

Example 12.3.1 In observation table 12.1 we can read:

- STA $= \{\lambda, \texttt{a}, \texttt{b}, \texttt{aa}, \texttt{ab}\}$ and among these RED $= \{\lambda, \texttt{a}\}$.
- OT$[\texttt{aa}][\lambda] = 0$ so $\texttt{aa} \notin L$.
- On the other hand we have OT$[\texttt{b}][\lambda] = 1$ which means that $\texttt{b} \in L$.
- Note also that the table is redundant: for example, OT$[\texttt{aa}][\lambda] =$ OT$[\texttt{a}][\texttt{a}] = 0$, and similarly OT$[\lambda][\texttt{a}] = [\texttt{a}][\lambda] = *$. This is only due to the fact that the table is an observation of the data. It does not compute or invent new information.

In the following, we will be only considering legal tables, i.e. those that are based on a set STA prefix-closed, a set EXP suffix-closed and a redundant table. Legality can be

Table 12.1. *An observation table.*

	λ	a	b
λ	0	*	1
a	*	0	0
b	1	1	*
aa	0	0	*
ab	0	*	*

Algorithm 12.4: GOLD-check legality.

```
Input: a table ⟨STA, EXP, OT⟩
Output: a Boolean indicating if the table is legal or not
OK ← true;
for s ∈ STA do                          /* check if STA is prefix-closed */
| if PREF(s) ⊄ STA then OK ← false
end
for e ∈ EXP do                          /* check if EXP is suffix-closed */
| if PREF(e) ⊄ EXP then OK ← false
end
for p ∈ STA do                                /* check if all is legal */
    for e ∈ EXP do
        for p ∈ PREF(e) do
        | if [sp ∈ STA ∧ OT[s][e] ≠ OT[sp][p^{-1}e] then OK ← false
        end
    end
end
return OK
```

checked with Algorithm 12.4. The complexity of the algorithm can be easily improved (see Exercise 12.1). In practice, a good policy is to stock the data in an association table, with the string as key and the label as value, and to manipulate in the observation table just the keys to the table. This makes the legality issue easy to deal with.

Definition 12.3.2 (Holes)
A **hole** in a table ⟨STA, EXP, OT⟩ is a pair (u, e) such that $\text{OT}[u][e] = *$.

A hole corresponds to a missing observation.

Definition 12.3.3 (Complete table)
The table ⟨STA, EXP, OT⟩ is **complete** (or has **no holes**) *if*$_{def}$ $\forall u \in \text{STA}, \forall e \in \text{EXP}, \ \text{OT}[u][e] \in \{0, 1\}$.

	λ	a
λ	0	1
a	1	0
b	1	0
aa	0	0
ab	1	0

(a) An observation table, which is not closed, because of row aa.

	λ	a
λ	0	1
a	1	0
b	0	1
aa	0	1
ab	1	0

(b) A complete and closed observation table.

Fig. 12.9.

Definition 12.3.4 (Rows)
We will refer to OT$[u]$ as the **row** indexed by u and will say that two rows OT$[u]$ and OT$[v]$ are **compatible** for OT (or u and v are consistent for OT) if_{def} $\nexists e \in$ EXP : $\big(\text{OT}[u][e] = 0$ and $\text{OT}[v][e] = 1\big)$ or $\big(\text{OT}[u][e] = 1$ and $\text{OT}[v][e] = 0\big)$. We denote this by $u \simeq_{\text{OT}} v$.

The goal is not going to be to detect if two rows are compatible, but if they are not.

Definition 12.3.5 (Obviously different rows)
Rows u and v are **obviously different** (OD) for OT (we also write OT$[u]$ and OT$[v]$ are obviously different) if_{def} $\exists e \in$ EXP : $\text{OT}[u][e], \text{OT}[v][e] \in \{0, 1\}$ and $\text{OT}[u][e] \neq \text{OT}[v][e]$.

Example 12.3.2 Table 12.1 is incomplete since it has holes. Rows OT[λ] and OT[a] are incompatible (and OD), but row OT[aa] is compatible with both OT[λ] and OT[a].

12.3.2 Complete tables

We now consider the ideal (albeit unrealistic) setting where there are no holes in the table.

Definition 12.3.6 (Closed table)
A table OT is **closed** if_{def} $\forall u \in$ BLUE, $\exists s \in$ RED : $\text{OT}[u] = \text{OT}[s]$.

The table presented in Figure 12.9(a) is not closed (because of row aa) but Table 12.9(b) is. Being closed means that every BLUE state can be matched with a RED one.

12.3.3 Building a DFA from a complete and closed table

Building an automaton from a table ⟨STA, EXP, OT⟩ can be done very easily as soon as certain conditions are met:

- The set of strings marking the states in STA is prefix-closed,
- The set EXP is suffix-closed,
- The table should be complete: holes correspond to undetermined pieces of information,
- The table should be closed.

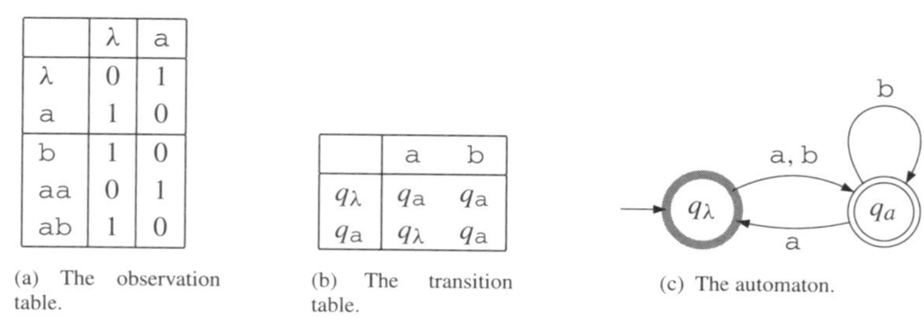

	λ	a
λ	0	1
a	1	0
b	1	0
aa	0	1
ab	1	0

(a) The observation table.

	a	b
q_λ	q_a	q_a
q_a	q_λ	q_a

(b) The transition table.

(c) The automaton.

Fig. 12.10. A table and the corresponding automaton.

Once these conditions hold we can use Algorithm GOLD-BUILDAUTOMATON (12.5) and convert the table into a DFA.

Algorithm 12.5: GOLD-BUILDAUTOMATON.

Input: a closed and complete observation table (STA, EXP, OT)
Output: a DFA $\mathcal{A} = \langle \Sigma, Q, q_\lambda, \mathbb{F}_\mathbb{A}, \mathbb{F}_\mathbb{R}, \delta \rangle$
$Q \leftarrow \{q_r : r \in \text{RED}\}$;
$\mathbb{F}_\mathbb{A} \leftarrow \{q_{we} : we \in \text{RED} \wedge \text{OT}[w][e] = 1\}$;
$\mathbb{F}_\mathbb{R} \leftarrow \{q_{we} : we \in \text{RED} \wedge \text{OT}[w][e] = 0\}$;
for $q_w \in Q$ **do**
| **for** $a \in \Sigma$ **do** $\delta(q_w, a) \leftarrow q_u : u \in \text{RED} \wedge \text{OT}[u] = \text{OT}[wa]$
end
return $\langle \Sigma, Q, q_\lambda, \mathbb{F}_\mathbb{A}, \mathbb{F}_\mathbb{R}, \delta \rangle$

Example 12.3.3 Consider Table 12.10(a). We can apply the construction from Algorithm 12.5 and obtain $Q = \{q_\lambda, q_a\}$, $\mathbb{F}_\mathbb{A} = \{q_a\}$, $\mathbb{F}_\mathbb{R} = \{q_\lambda\}$ and δ is given by the transition table 12.10(b). Automaton 12.10(c) can be built.

Definition 12.3.7 (Consistent table) Given an automaton $\mathcal{A}$ and an observation table $\langle$STA, EXP, OT$\rangle$, $\mathcal{A}$ is **consistent** with $\langle$STA, EXP, OT$\rangle$ when the following holds:

- $\text{OT}[u][e] = 1 \implies ue \in \mathbb{L}_{\mathbb{F}_\mathbb{A}}(\mathcal{A})$,
- $\text{OT}[u][e] = 0 \implies ue \in \mathbb{L}_{\mathbb{F}_\mathbb{R}}(\mathcal{A})$.

$\mathbb{L}_{\mathbb{F}_\mathbb{A}}(\mathcal{A})$ is the language recognised by $\mathcal{A}$ by accepting states, whereas $\mathbb{L}_{\mathbb{F}_\mathbb{R}}(\mathcal{A})$ is the language recognised by $\mathcal{A}$ by rejecting states.

Theorem 12.3.1 (Consistency theorem) *Let* $\langle$STA, EXP, OT$\rangle$ *be an observation table closed and complete. If* STA *is prefix-closed and* EXP *is suffix-closed then* GOLD-BUILDAUTOMATON*(*$\langle$STA, EXP, OT$\rangle$*) is consistent with the information in* $\langle$STA, EXP, OT$\rangle$.

Proof The proof is straightforward as GOLD-BUILDAUTOMATON builds a DFA directly from the data from ⟨STA, EXP, OT⟩. □

12.3.4 Building a table from the data

The second question is that of obtaining a table from a sample. At this point we want the table to be consistent with the sample, and to be just an alternative representation of the sample. Given a sample S and a set of states RED prefix-closed, it is always possible to build a set of experiments EXP such that the table ⟨STA, EXP, OT⟩ contains all the information in S (and no other information!). There can be many possible tables, one corresponding to each set of RED states we wish to consider. And, of course, in most cases, these tables are going to have holes.

Algorithm 12.6: GOLD-BUILDTABLE.

Input: a sample $S = \langle S_+,\ S_- \rangle$, a set RED prefix-closed
Output: table ⟨STA, EXP, OT⟩
EXP ← SUFF(S);
BLUE ← RED · Σ \ RED;
for $p \in$ RED ∪ BLUE **do**
  for $e \in$ EXP **do**
    if $p.e \in S_+$ **then** OT[p][e] ← 1
    else
      if $p.e \in S_-$ **then** OT[p][e] ← 0 **else** OT[p][e] ← $*$
    end
  end
end
return ⟨STA, EXP, OT⟩

Algorithm GOLD-BUILDTABLE (12.6) builds the table corresponding to a given sample and a specific set RED.

Example 12.3.4 Table 12.2 constructed for the sample $S = \{(\texttt{aa}, 1), (\texttt{bbaa}, 1), (\texttt{aba}, 0)\}$ and for the set of RED states $\{\lambda, \texttt{a}\}$ is given here. We have not entered the '$*$' symbols to increase readability (i.e. an empty cell denotes a symbol $*$).

12.3.5 Updating the table

We notice the following:

Proposition 12.3.2 *If* $\exists t \in$ BLUE *such that* OT[t] *is obviously different from any* OT[s] *(with* $s \in$ RED*), then whatever way we fill the holes in* ⟨STA, EXP, OT⟩*, the table will not be closed.*

Table 12.2. GOLD-BUILDTABLE
$(\{(\texttt{aa}, 1), (\texttt{bbaa}, 1), (\texttt{aba}, 0)\}), \{\lambda, \texttt{a}\})$.

	λ	a	aa	ba	aba	baa	bbaa
λ			1		0		1
a		1		0			
b						1	
aa	1						
ab		0					

In other words, if one BLUE state is obviously different from all the RED states, then even guessing each $*$ correctly is not going to be enough. This means that this BLUE state should be promoted before attempting to fill in the holes.

12.3.6 Algorithm GOLD

The general algorithm can now be described.

It is composed of four steps requiring four sub-algorithms:

(i) Given sample S, Algorithm GOLD-BUILDTABLE (12.6, page 248) builds an initial table with RED $= \{\lambda\}$, BLUE $= \Sigma$ and $E = \text{SUFF}(S)$.
(ii) Find a BLUE state obviously different (OD) with all RED states. Promote this BLUE state to RED and repeat.
(iii) Fill in the holes that are left (using Algorithm GOLD-FILLHOLES). If the filling of the holes fails, return the PTA (using Algorithm BUILD-PTA (12.1, page 239)).
(iv) Using Algorithm GOLD-BUILDAUTOMATON (12.5, page 247), build the automaton. If it is inconsistent with the original sample, return the PTA instead.

12.3.7 A run of the algorithm

Example 12.3.5 We provide an example run of Algorithm GOLD (12.7).

We use the following sample:

$$S_+ = \{\texttt{bb}, \texttt{abb}, \texttt{bba}, \texttt{bbb}\}$$
$$S_- = \{\texttt{a}, \texttt{b}, \texttt{aa}, \texttt{bab}\}.$$

We first build the observation table corresponding to RED $= \{\lambda\}$.

Now, Table 12.3 is not closed because of row OT[b]. So we promote $q_{\texttt{b}}$ and update the table, obtaining Table 12.4.

Table 12.3. *The table for* $S_+ = \{\texttt{bb}, \texttt{abb}, \texttt{bba}, \texttt{bbb}\}$
$S_- = \{\texttt{a}, \texttt{b}, \texttt{aa}, \texttt{bab}\}$ *and* RED $= \{\lambda\}$.

	λ	a	b	aa	ab	ba	bb	abb	bab	bba	bbb
λ		0	0	0			1	1	0	1	1
a	0	0					1				
b	0		1		0	1	1				

Algorithm 12.7: GOLD for DFA identification.

Input: a sample S

Output: a DFA consistent with the sample

RED $\leftarrow \{\lambda\}$;

BLUE $\leftarrow \Sigma$;

$\langle$STA, EXP, OT$\rangle \leftarrow$ GOLD-BUILDTABLE(S,RED);

while $\exists x \in$ BLUE *such that* OT[x] *is* OD **do**

    RED $\leftarrow$ RED $\cup \{x\}$;

    BLUE $\leftarrow$ BLUE $\cup \{xa : a \in \Sigma\}$;

    for $u \in$ STA **do**

        for $e \in$ EXP **do**

            if $ue \in S_+$ **then** OT[u][e] $\leftarrow 1$

            else if $ue \in S_-$ **then** OT[u][e] $\leftarrow 0$

            else OT[u][e] $\leftarrow *$

        end

    end

end

OT $\leftarrow$ GOLD-FILLHOLES(OT);

if fail then return BUILD-PTA(S)

else

    $\mathcal{A} \leftarrow$ GOLD-BUILDAUTOMATON($\langle$STA, EXP, OT$\rangle$);

    if CONSISTENT($\mathcal{A}$,S) **then return** $\mathcal{A}$

    else return BUILD-PTA(S)

end

Table 12.4. *The table for* $S_+ = \{$bb, abb, bba, bbb$\}$ $S_- = \{$a, b, aa, bab$\}$ *and* RED $= \{\lambda,$ b$\}$.

	λ	a	b	aa	ab	ba	bb	abb	bab	bba	bbb
λ		0	0	0			1	1	0	1	1
b	0		1		0	1	1				
a	0	0					1				
ba			0								
bb	1	1	1								

But Table 12.4 is not closed because of OT[bb]. Since q_{bb} is obviously different from both q_λ (because of experiment λ) and q_{b} (because of experiment b), we promote q_{bb} and update the table to Table 12.5.

At this point there are no BLUE rows that are obviously different from the RED rows. Therefore all that is needed is to fill the holes. Algorithm GOLD-FILLHOLES is now used to make the table complete.

Table 12.5. *The table for* $S_+ = \{$bb, abb, bba, bbb$\}$ $S_- = \{$a, b, aa, bab$\}$ *and* RED $= \{\lambda,$ b, bb$\}$.

	λ	a	b	aa	ab	ba	bb	abb	bab	bba	bbb
λ		0	0	0			1	1	0	1	1
b	0		1		0	1	1				
bb	1	1	1								
a	0	0					1				
ba			0								
bba	1										
bbb	1										

Table 12.6. *The table for* $S_+ = \{$bb, abb, bba, bbb$\}$ $S_- = \{$a, b, aa, bab$\}$ *after running the first phase of* GOLD-FILLHOLES.

	λ	a	b	aa	ab	ba	bb	abb	bab	bba	bbb
λ	1	0	0	0			1	1	0	1	1
b	0	0	1		0	1	1				
bb	1	1	1								
a	0	0					1				
ba			0								
bba	1										
bbb	1										

Table 12.7. *The table for* $S_+ = \{$bb, abb, bba, bbb$\}$ $S_- = \{$a, b, aa, bab$\}$ *after phase 2 of* GOLD-FILLHOLES.

	λ	a	b	aa	ab	ba	bb	abb	bab	bba	bbb
λ	1	0	0	0	1	1	1	1	0	1	1
b	0	0	1	1	0	1	1	1	1	1	1
bb	1	1	1	1	1	1	1	1	1	1	1
a	0	0					1				
ba			0								
bba	1										
bbb	1										

This algorithm first fills the rows corresponding to the RED rows by using the information contained in the BLUE rows which are compatible (in the sense of Definition 12.3.4). In this case there are a number of possibilities which may conflict. For example, we have a $\simeq_{\mathrm{OT}} \lambda$ but also a $\simeq_{\mathrm{OT}}$ b. And we are only considering pairs where the first prefix/state is BLUE and the second is RED.

We suppose that in this particular case, the algorithm has selected a $\simeq_{\mathrm{OT}}$ b, ba $\simeq_{\mathrm{OT}} \lambda$, bba $\simeq_{\mathrm{OT}} \lambda$ and bbb $\simeq_{\mathrm{OT}}$ bb. This results in building first Table 12.6. Then all the holes in the RED rows are filled by 1s (Table 12.7).

Table 12.8. *The complete table for* $S_+ = \{\text{bb, abb, bba, bbb}\}$ $S_- = \{\text{a, b, aa, bab}\}$ *after running* GOLD-FILLHOLES.

	λ	a	b	aa	ab	ba	bb	abb	bab	bba	bbb
λ	1	0	0	0	1	1	1	1	0	1	1
b	0	0	1	1	0	1	1	1	1	1	1
bb	1	1	1	1	1	1	1	1	1	1	1
a	0	0	1	1	0	1	1	1	1	1	1
ba	1	0	0	0	1	1	1	1	0	1	1
bba	1	0	0	0	1	1	1	1	0	1	1
bbb	1	1	1	1	1	1	1	1	1	1	1

Algorithm 12.8: GOLD-FILLHOLES.

Input: a table ⟨STA, EXP, OT⟩
Output: the table OT updated, with holes filled

```
for p ∈ BLUE do                          /* First fill in all the RED lines */
    if ∃r ∈ RED : p ≃_OT r then                    /* Find a compatible RED */
        for e ∈ EXP do
            if OT[p][e] ≠ * then OT[r][e] ← OT[p][e]
        end
    else
        return fail
    end
end
for r ∈ RED do
    for e ∈ EXP do if OT[r][e] = * then OT[r][e] ← 1
end
for p ∈ BLUE do                          /* Now fill in all the BLUE lines */
    if ∃r ∈ RED : p ≃_OT r then                    /* Find a compatible RED */
        for e ∈ EXP do
            if OT[p][e] = * then OT[p][e] ← OT[r][e]
        end
    else
        return fail
    end
end
return ⟨STA, EXP, OT⟩
```

The second part of the algorithm again visits the BLUE states, tries to find a compatible RED state and copies the corresponding RED row. This results in Table 12.8.

Finally, the third part of Algorithm 12.8 fills the remaining holes of the BLUE rows. This results in Table 12.8.

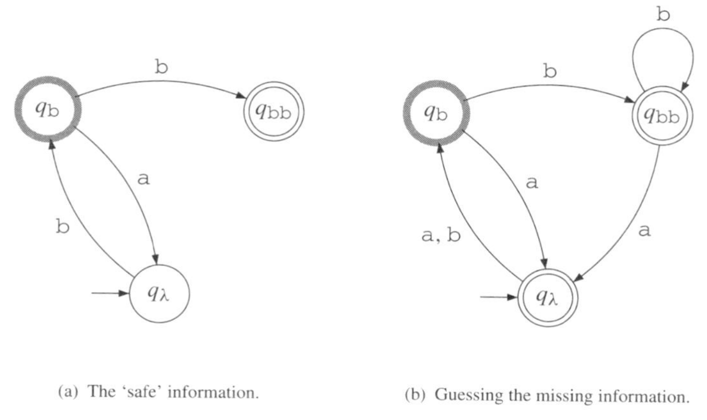

(a) The 'safe' information. (b) Guessing the missing information.

Fig. 12.11. The final filling of the holes for GOLD.

Next, Algorithm GOLD-BUILDAUTOMATON (12.5) is run on the table, with the resulting automaton depicted in Figure 12.11(b). The DFA accepts aa which is in S_+, therefore the PTA is returned instead.

One might consider that GOLD-FILLHOLES is far too trivial for such a complex task as that of learning automata. Indeed, when considering Table 12.5 a certain number of safe decisions about the automaton can be made. These are depicted in Figure 12.11(a). The others have to be guessed:

- equivalent lines for a could be λ and b (so q_a could be either q_λ or q_b),
- possible candidates for bba are λ and bb (so q_{bba} could be either q_λ or q_{bb}),
- possible candidates for bbb are λ and bb (so q_{bbb} could be either q_λ or q_{bb}).

In the choices above not only should the possibilities before merging be considered, but also the interactions between the merges. For example, even if in theory both states q_a and q_{bba} could be merged into state q_λ, they both cannot be merged together! We will not enter here into how this guessing can be done (greediness is one option).

Therefore the algorithm, having failed, returns the PTA depicted in Figure 12.12(a). If it had guessed the holes correctly an automaton consistent with the data (see Figure 12.12(b)) might have been returned.

12.3.8 Proof of the algorithm

We first want to prove that, alas, filling in the holes is where the real problems start. There is no tractable strategy that will allow us to fill the holes easily:

Theorem 12.3.3 (Equivalence of problems) *Let* RED *be a set of states prefix-closed, and S be a sample. Let* $\langle$STA, EXP, OT$\rangle$ *be an observation table consistent with all the data in S, with* EXP *suffix-closed.*

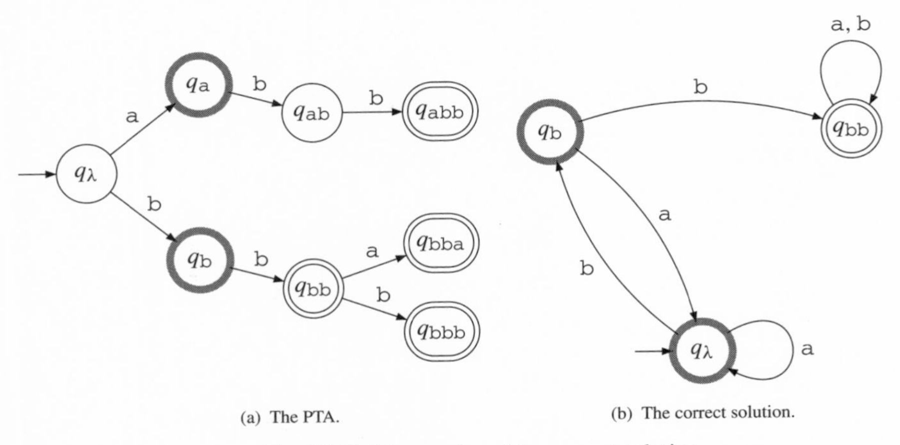

(a) The PTA. (b) The correct solution.

Fig. 12.12. The result and the correct solution.

The question:

> ***Name:*** *Consistent*
> ***Instance:*** *A sample S, a prefix-closed set* RED
> ***Question:*** *Does there exist a* DFA $= \langle \Sigma, Q, q_\lambda, \mathbb{F}_\mathbb{A}, \mathbb{F}_\mathbb{R}, \delta \rangle$ *with* $Q = \{q_u : u \in \text{RED}\}$, *and if* $ua \in \text{RED}$, $\delta(q_u, a) = q_{ua}$, *consistent with S?*

is equivalent to:

> ***Name:*** *Holes*
> ***Instance:*** *An observation table* $\langle$STA, EXP, OT$\rangle$
> ***Question:*** *Can we fill the holes in such a way as to have* $\langle$STA, EXP, OT$\rangle$ *closed?*

Proof If we can fill the holes and obtain a closed table, then a DFA can be constructed which is consistent with the data (by Theorem 12.3.1). If there is a DFA with the states of RED then we can use the DFA to fill the holes. □

From this we have:

Theorem 12.3.4 *The problem:*

> ***Name:*** *Minimum consistent* DFA *reachable from* RED
> ***Instance:*** *A sample S, a set* RED *prefix-closed such that each* OT[*s*] ($s \in$ RED) *is obviously different from all the others with respect to S, and a positive integer n*
> ***Question:*** *Is there a* DFA $= \langle \Sigma, Q, q_\lambda, \mathbb{F}_\mathbb{A}, \mathbb{F}_\mathbb{R}, \delta \rangle$ *with the conditions* $\{q_u : u \in \text{RED}\} \subseteq Q$, $|Q| = n$, *consistent with S?*

is $\mathcal{NP}$*-complete.*

And as a consequence:

Corollary 12.3.5 *Given S and a positive integer n, the question:*

> ***Name:*** *Minimum consistent* DFA
> ***Instance:*** *A sample S and a positive integer n*
> ***Question:*** *Is there a* DFA *with n states consistent with S?*

is $\mathcal{NP}$-complete.

Proof We leave the proofs that these problems are $\mathcal{NP}$-hard to Section 6.2 (page 119).

Proving that either of the problems belongs to $\mathcal{NP}$ is not difficult: simply producing a DFA and checking consistency is going to take polynomial time only as it consists of parsing the strings in S. □

On the positive side, identification in the limit can be proved:

Theorem 12.3.6 *Algorithm* GOLD, *given any sample $S = \langle S_+, S_- \rangle$:*

- *outputs a* DFA *consistent with S,*
- *admits a polynomial-sized characteristic sample,*
- *runs in time and space polynomial in $\|S\|$.*

Proof

- Remember that in the worst case the PTA is returned.
- The characteristic sample can be constructed in such a way as to make sure that all the states are found to be OD. The number of such strings is quadratic in the size of the target automaton. Furthermore it can be proved that none of these strings needs to be of length more than n^2. It should be noted that for this to work, the order in which the BLUE states are explored for promotion matters. If the size of the canonical acceptor of the language is n, then there is a characteristic sample CS_L with $\|\mathrm{CS}_L\| = 2n^2(|\Sigma| + 1)$, such that GOLD$(S)$ produces the canonical acceptor for all $S \supseteq \mathrm{CS}_L$.
- Space complexity is in $\mathcal{O}(\|S\| \cdot n)$ whereas time complexity is in $\mathcal{O}(n^2 \cdot \|S\|)$. We leave for Exercise 12.5 the question of obtaining a faster algorithm. □

Corollary 12.3.7 *Algorithm* GOLD *identifies $\mathcal{DFA}(\Sigma)$ in* POLY-CS *time.*

Identification in the limit in POLY-CS polynomial time follows from the previous remarks.

12.4 RPNI

In Algorithm GOLD, described in Section 12.3, there is more than a real chance that after many iterations the final problem of 'filling the holes' is not solved at all (and perhaps cannot be solved unless more states are added) and the PTA is returned. Even if this is mathematically admissible (since identification in the limit is ensured), in practice one would prefer an algorithm that does some sort of generalisation in all circumstances, and not just in the favourable ones.

This is what is proposed by algorithm RPNI (Regular Positive and Negative Inference). The idea is to greedily create clusters of states (by merging) in order to come up with a solution that is always consistent with the learning data. This approach ensures that some type of generalisation takes place and, in the best of cases (which we can characterise by giving sufficient conditions that permit identification in the limit), returns the correct target automaton.

12.4.1 The algorithm

We describe here a generic version of Algorithm RPNI. A number of variants have been published that are not exactly equivalent. These can be studied in due course. Basically, Algorithm RPNI (12.13) starts by building PTA(S_+) from the positive data (Algorithm BUILD-PTA (12.1, page 239)), then iteratively chooses possible merges, checks if a given merge is correct and is made between two compatible states (Algorithm RPNI-COMPATIBLE (12.10)), makes the merge (Algorithm RPNI-MERGE (12.11)) if admissible and promotes the state if no merge is possible (Algorithm RPNI-PROMOTE (12.9)).

The algorithm has as a starting point the PTA, which is a deterministic finite automaton. In order to avoid problems with non-determinism, the merge of two states is immediately followed by a folding operation: the merge in RPNI always occurs between a RED state and a BLUE state. The BLUE states have the following properties:

- If q is a BLUE state, it has exactly one predecessor, i.e. whenever $\delta(q_1, a_1) = \delta(q_2, a_2) = q$, then necessarily $q_1 = q_2$ and $a_1 = a_2$.
- q is the root of a tree, i.e. if $\delta(q, u \cdot a)$=$\delta(q, v \cdot b)$ then necessarily $u = v$ and $a = b$.

Algorithm 12.9: RPNI-PROMOTE.

Input: a DFA $\mathcal{A} = \langle \Sigma, Q, q_\lambda, \mathbb{F}_\mathbb{A}, \mathbb{F}_\mathbb{R}, \delta \rangle$, sets RED, BLUE $\subseteq Q$, $q_u \in$ BLUE
Output: $\mathcal{A}$, RED, BLUE updated
RED $\leftarrow$ RED $\cup \{q_u\}$;
BLUE $\leftarrow$ BLUE $\cup \{\delta(q_u, a), a \in \Sigma\}$;
return $\mathcal{A}$, RED, BLUE

Algorithm RPNI-PROMOTE (12.9), given a BLUE state q_u, promotes this state to RED and all the successors in $\mathcal{A}$ of this state become BLUE.

Algorithm RPNI-COMPATIBLE (12.10) returns YES if the current automaton cannot parse any string from S_- but returns NO if some counter-example is accepted by the current automaton. Note that the automaton $\mathcal{A}$ is deterministic.

Algorithm 12.10: RPNI-COMPATIBLE.

Input: $\mathcal{A}$, S_-
Output: a Boolean, indicating if $\mathcal{A}$ is consistent with S_-
for $w \in S_-$ **do**
| **if** $\delta_{\mathcal{A}}(q_\lambda, w) \cap \mathbb{F}_{\mathbb{A}} \neq \emptyset$ **then return false**
end
return true

Algorithm RPNI-MERGE (12.11) takes as arguments a RED state q and a BLUE state q'. It first finds the unique pair (q_f, a) such that $q' = \delta_{\mathcal{A}}(q_f, a)$. This pair exists and is unique because q' is a BLUE state and therefore the root of a tree. RPNI-MERGE then redirects $\delta(q_f, a)$ to q. After that, the tree rooted in q' (which is therefore disconnected from the rest of the DFA) is folded (RPNI-FOLD) into the rest of the DFA. The possible intermediate situations of non-determinism (see Figure 12.7, page 242) are dealt with during the recursive calls to RPNI-FOLD. This two-step process is shown in Figures 12.13 and 12.14.

Algorithm 12.11: RPNI-MERGE.

Input: a DFA $\mathcal{A}$, states $q \in$ RED, $q' \in$ BLUE
Output: $\mathcal{A}$ updated
Let (q_f, a) be such that $\delta_{\mathcal{A}}(q_f, a) = q'$;
$\delta_{\mathcal{A}}(q_f, a) \leftarrow q$;
return RPNI-FOLD($\mathcal{A}, q, q'$)

Algorithm RPNI (12.13) depends on the choice of the function CHOOSE. Provided it is a deterministic function (such as one that chooses the minimal $\langle u, a \rangle$ in the lexicographic order), convergence is ensured.

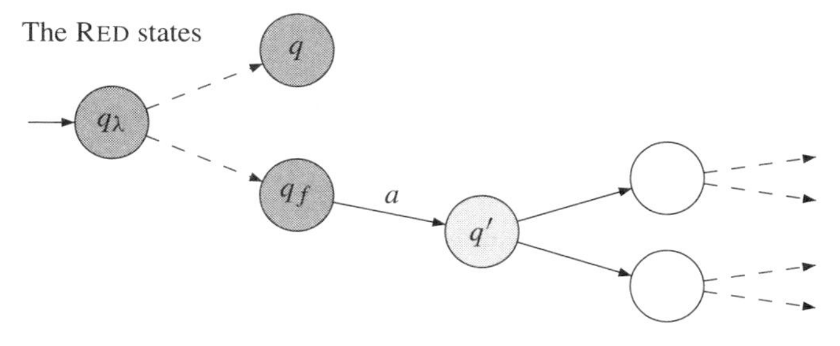

Fig. 12.13. The situation before merging.

Algorithm 12.12: RPNI-FOLD.

Input: a DFA $\mathcal{A}$, states $q, q' \in Q$, q' being the root of a tree
Output: $\mathcal{A}$ updated, where subtree in q' is folded into q
if $q' \in \mathbb{F}_{\mathbb{A}}$ **then** $\mathbb{F}_{\mathbb{A}} \leftarrow \mathbb{F}_{\mathbb{A}} \cup \{q\}$;
for $a \in \Sigma$ **do**
  if $\delta_{\mathcal{A}}(q', a)$ *is defined* **then**
    if $\delta_{\mathcal{A}}(q, a)$ *is defined* **then**
      $\mathcal{A} \leftarrow$RPNI-FOLD$(\mathcal{A}, \delta_{\mathcal{A}}(q, a),\ \delta_{\mathcal{A}}(q', a))$
    else
      $\delta_{\mathcal{A}}(q, a) \leftarrow \delta_{\mathcal{A}}(q', a)$;
    end
  end
end
return $\mathcal{A}$

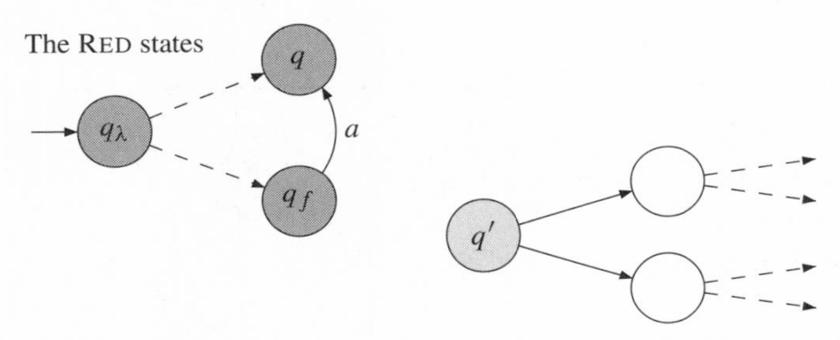

Fig. 12.14. The situation after merging and before folding.

12.4.2 The algorithm's proof

Theorem 12.4.1 *Algorithm* RPNI *(12.13) identifies in the limit* $\mathcal{DFA}(\Sigma)$ *in* POLY-CS *time.*

Proof We use here Definition 7.3.3 (page 154). We first prove that the algorithm computes a consistent solution in polynomial time. First note that the size of the PTA ($\|\text{PTA}\|$ being the number of states) is polynomial in $\|S\|$. The function CHOOSE can only be called at most $\|\text{PTA}\|$ number of times. At each call, compatibility of the running BLUE state will be checked with each state in RED. This again is bounded by the number of states in the PTA. And checking compatibility is also polynomial.

Then to prove that there exists a polynomial characteristic set we constructively add the examples sufficient for identification to take place. Let $\mathcal{A} = \langle \Sigma, Q, q_\lambda, \mathbb{F}_{\mathbb{A}}, \mathbb{F}_{\mathbb{R}}, \delta \rangle$ be the complete minimal canonical automaton for the target language L. Let $<_{\text{CHOOSE}}$ be the order relation associated with function CHOOSE. Then compute the minimum

Algorithm 12.13: RPNI.

Input: a sample $S = \langle S_+, S_- \rangle$, functions COMPATIBLE, CHOOSE
Output: a DFA $\mathcal{A} = \langle \Sigma, Q, q_\lambda, \mathbb{F}_\mathbb{A}, \mathbb{F}_\mathbb{R}, \delta \rangle$
$\mathcal{A} \leftarrow$ BUILD-PTA(S_+);
RED $\leftarrow \{q_\lambda\}$;
BLUE $\leftarrow \{q_a : a \in \Sigma \cap \text{PREF}(S_+)\}$;
while BLUE $\neq \emptyset$ **do**
  CHOOSE($q_b \in$ BLUE);
  BLUE $\leftarrow$ BLUE $\setminus \{q_b\}$;
  if $\exists q_r \in$ RED *such that* RPNI-COMPATIBLE(RPNI-MERGE($\mathcal{A}$, q_r, q_b), S_-) **then**
    $\mathcal{A} \leftarrow$ RPNI-MERGE($\mathcal{A}$, q_r, q_b);
    BLUE $\leftarrow$ BLUE $\cup \{\delta(q, a) : q \in$ RED $\wedge\, a \in \Sigma \wedge \delta(q, a) \notin$ RED$\}$;
  else
    $\mathcal{A} \leftarrow$ RPNI-PROMOTE(q_b, $\mathcal{A}$)
  end
end
for $q_r \in$ RED **do** `/* mark rejecting states */`
  if $\lambda \in (\mathbb{L}(\mathcal{A}_{q_r})^{-1} S_-)$ **then** $\mathbb{F}_\mathbb{R} \leftarrow \mathbb{F}_\mathbb{R} \cup \{q_r\}$
end
return $\mathcal{A}$

distinguishing string between two states q_u, q_v (MD), and the shortest prefix of a state q_u (SP):

- $\text{MD}(q_u, q_v) = \min_{<\text{CHOOSE}}\{w \in \Sigma^\star : \big(\delta(q_u, w) \in \mathbb{F}_\mathbb{A} \wedge \delta(q_v, w) \in \mathbb{F}_\mathbb{R}\big) \vee \big(\delta(q_u, w) \in \mathbb{F}_\mathbb{R} \wedge \delta(q_v, w) \in \mathbb{F}_\mathbb{A}\big)\}$.
- $\text{SP}(q_u) = \min_{<\text{CHOOSE}}\{w \in \Sigma^\star : \delta(q_\lambda, w) = q_u\}$.

RPNI-CONSTRUCTCS($\mathcal{A}$) (Algorithm 12.14) uses these definitions to build a characteristic sample for RPNI, for the order $<_{\text{CHOOSE}}$, and the target language.

$\text{MD}(q_u, q_v)$ represents the minimum suffix allowing us to establish that states q_u and q_v should never be merged. For example, if we consider the automaton in Figure 12.15, in which the states are numbered in order to avoid confusion, this string is aa for q_1 and q_2.

$\text{SP}(q_u)$ is the ***shortest prefix*** in the chosen order that leads to state q_u. Normally this string should be u itself. For example, for the automaton represented in Figure 12.15, $\text{SP}(q_1) = \lambda$, $\text{SP}(q_2) =$ a, $\text{SP}(q_3) =$ aa and $\text{SP}(q_4) =$ b. □

12.4.3 A run of the algorithm

To run RPNI we first have to select a function CHOOSE. In this case we use the lex-length order over the prefixes leading to a state in the PTA. This allows us to mark the states once

Algorithm 12.14: RPNI-CONSTRUCTCS.

Input: $\mathcal{A} = \langle \Sigma, Q, q_\lambda, \mathbb{F}_\mathbb{A}, \mathbb{F}_\mathbb{R}, \delta \rangle$
Output: $S = \langle S_+, S_- \rangle$
$S_+ \leftarrow \emptyset$;
$S_- \leftarrow \emptyset$;
for $q_u \in Q$ **do**
  for $q_v \in Q$ **do**
    for $a \in \Sigma$ *such that* $\mathbb{L}(\mathcal{A}_{q_v}) \cap a\,\Sigma^\star \neq \emptyset$ *and* $q_u \neq \delta(q_v, a)$ **do**
      $w \leftarrow \mathrm{MD}(q_u, q_v)$;
      if $\delta(q_\lambda, u \cdot w) \in \mathbb{F}_\mathbb{A}$ **then**
        $S_+ \leftarrow S_+ \cup \mathrm{SP}(q_u) \cdot w$; $S_- \leftarrow S_- \cup \mathrm{SP}(q_v)a \cdot w$
      else $S_- \leftarrow S_- \cup \mathrm{SP}(q_u) \cdot w$; $S_+ \leftarrow S_+ \cup \mathrm{SP}(q_v)a \cdot w$
    end
  end
end
return $\langle S_+, S_- \rangle$

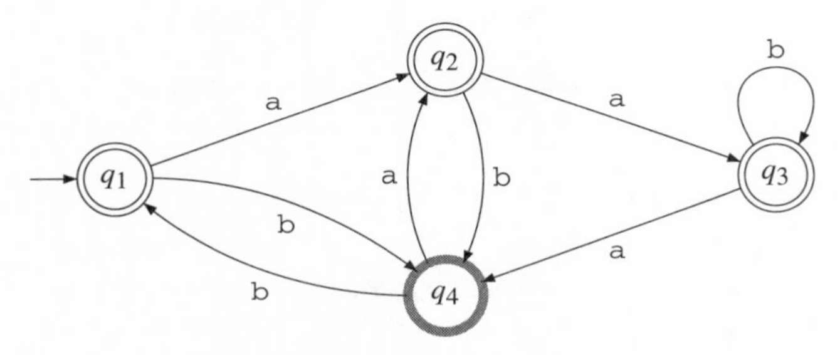

Fig. 12.15. Shortest prefixes of a DFA.

and for all. With this order, state q_1 corresponds to q_λ in the PTA, q_2 to q_{a}, q_3 to q_{b} ,q_4 to q_{aa}, and so forth.

The data for the run are:

$$S_+ = \{\mathtt{aaa}, \mathtt{aaba}, \mathtt{bba}, \mathtt{bbaba}\}$$
$$S_- = \{\mathtt{a}, \mathtt{bb}, \mathtt{aab}, \mathtt{aba}\}$$

From this we build PTA(S_+), depicted in Figure 12.16.

We now try to merge states q_1 and q_2, by using Algorithm RPNI-MERGE with values $\mathcal{A}$, q_1, q_2. Once transition $\delta_\mathcal{A}(q_1, \mathtt{a})$ is redirected to q_1, we reach the situation represented in Figure 12.17.

This is the point where Algorithm RPNI-FOLD is called, in order to fold the subtree rooted in q_2 into the rest of the automaton; the result is represented in Figure 12.18.

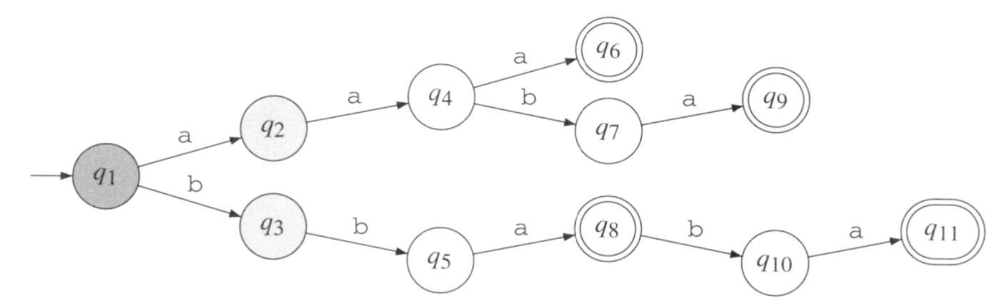

Fig. 12.16. PTA(S_+).

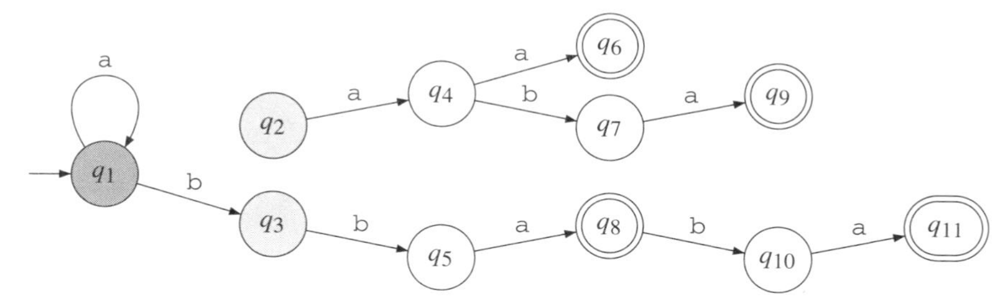

Fig. 12.17. After $\delta_{\mathcal{A}}(q_1, \mathrm{a}) = q_1$.

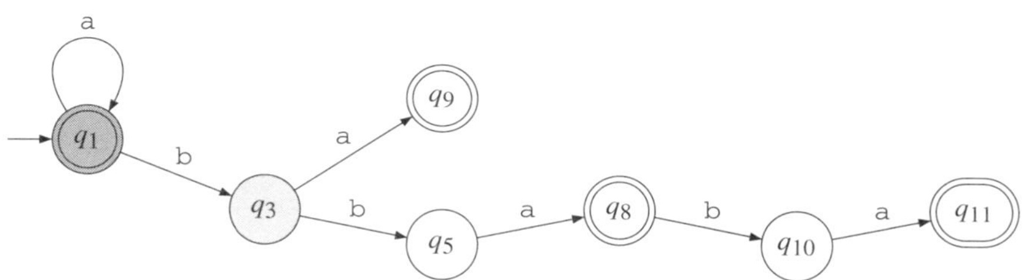

Fig. 12.18. After merging q_2 and q_1.

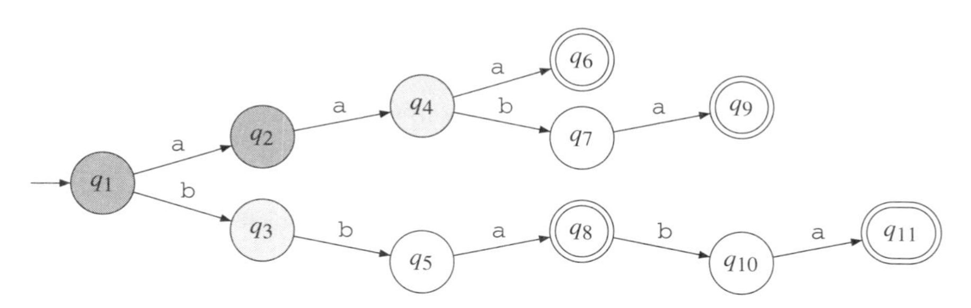

Fig. 12.19. The PTA with q_2 promoted.

The resulting automaton can now be tested for compatibility but if we try to parse the negative examples we notice that counter-example a is accepted. The merge is thus abandoned and we return to the PTA.

State q_2 is now promoted to RED, and its successor q_4 is BLUE (Figure 12.19).

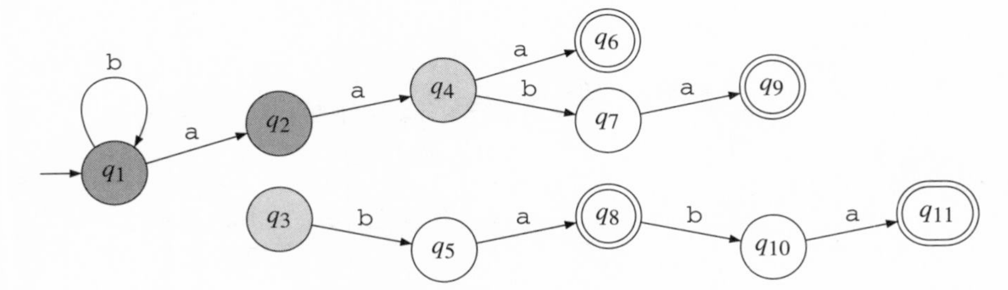

Fig. 12.20. Trying to merge q_3 and q_1.

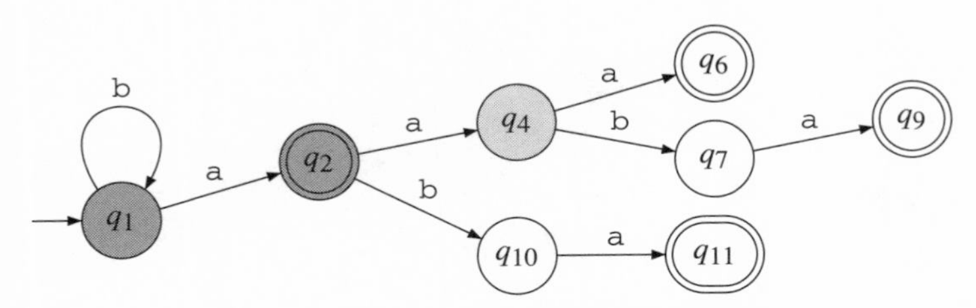

Fig. 12.21. After the folding.

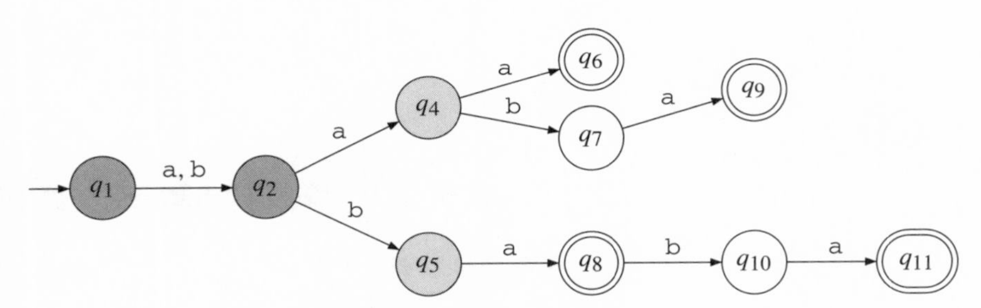

Fig. 12.22. Trying to merge q_3 and q_2.

So the next BLUE state is q_3 and we now try to merge q_3 with q_1. The automaton in Figure 12.20 is built by considering the transition $\delta_{\mathcal{A}}(q_1, \mathtt{b}) = q_1$. Then RPNI-FOLD($\mathcal{A}$, q_1, q_3) is called.

After folding, we get an automaton (see Figure 12.21) which again parses counter-example a as positive.

Therefore the merge $\{q_1, q_3\}$ is abandoned and we must now check the merge between q_3 and q_2. After folding, we are left with the automaton in Figure 12.22 which this time parses the counter-example aba as positive.

Since q_3 cannot be merged with any RED state, there is again a promotion: RED $=$ $\{q_1, q_2, q_3\}$, and BLUE $= \{q_4, q_5\}$. The updated PTA is depicted in Figure 12.23.

The next BLUE state is q_4 and the merge we try is q_4 with q_1. But a (which is the distinguishing suffix) is going to be accepted by the resulting automaton (Figure 12.24).

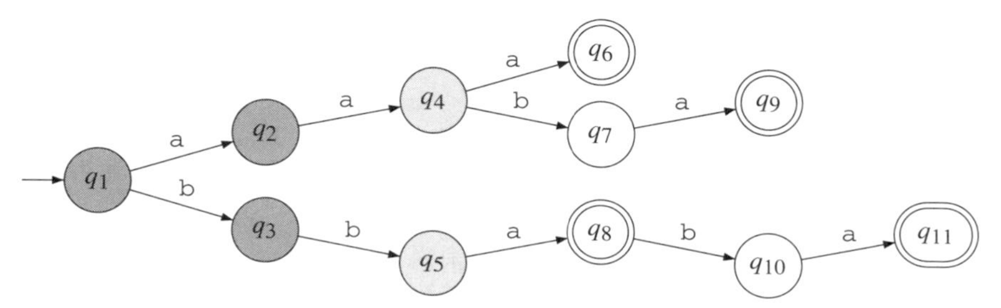

Fig. 12.23. The PTA with q_3 promoted.

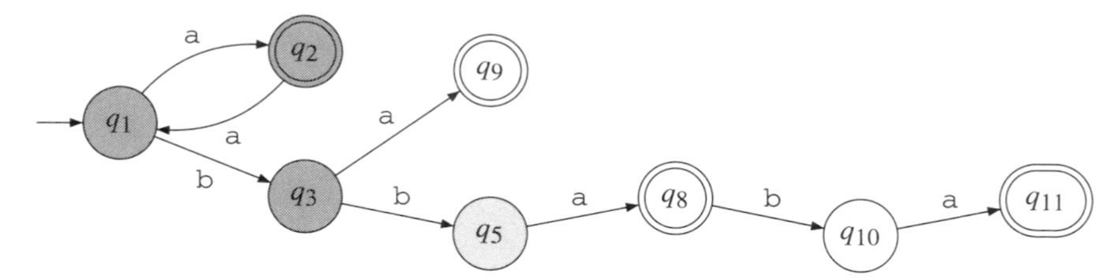

Fig. 12.24. Merging q_4 with q_1 and folding.

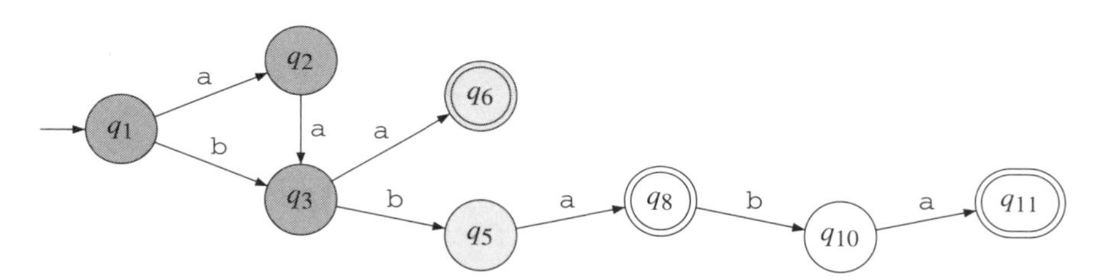

Fig. 12.25. Automaton after merging q_4 with q_3.

The merge between q_4 and q_2 is then tested and fails because of a now being parsed. The next merge (q_4 with q_3) is accepted. The resulting automaton is shown in Figure 12.25.

The next BLUE state is q_5; notice that state q_6 has the shortest prefix at that point, but what counts is the situation in the original PTA. The next merge to be tested is q_5 with q_1: it is rejected because of string a which is a counter-example that would be accepted by the resulting automaton (represented in Figure 12.26). Then the algorithm tries merging q_5 with q_2: this involves folding in the different states q_8, q_{10} and q_{11}. The merge is accepted, and the automaton depicted in Figure 12.27 is constructed.

Finally, BLUE state q_6 is merged with q_1. This merge is accepted, resulting in the automaton represented in Figure 12.28.

Last, the states are marked as final (rejecting). The final accepting ones are correct, but by parsing strings from S_-, state q_2 is marked as rejecting (Figure 12.29).

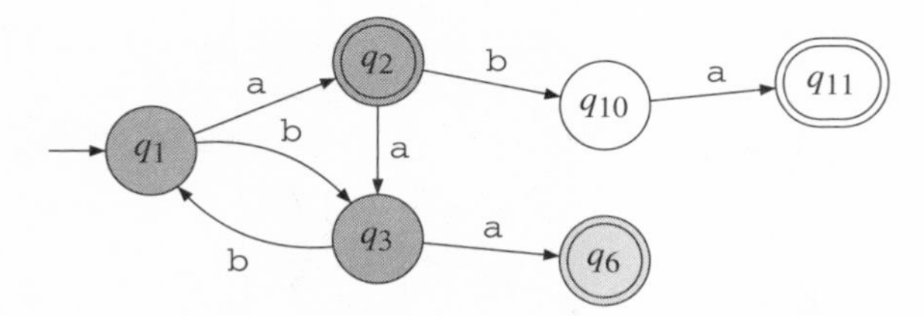

Fig. 12.26. Automaton after merging q_5 with q_1, and folding.

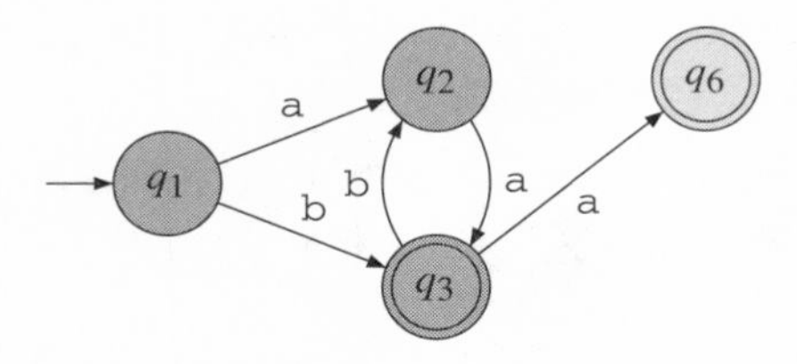

Fig. 12.27. Automaton after merging q_5 with q_2, and folding.

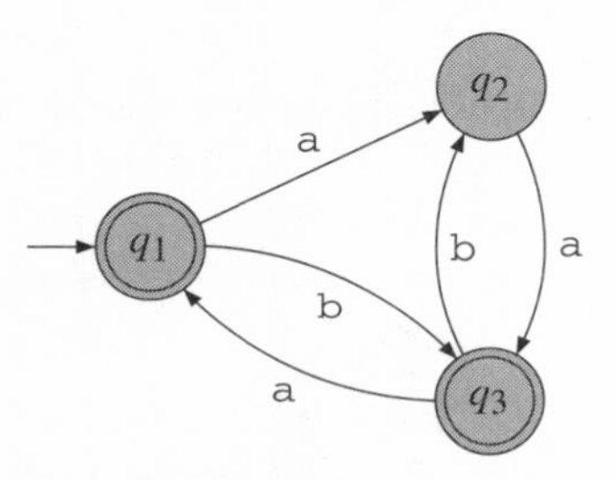

Fig. 12.28. Automaton after merging q_6 with q_1.

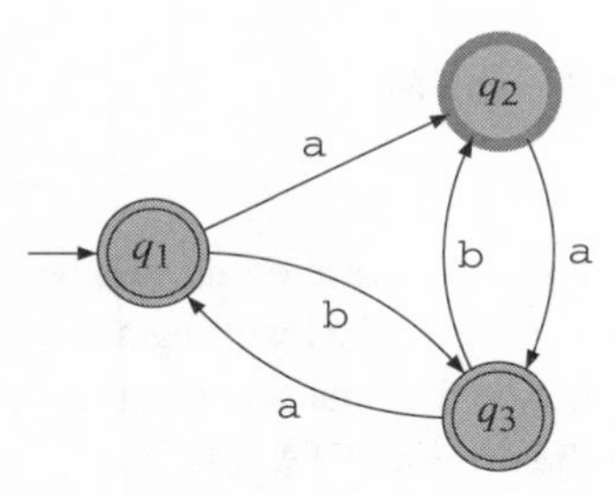

Fig. 12.29. Automaton after marking the final rejecting states.

12.4.4 Some comments about implementation

The RPNI algorithm, if implemented as described above, does not scale up. It needs a lot of further work done to it before reaching a satisfactory implementation. It will be necessary to come up with a correct data structure for the PTA and the intermediate automata. One should consider the solutions proposed by the different authors working in the field.

The presentation we have followed here avoids the heavy non-deterministic formalisms that can be found in the literature and that add an extra (and mostly unnecessary) difficulty to the implementation.

12.5 Exercises

12.1 Algorithm 12.4 (page 245) has a complexity in $\mathcal{O}(|\text{STA}|^2 + |\text{STA}| \cdot |E|^2)$. Find an alternative algorithm, with a complexity in $\mathcal{O}(|\text{STA}| \cdot |E|)$.

12.2 Run Gold's algorithm for the following data:

$$S_+ = \{\texttt{a}, \texttt{abb}, \texttt{bab}, \texttt{babbb}\}$$
$$S_- = \{\texttt{ab}, \texttt{bb}, \texttt{aab}, \texttt{b}, \texttt{aaaa}, \texttt{babb}\}$$

12.3 Take RED $= \{q_\lambda\}$, EXP $= \{\lambda, \texttt{b}, \texttt{bbb}\}$. Suppose $S_+ = \{\lambda, \texttt{b}\}$ and $S_- = \{\texttt{bbb}\}$. Construct the corresponding DFA, with Algorithm GOLD-BUILDAUTOMATON 12.5 (page 247). What is the problem?

12.4 Build an example where RED is not prefix-closed and for which Algorithm GOLD-BUILDAUTOMATON 12.5 (page 247) fails.

12.5 In Algorithm GOLD the complexity seems to depend on revisiting each cell in OT various times in order to decide if two lines are obviously different. Propose a data structure which allows the first phase of the algorithm (the deductive phase) to be in $\mathcal{O}(n \cdot \|S\|)$ where n is the size (number of states) of the target DFA.

12.6 Construct the characteristic sample for the automaton depicted in Figure 12.30(a) with Algorithm RPNI-CONSTRUCTCS (12.14).

12.7 Construct the characteristic sample for the automaton depicted in Figure 12.30(b), as defined in Algorithm RPNI-CONSTRUCTCS (12.14).

12.8 Run Algorithm RPNI for the order relations $\leq_{alpha}$ and $\leq_{lex\text{-}length}$ on

$$S_+ = \{\texttt{a}, \texttt{abb}, \texttt{bab}, \texttt{babbb}\}$$
$$S_- = \{\texttt{ab}, \texttt{bb}, \texttt{aab}, \texttt{b}, \texttt{aaaa}, \texttt{babb}\}.$$

12.9 We consider the following definition in which a learning algorithm is supposed to learn from its previous hypothesis and the new example:

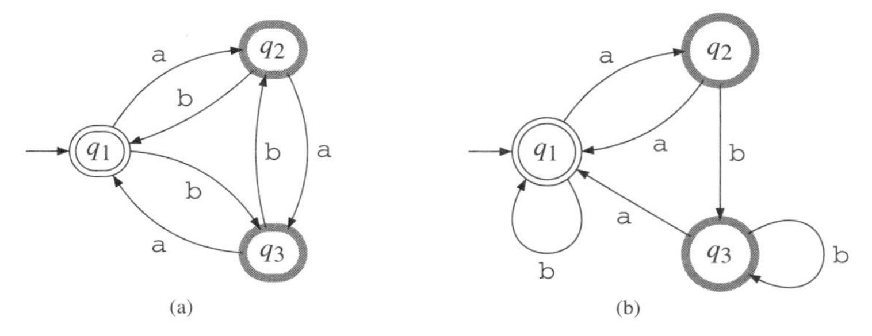

Fig. 12.30. Target automata for the exercises.

Definition 12.5.1 An algorithm **A incrementally** identifies grammar class $\mathcal{G}$ in the limit if_{def} given any T in $\mathcal{G}$, and any presentation ϕ of $\mathbb{L}(T)$, there is a rank n such that if $i \geq n$, $\mathbf{A}(\phi(i), G_i) \equiv T$.

Can we say that RPNI is incremental? Can we make it incremental?

12.10 What do you think of the following conjecture?
Conjecture of non-incrementality of the regular languages. There exists no incremental algorithm that identifies the regular languages in the limit from an informant. More precisely, let **A** be an algorithm that given a DFA $\mathcal{A}_k$, a current hypothesis, and a labelled string w_k (hence a pair $\langle w_k, l(w_k)\rangle$), returns an automaton $\mathcal{A}_{k+1}$. In that case we say that **A** identifies in the limit an automaton T if_{def} for no rank k above n, $\mathcal{A}_k \not\equiv T$.
Note that $l(w)$ is the label of string w, i.e. 1 or 0.

12.11 Devise a collusive algorithm to identify DFAs from an informant. The algorithm should rely on an encoding of DFAs over the intended alphabet Σ. The algorithm checks the data, and, if some string corresponds to the encoding of a DFA, this DFA is built and the sample is reconsidered: is the DFA minimal and compatible for this sample? If so, the DFA is returned. If not, the PTA is returned. Check that this algorithm identifies DFAs in POLY-CS time.

12.6 Conclusions of the chapter and further reading

12.6.1 Bibliographical background

In Section 12.1 we have tried to define in a uniform way the problem of learning DFAs from an informant. The notions developed here are based on the common notion of the prefix tree acceptor, sometimes called the augmented PTA, which has been introduced by various authors (Alquézar & Sanfeliu, 1994, Coste & Fredouille, 2003). It is customary to present learning in a more asymmetric way as generalising from the PTA and controlling the generalisation (i.e. avoiding over-generalisation) through the negative examples. This approach is certainly justified by the capacity to define the search space neatly (Dupont, Miclet & Vidal, 1994): we will return to it in Chapters 6 and 14.

Here the presentation consists of viewing the problem as a classification question and giving no advantage to one class over another. Among the number of reasons for preferring this idea, there is a strong case for manipulating three types of states, some of which are of unknown label. There is a point to be made here: if in ideal conditions, and when convergence is reached, the hypothesis DFA (being exactly the target) will only have final states (some accepting and some rejecting), this will not be the case when the result is incorrect. In that case deciding that all the non-accepting states are rejecting is bound to be worse than leaving the question unsettled.

The problem of identifying DFAs from an informant has attracted a lot of attention: E. Mark Gold (Gold, 1978) and Dana Angluin (Angluin, 1978) proved the intractability of finding the smallest consistent automaton. Lenny Pitt and Manfred Warmuth extended this result to non-approximability (Pitt & Warmuth, 1993). Colin de la Higuera (de la Higuera, 1997) noticed that the notion of polynomial samples was non-trivial.

E. Mark Gold's algorithm (GOLD) (Gold, 1978) was the first grammatical inference algorithm with strong convergence properties. Because of its incapacity to do better than return the PTA the algorithm is seldom used in practice. There is nevertheless room for improvement. Indeed, the first phase of the algorithm (the deductive step) can be implemented with a time complexity of $\mathcal{O}(\|S\| \cdot n)$ and can be used as a starting point for heuristics.

Algorithm RPNI is described in Section 12.4. It was developed by Jose Oncina and Pedro García (Oncina & García, 1992). Essentially this presentation respects the original algorithm. We have only updated the notations and somehow tried to use a terminology common to the other grammatical inference algorithms. There have been other alternative approaches based on similar ideas: work by Boris Trakhtenbrot and Ya Barzdin (Trakhtenbrot & Bardzin, 1973) and by Kevin Lang (Lang, 1992) can be checked for details.

A more important difference in this presentation is that we have tried to avoid the non-deterministic steps altogether. By replacing the symmetrical merging operation, which requires determinisation (through a *cascade of merges*), by the simpler asymmetric folding operation, NFAs are avoided.

In the same line, an algorithm that doesn't construct the PTA explicitly is presented in (de la Higuera, Oncina & Vidal, 1996). The RED, BLUE terminology was introduced in (Lang, Pearlmutter & Price, 1998), even if does not coincide exactly with previous definitions: in the original RPNI the authors use *shortest prefixes* to indicate the RED states and elements of the *kernel* for some prefixes leading to the BLUE states. Another analysis of merging can be found in (Lambeau, Damas & Dupont, 2008).

Algorithm RPNI has been successfully adapted to tree automata (García & Oncina, 1993), and infinitary languages (de la Higuera & Janodet, 2004).

An essential reference for those wishing to write their own algorithms for this task is the datasets. Links about these can be found on the grammatical inference webpage (van Zaanen, 2003). One alternative is to generate one's own targets and samples. This can be done with the GOWACHIN machine (Lang, Pearlmutter & Coste, 1998).

12.6.2 Some alternative lines of research

Both GOLD and RPNI have been considered as good starting points for other algorithms. During the ABBADINGO competition, state merging was revisited, the order relation being built during the run of the algorithm. This led to heuristic EDSM (evidence driven state merging), which is described in Section 14.5.

More generally the problem of learning DFAs from positive and negative strings has been tackled by a number of other techniques (some of which are presented in Chapter 14).

12.6.3 Open problems and possible new lines of research

There are a number of questions that still deserve to be looked into concerning the problem of learning DFAs from an informant, and the algorithms GOLD and RPNI:

Concerning the problem of learning DFAs from an informant. Both algorithms we have proposed are incapable of adapting correctly to an incremental setting. Even if a first try was made by Pierre Dupont (Dupont, 1996), there is room for improvement. Moreover, one aspect of incremental learning is that we should be able to *forget* some of the data we are learning from during the process. This is clearly not the case with the algorithms we have seen in this chapter.

Concerning the GOLD algorithm. There are two research directions one could recommend here. The first corresponds to the deductive phase. As mentioned in Exercise 12.5, clever data structures should accelerate the construction of the table with as many obviously different rows as possible.

The second line of research corresponds to finding better techniques and heuristics to fill the holes.

Concerning the RPNI algorithm. The complexity of the RPNI algorithm remains loosely studied. The actual computation which is proposed is not convincing, and empirically, those that have consistently used the algorithm certainly do not report a cubic behaviour. A better analysis of the complexity (joined with probably better data structures) is of interest in that it would allow us to capture the parts where most computational effort is spent.

Other related topics. A tricky open question concerns the collusion issues. An alternative 'learning' algorithm could be to find in the sample a string which would be the encoding of a DFA, decode this string and check if the characteristic sample for this automaton is included. This algorithm would then rely on collusion: it needs a teacher to encode the automaton. Collusion is discussed in (Goldman & Mathias, 1996): for the learner-teacher model to be able to resist collusion, an adversary is introduced. This, here, corresponds to the fact that the characteristic sample is to be included in the learning sample for identification to be mandatory. But the fact that one can encode the target into a unique string, which is then correctly labelled and passed to the learner together with a proof that the number of states is at least n, which has been remarked in a number of papers (Castro & Guijarro, 2000, Denis & Gilleron, 1997, Denis, d'Halluin & Gilleron, 1996) remains troubling.

13
Learning with queries

> Among the more interesting remaining theoretical questions are: inference in the presence of noise, general strategies for interactive presentation and the inference of systems with semantics.
>
> ***Jerome Feldman*** *(Feldman, 1972)*

> La simplicité n'a pas besoin d'être simple, mais du complexe resserré et synthétisé.
>
> ***Alfred Jarry***

We describe algorithm LSTAR, introduced by Dana Angluin, which has inspired several variants and adaptations to other classes of languages.

13.1 The minimally adequate teacher

A minimally adequate teacher (MAT) is an Oracle that can give answers to membership queries and strong equivalence queries. We analysed in Section 9.2 the case where you want to learn with less.

The main algorithm that works in this setting is called LSTAR. The general idea of LSTAR is:

- find a consistent observation table (representing a DFA),
- submit it as an equivalence query,
- use the counter-example to update the table,
- submit membership queries to make the table *closed and complete*,
- iterate until the Oracle, upon an equivalence query, tells us that the correct language has been reached.

The observation table we use is analogous to that described in Section 12.3, so we will use the same formalism here.

13.1.1 An observation table

An observation table is a specific tabular representation of an automaton. An example is given in Table 13.1(a).

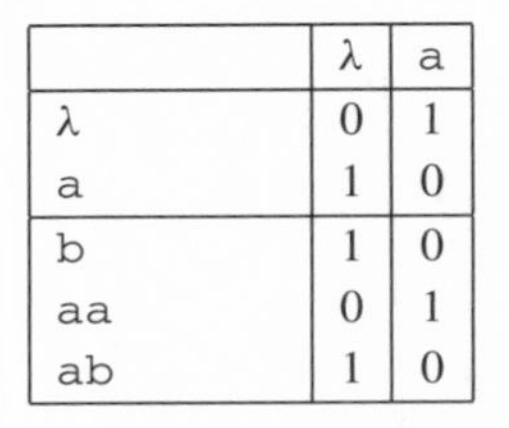

	λ	a
λ	0	1
a	1	0
b	1	0
aa	0	1
ab	1	0

(a) An observation table.

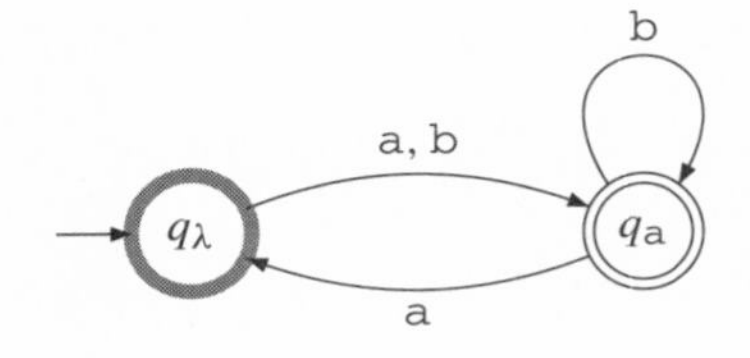

(b) The corresponding automaton.

Fig. 13.1. The observation table and the corresponding automaton.

The meaning of the table can be made reasonably clear. By concatenating the name of a row r with the name of a column c we get a string rc. The string is in the language if the corresponding cell $\text{OT}[r][c]$ contains a 1, and is not if it is a 0. If the table complies with certain conditions of consistency, an automaton can easily be extracted from the table. The automaton corresponding to Table 13.1(a) is depicted in Figure 13.1(b). We formally describe a procedure allowing us to extract a DFA from a table (when possible) in Section 13.1.2.

An observation table is a triple $\langle \text{STA}, \text{EXP}, \text{OT} \rangle$ where:

- $\text{STA} = \text{RED} \cup \text{BLUE}$ is a set of strings, denoting labels of states,
- $\text{RED} \subset \Sigma^*$ is a finite set of states,
- $\text{EXP} \subset \Sigma^*$ is the experiment set,
- $\text{BLUE} = \text{RED} \cdot \Sigma \setminus \text{RED}$ is the set of states successors of RED that are not RED,
- $\text{OT} : \text{STA} \times \text{EXP} \to \{0, 1, *\}$ is a function such that:
- $\text{OT}[u][e] = \begin{cases} 1 & \text{if } ue \in L \\ 0 & \text{if } ue \notin L \\ * & \text{otherwise (not known).} \end{cases}$

Again, to simplify, RED and BLUE will be sets of strings also used to label the states.

There are a number of key ideas one wants to understand in order to grasp this algorithm.

Definition 13.1.1 (Holes) A **hole** in a table $\langle \text{STA}, \text{EXP}, \text{OT} \rangle$ is a pair (u, e) such that $\text{OT}[u][e] = *$.

A table is **complete** if_{def} it has no holes.

The problem with incomplete tables is that we do not have all the information needed to extract a DFA from a table. In Section 12.3 (page 243) this was the key problem and no satisfying solution was given. Consider for instance Table 13.2(a). There are several holes that could be filled in various manners. For example, the hole corresponding to $\text{OT}[\texttt{ab}][\texttt{b}]$ indicates that there is no fixed or known value for $\delta(q_{\texttt{ab}}, \texttt{b})$. In order to build a DFA from this incomplete table, we notice that $\delta(q_{\texttt{ab}}, \texttt{b})$ could just as well be q_λ as $q_{\texttt{ab}}$. The situation is represented Figure 13.2(b).

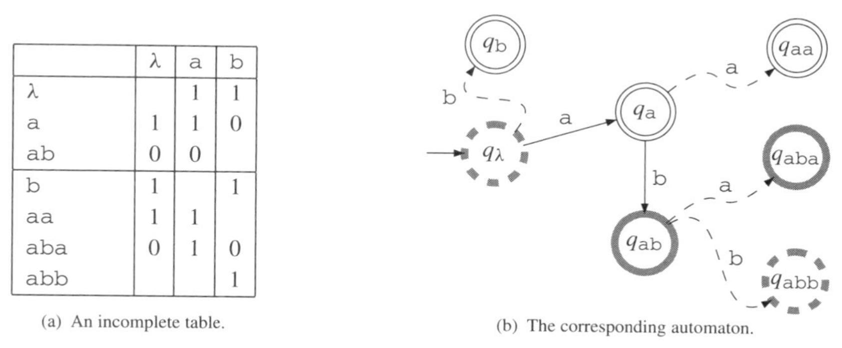

	λ	a	b
λ		1	1
a	1	1	0
ab	0	0	
b	1		1
aa	1	1	
aba	0	1	0
abb			1

(a) An incomplete table.

(b) The corresponding automaton.

Fig. 13.2. The automaton corresponding to an incomplete table.

	λ	a
λ	0	0
a	0	0
aa	0	0
aab	1	0
b	1	0
ab	1	0
aaba	0	0
aabb	1	0

(a) The observation table.

	a	b
q_λ	q_λ	q_{b}
q_{b}	q_λ	q_{b}

(b) The transition table.

(c) Automaton.

Fig. 13.3. An automaton and a table.

13.1.2 Building a DFA from a complete and closed table

Building an automaton from a table ⟨STA, EXP, OT⟩ can be done very easily if certain conditions are met:

- The set of strings marking the states in STA must be prefix-closed,
- The set EXP is suffix-closed,
- The table must be complete and therefore have no holes,
- The table must be closed.

If these conditions hold we can use Algorithm LSTAR-BUILDAUTOMATON (13.1, similar to Algorithm 12.5, page 247).

Example 13.1.1 Consider Table 13.3(a). We can apply the construction from Algorithm 13.1 and obtain $Q = \{q_\lambda, q_{\text{a}}\}$, $\mathbb{F}_{\mathbb{A}} = \{q_{\text{a}}\}$, $\mathbb{F}_{\mathbb{R}} = \{q_\lambda\}$ and δ given by the transition table 13.3(b). Then the automaton 13.3(c) can be built.

Algorithm 13.1: LSTAR-BUILDAUTOMATON.

Input: a closed and complete observation table (STA, EXP, OT)
Output: DFA $\langle \Sigma, Q, q_\lambda, \mathbb{F}_\mathbb{A}, \mathbb{F}_\mathbb{R}, \delta \rangle$
$Q \leftarrow \{q_u : u \in \text{RED} \wedge \forall v < u \ \text{OT}[v] \neq \text{OT}[u]\}$;
$\mathbb{F}_\mathbb{A} \leftarrow \{q_u \in Q : \text{OT}[u][\lambda] = 1\}$;
$\mathbb{F}_\mathbb{R} \leftarrow \{q_u \in Q : \text{OT}[u][\lambda] = 0\}$;
for $q_u \in Q$ **do**
| **for** $a \in \Sigma$ **do** $\delta(q_u, a) \leftarrow q_w \in Q : \text{OT}[ua] = \text{OT}[w]$
end
return $\langle \Sigma, Q, q_\lambda, \mathbb{F}_\mathbb{A}, \mathbb{F}_\mathbb{R}, \delta \rangle$

13.1.3 Consistency

Definition 13.1.2 (Consistent table) Given an automaton $\mathcal{A}$ and an observation table $\langle \text{STA}, \text{EXP}, \text{OT} \rangle$, $\mathcal{A}$ is **consistent** with $\langle \text{STA}, \text{EXP}, \text{OT} \rangle$ when the following holds:

- $\text{OT}[u][e] = 1 \implies ue \in \mathbb{L}_{\mathbb{F}_\mathbb{A}}(\mathcal{A})$,
- $\text{OT}[u][e] = 0 \implies ue \in \mathbb{L}_{\mathbb{F}_\mathbb{R}}(\mathcal{A})$.

Remember that $\mathbb{L}_{\mathbb{F}_\mathbb{A}}(\mathcal{A})$ is the language recognised by $\mathcal{A}$ by accepting states, whereas $\mathbb{L}_{\mathbb{F}_\mathbb{R}}(\mathcal{A})$ is the language recognised by $\mathcal{A}$ by rejecting states.

Theorem 13.1.1 (Consistency theorem) Let $\langle \text{STA}, \text{EXP}, \text{OT} \rangle$ be an observation table, closed and complete. If STA is prefix-closed and EXP is suffix-closed then LSTAR-BUILDAUTOMATON($\langle \text{STA}, \text{EXP}, \text{OT} \rangle$) is consistent with the data in $\langle \text{STA}, \text{EXP}, \text{OT} \rangle$.

Proof LSTAR-BUILDAUTOMATON($\langle \text{STA}, \text{EXP}, \text{OT} \rangle$) is built from the data from $\langle \text{STA}, \text{EXP}, \text{OT} \rangle$. □

Completing the table by submitting membership queries should allow us to solve some of the ambiguity issues, even if not all of them.

13.1.4 Tables with no holes

We consider here the case where there are no holes in the table. To reach this situation, we will have filled the holes by using membership queries.

Definition 13.1.3 (Equivalent prefixes and rows) Two prefixes u and v are **equivalent** *if$_{def}$* $\text{OT}[u] = \text{OT}[v]$. We will denote this by $u \equiv_{\text{EXP}} v$.

The next definition is similar to Definition 12.3.6 (page 246).

	λ	a
λ	0	1
a	1	0
b	1	0
aa	0	1
ab	1	1

(a) A table that is not closed, because of row ab.

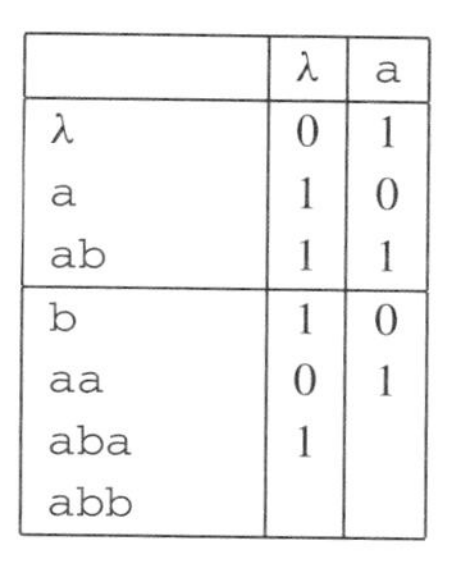

	λ	a
λ	0	1
a	1	0
ab	1	1
b	1	0
aa	0	1
aba	1	
abb		

(b) Closing the table.

Fig. 13.4. Closing a table.

Definition 13.1.4 (Closed table) A table ⟨STA, EXP, OT⟩) is **closed** if_{def} given any row u of BLUE there is some row v in RED such that $u \equiv_{\text{EXP}} v$.

Checking if the table is closed is straightforward. But what can the algorithm do once it has found that the table is not closed? Let s be the row (of BLUE) that does not appear in RED, add s to RED, and $\forall a \in \Sigma$, add sa to BLUE. This corresponds to the ***promotion*** introduced in Section 12.2.3 (page 242).

By repeating this until the table is closed, we are done. Notice that the number of iterations is bounded by the size of the automaton.

Example 13.1.2 In Gold's algorithm (Section 12.3) all the RED states were obviously different from each other. Moreover, a state was moved to the upper part of the table only when this condition was met. Because of the lack of control the learner has over the Oracle, this is not the case here.

An ***inconsistent*** table is one from which an automaton cannot be extracted. This is different from Section 12.3: in this case it is possible to have RED prefixes/states that seem equivalent and need separating.

Definition 13.1.5 A table is **consistent** if_{def} every equivalent pair of rows in RED remains equivalent in STA after appending any symbol.
$\text{OT}[s_1] = \text{OT}[s_2] \implies \forall a \in \Sigma, \text{OT}[s_1 a] = \text{OT}[s_2 a]$.

What do we do when we have an inconsistent table? If it is inconsistent, then let $a \in \Sigma$ be the symbol for which $\text{OT}[s_1] = \text{OT}[s_2]$ but $\text{OT}[s_1 a] \neq \text{OT}[s_2 a]$. Let e be the experiment for which the inconsistency has been found ($\text{OT}[s_1 a][e] \neq \text{OT}[s_2 a][e]$). Then by adding experiment ae to the table, rows $\text{OT}[s_1]$ and $\text{OT}[s_2]$ are different. Indeed, $\text{OT}[s_1][ae] \neq \text{OT}[s_2][ae]$.

Example 13.1.3 Table 13.5(a) is inconsistent: rows a and ab look the same, but, upon experiment a they fail to be equivalent, since rows aa and aba are different. Therefore column (and experiment) aa is added, resulting in Table 13.5(b).

	λ	a
λ	0	1
a	1	0
ab	1	0
b	1	0
aa	0	1
aba	0	0
abb	1	0

(a) An inconsistent table (because of a and ab).

	λ	a	aa
λ	0	1	0
a	1	0	1
ab	1	0	0
b	1	0	
aa	0	1	
aba	0	1	
abb	1	0	

(b) The table has become consistent.

	λ	a
λ	0	1
a	1	0
ab	1	0
b	1	0
aa	0	1
aba	0	1
abb	1	0

(c) A consistent table.

Fig. 13.5. Consistency.

On the other hand, Table 13.5(c) is consistent, since we have not only OT[a] = OT[ab], but also OT[aa] = OT[aba] and OT[ab] = OT[abb].

Once the learner has built a complete, closed and consistent table, it can construct the DFA using Algorithm LSTAR-BUILDAUTOMATON and make an equivalence query!

Obviously, if the Oracle returns a positive answer to the algorithm's equivalence query, it can halt. If she returns a counter-example (u), then the learner should add as RED states all the prefixes of u, and complete the BLUE section accordingly (with all strings pa ($a \in \Sigma$) such that p is a prefix of u but pa is not). In this way at least one new RED line *obviously different* from all the others will have been added.

13.2 The algorithm

Algorithm LSTAR (13.2) can now be described. First the observation table is initialised by Algorithm LSTAR-INITIALISE (13.3). This consists of building one RED row (λ) and as many BLUE rows as there are symbols in the alphabet. Then the iterative construction begins. When the table is not closed an extra row is added (Algorithm LSTAR-CLOSE (13.4)), when it is inconsistent an extra column is added (Algorithm LSTAR-CONSISTENT (13.5)). At every moment, membership queries are made to fill in the holes. When we are ready and the table is closed and consistent, an equivalence query is made and if unsuccessful, new rows are added (Algorithm LSTAR-USEEQ (13.6)).

13.2.1 A run of LSTAR

We run LSTAR over an example. We start with the empty table, in which RED $= \{\lambda\}$ and EXP $= \{\lambda\}$. A first membership query is made with string λ. The answer is YES. Then a and b are added to BLUE and the membership queries a and b are made. Suppose again the answers are YES. The corresponding Table 13.6(a) is closed and complete, so an equivalence query is made for the automaton depicted in Figure 13.6(b).

Algorithm 13.2: LSTAR Learning Algorithm.

Input: –
Output: DFA $\mathcal{A}$
LSTAR-INITIALISE;
repeat
  while ⟨STA, EXP, OT⟩ *is not closed or not consistent* **do**
    if ⟨STA, EXP, OT⟩ *is not closed* **then**
    ⟨STA, EXP, OT⟩ ← LSTAR-CLOSE(⟨STA, EXP, OT⟩);
    if ⟨STA, EXP, OT⟩ *is not consistent* **then**
    ⟨STA, EXP, OT⟩ ← LSTAR-CONSISTENT(⟨STA, EXP, OT⟩)
  end
  Answer← EQ(⟨STA, EXP, OT⟩);
  if Answer≠ YES **then**
  ⟨STA, EXP, OT⟩ ← LSTAR-USEEQ(⟨STA, EXP, OT⟩, Answer)
until Answer= YES ;
return LSTAR-BUILDAUTOMATON(⟨STA, EXP, OT⟩)

Algorithm 13.3: LSTAR-INITIALISE.

Input: –
Output: table ⟨STA, EXP, OT⟩
RED ← $\{q_\lambda\}$;
BLUE ← $\{q_a : a \in \Sigma\}$;
EXP ← $\{\lambda\}$;
OT[λ][λ] ← MQ(λ);
for $a \in \Sigma$ **do** OT[a][λ] ← MQ(a);
return ⟨STA, EXP, OT⟩

Let abb be the negative counter-example returned by the Oracle. The table is updated to Table 13.7(a). The table now closed and the holes can be filled through membership queries, yielding Table 13.7(a).

But Table 13.6(a) is not closed as rows OT[a] and OT[ab] coincide whereas rows OT[a · b] and OT[ab · b] do not. Experiment b is the reason for this, so it is added as an experiment and the new Table 13.7(c) has to be completed. Table 13.7(d) is therefore obtained, which is now closed and complete and can be transformed into an automaton that will be proposed as an equivalence query (Figure 13.8).

We suppose this time the equivalence query is met with a positive answer so we halt.

Algorithm 13.4: LSTAR-CLOSE.

Input: a table $\langle \text{STA}, \text{EXP}, \text{OT} \rangle$
Output: table $\langle \text{STA}, \text{EXP}, \text{OT} \rangle$ updated
for $s \in$ BLUE *such that* $\forall u \in$ RED $\text{OT}[s] \neq \text{OT}[u]$ **do**
  RED $\leftarrow$ RED $\cup \{s\}$;
  BLUE $\leftarrow$ BLUE $\setminus \{s\}$;
  for $a \in \Sigma$ **do** BLUE $\leftarrow$ BLUE $\cup \{s \cdot a\}$;
  for $u, e \in \Sigma^\star$ *such that* $\text{OT}[u][e]$ *is a hole* **do** $\text{OT}[u][e] \leftarrow \text{MQ}(ue)$
end
return $\langle \text{STA}, \text{EXP}, \text{OT} \rangle$

Algorithm 13.5: LSTAR-CONSISTENT.

Input: a table $\langle \text{STA}, \text{EXP}, \text{OT} \rangle$
Output: table $\langle \text{STA}, \text{EXP}, \text{OT} \rangle$ updated
find $s_1, s_2 \in$ RED, $a \in \Sigma$ and $e \in$ EXP such that $\text{OT}[s_1] = \text{OT}[s_2]$ and
$\text{OT}[s_1 \cdot a][e] \neq \text{OT}[s_2 \cdot a][e]$;
EXP $\leftarrow$ EXP $\cup \{a \cdot e\}$;
for $u, e \in \Sigma^\star$ *such that* $\text{OT}[u][e]$ *is a hole* **do** $\text{OT}[u][e] \leftarrow \text{MQ}(ue)$;
return $\langle \text{STA}, \text{EXP}, \text{OT} \rangle$

Algorithm 13.6: LSTAR-USEEQ.

Input: a table $\langle \text{STA}, \text{EXP}, \text{OT} \rangle$, string Answer
Output: table $\langle \text{STA}, \text{EXP}, \text{OT} \rangle$ updated
for $p \in$ PREF(Answer) **do**
  RED $\leftarrow$ RED $\cup \{p\}$;
  for $a \in \Sigma : pa \notin$ PREF(Answer) **do** BLUE $\leftarrow$ BLUE $\cup \{pa\}$
end
for $u, e \in \Sigma^\star$ *such that* $\text{OT}[u][e]$ *is a hole* **do** $\text{OT}[u][e] \leftarrow \text{MQ}(ue)$;
return $\langle \text{STA}, \text{EXP}, \text{OT} \rangle$

13.2.2 Proof of the algorithm

The algorithm LSTAR clearly terminates. Since every regular language admits a unique minimal DFA, let us suppose, without loss of generality, that the target is this minimum DFA with n states. But since any DFA consistent with a table has at least as many states

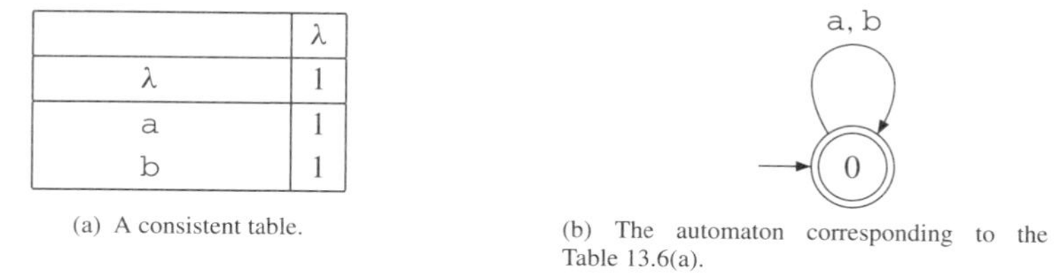

	λ
λ	1
a	1
b	1

(a) A consistent table.

(b) The automaton corresponding to the Table 13.6(a).

Fig. 13.6. Consistency.

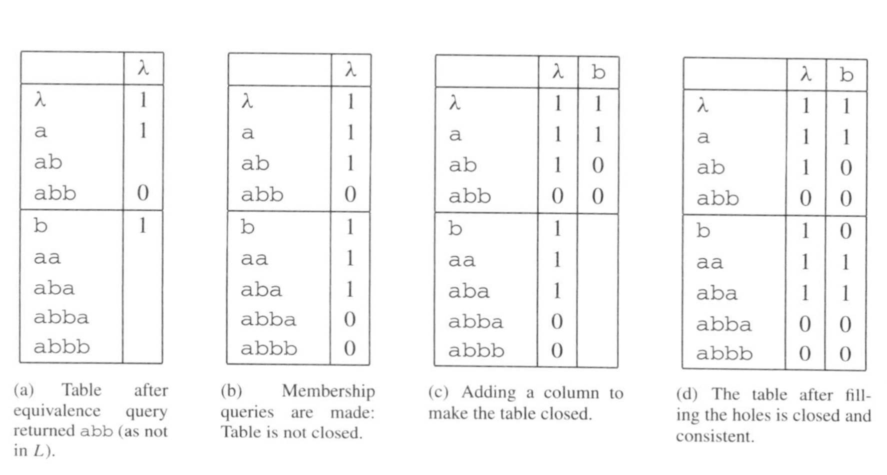

	λ
λ	1
a	1
ab	
abb	0
b	1
aa	
aba	
abba	
abbb	

(a) Table after equivalence query returned abb (as not in L).

	λ
λ	1
a	1
ab	1
abb	0
b	1
aa	1
aba	1
abba	0
abbb	0

(b) Membership queries are made: Table is not closed.

	λ	b
λ	1	1
a	1	1
ab	1	0
abb	0	0
b	1	
aa	1	
aba	1	
abba	0	
abbb	0	

(c) Adding a column to make the table closed.

	λ	b
λ	1	1
a	1	1
ab	1	0
abb	0	0
b	1	0
aa	1	1
aba	1	1
abba	0	0
abbb	0	0

(d) The table after filling the holes is closed and consistent.

Fig. 13.7. Running LSTAR.

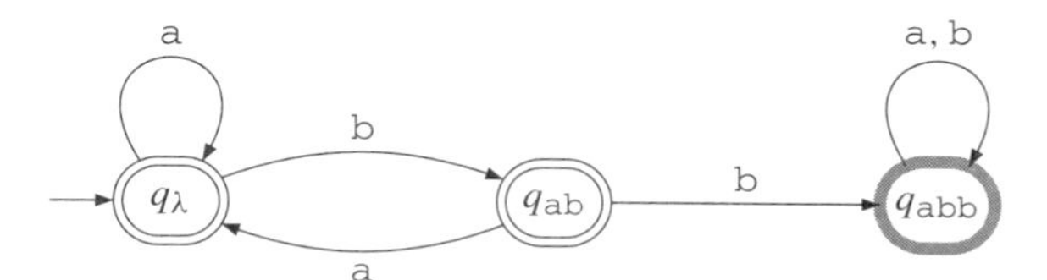

Fig. 13.8. Automaton after running LSTAR.

as different rows in RED, and if a table is closed and consistent then the construction of a consistent DFA is unique, therefore the table can only grow 'vertically' until it has n different rows.

Now, each closure failure adds one different row to RED, and each inconsistency failure adds one experiment, which also creates a new row in RED. Each counter-example also adds one different row to RED (notice that many rows can appear because of the prefixes, but what matters is that at least one appears that is different from the others). Every time the table is not consistent or an equivalence query is met by a counter-example, at least one new row is introduced. Furthermore, the number of steps between two such events is also finite. So the total number of these operations is bounded.

Now, for correction, if the algorithm has built a table with n obviously different rows in RED, and n is the size of the minimal DFA for the target, then it is the target.

The algorithm therefore is correct and terminates.

Let us now study the complexity of LSTAR:

- At most n experiments will be made (including λ), since an experiment necessarily introduces a new different row. So $|\text{EXP}| \leq n$.
- For the same reasons at most n equivalence queries are made.
- The number of membership queries is bounded by the total size of the table, which is at most n (the number of experiments/columns), multiplied by the number of lines ($\leq nm$ where m is the length of the longest counter-example returned by the Oracle).

Therefore the total number of queries made is at most n^2m. A computation of this number at each step of the algorithm is possible and gives similar results: the table only grows with the size of the counter-examples returned by the Oracle.

13.2.3 About implementation issues

One difficulty with the implementation of LSTAR comes from maintaining the redundancy. Actually it is not necessary to implement the actual table. A better idea is to manage three association tables:

- A first table MQ contains the result of the membership queries. It can be consulted in constant time to know whether a particular string has been queried or not, and if it has, whether or not it belongs to the language.
- A second table PREF contains the different names of the rows, and for each row, the status: is it RED or BLUE?
- A third table just contains the different experiments.

The actual observation table is only simulated by a function $\text{OT}(u, v)$ which will return the value $\text{MQ}[uv]$.

13.3 Exercises

13.1 Run algorithm LSTAR on the automaton from Figure 13.9. You will need to simulate the Oracle too!

13.2 Replace the equivalence queries with the use of sampling query EX() in the above. What values of m_i should you consider?

13.3 It was suggested in Section 9.4 (page 191) that equivalence queries can be replaced by sampling. Write the algorithm allowing us to actually simulate the equivalence query.

13.4 If one chooses to sample instead of making equivalence queries, one problem is: what do we do with all the examples that did fit the hypothesis until a counter-example was found? One alternative is to enter all this information into the observation table. Another is to ignore it. Which is better? Why?

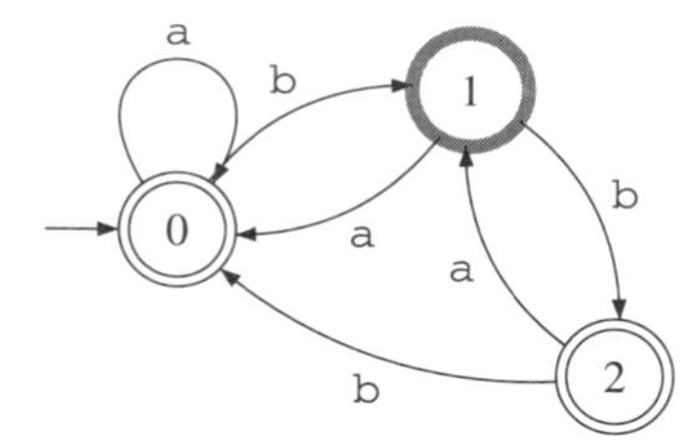

Fig. 13.9. A target automaton.

13.4 Conclusions of the chapter and further reading

In Chapter 9 we discussed a number of implications of active learning. In particular, negative results were given.

13.4.1 Bibliographical background

Algorithm LSTAR was is due to Dana Angluin (Angluin, 1990) and was later adapted for a robotics scenario by Ron Rivest and Robert Schapire (Rivest & Schapire, 1993), and has led to a number of variants (both in description and in the combination between the membership and equivalence queries (Balcázar *et al.*, 1994b, Kearns & Vazirani, 1994). We have concentrated in this chapter on DFAs. In the case of context-free grammars the negative proofs by Dana Angluin and Michael Kharitonov (Angluin & Kharitonov, 1991) with MATs are related to cryptographic assumptions. On the other hand, if structural information is available, Yasubumi Sakakibara proves the learnability of the class of context-free grammars in this model (Sakakibara, 1990). Learning balls of strings from different types of queries has also been studied (de la Higuera, Janodet & Tantini, 2008).

Returning to the DFA case, it should be noticed that the Oracle has no reason to return the counter-example the learner really needs. A more helpful setting was studied in (Birkendorf, Boeker & Simon, 2000).

Other studies contemplate the fact that the Oracle is somehow bounded. In practice, it may be difficult to imagine that the Oracle has the resources to return an exponentially long example; furthermore, if the learner can find a hypothesis correct only over the strings of reasonable length, this may be sufficient. These questions are discussed in (Castro, 2001, Watanabe, 1994).

13.4.2 Some alternative lines of research

The model has received considerable attention and there are many papers on learning with different sorts of queries. Yasubumi Sakakibara (Sakakibara, 1987) learns context-free grammars from queries. Takashi Yokomori (Yokomori, 1996) learns two-tape automata from both queries and counter-examples, and also non-deterministic finite automata from queries in polynomial time but depending on the size of the associated DFA (Yokomori, 1994). Juan-Manuel Vilar extends queries to translation tasks in (Vilar, 1996), and Oded Maler and Amir Pnueli (Maler & Pnueli, 1991) learn Büchi automata from queries over infinite strings.

A recent idea is that of combining membership queries and equivalence queries in some way. Correction queries (Beccera-Bonache, Bibire & Horia Dediu, 2005, Becerra-Bonache, Horia Dediu & Tîrnauca, 2006) correspond to strings that the learner hypothesises as being in the language. The Oracle then presents some correction of the string if the string does not belong to the language. There are many ways of defining such corrections, some being more theoretical and others (using the edit distance) closer to possible applications (Becerra-Bonache *et al.*, 2008, Kinber, 2008, Tirnauca, 2008). Again, as in other questions, the problem of correctly defining a topology over the languages is of crucial importance, as the sort of correction one would expect is one of a string *close* to the queried string.

In practice, getting hold of equivalence queries is considered to be the hardest of problems. One way around this is to sample. An evolutionary algorithm following this line is proposed in (Bongard & Lipson, 2005).

One can also consider the case where the Oracle can answer probabilistic queries. We visit this question in the corresponding chapters (Chapter 10 for some negative results, and Chapter 16 for some positive ones). A typical idea is to introduce *specific sampling queries* in order to learn probabilistic machines (de la Higuera & Oncina, 2004).

13.4.3 Open problems and possible new lines of research

We proposed in Section 9.6.3 some problems relating to learning with queries. More generally, there are a number of reasons for which inventing new query-learning algorithms or making the existing ones more efficient (not just in time, but also in number of queries) are important issues, and our feeling is that more research in this area should be encouraged.

14

Artificial intelligence techniques

Si al cabo de tres partidas de póquer no sabes todavía quien es el tonto, es que el tonto eres tú.

Manuel Vicent

The training program of an artificial intelligence can certainly include an informant, whether or not children receive negative instances.

E. Mark Gold *(Gold, 1967)*

In the different combinatorial methods described in the previous chapters, the main engine to the generalisation process is something like:

> 'If nothing tells me not to generalise, do it.'

For example, in the case where we are learning from an informant, the negative data provide us with the reason why one should not generalise (which is usually performed by merging two states). In the case of learning from text, the limitations exercised by the extra bias on the grammar class are what avoid over-generalisation.

But as a principle, the 'do it if you are allowed' idea is surely not the soundest. Since identification in the limit is achieved through elimination of alternatives, and an alternative can only be eliminated if there are facts that prohibit it, the principle is mathematically sound but defies *common sense*.

There are going to be both advantages and risks to considering a less *optimistic* point of view, which could be expressed somehow like:

> 'If there are good reasons to generalise, then do it.'

On one hand, generalisations will be justified through the fact that there is some positive ground, some good reason to make them. But on the other hand we will most often lose the mathematical performance guarantees. It will then be a matter of expertise to decide if some extra bias has been added to the system, and then to know if this bias is desired or not.

Let us discuss this point a little further. Suppose the task is learning DFAs from text, and the algorithm takes as a starting point the prefix tree acceptor, then performs different merges between states whenever this seems like a good idea. In this case the 'good idea' might be something like 'if it contributes to diminishing the size of the automaton'. Then you will have added such a heavy bias that your learning algorithm will always return the universal automaton that recognises any string! Obviously this is a simple example that would not fool anyone for long. But if we follow on with the idea, it is not difficult to come up with algorithms whose task is to learn context-free grammars, but whose construction rules are such that only grammars that generate regular languages can be effectively learnt!

As, when dealing with these heuristics, the option of identification does not make sense, it is going to be remarkably difficult to decide when a given algorithm has such a hidden bias and when it does not.

On the other hand, there are a number of reasons for which researchers in artificial intelligence have worked thoroughly in searching for new heuristics for grammatical inference problems:

- The sheer size of the search spaces makes the task of interest. When describing the set of admissible solutions as a partition lattice (see Section 6.3.4), the size of this lattice increases in a dramatic way (defined by the Bell formula) with the size of the positive data.
- We also saw in Chapter 6 that the basic operations related with the different classes of grammars and automata were intractable: the equivalence problem, the 'smallest consistent' problem.
- Furthermore the challenging nature of the associated $\mathcal{NP}$-hard problems is also something to be taken into account. Even if $\mathcal{NP}$ is in theory a unique class, there are degrees of approximation that can be different from problem to problem, and it is well known by scientists working on the effective resolution of $\mathcal{NP}$-hard problems that some can be tackled easier than others, at least in particular instances, or that the size of the tractable instances of the problem may vary from one problem to another. In the case of learning DFAs, the central problem concerns finding the 'minimum consistent DFA', a problem for which only small instances seem to be tractable by exhaustive algorithms, 'small' corresponding to less than about 30 states in the target.
- A fourth reason is that the point of view we have defended since the beginning, that there is a target language to be found, can in many situations not be perfectly adapted. In the case where we are given a sample and the problem is to find the smallest consistent grammar, this corresponds to trying to solve an intractable but combinatorially well-defined problem; there are finally cases where the data are noisy or where the target is moving.

The number of possible heuristics is very large, and we will only survey here some of the main ideas that have been tried in the field.

14.1 A survey of some artificial intelligence ideas

We will comment at the end of the chapter that more techniques could be tested; indeed one could almost systematically take an artificial intelligence text book, choose some metaheuristic method for solving hard problems, and then try to adapt it to the task of learning grammars or automata. We only give the flavour of some of the better studied ideas here. In the next sections we will describe the use of the following techniques in grammatical inference:

- genetic algorithms,
- Tabu search,
- using the MDL principle,
- heuristic greedy search,
- constraint satisfaction.

We aim here to recall the key ideas of the technique and to show, in each case, through a very brief example, how the technique can be used in grammatical inference. The goal is certainly not to be technical nor to explain the finer tuning explanations necessary in practice, but only to give the idea and to point, in the bibliographical section, to further work with the technique.

14.2 Genetic algorithms

The principle of genetic algorithms is to simulate biological modifications of genes and hope that, via evolutionary mechanisms, nature increases the quality of its population. The fact that both bio-computing and grammatical inference deal with languages and strings adds a specific flavour to this approach here.

14.2.1 Genetic algorithms: general approach

Genetic algorithms maintain a population of strings that each encode a given solution to the learning problem, then by defining the specific genetic operators, allow this population to evolve and better itself (through an adequacy to a given fitness function).

In the case where the population is made of grammars or automata supposed to somehow better describe a learning sample, a certain number of issues should be addressed:

(i) What is the search space? Does it comprise all strings describing grammars or only those that correspond to correct ones?
(ii) How do we build the first generation?
(iii) What are the genetic operators? Typically some sort of mutation should exist: a symbol in the string can mutate into another symbol. Also a crossing-over operator is usually necessary: this operation takes two strings and mixes them together in some way to obtain the siblings for the next generation.
(iv) What happens when, after an evolution, the given string does not encode a solution any more? One may consider having *stopping sequences* in the string so that non-encoding bits can be blocked between these special sequences. This is quite a nice idea leading to interesting interpretations about what is known as *junk* DNA.
(v) What fitness function should be used? How do we compare the quality of two solutions?

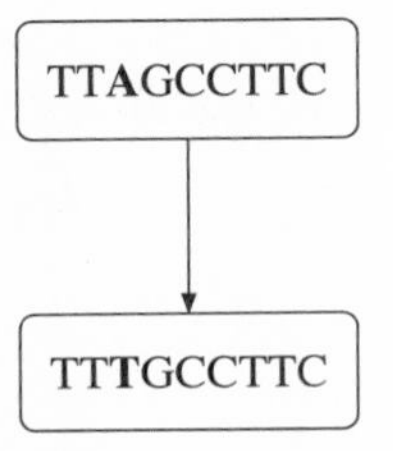

Fig. 14.1. Mutation.

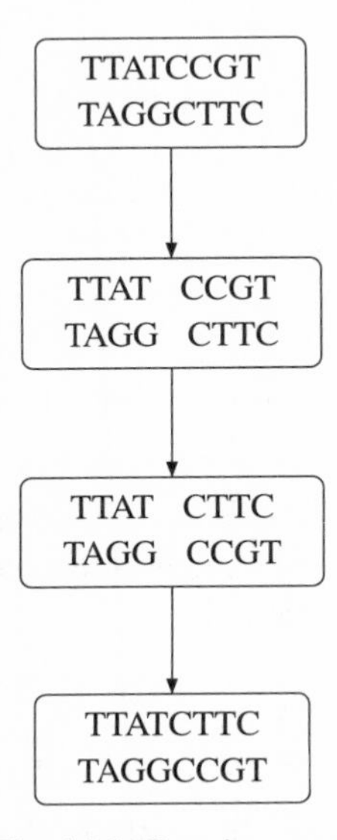

Fig. 14.2. Crossing-over.

There are also many other parameters that need tuning, such as the number of generations, the number of elements of a generation that should be kept, quantities of genetic operations that should take place, etc.

Mechanisms of evolution are essentially of two types (at the gene level): mutation and crossing-over.

Mutation consists of taking a string and letting one of the letters be modified. For example, in Figure 14.1, the third symbol is substituted. When implemented, the operation consists of randomly selecting a position, and (again randomly) modifying this symbol.

Crossing-over is more complex and involves two strings. They both get cut into two substrings, and then the crossing-over position takes place, with an exchange of the halves. In some cases the position where the strings is cut has to be the same. We give an example of this in Figure 14.2: two strings are divided at the same position (here after the fourth symbol), then crossing-over takes place.

14.2.2 A genetic algorithm in grammatical inference

We show here a simple way of implementing the different points put forward in the previous section for the specific task of learning NFAs from positive and negative examples. We are given a sample $\langle S_+, S_-\rangle$.

(i) *What is the search space?* We consider the lattice as defined in Section 6.3, containing all NFAs strongly structurally complete with $\langle S_+, S_-\rangle$. Each NFAs can therefore be represented by a partition of the states over MCA(S_+). There are several ways to represent partitions as strings. We use the following formalism: let $|E| = n$, and Π be a partition, $\Pi = \{B_1, B_2, \ldots, B_k\}$, of E. We associate with the partition Π the string w_Π : $w_\Pi(j) = m \iff j \in B_m$.

   For example, the partition $\{\{1, 2, 6\}, \{3, 7, 9, 10\}, \{4, 8, 12\}, \{5\}, \{11\}\}$ is encoded by the string w_Π=(112341232253).

(ii) *How do we generate the first generation?* We randomly start from MCA(S_+) (see Definition 6.3.2, page 125), make a certain number of merges, obtaining in that way a population of NFAs, all in the lattice and all (weakly) structurally complete.

(iii) *What are the genetic operators?* Structural operators are structural mutation and structural crossing-over over the strings w_Π. We illustrate this in Example 14.2.1.

(iv) *What happens when, after an evolution, the given string does not encode a grammar any more?* This problem does not arise here, as the operators are built to remain inside the lattice.

(v) *What fitness function should be used?* Here the two key issues are the number of strings from S_- accepted and the size of the NFA. We obviously want both as low as possible. One possibility is even to discard any NFA such that $\mathbb{L}(\mathcal{A}) \cap S_- \neq \emptyset$.

Example 14.2.1 We start with an MCA for the sample $S_+ = \{\texttt{aaaa}, \texttt{abba}, \texttt{baa}\}$ as represented in Figure 14.3. Consider an element of the initial population obtained by using the partition represented by string $w_\Pi = (112341232253)$. This NFA is drawn in Figure 14.4(a). Then if the second position is selected and the 1 is substituted by a 4 in string w_Π, we obtain string (142341232253). Building the corresponding partitions is straightforward.

Now a crossing-over between two partitions is shown in Figure 14.5. The partitions are encoded as strings, which are cut and mixed, resulting in two different partitions.

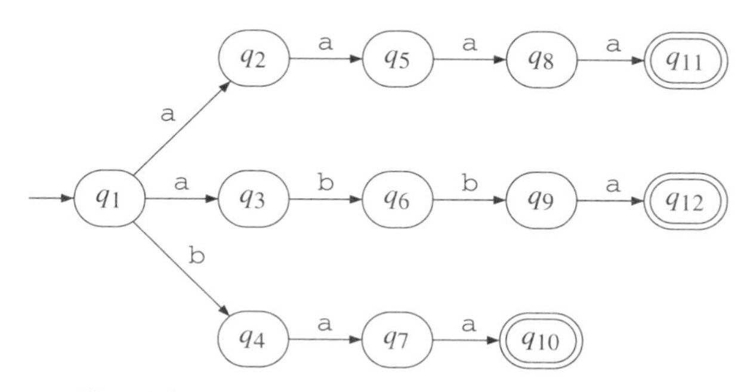

Fig. 14.3. The MCA for $S_+ = \{\texttt{aaaa}, \texttt{abba}, \texttt{baa}\}$.

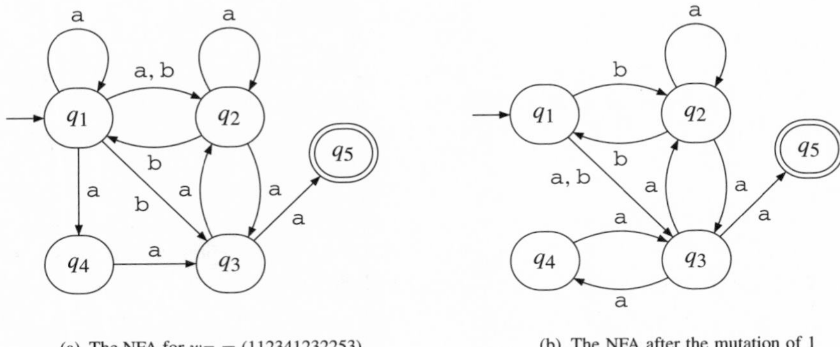

(a) The NFA for $w_\Pi = (112341232253)$.

(b) The NFA after the mutation of 1 to 4 in position 2.

Fig. 14.4. A mutation operation for an NFA.

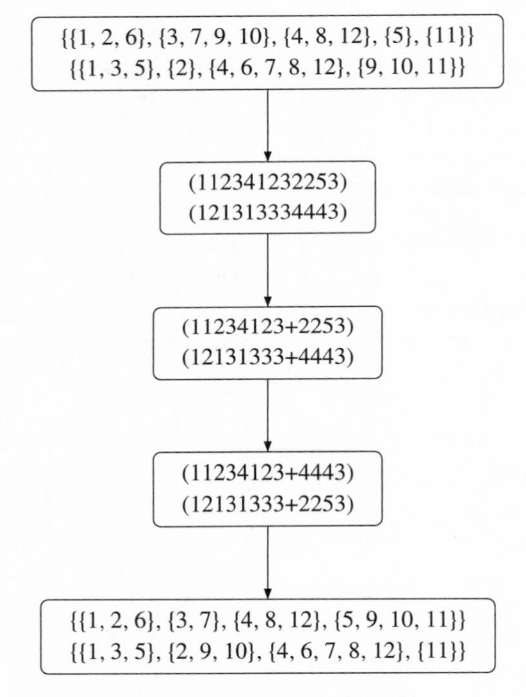

Fig. 14.5. Crossing-over with partitions.

14.3 Tabu search

When searching in large spaces, *hill-climbing* techniques try to explore the space progressively from a starting point to a local optimum, where quality is measured through a fitness function. As the optimum is only local, to try to explore the space further, different ideas

Algorithm 14.1: TABU.

Input: a sample $S = \langle S_+, S_- \rangle$, a fitness function v, an integer kmax, an initial NFA $\mathcal{A}$
Output: an NFA
$k \leftarrow 0$;
$T \leftarrow \emptyset$;
$\mathcal{A}* \leftarrow \mathcal{A}$;
while $k \neq$ kmax **do**
  build R, the set of admissible transitions that can be added or removed;
  select r in $R \setminus T$, such that the addition or deletion of r to or from $\mathcal{A}$ realises the maximum of v on S;
  add or delete r from $\mathcal{A}$;
  if $v(\mathcal{A}) > v(\mathcal{A}*)$ **then** $\mathcal{A}* \leftarrow \mathcal{A}$;
  TABU-UPDATE(T, r);
  $k \leftarrow k + 1$
end
return $\mathcal{A}*$

Algorithm 14.2: TABU-UPDATE(T, r).

Input: the Tabu list T, its maximal size m, the new element r
Output: T
if CARD(T) $= m$ **then** delete its last element;
Add r as the first element of T;
return T

have been proposed, one of which corresponds to using *Tabu lists*, or lists of operations that are forbidden, at least for a while.

14.3.1 What is Tabu search?

Tabu search also requires the definition of a search space, then the definition of some local operators to move around this space: each solution has neighbours and all these neighbours should be measured, the best (for a given fitness function) being kept for the next iteration. The idea is to iteratively try to find a neighbour of the current solution, that betters it. In order to avoid going through the same elements over and over, and getting out of a local optimum, a *Tabu* list of the last few moves is kept, and the algorithm chooses an element outside this list.

There is obviously an issue in reaching a local optimum for the fitness function. To get out of this situation (i.e. where all the neighbours are worse than the current solution) some sort of major change has to be made.

This sort of heuristic depends strongly on the tuning of a number of parameters. We will not discuss these here as they require us to take into account a large number of factors (size of the alphabet, the target, ...).

14.3.2 A Tabu search algorithm for grammatical inference

The goal is to learn regular languages, defined here by NFAs, from an informant. An inductive bias is proposed: the number of states in the automaton (or at least an upper bound of this number) is fixed. The search space is the set of all NFAs with n states, and the neighbour relation is given by, from one NFA to another, the addition or the removal of just one transition.

(i) *What is the search space?* The search space is made of λ-NFAs with n states, of which one (q_A) is accepting and all the others are rejecting. Furthermore q_A is reachable by λ-transitions only, and λ-transitions can only be used for this. There is no transition from q_A. An element in this space is represented in Figure 14.6.

(ii) *What are the local operators?* Adding and removing a transition. If a transition is added, it has to comply with the above rules. For example, in the automaton represented in Figure 14.6, any of the transitions could be removed, and transitions could be added connecting states q_1, q_2 and q_3, or λ-transitions leading to state q_A.

(iii) *What fitness function should be used?* We count here the number of strings in S correctly labelled by the NFA. We introduce a very simple fitness function v which will just parse the sample and return the number of errors.

(iv) *How do we initialise?* In theory, any n-state automaton complying with the imposed rules would be acceptable.

We denote:

- $\mathcal{A}*$ is the best solution reached so far,
- T is the Tabu list of transitions that have been added or removed in the last m moves,
- kmax is an integer bounding the number of iterations of the algorithm,
- Q is a set of $n-1$ states, $q_A \notin Q$ is the unique accepting state,
- the rules by triples: $R = Q \times \Sigma \times Q \cup Q \times \{\lambda\} \times \{q_A\}$. Therefore $(q, a, q') \in R \iff q' \in \delta_N(q, a)$.

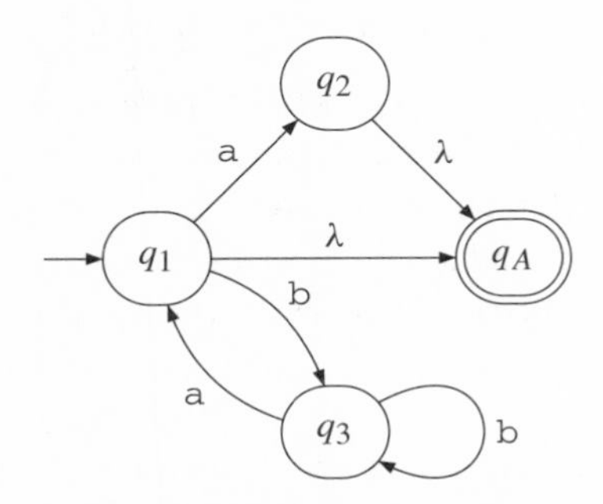

Fig. 14.6. A Tabu automaton.

14.4 MDL principle in grammatical inference

The minimum description length (MDL) principle states that the best solution is one that minimises the combination of the encoding of the grammar and the encoding of the data when parsed by the grammar.

But the principle obeys to some very strict rules. To be more exact, the way both the grammar and the data should be encoded depends on a universal Turing machine, and has to be handled carefully.

14.4.1 What is the MDL principle?

The MDL principle is a refinement of the Occam principle. Remember the Occam principle tells us that between various hypotheses, one should choose the simplest. The notion of 'simplest' here refers to some fixed notation system. The MDL principle adds the fact that simplicity should also be measured in the way the data are explained by the hypothesis. That means that the simplicity of a hypothesis (with respect to some data S) is the sum of the size of the encoding of the hypothesis and the size of the encoding of the data where the encoding of the data can be dependent on the hypothesis.

As a motivating example, take the case of a sample containing strings abaa, abaaabaa, and abaaabaaabaaabaa. A learning algorithm may come up with either of the two automata depicted in Figure 14.7. Obviously the left-hand one (Figure 14.7(a)) is easier to encode than the right-hand one (Figure 14.7(b)). But on the other hand the first automaton does not help us reduce the size of the encoding of the data, whereas using the second one, the data can easily be encoded as something like $\{1, 2, 4\}$, denoting the number of cycles one should make to generate each string.

14.4.2 A simple MDL algorithm for grammatical inference

To illustrate these ideas let us try to learn a DFA from text. We are given a positive sample S_+.

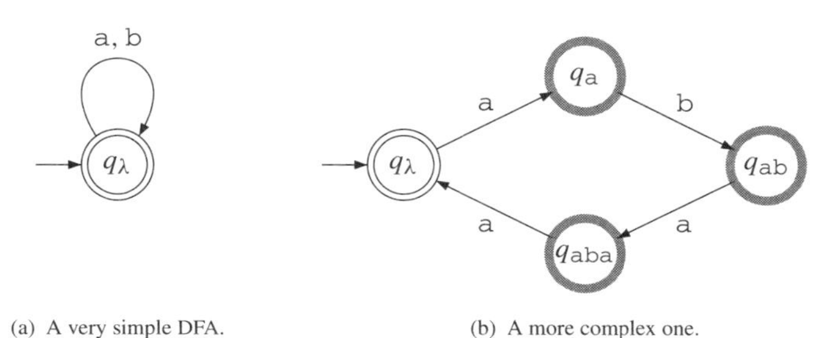

(a) A very simple DFA. (b) A more complex one.

Fig. 14.7. Two candidates for the sample $S_+ = \{$abaa, abaaabaa, abaaabaaab aaabaa$\}$.

Algorithm 14.3: MDL.

Input: S_+ function CHOOSE
Output: $\mathcal{A} = \langle \Sigma, Q, q_\lambda, \mathbb{F}_\mathbb{A}, \mathbb{F}_\mathbb{R}, \delta \rangle$
$\mathcal{A} \leftarrow$ Build-PTA(S_+);
RED $\leftarrow \{q_\lambda\}$;
current_score $\leftarrow \infty$;
BLUE $\leftarrow \{q_a : a \in \Sigma$ and $S_+ \cap a\,\Sigma^\star \neq \emptyset\}$;
while BLUE $\neq \emptyset$ **do**
    $q_b \leftarrow$ CHOOSE(BLUE);
    BLUE $\leftarrow$ BLUE $\setminus \{q_b\}$;
    if $\exists q_r \in$ RED : $\text{sc}(\text{MERGE}(q_r, q_b, \mathcal{A}), S_+) < \text{current\_score}$ **then**
        $\mathcal{A} \leftarrow$MERGE($q_r, q_b, \mathcal{A}$);
        current_score $\leftarrow$ sc($\mathcal{A}$,S_+)
    else
        $\mathcal{A} \leftarrow$ PROMOTE($q_b, \mathcal{A}$)
    end
end
for $q_r \in$ RED **do**
    if $\mathbb{L}(\mathcal{A}_{q_r}) \cap S_+ \neq \emptyset)$ **then** $\mathbb{F}_\mathbb{A} \leftarrow \mathbb{F}_\mathbb{A} \cup \{q_r\}$
end
return $\mathcal{A}$

Let us define the score of an automaton as the number of states of the DFA multiplied by the size of the alphabet. This is of course questionable, and should only be considered for a first approach; since this size is supposed to be compared with the size of the data, it is essential that the size is fairly computed. Ideally, the size should be that of the smallest Turing machine whose output is the automaton...

Then, given a string w and a DFA $\mathcal{A}$, we can encode the string depending on the number of choices we have at each stage. For example, using the arguments discussed above for the automata from Figure 14.7, we just have to encode a string by the numbers of its choices every time a choice has to be made. We therefore associate with each state of $\mathcal{A}$ the value $\text{ch}(q) = \log\big(|\{a \in \Sigma : \delta(q,a) \text{ is defined}\}|\big)$ if $q \notin \mathbb{F}_\mathbb{A}$. If $q \in \mathbb{F}_\mathbb{A}$ then $\text{ch}(q) = \log\big(1 + |\{a \in \Sigma : \delta(q,a) \text{ is defined}\}|\big)$, since one more choice is possible. The value ch corresponds to the size of the encoding of the choices a parser would have in that state. So in the automaton 14.7(b), we have $\text{ch}(q_\lambda) = \log 2$, $\text{ch}(q_{\mathtt{a}}) = \text{ch}(q_{\mathtt{ab}}) = \text{ch}(q_{\mathtt{aba}}) = \log 1 = 0$. The fact that no cost is counted corresponds to the idea that no choice has to be made and is also consistent with $\log 1 = 0$.

From this we define the associated value $\text{ch}(w) = \text{ch}(q_\lambda, w)$ which determines the size of the encoding of the path followed to parse string w, which depends on the recursive definition: $\text{ch}(q, \lambda) = \text{ch}(q_\lambda)$ and $\text{ch}(q, a \cdot w) = \text{ch}(q) + \text{ch}(\delta_\mathcal{A}(q,a), w)$.

Table 14.1. *Computations of all the* ch(x), *for the* PTA *and the different automata*

$\|\mathcal{A}\|$	a	a^2	b^2	a^3
12	$1 + \log 3$	$2 + \log 3$	$1 + \log 3$	$3 + \log 3$
6	$2 \log 3$	$3 \log 3$	$2 \log 3$	$4 \log 3$
2	$2 \log 3$	$3 \log 3$	$2 \log 3$	$4 \log 3$
1	2	3	3	4
$\|\mathcal{A}\|$	b^2a	a^4	ab^2a	b^4
12	$1 + \log 3$	$3 + \log 3$	$1 + \log 3$	$1 + \log 3$
6	$2 \log 3$	$5 \log 3$	$3 \log 3$	$2 \log 3$
2	$3 \log 3$	$5 \log 3$	$4 \log 3$	$3 \log 3$
1	4	5	5	5

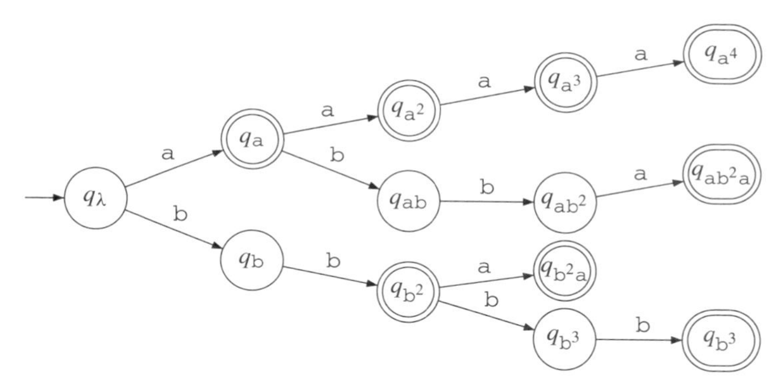

Fig. 14.8. PTA(S_+) where $S_+ = \{a, a^2, b^2, a^3, b^2a, a^4, ab^2a, b^4\}$.

We can now, given a sample S_+ and a DFA $\mathcal{A}$, measure the score sc of $\mathcal{A}$ and S_+, (denoted by sc($\mathcal{A}$, S_+)) as $\|\mathcal{A}\| \cdot |\Sigma| + \sum_{w \in S_+} \text{ch}(w)$, where $\|\mathcal{A}\|$ is the number of states of $\mathcal{A}$.

We can build a simple state-merging algorithm (Algorithm MDL, 14.3) which will merge states until it can no longer lower the score. The operations MERGE and PROMOTE are as in Chapter 12.

The training sample is $S_+ = \{a, a^2, b^2, a^3, b^2a, a^4, ab^2a, b^4\}$. From this we build PTA(S_+), depicted in Figure 14.8. We compute the score of the running solution and we get $13 \log(2) + 8 \log(3)$ for the derivations, and 26 for the PTA. The total is therefore $\text{sc}(\mathcal{A}, S_+) = 39 + 8 \log(3) \approx 51.68$.

The exact computations of the ch(x) can be found in Table 14.1. Merging q_a with q_λ is tested; this requires recursive merging (for determinisation). The resulting automaton is represented in Figure 14.9. The new automaton has values $\text{ch}(q_\lambda) = \text{ch}(q_{b^2}) = \log 3$,

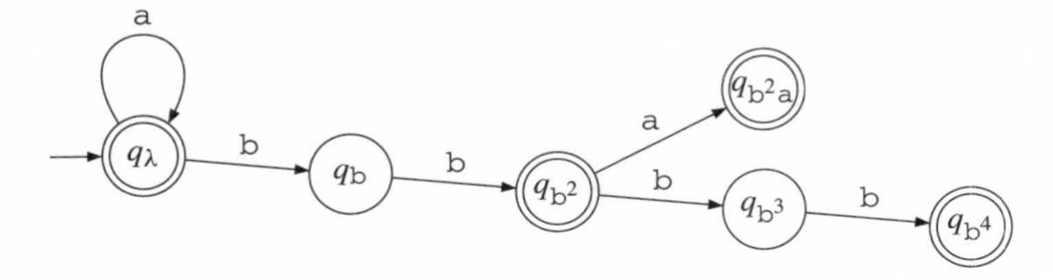

Fig. 14.9. We merge q_a with q_λ.

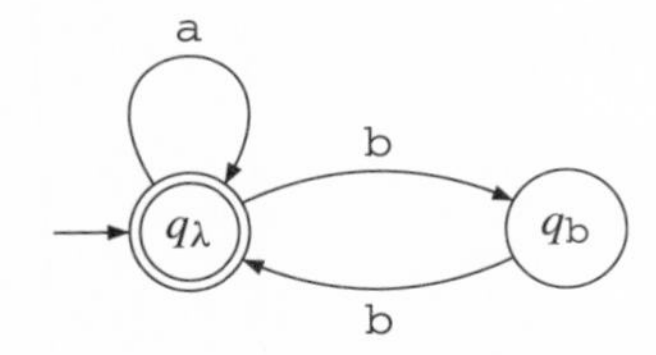

Fig. 14.10. The returned solution.

and $\text{ch}(q_b) = \text{ch}(q_{b^2a}) = \text{ch}(q_{b^3}) = \text{ch}(q_{b^4}) = \log 1 = 0$. So $\text{sc}(\mathcal{A}, S_+)$ can be computed as: $6 \cdot 2 + 23 \log(3) < 36 + 11 \log(3)$ (roughly 48.45 against 51.68), so the merge is accepted.

We try to merge q_b with q_λ and obtain the universal automaton whose score is 2 for the DFA + 31 log(3). This gives a score of 51.13, which is more than the score of our current solution (with six states). Therefore the merge is rejected. q_b is promoted and we test merging q_{ab} with q_λ. After determinisations we obtain the two-state automaton depicted in Figure 14.10. The score of this DFA is 43.68, which is better than the current best. Since no more merges are possible, the algorithm halts with this DFA as the solution.

Note that by taking a different scoring scheme, the result would have been very different (see Exercises 14.7 and 14.8).

14.5 Heuristic greedy state merging

When we described RPNI (Section 12.4), we presented it as a deterministic algorithm. Basically, the order in which the compatibilities are checked is defined right from the start. Moreover, as soon as two states are mergeable, they are merged. This is clearly an optimistic point of view, and there may be another, based on choosing the *best* merge.

But one should remember that RPNI identifies in the limit. This may well no longer be the case if we use a heuristic to define the best possible merge: one can usually imagine a (luckily counter-intuitive) distribution that will make us explore the lattice the wrong way.

14.5.1 How do greedy state-merging algorithms work?

The general idea of a *greedy* state-merging algorithm is as follows:

- choose two states,
- perform a cascade of forced merges until the automaton is deterministic,

- if this automaton accepts some sentences from S_-, backtrack and choose another couple,
- if not, loop until no merging is possible.

Now how are the moves chosen? Consider the current automaton for which RPNI has to make a decision: what moves are allowed?

There are two possibilities:

- merging a BLUE with a RED,
- promoting a BLUE to RED and changing all its successors that are not RED to BLUE.

Promotion takes place when a BLUE state can be merged with no RED state. This means that this event (similar to having a row obviously different in Gold's algorithm) has to be systematically checked.

But once there are no possible promotions, the idea is to do better than RPNI and, instead of greedily checking in order to find the first admissible merge, to check all possible legal merges between a BLUE state and a RED state, compute a score for each merge, and then choose the merge with the highest score.

14.5.2 Evidence driven state merging (EDSM)

The evidence-driven approach (see Algorithm 14.4) consists of computing for every pair of states (one BLUE, the other RED) the score of that merge as the number of strings that end in the same state if that merge is done. To do that, the strings from S_+ and S_- have

---

Algorithm 14.4: EDSM-COUNT.

Input: $\mathcal{A}$, S_+, S_-
Output: the score sc of $\mathcal{A}$
for $q \in Q$ **do** tp[q] $\leftarrow 0$; tn[q] $\leftarrow 0$;
for $w \in S_+$ **do** tp[$\delta_{\mathcal{A}}(q_\lambda, w)$] $\leftarrow$ tp[$\delta_{\mathcal{A}}(q_\lambda, w)$] $+ 1$;
for $w \in S_-$ **do** tn[$\delta_{\mathcal{A}}(q_\lambda, w)$] $\leftarrow$ tn[$\delta_{\mathcal{A}}(q_\lambda, w)$] $+ 1$;
sc$\leftarrow$ 0;
for $q \in Q$ **do**
  if sc$\neq -\infty$ **then**
    if tn[q] > 0 **then**
      if tp[q] > 0 **then** sc$\leftarrow -\infty$ **else** sc$\leftarrow$ sc+tn[q]-1
    else
      if tp[q] > 0 **then** sc$\leftarrow$ sc+tp[q]-1
    end
  end
end
return sc

Algorithm 14.5: EDSM $\mathcal{A}$.

Input: $S = \langle S_+, S_- \rangle$, functions COMPATIBLE, CHOOSE
Output: $\mathcal{A} = \langle \Sigma, Q, q_\lambda, \mathbb{F}_\mathbb{A}, \mathbb{F}_\mathbb{R}, \delta \rangle$
$\mathcal{A} \leftarrow$ BUILD-PTA(S_+); RED $\leftarrow \{q_\lambda\}$; BLUE $\leftarrow \{q_a : a \in \Sigma$ and $S_+ \cap a\,\Sigma^\star \neq \emptyset\}$;
while BLUE $\neq \emptyset$ **do**
  promotion $\leftarrow$ **false**;
  for $q_b \in$ BLUE **do**
    if ***not*** promotion **then**
      $bs \leftarrow -\infty$;
      atleastonemerge $\leftarrow$ **false**;
      for $q_r \in$ RED **do**
        $s \leftarrow$EDSM-COUNT(MERGE($q_r, q_b, \mathcal{A}$),S_+, S_-);
        if $s > -\infty$ **then** atleastonemerge $\leftarrow$ **true**
        if $s > bs$ **then** $bs \leftarrow s$; $\overline{q_r} \leftarrow q_r$; $\overline{q_b} \leftarrow q_b$
      end
      if ***not*** atleastonemerge **then** `/* no merge is possible */`
        PROMOTE($q_b, \mathcal{A}$); promotion $\leftarrow$ **true**;
      end
    end
  end
  if ***not*** promotion **then** `/* we can merge */`
    BLUE $\leftarrow$ BLUE $\setminus \{\overline{q_b}\}$; $\mathcal{A} \leftarrow$MERGE($\overline{q_r}, \overline{q_b}, \mathcal{A}$)
  end
end
for $x \in S_+$ **do** $\mathbb{F}_\mathbb{A} \leftarrow \mathbb{F}_\mathbb{A} \cup \{\delta(q_\lambda, x)\}$;
for $x \in S_-$ **do** $\mathbb{F}_\mathbb{R} \leftarrow \mathbb{F}_\mathbb{R} \cup \{\delta(q_\lambda, x)\}$;
return $\mathcal{A}$

to be parsed. If, by doing that merge, a conflict arises (a negative string is accepted or a positive string is rejected) the score is $-\infty$. The merge with the highest score is chosen.

The corresponding algorithm (Algorithm EDSM, 14.5) is given with a specific counting scheme (Algorithm EDSM-COUNT, 14.4). The merging function is exactly the one (Algorithm 12.11) introduced in Section 12.4.

Example 14.5.1

$$S_+ = \{\mathtt{a}, \mathtt{aaa}, \mathtt{bba}, \mathtt{abab}\}$$
$$S_- = \{\mathtt{ab}, \mathtt{bb}\}$$

Consider the DFA represented in Figure 14.11. States q_λ and $q_\mathtt{a}$ are RED, whereas states $q_\mathtt{b}$ and $q_\mathtt{ab}$ are BLUE.

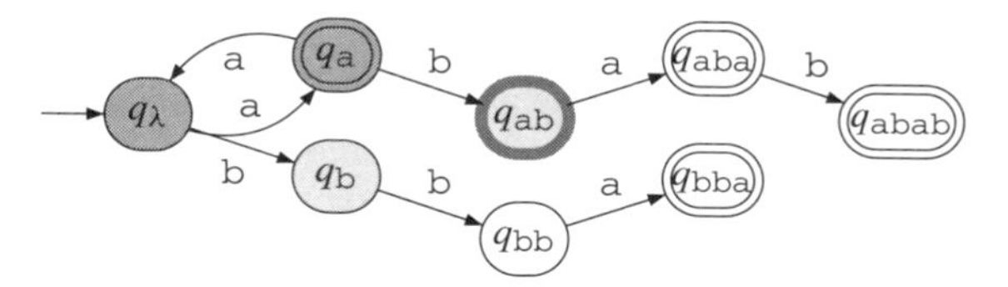

Fig. 14.11. The DFA before attempting a merge.

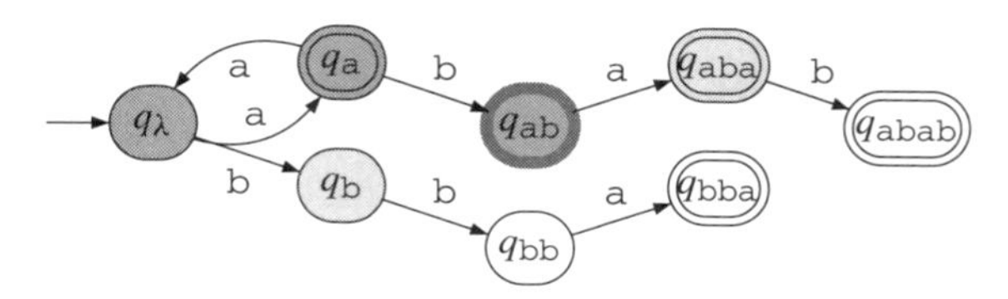

Fig. 14.12. The DFA after the promotion of q_{ab}.

State q_{ab} is now selected and the counts are computed:

- EDSM-COUNT(MERGE($q_\lambda, q_{\mathrm{ab}}, \mathcal{A}$)) $= -\infty$, because this consists of merging q_{ab} with q_{abab}.
- EDSM-COUNT(MERGE($q_{\mathrm{a}}, q_{\mathrm{ab}}, \mathcal{A}$)) $= -\infty$, because this consists of merging q_{a} with q_{ab}.

Therefore a promotion takes place: since we have a BLUE which can be merged, the DFA is updated with state q_{ab} promoted to RED (see Figure 14.12).

Suppose instead state b was selected. The counts are now different:

- EDSM-COUNT(MERGE($q_\lambda, q_{\mathrm{b}}, \mathcal{A}$)) $= 2$
- EDSM-COUNT(MERGE($q_{\mathrm{a}}, q_{\mathrm{b}}, \mathcal{A}$)) $= 3$

In this case, the merge between q_{a} and q_{b} would be selected.

There are different ways to perform the possible operations but what characterises the evidence-driven state-merging techniques is that one should be careful to always check first if some promotion is possible.

A different idea is to use a heuristic to decide before testing consistency in what order the merges should be checked. This is of course cheaper, but the problem is that promotion is then hard to detect. In practice this approach (called ***data-driven***) has not proven to be successful, at least in the deterministic setting. When trying to learn probabilistic automata, it seems that data-driven state merging is a good option.

14.6 Graph colouring and constraint satisfaction

The PTA can be seen as a graph for which the goal is to obtain a colouring of the nodes respecting a certain number of conditions. These conditions can be described as constraints, and again the nodes of the graph have to be valued in a way satisfying a set of constraints. Alas these constraints are dynamic (some constraints will depend on others).

There are too many options to mention them all here; we just describe briefly how we can convert the problem of learning a DFA from an informed sample $\langle S_+, S_- \rangle$ into a constraint satisfaction question.

We first build the complete prefix tree acceptor PTA(S_+, S_-) using Algorithm 12.1, page 239. Let us suppose the PTA has m states.

Now consider the graph whose nodes are the states of the PTA and where there is an edge between two nodes/states q and q' if they are incompatible, i.e. they cannot be merged. Then the problem is to find a colouring of the graph (no two adjacent nodes can take the same colour) with a minimum number of colours.

An alternative problem whose resolution can provide us with a partition is that of building cliques in the graph of consistency.

Technically things are a little more complex, since the constraints are dynamic: choosing to colour two nodes with a given colour corresponds to merging the states, with the usual problems relating to determinism.

Nevertheless there are many heuristics that have been tested for these particular well-known problems.

Let us model the problem further. We consider (given the PTA) m variables $X_1, \ldots, X_m$, and n possible values $1,\ldots,n$, corresponding to the n states of the target automaton (which supposes we take an initial gamble on the size of the intended target).

One can describe three types of constraints:

- global constraints: $q_i \in \mathbb{F}_{\mathbb{A}}, q_j \in \mathbb{F}_{\mathbb{R}} \implies X_i \neq X_j$,
- propagation constraints: $X_k \neq X_l \wedge \delta(q_i, a) = q_k \wedge \delta(q_j, a) = q_l \implies X_i \neq X_j$,
- deterministic constraints: $\delta(q_i, a) = q_j \wedge \delta(q_k, a) = q_l \implies [X_i = X_k \implies X_j = X_l]$.

Note that the deterministic constraints are dynamic: they will only be used when the algorithm starts deciding to effectively colour some states.

Among the different systems used to get around such constraints, conflict diagnosis (using intelligent backtracking) has been used.

Example 14.6.1 Consider the PTA represented in Figure 14.13. Then a certain number of initial global constraints can be established. If we denote by $\langle X_i, X_j \rangle$ the constraint: "q_i

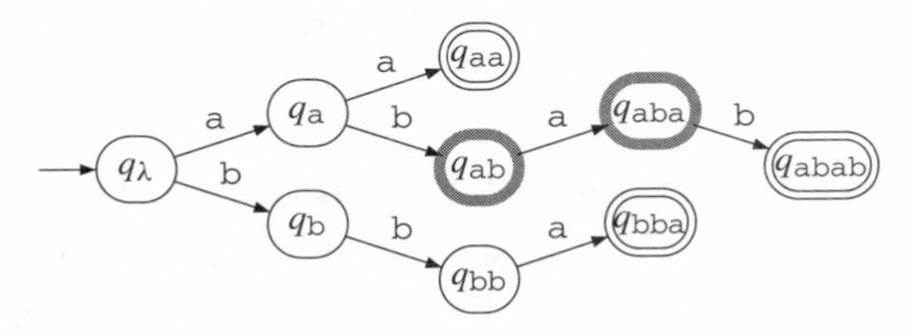

Fig. 14.13. PTA ({(aa, 1) (aba, 0) (bba, 1) (ab, 0) (abab, 1)}).

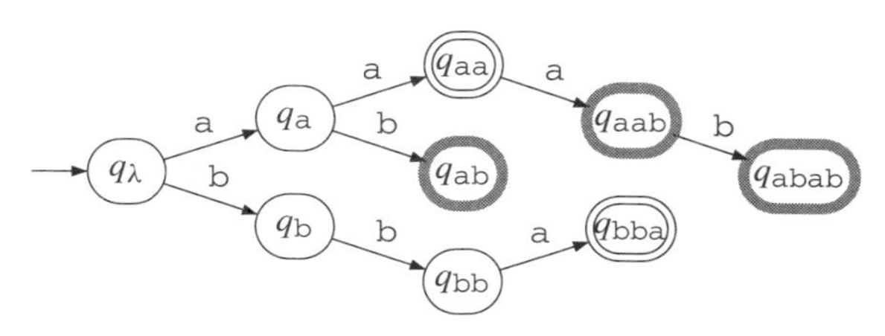

Fig. 14.14. PTA ({(aa, 1) (aab, 0) (bba, 0) (ab, 0) (abab, 1)}).

and q_j cannot be merged", we have by taking all pairs of states, one being accepting and the other rejecting:

Initial constraints: $\langle X_{\mathtt{aa}}, X_{\mathtt{ab}}\rangle$, $\langle X_{\mathtt{aa}}, X_{\mathtt{aba}}\rangle$, $\langle X_{\mathtt{abab}}, X_{\mathtt{ab}}\rangle$, $\langle X_{\mathtt{abab}}, X_{\mathtt{aba}}\rangle$, $\langle X_{\mathtt{bba}}, X_{\mathtt{ab}}\rangle$, $\langle X_{\mathtt{bba}}, X_{\mathtt{aba}}\rangle$.

We can now represent some of the propagation constraints, where we use the rule $\langle X_{ua}, X_{va}\rangle \implies \langle X_u, X_v\rangle$.

This, when propagated, gives us $\langle X_{\mathtt{a}}, X_{\mathtt{ab}}\rangle$, $\langle X_{\mathtt{ab}}, X_{\mathtt{bb}}\rangle$, $\langle X_{\mathtt{a}}, X_{\mathtt{b}}\rangle$, $\langle X_{\mathtt{a}}, q_{\mathtt{aba}}\rangle$, $\langle X_\lambda, X_{\mathtt{ab}}\rangle$.

14.7 Exercises

14.1 Write the different missing algorithms for the genetic algorithms.

14.2 Find a difficult language to identify with a genetic algorithm.

14.3 The definition of the value function v for the Tabu search method is very naive. Can we do better?

14.4 Find a difficult language to identify with the Tabu search algorithm, using a fixed k.

14.5 If the target is a complete automaton, then the MDL algorithm will perform poorly. Why?

14.6 What would be a good class of DFAs for the MDL algorithm? Hint: one may want to have large alphabets but only very few transitions. A definition in the spirit of Definition 4.4.1 (page 84) may be a good idea.

14.7 Using the same data as in Section 14.4, run the MDL algorithm with a score function that ignores the size of the alphabet, i.e. $\mathrm{sc}(\mathcal{A}, S_+)) = \|\mathcal{A}\| + \sum_{w \in S_+} \mathrm{ch}(w)$.

14.8 Conversely, let us suppose we intend to represent the automaton as a table with three entries (one for the alphabet and two for the states). Therefore we could choose $\mathrm{sc}(\mathcal{A}, S_+)) = \|\mathcal{A}\|^2 \cdot |\Sigma| + \sum_{w \in S_+} \mathrm{ch}(w)$.

14.9 In Algorithm EDSM, the computation of the scores is very expensive. Can we combine the data-driven and evidence-driven approaches in order to not have to compute all the scores but to be able to discover the promotion situations?

14.10 In Algorithm EDSM, once $\mathrm{sc}(q,q',\mathcal{A})$ is computed as $-\infty$, does it need to be recomputed? Is there any way to avoid such expensive recomputations?

14.11 Build the set of constraints corresponding to the PTA represented in Figure 14.14.

14.8 Conclusions of the chapter and further reading

14.8.1 Bibliographical background

Some of the key ideas presented in the first section of this chapter have been discussed in (de la Higuera, 2006a).

The question of the size of the DFA is a real issue. Following along the lines of the ABBADINGO competition (Lang & Pearlmutter, 1997), many authors were keen on finding algorithms working with targets of a few hundred states, but there has also been a group of researchers interested in considering the purely combinatoric problem of finding the smallest DFA and coming up with some heuristics for this case (de Oliveira & Silva, 2001).

Section 14.2 on genetic algorithms is based on work by many researchers but especially (Dupont, 1994, Sakakibara & Kondo, 1999, Sakakibara and Muramatsu, 2000). Research in this area has taken place both from the grammatical inference perspective and from the genetic algorithms perspective. It is therefore essential, when wanting to study this particular topic, to look at the bibliography from both areas. An attempt to use genetic algorithms (using the term coevolutionary learning) in an active setting was made by Josh Bongard and Hod Lipson (Bongard & Lipson, 2005).

The MDL principle is known under different names (Rissanen, 1978, Wallace & Ball, 1968). In grammatical inference some key ideas were introduced by Gerard Wolff (Wolff, 1978). More recently his work was pursued by Pat Langley and Sean Stromsten (Langley, 1995, Langley & Stromsten, 2000) and by George Petasis *et al.* (Petasis *et al.*, 2004a). New ideas, in the case of learning DFAs with MDL, are by Pieter Adriaans and Ceriel Jacobs (Adriaans & Jacobs, 2006).

Section 14.3, concerning Tabu search, is based on work by Jean-Yves Giordano (Giordano, 1996). A more general presentation of Tabu search can be found in Fred Glover's book (Glover & Laguna, 1997).

The main results concerning algorithm EDSM correspond to work done by Nick Price and Kevin Lang during or after the ABBADINGO competition (Lang, 1999, Lang, Pearlmutter & Price, 1998). The cheaper (but also worse) data-driven approach was used by Colin de la Higuera *et al.* (de la Higuera, Oncina & Vidal, 1996)

Section 14.6 is based on work by Alan Biermann (Biermann, 1971), Arlindo de Oliveira and João Marques Silva (de Oliveira & Silva, 1998), and François Coste (Coste & Nicolas, 1998a, 1998b).

Pure heuristics are problematic in that they can introduce an unwanted, undeclared added bias. Typically if the intended class of grammars is different from the one that is really going to be learnt, something is wrong.

An interesting alternative is to base a heuristic on a provably convergent algorithm. There is still the necessity to study what is really happening in this case, but a prudent guess is that somehow one is keeping control of the bias.

14.8.2 Some alternative lines of research

- A very different approach to learning context-free grammars has been followed in system SYNAPSE by Katsuhiko Nakamura and his colleagues (Nakamura & Matsumoto, 2005): the goal there is to learn the CYK parser directly and inductively. The system is in part incremental. Along similar lines, genetic algorithms have been tried, with the same goal of learning the parser (Sakakibara & Kondo, 1999, Sakakibara & Muramatsu, 2000).
- Neural networks have been used in grammatical inference with varying degrees of success. Among the best known papers are those by René Alquézar and Alberto Sanfeliu, Lee Giles, Mikel Forcada and their colleagues (Alquézar & Sanfeliu, 1994, Carrasco, Forcada & Santamaria, 1996, Giles, Lawrence & Tsoi, 2001). In some cases a mixture of numerical and symbolic techniques were used, the symbolic grammatical inference allowing us to better find the parameters for the recurrent network and conversely the network allowing us to decide compatibility of states. An important question is that of extracting an automaton from a learnt neural network, so as to avoid the black-box effect. In this case, we can be facing an interesting task of interactive learning in which the neural network can play the part of the Oracle.

14.8.3 Open problems and possible new lines of research

There is obviously a lot of work possible, once the limits of provable methods are well understood. Let us discuss a certain number of elements of reflection:

- The GOLD algorithm gives an interesting basis for learning as it redefines the search space. This should be considered as a good place to start from. Moreover, the complexity of the algorithm can be considerably reduced through a careful use of good data structures.
- Use of semantic information in real tasks should be encouraged. This semantic information needs to be translated into syntactic constraints that in turn could improve the algorithms.
- EDSM is a good example of what should be done: the basis is a provable algorithm (RPNI) in which the greediness is controlled by a *common sense* function instead of by an arbitrary order.

15

Learning context-free grammars

Too much faith should not be put in the powers of induction, even when aided by intelligent heuristics, to discover the right grammar. After all, stupid people learn to talk, but even the brightest apes do not.

***Noam Chomsky**, 1963*

It seems a miracle that young children easily learn the language of any environment into which they were born. The generative approach to grammar, pioneered by Chomsky, argues that this is only explicable if certain deep, universal features of this competence are innate characteristics of the human brain. Biologically speaking, this hypothesis of an inheritable capability to learn any language means that it must somehow be encoded in the DNA of our chromosomes. Should this hypothesis one day be verified, then linguistics would become a branch of biology.

***Niels Jerne**, Nobel Lecture, 1984*

Context-free languages correspond to the second 'easiest' level of the Chomsky hierarchy. They comprise the languages generated by context-free grammars (see Chapter 4).

All regular languages are context-free but the converse is not true. Among the languages that are context-free but not regular, some 'typical' ones are:

- $\{\mathtt{a}^n\mathtt{b}^n : n \geq 0\}$. This is the classical text book language used to show that automata cannot count in an unrestricted way.
- $\{w \in \{\mathtt{a}, \mathtt{b}\}^* : |w|_{\mathtt{a}} = |w|_{\mathtt{b}}\}$. This language is a bit more complicated than the previous one. But the same argument applies: You cannot count the a's and the b's nor the difference between the number of occurrences of each letter.
- The language of *palindromes*: $\{w \in \{\mathtt{a}, \mathtt{b}\}^* : |w| = n, \ \wedge \forall i \in [n] \ w(i) = w(n-i+1)\} = \{w \in \{\mathtt{a}, \mathtt{b}\}^* : w = w^R\}$.
- *Dyck*, or the language of well-formed brackets. The language of all bracketed strings or balanced parentheses is classical in formal language theory. If just working on one pair of brackets (denoted by a and b), it is defined by the rewriting system $\langle\{\mathtt{ab} \vdash \lambda\}, \lambda\rangle$, i.e. by those strings whose brackets all disappear by deleting iteratively every substring ab. The language is context-free and can be generated by the grammar $\langle\{\mathtt{a}, \mathtt{b}\}, \{N_1\}, R, N_1\rangle$ with $R = \{N_1 \rightarrow \mathtt{a}N_1\mathtt{b}N_1; N_1 \rightarrow \lambda\}$. This known as $Dyck_1$, because it uses only one pair of brackets. And for each n, the language $Dyck_n$, over n pairs of brackets, is also context-free.

- The language generated by the grammar $\langle\{\mathtt{a}, \mathtt{b}\}, \{N_1\}, R, N_1\rangle$ with $R = \{N_1 \rightarrow \mathtt{a}N_1N_1; N_1 \rightarrow \mathtt{b}\}$ is called the *Lukasiewicz* language.

It has been suggested by many authors that context-free grammars are a better model for natural language than regular grammars, even if it is also admitted that a certain number of constructs can only be found in context-sensitive languages.

There are other reasons for wanting to learn context-free grammars: these appear in computational linguistics or in the analysis of web documents where the tag languages need opening and closing tags. In bio-informatics, also, certain constructs of the secondary structure are context-free.

15.1 The difficulties

When moving up from the regular world to the context-free world we are faced with a whole set of new difficulties.

Do we learn context-free languages or context-free grammars? This is going to be the first (and possibly most crucial) question. When dealing with regular languages (and grammars) the issue is much less troublesome, as the Myhill-Nerode theorem provides us with a nice one-to-one relationship between the languages and the automata. In the case of context-freeness, there are several reasons that make it difficult to consider grammars instead of languages:

- The first very serious issue is that equivalence of context-free grammars is undecidable. This is also the case for the subclass of the linear grammars. As an immediate consequence there will be the fact that canonical forms will be unavailable, at least in a constructible way. Moreover, the critical importance of the undecidability issue can be seen in the following problem:

  Suppose class $\mathcal{L}$ is learnable, and we are given two grammars G_1 and G_2 for languages in $\mathcal{L}$. Then we could perhaps generate examples from $\mathbb{L}(G_1)$ and learn some grammar H_1 and do the same from $\mathbb{L}(G_2)$ obtaining H_2. Checking the syntactic equality between H_1 and H_2 corresponds in an intuitive way to solving the equivalence between G_1 and G_2. Moreover the fact that the algorithm constructs in some way a normal and canonical form, since it depends on the examples, is puzzling. The question we raise here is: 'Can we use a grammatical inference algorithm to solve the equivalence problem?'.

  This is obviously not a tight argument. But if one requires 'learnable' to mean 'have characteristic samples' then the above reasoning at least proves that the so-called characteristic samples have to be uncomputable.
- A second troubling issue is that of 'expansiveness': in certain cases the grammar can be exponentially smaller than any string in the language. Consider for instance the grammar $G_n = \langle\{\mathtt{a}\}, \{N_k : k \in [n]\}, R_n, N_1\rangle$ with $R_n = \bigcup_{i<n}\{N_i \rightarrow N_{i+1}N_{i+1}\} \cup \{N_n \rightarrow \mathtt{a}\}$. Then the only string in the language $\mathbb{L}(G_n)$ is $\mathtt{a}^{2^{n-1}}$ which is of length 2^{n-1}. There is therefore an exponential relation between the size of the grammar and the length of even the shortest strings the grammar can produce. If we take a point of view where learning is seen as a compression question, then compressing into logarithmic size is surely not a problem and is most recommendable. But on the other hand if the question we ask is "what examples are needed to learn?", then

we face a problem. In the terms we have been using so far, the characteristic samples would be exponential.

- When studying the learnability of regular languages, there is an important difference between learning deterministic representations and non-deterministic ones. In the case of context-freeness, things are even more complex as there are two different notions related to determinism.

  The first possible notion corresponds to *ambiguity*: a grammar is ambiguous if it admits ambiguous strings, i.e. strings that have two different derivation trees associated. It is well known that there exist inherently ambiguous languages, i.e. languages for which all grammars have to be ambiguous. All reasonable questions relating to ambiguity are undecidable, so one cannot limit oneself to the class of the unambiguous languages, nor check the ambiguity of an individual string.

  The second possible notion is used by *deterministic languages*. Here, determinism refers to the determinism of the pushdown automaton that recognises the language. There is a well-represented subclass of the deterministic languages for which, furthermore, the equivalence problem is decidable. There have been no serious attempts to learn deterministic pushdown automata, so we will not enter this subject here.

- *Intelligibility* is another issue that becomes essential when dealing with context-free grammars. A context-free language can be generated by many very different grammars, some of which fit the structure of the language better than others. Take for example the grammar (based on the Lukasiewicz grammar) $\langle\{\texttt{a},\texttt{b}\},\{N_1,N_2\},R,N_1\rangle$ with $R=\{N_1\rightarrow \texttt{a}N_2N_2;\ N_2\rightarrow \texttt{b};\ N_1\rightarrow \texttt{a}N_2;\ N_1\rightarrow\lambda\}$. Is this a better grammar to generate the single-bracket language? An equivalent grammar would be the grammar in Chomsky (quadratic) normal form: $\langle\{\texttt{a},\texttt{b}\},\{N_1,N_2,N_3,A,B\},R,N_1\rangle$ with $R=\{N_1\rightarrow\lambda+N_2N_3;\ N_2\rightarrow AN_1;\ N_3\rightarrow BN_1;\ A\rightarrow \texttt{a};\ B\rightarrow\texttt{b}\}$.

  The point we are raising is that there is really a lot of semantics hidden in the structure defined by the grammar. This gives yet another reason for considering that the problem is about learning context-free grammars rather than context-free languages!

15.1.1 Dealing with linearity

As regular languages, linear languages and context-free languages all share the curse of not being learnable from positive examples, an alternative is to reduce the class of languages in order to obtain a family that would not be super-finite, but on the other hand would be identifiable.

Definition 15.1.1 (Linear context-free grammars) A context-free grammar $G=(\Sigma,V,R,N_1)$ is **linear** *if$_{def}$* $R\subset V\times(\Sigma^*V\Sigma^*\cup\Sigma^*)$.

Definition 15.1.2 (Even linear context-free grammars) A context-free grammar $G=(\Sigma,V,R,N_1)$ is an **even linear** grammar *if$_{def}$* $R\subset V\times(\Sigma V\Sigma\cup\Sigma\cup\{\lambda\})$.

Thus languages like $\{a^nb^n : n\in\mathbb{N}\}$, or the set of all palindromes, are even linear without being regular. But using reduction techniques from Section 7.4, we find a clear relationship with the regular languages. Indeed the operation allowing us to simulate an even linear grammar by a finite automaton is called a ***regular reduction***:

Definition 15.1.3 (Regular reduction) Let $G = (\Sigma, V, R, N_1)$ be an even linear grammar. We say that the NFA $\mathcal{A} = \langle \Sigma_R, Q, q_1, q_F, \emptyset, \delta_R \rangle$ is the **regular reduction** of G *if*$_{def}$

- $\Sigma_R = \{\langle ab \rangle : a, b \in \Sigma\} \cup \Sigma$,
- $Q = \{q_i : N_i \in V\} \cup \{q_F\}$,
- $\delta_R(q_i, \langle ab \rangle) = \{q_j : (N_i, aN_jb) \in R\}$,
- $\forall a \in \Sigma,\ \delta_R(q_i, a) = \{q_F : (N_i, a) \in R\}$,
- $\forall i$ such that $(N_i, \lambda) \in R, q_F \in \delta_R(q_i, \lambda)$.

Theorem 15.1.1 *Let G be an even linear grammar and let R be its regular reduction. Then $a_1 \cdots a_n \in \mathbb{L}(G)$ if and only if $\langle a_1 a_n \rangle \langle a_2 a_{n-1} \rangle \cdots \in \mathbb{L}(R)$.*

Proof This is clear by the construction of the regular reduction, but more detail can be found in the construction presented in Section 7.4.3 (page 160). □

The corollary of the above construction is that any technique based on learning the class of all regular languages or subclasses of regular languages can be transposed to subclasses of even linear languages. For instance, in the setting of learning from positive examples only, positive results concerning subclasses of even linear languages have been obtained.

Very simple grammars are a very restricted form of grammar that are not linear but are strongly deterministic. They constitute another class of context-free grammars for which positive learning results have been obtained. They are context-free grammars in a restricted Greibach normal form:

Definition 15.1.4 (Very simple grammars) A context-free grammar $G = (\Sigma, V, R, N_1)$ is a **very simple grammar** *if*$_{def}$ $R \subset (V \times \Sigma V^*)$ and for any $a \in \Sigma$ $(A, a\alpha) \in R \wedge (B, a\beta) \in R \implies [A = B \wedge \alpha = \beta]$.

Lemma 15.1.2 (Some properties of very simple grammars)
Let $G = (\Sigma, V, R, N_1)$ be a very simple grammar, let $\alpha, \beta \in V^+$ and let $x \in \Sigma^+, u, u_1, u_2 \in \Sigma^*$. Then:

- $N_1 \stackrel{*}{\Longrightarrow} x\alpha \wedge N_1 \stackrel{*}{\Longrightarrow} x\beta \Rightarrow \alpha = \beta$ *(forward determinism),*
- $\alpha \stackrel{*}{\Longrightarrow} x \wedge \beta \stackrel{*}{\Longrightarrow} x \Rightarrow \alpha = \beta$ *(backward determinism),*
- $N_1 \stackrel{*}{\Longrightarrow} u_1\alpha \stackrel{*}{\Longrightarrow} u_1 x \ \wedge N_1 \stackrel{*}{\Longrightarrow} u_2\beta \stackrel{*}{\Longrightarrow} u_2 x \Rightarrow u_1^{-1}L = u_2^{-1}L.$

Very simple grammars are therefore deterministic both for a top-down and a bottom-up parse. Moreover, a nice congruence can be extracted, which will prove to be the key to building a succesful identification algorithm. One should point out that they are nevertheless quite limited: each symbol in the final alphabet can only appear once in the entire grammar.

Example 15.1.1 The grammar $G = (\Sigma, V, R, N_1)$ (with $\Sigma = \{\mathtt{a}, \mathtt{b}, \mathtt{c}, \mathtt{d}, \mathtt{e}, \mathtt{f}\}$) is a very simple grammar:

$$N_1 \to \mathtt{a}N_1N_2 + \mathtt{f}$$
$$N_2 \to \mathtt{b}N_2 + \mathtt{c} + \mathtt{d}N_3N_3$$
$$N_3 \to \mathtt{e}$$

The language generated by G can be represented by the following extended regular expression: $\mathtt{a}^n\mathtt{f}\big(\mathtt{b}^*(\mathtt{c} + \mathtt{dee})\big)^n$.

Theorem 15.1.3 *The class of very simple grammars can be identified in the limit from text by an algorithm that*

- *has polynomial update time,*
- *makes a polynomial number of implicit prediction errors and mind changes.*

Proof [sketch] Let us describe the algorithm. In a very simple grammar for any $a \in \Sigma$ there is exactly one rule of shape $(N, a\alpha) \in R$, so the number of rules in the grammar is exactly $|\Sigma|$ and there are at most $|\Sigma|$ non-terminals. The algorithm goes through three steps:

Step 1 For each $a \in \Sigma$ and making use of equations in Lemma 15.1.2, determine the left part of the only rule in which a appears.

Step 2 As there is exactly one rule for each terminal symbol, the rules applied in the parsing of any string are known. Then, we can construct an equation, for each training string, that relates the length of the string and the lengths of the right part of the rules used in the derivation of the string. We now solve the system of equations to determine the length of the right-hand side of each rule.

Step 3 Simulating the parse for each training string, we determine the order in which the rules are applied and the non-terminals that appear on the right-hand side of the rules.

□

We run the sketched algorithm on a simple example. Suppose the data consists of the strings $\{\mathtt{afbc}, \mathtt{f}, \mathtt{afbbc}, \mathtt{aec}, \mathtt{afbdee}\}$. Step 1 will allow us to cluster the letters into three groups: $\{\mathtt{b}, \mathtt{c}, \mathtt{d}\}$, $\{\mathtt{a}, \mathtt{f}\}$ and $\{\mathtt{e}, \mathtt{g}\}$. Indeed since we have $N_1 \overset{*}{\Longrightarrow} \mathtt{afb}\alpha \overset{*}{\Longrightarrow} \mathtt{afbc} \wedge N_1 \overset{*}{\Longrightarrow} \mathtt{afb}\beta \overset{*}{\Longrightarrow} \mathtt{afbbc}$ we deduce that $\alpha = \beta$ and that the left-hand side of the rules for $\mathtt{b}$ and $\mathtt{c}$ are identical. Now for step 2 and simplifying we can deduce that the rules corresponding to $\mathtt{c}$, $\mathtt{e}$ and $\mathtt{f}$ are all of length 1 (so $N_e \leftarrow \mathtt{e}$). It follows that the rule for letter $\mathtt{b}$ is of length 2 and those for $\mathtt{a}$ and $\mathtt{d}$ are of length 3. We can now materialise this by bracketing the strings in the learning sample:

$$\{(\mathtt{af}(\mathtt{bc})), (\mathtt{f}), (\mathtt{af}(\mathtt{b}(\mathtt{bc}))), (\mathtt{aec}), (\mathtt{af}(\mathtt{b}(\mathtt{dee})))\}$$

And by reconstruction we get the fact that the rules are:

$$N_1 \rightarrow \mathtt{a}N_1N_2 + \mathtt{f}$$
$$N_2 \rightarrow \mathtt{b}N_2 + \mathtt{c} + \mathtt{d}N_3N_3$$
$$N_3 \rightarrow \mathtt{e}$$

One should note that the complexity will rise exponentially with the size of the alphabet.

15.1.2 Dealing with determinism

It might seem from the above that the key to success is to limit ourselves to linear grammars, but if we consider Definition 7.3.3 the results are negative:

Theorem 15.1.4 *For any alphabet Σ of size at least two, $\mathcal{LIN}(\Sigma)$ cannot be identified in the limit by polynomial characteristic samples from an informant.*

Proof Consider two linear languages. At least one string of their symmetric difference should appear in the characteristic sample in order to be able to distinguish them. But the length of the smallest string in the symmetric difference cannot be bounded by any polynomial in the size of the grammar since deciding if two linear grammars are equivalent is undecidable.

It should be noted that this result is independent of the sort of representation that is used. Further elements concerning this issue are discussed in Section 6.4. □

Corollary 15.1.5 *For any alphabet Σ of size at least two, $\mathcal{CFG}(\Sigma)$ cannot be identified in the limit by polynomial characteristic samples from an informant.*

We saw in Chapter 12 that DFAs were identifiable in the limit by polynomial characteristic samples (POLY-CS polynomial time) from an informant. So if we want to get positive results in this setting, we need to restrict the class of linear grammars further.

Deterministic linear grammars provide a non-trivial extension of the regular grammars:

Definition 15.1.5 (Deterministic linear grammars) A **deterministic linear context-free grammar** $G = (\Sigma, V, R, N_1)$ is a (linear) grammar where $R \subset \ \times (\Sigma V \Sigma^* \cup \{\lambda\})$ and $(N, a\alpha), (N, a\beta) \in R \Rightarrow \alpha = \beta$.

Definition 15.1.6 (Deterministic linear grammar normal form)
A deterministic linear grammar $G = (\Sigma, V, R, N_1)$ is in **normal form** *if*$_{def}$

(i) G has no useless non-terminals,
(ii) $\forall (N, aN'w) \in R, w = \text{lcs}(a^{-1}L_G(N))$,
(iii) $\forall N, N' \in R,\ L_G(N) = L_G(N') \Rightarrow N = N'$.

Remember that $\text{lcs}(L)$ is the least common suffix of language L. Having a *nice* normal form allows us to claim:

Theorem 15.1.6 *The class of deterministic linear grammars can be identified in the limit in polynomial time and data from an informant.*

Proof [sketch] The algorithm works by an incremental (by levels) construction of the canonical grammar.

The algorithm maintains a queue of non-terminals to explore. At the beginning, the start symbol is added to the grammar and to the exploration queue. At each step, a non-terminal (N) is extracted from the queue and a terminal symbol (a) is chosen in order to further parse the data. From these a new rule is proposed, based on the second condition of Definition 15.1.6 of the normal form for deterministic linear grammars: $N \rightarrow aN_?w$. Each time a new rule is proposed the only non-terminal that appears on its right-hand side ($N_?$) is checked for equivalence with a non-terminal in the grammar. We denote this non-terminal by $N_?$ in order to indicate that it is still to be named.

If a compatible non-terminal is found, the non-terminal in the rule is named after it. If no non-terminal is found, a new non-terminal is added to the grammar (corresponding to a promotion) and to the exploration list. In both cases the rule is added to the grammar.

By simulating the run of this algorithm over a particular grammar, a characteristic sample can be constructed. □

We run the sketched algorithm on an example. Let the learning sample consist of the sets $S_+ = \{\mathtt{abbabb}, \mathtt{bba}, \mathtt{babbaaa}, \mathtt{aabbabbbb}, \mathtt{baabbabbaa}\}$ and $S_- = \{\mathtt{b}\}$.

The first non-terminal is N_1 and the first terminal symbol we choose is a. Therefore a rule with profile $N_1 \rightarrow \mathtt{a}N_?w$ is considered. First, string w is sought by computing lcs({bbabb, abbabbbb})= bb.

Therefore rule $N_1 \rightarrow \mathtt{a}N_?\mathtt{bb}$ is suggested as a starting point (N_1 being the axiom). Can non-terminal $N_?$ be merged with N_1? Since rule $N_1 \rightarrow \mathtt{a}N_1\mathtt{bb}$ does not create a conflict, the merge is accepted and the rule is kept. Thus the current set of rules is $N_1 \rightarrow \mathtt{a}N_1\mathtt{bb}$; $N_1 \stackrel{*}{\Longrightarrow} \mathtt{bba} + \mathtt{babbaaa} + \mathtt{baabbabbaa}$.

Now terminal symbol b is brought forward and the different elements of the corresponding rule $N_1 \rightarrow \mathtt{b}N_?w$ have to be identified. We start with w which is lcs({ba, abbaaa, aabbabbaa}) so a. Again adding rule $N_1 \rightarrow \mathtt{b}N_1\mathtt{a}$ is considered but the resulting grammar (with rules $N_1 \rightarrow \mathtt{a}N_1\mathtt{bb}$; $N_1 \rightarrow \mathtt{b}N_1\mathtt{a}$; $N_1 \stackrel{*}{\Longrightarrow} \mathtt{b} + \mathtt{aabbabba}$) would generate string b which is in S_-.

So the current grammar is $\{N_1 \rightarrow \mathtt{a}N_1\mathtt{bb};\ N_1 \rightarrow \mathtt{b}N_2\mathtt{a};\ N_2 \stackrel{*}{\Longrightarrow} \mathtt{b} + \mathtt{abbaaa} + \mathtt{aabbabba}\}$. We compute lcs({bbaaa, aabbabba}) $=$ a. Therefore $N_2 \rightarrow \mathtt{a}N_?\mathtt{a}$ is accepted. We are left with b (this leads to the rule $N_2 \rightarrow \mathtt{b}$) and finally the grammar contains the following rules:

$$N_1 \rightarrow \mathtt{a}N_1\mathtt{bb} + \mathtt{b}N_2\mathtt{a}$$
$$N_2 \rightarrow \mathtt{a}N_1\mathtt{a} + \mathtt{b}$$

15.1.3 Dealing with sparseness

A string is the result of many rules that have all got to be learnt in 'one shot'. In a certain sense, there is an *all or nothing* issue here: hill climbing seems to be impossible, and the number of examples needed to justify the set of all rules can easily seem too large.

Moreover, from string to string (in the language) local modifications may not work. This can be measured in the following way: given two strings u and v in L, the number of modifications one needs to make to string u in order to obtain string v is going to be such that one will not be able to use the couple (u, v) to learn an isolated rule which allows us to get from u to v.

15.2 Learning reversible context-free grammars

In Section 11.2 (page 223), we introduced a class of look-ahead languages. Learning could take place by eliminating the sources of ambiguity through merging. This was done in the context of the regular languages. We now show how this idea can also lead to an algorithm for context-free languages, even if we will need some extra information about the structure of the strings.

15.2.1 Unlabelled trees or skeletons

In practice the positive data from which we will usually be learning from cannot be trees, labelled at the internal nodes. It will either just consist of the strings themselves or, in a more helpful setting, of bracketed strings. As explained in Section 3.3.1 (page 52), these correspond to trees with unlabelled internal nodes.

Definition 15.2.1 Let $G = \langle \Sigma, V, R, N_1 \rangle$ be a context-free grammar. A skeleton for string α (over $\Sigma \cup V$) is a derivation tree with frontier α and in which all internal nodes are labelled by a new symbol '?'.

In Figure 15.1(a) we represent a parse tree for `aaba`, and in Figure 15.1(b) the corresponding skeleton.

15.2.2 K-contexts

Definition 15.2.2 Let $G = \langle \Sigma, V, R, N_1 \rangle$ be a context-free grammar. A k**-deep** derivation in G is a derivation

$$\begin{aligned} N_{i_0} &\Rightarrow \alpha_1 N_{i_1} \beta_1 \\ &\Rightarrow \alpha_1 \alpha_2 N_{i_2} \beta_2 \beta_1 \\ &\stackrel{k}{\Longrightarrow} \alpha_1 \alpha_2 \cdots \alpha_{k-1} \alpha_k N_{i_k} \beta_k \beta_{k-1} \cdots \beta_2 \beta_1 \end{aligned}$$

where $\forall l \leq k,\ \alpha_l, \beta_l \in (\Sigma \cup V)^*$.

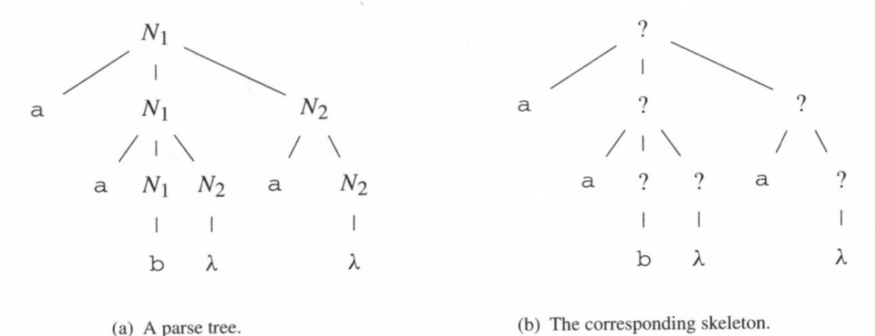

(a) A parse tree. (b) The corresponding skeleton.

Fig. 15.1. A parse tree for aaba and the corresponding skeleton. Some of the grammar rules are $N_1 \rightarrow \mathtt{a}N_1N_2$, $N_1 \rightarrow \mathtt{b}$, $N_2 \rightarrow \mathtt{a}N_2$ and $N_2 \rightarrow \lambda$.

Intuitively, this corresponds to a tree with just one long branch of length k.

Definition 15.2.3 Let $G = \langle \Sigma, V, R, N_1 \rangle$ be a context-free grammar and N_i, N_j be two non-terminal symbols in V. N_j is a k-**ancestor** of N_i if_{def} there exists a k-deep derivation of N_j into $\alpha N_i \beta$, $(\alpha, \beta \in (\Sigma \cup V)^*)$.

We define these as sets, i.e. k-ancestors(N) is the set of all k-ancestors of non-terminal N.

Example 15.2.1 From Figure 15.1(a), we can compute:

- 2-ancestors$(N_1) = \{N_1\}$, 2-ancestors$(N_2) = \{N_1, N_2\}$,
- 1-ancestors$(N_1) = \{N_1\}$, 1-ancestors$(N_1) = \{N_1, N_2\}$.

Now a k-context is defined as follows:

Definition 15.2.4 Let $G = \langle \Sigma, V, R, N_1 \rangle$ be a context-free grammar with a specific non-terminal N_i in V. The k-**contexts** of N_i are all trees built as follows:

(i) Let t be the derivation tree for a derivation $N_j \overset{k}{\Longrightarrow} \alpha N_i \beta \overset{*}{\Longrightarrow} u N_i v$ where derivation $N_j \overset{k}{\Longrightarrow} \alpha N_i \beta$ is a k-deep derivation and $\alpha \overset{*}{\Longrightarrow} u$ and $\beta \overset{*}{\Longrightarrow} v$, with $u, v \in \Sigma^\star$.
(ii) Let $z_\$$ be the address of N_i in t $(t(z_\$) = N_i, |z_\$| = k)$.
(iii) We build the k-context $c[t, z_\$]$ as a tree of domain $\mathrm{Dom}(t) \setminus \{z_\$ a u : a \in \mathbb{N}, u \in \mathbb{N}^*\}$ and such that $c[t, z_\$] : \mathrm{Dom}(t) \rightarrow \Sigma \cup V \cup \{\lambda, \$, ?\}$, with:
   - $c[t, z_\$] = \$$
   - $c[t, u] = ?$ if $u1 \in \mathrm{Dom}(t)$ (i.e. if u is an internal node of the tree)
   - $c[t, u] = t(u)$ if not.

Example 15.2.2 Consider the grammar with rules $N_1 \rightarrow \mathtt{a}N_1N_2$, $N_1 \rightarrow \mathtt{b}$, $N_2 \rightarrow \mathtt{a}N_2$ and $N_2 \rightarrow \lambda$. We show in Figure 15.2(a) a parse tree t, and the corresponding 2-context $c[t, 11]$ in Figure 15.2(b).

Note that a non-terminal can have an infinity of k-contexts, but only one 0-context, which is always \$.

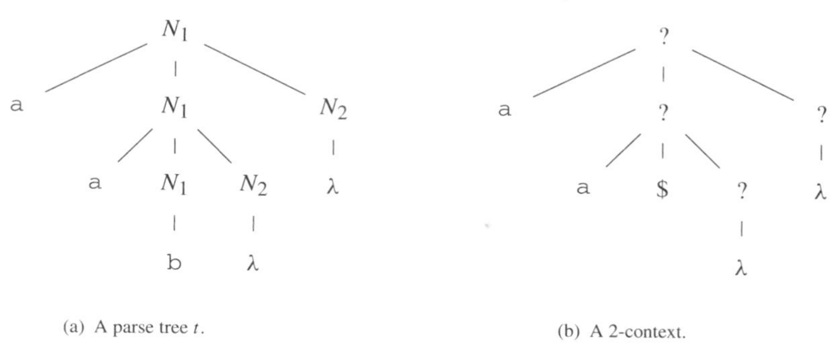

(a) A parse tree t.

(b) A 2-context.

Fig. 15.2. A parse tree t and the 2-context $c[t, 11]$.

In practice we will not be given a grammar from which one would compute the k-contexts. Instead, a basic grammar is constructed from a learning sample. This will allow us to be sure that the number of k-contexts remains finite. We thus denote by k-contexts(S_+, N_i) the set of all k-contexts of the non-terminal N_i with respect to the sample S_+.

15.2.3 *K*-reversible grammars

From the above we now define k-reversible grammars:

Definition 15.2.5 A context-free grammar G is k-**reversible** *if$_{def}$* the following two properties hold:

(i) If there exist two rules $N_l \rightarrow \alpha N_i \beta$ and $N_l \rightarrow \alpha N_j \beta$ then $N_i = N_j$ (invertibility condition).
(ii) If there exist two rules $N_i \rightarrow \alpha$ and $N_j \rightarrow \alpha$ and there is a k-context common to N_i and N_j, then $N_i = N_j$ (reset-free condition).

To say that a language is k-reversible (for some k) is nevertheless not that informative; if being k-reversible implies strong rules over the type of grammars, this is not true for the languages:

Theorem 15.2.1 *For any context-free language L, there exists a k-reversible grammar G such that $\mathbb{L}(G) = L$.*

Proof [sketch] The above result is already true even for fixed $k = 0$. One can transform any context-free grammar into a 0-reversible one, even if the transformation process can be costly. It should be noted that the above theorem says nothing about sizes. The corresponding grammar can in fact be of exponential size in the size of the original one. □

15.2.4 The algorithm

The first step consists of building, from a sample of unlabelled trees, an initial grammar. Basically it consists of converting the unlabelled trees into derivation trees where a unique

Algorithm 15.1: INITIALISE-K-REV-CF.

```
Data: A positive sample of unlabelled strings S+, k ∈ ℕ
Result: A grammar G = (Σ, V, R, N1) such that 𝕃(G) = ⋃_{t∈S+} Frontier(t)
V ← ∅;
for t ∈ S+ do
    t(λ) ← N1;
    for u ∈ Dom(t) do
        if t(u) = '?' then t(u) ← N_t^u; V ← V ∪ {N_t^u}
    end
    for u ∈ Dom(t) do
        if u1 ∈ Dom(t) then                    /* u is an internal node */
            m ← max{i ∈ ℕ : ui ∈ Dom(t)};
            R ← R ∪ {t(u) → t(u1)...t(um)}
        end
    end
end
return G
```

non-terminal (the axiom) is used to label every root of the trees in the sample S_+. All other internal nodes are labelled by non-terminals that are used exactly once. Algorithm K-REV-CF (15.2) first calls Algorithm INITIALISE-K-REV-CF (15.1) and then uses the labelled trees to merge the non-terminals until a k-reversible grammar is obtained.

In Algorithm K-REV-CF (15.2), the MERGE function is very simple: it consists of taking two non-terminals and merging them into just one. All the occurrences of each non-terminal in all the rules of the grammar are then replaced by the new variable.

Algorithm 15.2: K-REV-CF.

```
Data: A positive sample of unlabelled strings S+, k ∈ ℕ
Result: A k reversible grammar G = (Σ, V, R, N1)
INITIALISE-K-REV-CF(S+, k);
while G not reset-free and G not invertible do
    if ∃(N_l → αN_iβ) ∈ R ∧ (N_l → αN_jβ) ∈ R
    then MERGE(N_i, N_j);                              /* not reset-free */
    if ∃i, j ∈ [|V|], ∃α ∈ (Σ ∪ V)* : (N_i → α) ∈ R∧
    (N_j → α) ∈ R ∧ k-context(S+, N_i) ∩ k-contexts(S+, N_j) ≠ ∅
    then MERGE(N_i, N_j);                              /* not invertible */
end
return G
```

15.2.5 Running the algorithm

Consider the learning sample containing trees:

- ?(?(b b ?(a b)a)?(b a))
- ?(?(a b)a)
- ?(b a b ?(a b))
- ?(b b ?(a b)a)
- ?(?(b a)?(b a))

The first step (running Algorithm 15.1) leads to renaming the nodes labelled by '?':

- N_1(N_2(b b N_3(a b)a)N_4(b a))
- N_1(N_5(a b)a)
- N_1(b a b N_6(a b))
- N_1(b b N_7(a b)a)
- N_1(N_8(b a)N_9(b a))

The corresponding trees are represented in Figure 15.3.

There are six 1-contexts, shown in Figure 15.4(a–f).

With each non-terminal are associated its k-contexts. Here, and with k=1,

- 1-contexts(N_1)= ∅
- 1-contexts(N_2)={?($?(b a))} (1-context (e))

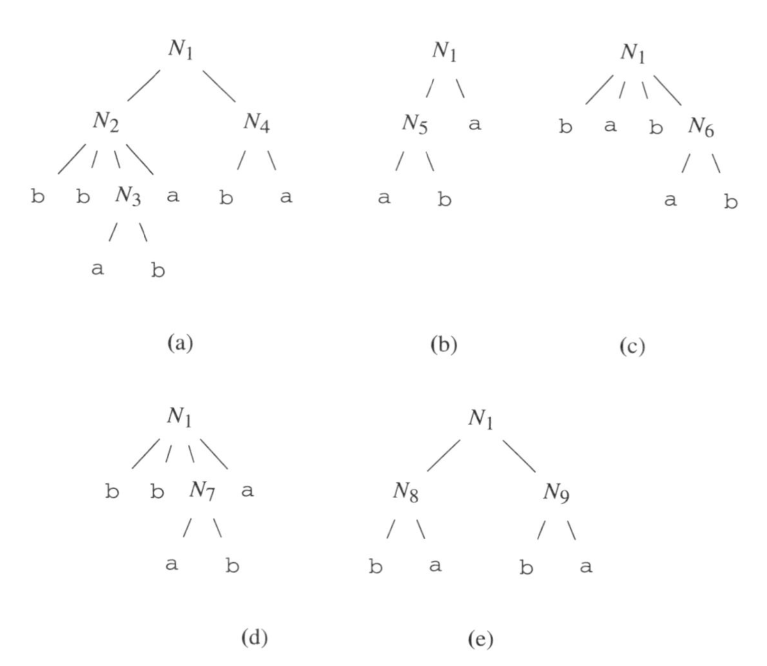

Fig. 15.3. After running Algorithm INITIALISE-K-REV-CF.

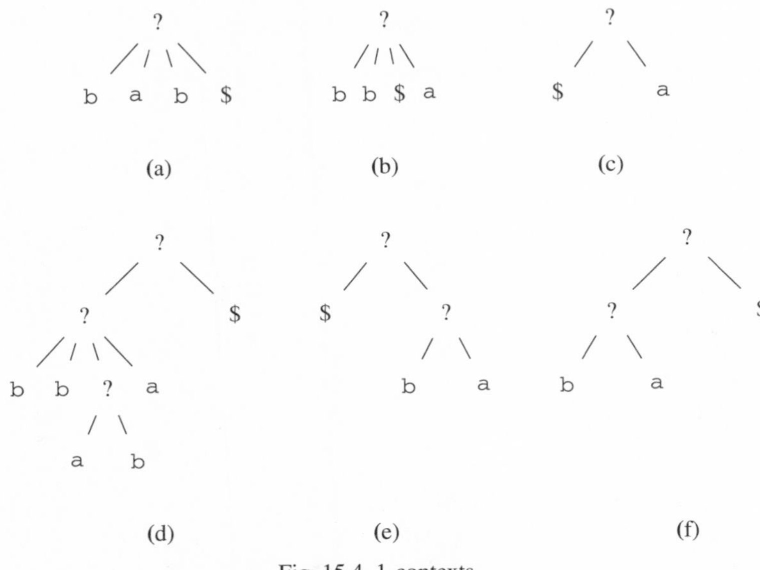

Fig. 15.4. 1-contexts.

- 1-contexts(N_3)={?(b b $ a)} (1-context (b))
- 1-contexts(N_4)={?(?(b b ?(a b)a)$)} (1-context (d))
- 1-contexts(N_5)={?($ a)} (1-context (c))
- 1-contexts(N_6)={?(b a b $)} (1-context (a))
- 1-contexts(N_7)={?(b b $ a)} (1-context (b))
- 1-contexts(N_8)={?($?(b a))} (1-context (e))
- 1-contexts(N_9)={?(?(b a)$)} (1-context (f))

Now suppose we are running Algorithm 15.2 with $k = 1$.

- The initial grammar is
  $N_1 \rightarrow N_2N_4 + N_5\text{a} + \text{bab}N_6 + \text{bb}N_7\text{a} + N_8N_9$; $N_2 \rightarrow \text{bb}N_3\text{a}$; $N_3 \rightarrow \text{ab}$; $N_4 \rightarrow \text{ba}$; $N_5 \rightarrow \text{ab}$; $N_6 \rightarrow \text{ab}$; $N_7 \rightarrow \text{ab}$; $N_8 \rightarrow \text{ba}$; $N_9 \rightarrow \text{ba}$.
- Since we have the rules $N_3 \rightarrow \text{ab}$ and $N_7 \rightarrow \text{ab}$, and N_3 and N_7 share a common 1-context, the grammar is not reset-free, so N_3 and N_7 are merged (into N_3). At this point our running grammar is
  $N_1 \rightarrow N_2N_4 + N_5\text{a} + \text{bab}N_6 + \text{bb}N_3\text{a} + N_8N_9$; $N_2 \rightarrow \text{bb}N_3\text{a}$; $N_3 \rightarrow \text{ab}$; $N_4 \rightarrow \text{ba}$; $N_5 \rightarrow \text{ab}$; $N_6 \rightarrow \text{ab}$; $N_8 \rightarrow \text{ba}$; $N_9 \rightarrow \text{ba}$.
- The algorithm then discovers that for N_1 and N_2 the invertibility condition doesn't hold, so they are merged, resulting in the grammar
  $N_1 \rightarrow N_1N_4 + N_5\text{a} + \text{bab}N_6 + \text{bb}N_3\text{a} + N_8N_9$; $N_3 \rightarrow \text{ab}$; $N_4 \rightarrow \text{ba}$; $N_5 \rightarrow \text{ab}$; $N_6 \rightarrow \text{ab}$; $N_8 \rightarrow \text{ba}$; $N_9 \rightarrow \text{ba}$.
- At this point the algorithms halts. One can notice that the grammar can be simplified without modifying the language.

15.2.6 Why the algorithm works

The complexity of the algorithm is clearly polynomial in the size of the learning sample. Moreover,

Theorem 15.2.2 *Algorithm 15.1 identifies context-free grammars in the limit from structured text and admits polynomial characteristic samples.*

We do not give the proof here. As usual, the tree notations make things extremely cumbersome. But some of the key properties are as follows:

Properties 15.2.3

- The order in which the merges are done doesn't matter. The resulting grammar is the same.
- Complexity is polynomial (for fixed k) with the size of the sample.
- The algorithm admits polynomial characteristic samples.

15.3 Constructive rewriting systems

An altogether different way of generating a language is through rewriting systems. It is possible to define special systems by giving a base and a rewrite mechanism, such that $b \in L$, and if $w \in L$ then $R(w) \in L$.

These systems can be learnt from text or from an informant depending on the richness of the class of rewriting systems considered.

15.3.1 The general mechanism

Let Σ be an alphabet, B a finite subset of $\Sigma^\star$ called the base, and R a set of rules: $(\Sigma^\star)^n \rightarrow \Sigma^\star$ which is some *constructive* function.

We expect that a certain number of properties hold. Informally,

- the smallest string(s) in L should be in B,
- from two strings u and v such that $R(u) = v$ one should be able to deduce something about R,
- if the absence in L of a string is needed to deduce rules from R then an informant will be needed.

Obviously other issues are raised here that we have seen in previous sections: when attempting to learn a rule, this rule should not be masked by a different rule that is somehow learnt before. The question behind this remark is of the existence of a normal form.

It should be noted that this type of mechanism avoids the difficult question of non-linearity. The difference in size between two positive examples (strings) u and v, such that v is obtained by applying R once to v, is going to be small.

15.3.2 Pure grammars

A typical case of learning constructive rewriting systems concerns inferring *pure grammars* from text. Pure grammars are basically context-free grammars where there is just one alphabet: the non-terminal and terminal symbols are interchangeable.

Definition 15.3.1 (Pure grammars) A **pure grammar** $G = (\Sigma, R, u)$ is a triple where Σ is an alphabet, $R \subset \Sigma \times \Sigma^\star$ is the set of rules and u is a string from $\Sigma^\star$ called the axiom.

Derivation is expressed as in usual context-free grammars. The only difference is that the set of variables and the set of terminal symbols coincide.

Example 15.3.1 Let $G = (\Sigma, R, \texttt{b})$ with $\Sigma = \{\texttt{a}, \texttt{b}\}$, $R = \{\texttt{b} \rightarrow \texttt{abb}\}$ and the axiom is b. The smallest strings in $\mathbb{L}(G)$ are b, abb, aabbb, ababb. It is easy to see that $\mathbb{L}(G)$ is the Lukasiewicz language.

The fact that there is only one alphabet means that in the long term, the different strings involved in a derivation will appear in the sample. This (avoiding the curse of expansiveness) allows learning pure grammars to become feasible, even from text. Some restrictions to the class of grammars nevertheless have to be added in order to obtain stronger results, like those involving polynomial bounds.

Definition 15.3.2
A pure grammar $G = (\Sigma, R, u)$ is **monogenic** *if*$_{def}$ $u \stackrel{*}{\Longrightarrow} w \rightarrow w'$ means that there are unique strings v_1 and v_2 such that $w = v_1 x v_2$, $w' = v_1 y v_2$ and $(xy) \in R$.

A pure grammar G is **deterministic** *if*$_{def}$ for each symbol a in Σ there is at most one production with a on the left-hand side.

A pure grammar G is k-**uniform** *if*$_{def}$ all rules (l, r) in R have $|r| = k$.

A language is pure if there is a pure grammar that generates it. It is deterministic if there is a pure deterministic grammar that generates it. And it is k-uniform if there is a pure k-uniform grammar that generates it.

Let us denote, for an alphabet Σ, by $\mathcal{PURE}(\Sigma)$, $\mathcal{PURE} - \mathcal{DET}(\Sigma)$ and $\mathcal{PURE}$-k-$\mathcal{UNIFORM}(\Sigma)$ the classes respectively of pure, pure deterministic and k-uniform languages over Σ.

Example 15.3.2 The Lukasiewicz language, if we consider the pure grammar with the unique rule $\texttt{b} \rightarrow \texttt{abb}$, is clearly monogenic and deterministic. It also is 3-uniform, trivially.

Theorem 15.3.1 *The class $\mathcal{PURE}(\Sigma)$ of all pure languages over the alphabet Σ is not identifiable in the limit from text.*

Proof It is easy to notice that with any non-empty finite language L over Σ we can associate a pure grammar of the form $G = (\Sigma \cup \{\texttt{a}\}, R, \texttt{a})$, with as many rules $(\texttt{a}, w)$ as there are strings w in L. Notice that symbol a does not belong to Σ. In this case we have $L = \mathbb{L}(G) \setminus \{\texttt{a}\}$. And since one can also generate infinite languages, the theorem follows easily using Gold's results (Theorem 7.2.3, page 151). □

Theorem 15.3.2 *The class $\mathcal{PURE}$-k-$\mathcal{UNIFORM}(\Sigma)$ of all k-uniform pure languages over the alphabet Σ is identifiable in the limit from text.*

Proof We provide a non-constructive proof. Finding the axiom is easy (the smallest string) and so is finding the k (i.e. by looking at the differences between the lengths of the strings). Then since the number of possible rules is finite, the class therefore has finite thickness and is learnable from text. □

The above result does not directly give us an algorithm for learning (such an algorithm is to be built in Exercise 15.9). To give a flavour of the type of algorithmics involved in this setting, let us run an intuitive version of the intended algorithm. Given a learning sample S_+, we can build a pure grammar as follows:

Suppose the learning data are {a, bab, cac, bccaccb, cbcacbc}. The axiom is found immediately and is a since it is the shortest string. k is necessarily 3. Then bab is chosen and is obtained from the axiom by applying the rule a $\rightarrow_R$ bab. The sample is simplified and is now {cac, bccaccb, cbcacbc}. The rule a $\rightarrow_R$ cac is invented to cope with cac. The set of rules is now capable of generating the entire sample.

15.4 Reducing rewriting systems

An alternative to using context-free grammars, which are naturally expanding (starting with the axiom), is to use reducing rewriting systems. The idea is that the rewriting system should eventually halt, so, in some sense, the left-hand sides of the rules should be larger than the right-hand sides. With just the length, this is not too difficult, but the class of languages is then of little interest.

In order to study a class containing all the regular languages, but also some others, we introduce delimited string-rewriting systems (SRS). This class, since it contains all the regular languages, will require more than text to be learnable. We therefore study the learnability of this class from an informant.

The rules of delimited string-rewriting systems allow us to replace substrings in strings. There are variants where variables are allowed, but these usually give rise to extremely powerful classes of languages, so for grammatical inference purposes we concentrate on simple variable free rewriting systems.

15.4.1 Strings, terms and rewriting systems

Let us introduce two new symbols \$ and £ that do not belong to the alphabet Σ and will respectively mark the beginning and the end of each string. The languages we are concerned with are thus subsets of $\$\Sigma^*£$. As for the rewrite rules, they will be made of pairs of ***terms** partially* marked; a term is a string over alphabet $\{\$, £\} \cup \Sigma$. Such strings have the restriction that the symbol \$ may only appear in first position whereas symbol £ may only appear in last position. Each term therefore belongs to $(\lambda + \$)\ \Sigma^\star(\lambda + £)$. We denote this set by $\mathbf{T}(\Sigma)$.

Formally, $\mathbf{T}(\Sigma) = \$\ \Sigma^\star\ £ \cup \$\ \Sigma^\star \cup \Sigma^\star\ £ \cup \Sigma^\star = (\$ + \lambda)\ \Sigma^\star(£ + \lambda)$.

The forms of the terms will constrain their use either to the beginning, or to the end, or to the middle, or even to the string taken as a whole.

Terms in $\mathbf{T}(\Sigma)$ can be of one of the following types:

- (Type 1.) $w \in \Sigma^\star$ (used to denote substrings) or
- (Type 2.) $w \in \$\,\Sigma^\star$ (used to denote prefixes) or
- (Type 3.) $w \in \Sigma^\star\,£$ (used to denote suffixes) or
- (Type 4.) $w \in \$\,\Sigma^\star\,£$ (used to denote whole strings).

Given a string w in $\mathbf{T}(\Sigma)$, the ***root*** of w is the string $\$^{-1}w£^{-1}$, $\$^{-1}w$, $w£^{-1}$ and w, respectively.

We define a specific order relation over $\mathbf{T}(\Sigma)$:

$$u <_{\text{DSRS}} v if_{def} \text{root}(u) <_{lex\text{-}length} \text{root}(v) \vee \\ \left[\text{root}(u) = \text{root}(v) \wedge \text{type}(u) < \text{type}(v)\right]$$

Example 15.4.1 $\Sigma = \{\mathtt{a}, \mathtt{b}\}$. Then \$a£, \$aab£ and \$£ are strings in $\$\Sigma^*£$. aa, \$b£ and \$baa£ are terms (elements of $\mathbf{T}(\Sigma)$, of respective types 1, 2, 3 and 4). The root of both \$aab£ and \$aab£ is aab.

Finally, we have ab $<_{\text{DSRS}}$ \$ab $<_{\text{DSRS}}$ ab£ $<_{\text{DSRS}}$ \$ab£ $<_{\text{DSRS}}$ ba.

We can now define the rewriting systems we are considering:

Definition 15.4.1 (Delimited string-rewriting system)

- A *rewrite rule* ρ is an ordered pair of terms $\rho = (l, r)$, generally written as $\rho = l \vdash r$. l is called the left-hand side of R and r its right-hand side.
- We say that $\rho = l \vdash r$ is a *delimited rewrite rule* if_{def} l and r are of the same type.
- By a *delimited string-rewriting system* (DSRS), we mean any finite set $\mathcal{R}$ of delimited rewrite rules.

The order $<_{\text{DSRS}}$ extends to rules: $(l_1, r_1) <_{\text{DSRS}} (l_2, r_2)$ if_{def} $l_1 <_{\text{DSRS}} l_2 \vee \left[l_1 = l_2 \wedge r_1 <_{lex\text{-}length} r_2\right]$.

A system is ***deterministic*** if_{def} no two rules share a common left-hand side.

Given a system $\mathcal{R}$ and a string w, there may be several rules that seem to be applicable upon w. Nevertheless only one rule is eligible. This is the rule having the smallest left-hand side, for the order $<_{\text{DSRS}}$. Formally, a rule $\rho = l \vdash r$ is eligible for string w if

$$\exists u, v \in \mathbf{T}(\Sigma) : \; w = ulv \\ \forall u'l'v' : \exists \rho' = l' \vdash r', l <_{\text{DSRS}} l'$$

One should note that the same rule might be eligible in different places. We systematically privilege the leftmost position.

Example 15.4.2 With system $(\{\mathtt{ab} \vdash \lambda; \mathtt{ba} \vdash \lambda\}, \$£)$, if we consider string \$ababbaba£, both rules ab $\vdash \lambda$ and ab $\vdash \lambda$ can be used, and each in various positions. The eligible rule is the first and it must be used in the leftmost position, therefore:

$$\$\underline{\mathtt{ab}}\mathtt{abbaba}£ \vdash \$\mathtt{abbaba}£$$

Given a DSRS $\mathcal{R}$ and two strings $w_1, w_2 \in \mathbf{T}(\Sigma)$, we say that w_1 ***rewrites in one step into*** w_2, written $w_1 \vdash_{\mathcal{R}} w_2$ or simply $w_1 \vdash w_2$, if_{def} there exists an eligible rule $(l \vdash r) \in \mathcal{R}$ for w_1, and there are two strings $u, v \in (\lambda + \$)\ \Sigma^\star(\lambda + \pounds)$ such that $w_1 = ulv$ and $w_2 = urv$, and furthermore u is shortest for this rule.

A string w is ***reducible*** if_{def} there exists w' such that $w \vdash w'$, and ***irreducible*** otherwise.

Example 15.4.3 Again for system $(\{\texttt{ab} \vdash \lambda;\ \texttt{ba} \vdash \lambda\}, \$\pounds)$, string $\$\texttt{ababa}\pounds$ is reducible whereas $\$\texttt{bbb}\pounds$ is not.

The constraints on \$ and £ are such that these symbols always remain in their typical positions during the reductions at the beginning and the end of the string.

Let $\vdash^*_{\mathcal{R}}$ (or simply $\vdash^*$) denote the reflexive and transitive closure of $\vdash_{\mathcal{R}}$. We say that w_1 ***reduces to*** w_2 or that w_2 ***is derivable from*** w_1 if_{def} $w_1 \vdash^*_{\mathcal{R}} w_2$.

Definition 15.4.2 (Language induced by a DSRS**)** Given a DSRS $\mathcal{R}$ and an irreducible string $e \in \Sigma^\star$, we define the language $\mathbb{L}(\mathcal{R}, e)$ as the set of strings that reduce to e using the rules of $\mathcal{R}$:

$$\mathbb{L}(\mathcal{R}, e) = \{w \in \Sigma^\star : \$w\pounds \vdash^*_{\mathcal{R}} \$e\pounds\}$$

Deciding whether a string w belongs to a language $\mathbb{L}(\mathcal{R}, e)$ or not consists of trying to obtain e from w by a rewriting derivation. We will denote by $\text{APPLY}_{\mathcal{R}}(\mathcal{R}, w)$ the string obtained by applying the different rules in $\mathcal{R}$ until no more rules can be applied. We extend the notation to a set of strings:

$$\text{APPLY}_{\mathcal{R}}(\mathcal{R}, S) = \{\mathcal{R}(w) :\ w \in S\}$$

Example 15.4.4 This time let us consider the Lukasiewicz language which can be represented by the system $(\{\texttt{abb} \vdash\ \texttt{b}\}, \$\texttt{b}\pounds)$. But there is an alternative system: $(\{\$\texttt{ab} \vdash \$;\ \texttt{aab} \vdash \texttt{a}\}, \$\texttt{b}\pounds)$.

Let us check that for either system, string aababbabb can be obtained as an element of the language:

$$\$\texttt{aab}\underline{\texttt{abb}}\texttt{abb}\pounds \vdash \$\texttt{a}\underline{\texttt{abb}}\texttt{abb}\pounds \vdash \$\texttt{ab}\underline{\texttt{abb}}\pounds \vdash \$\underline{\texttt{abb}}\pounds \vdash \$\texttt{b}\pounds$$

$$\$\underline{\texttt{aab}}\texttt{abbabb}\pounds \vdash \$\underline{\texttt{aab}}\texttt{babb}\pounds \vdash \underline{\$\texttt{ab}}\texttt{abb}\pounds \vdash \underline{\$\texttt{ab}}\texttt{b}\pounds \vdash \$\texttt{b}\pounds$$

Let $|\mathcal{R}|$ be the number of rules of $\mathcal{R}$ and $\|\mathcal{R}\|$ be the sum of the lengths of the strings $\mathcal{R}$ is involved in: $\|\mathcal{R}\| = \sum_{(l \vdash r) \in \mathcal{R}} |lr|$.

Here are some examples of DSRS and associated languages:

Example 15.4.5 Let $\Sigma = \{\texttt{a}, \texttt{b}\}$.

- $\mathbb{L}(\{\texttt{ab} \vdash \lambda\}, \lambda)$ is the Dyck language. Indeed, since this single rule erases substring ab, we get the following example of a derivation:

$$\$\texttt{aabb}\underline{\texttt{ab}}\pounds \vdash \$\texttt{a}\underline{\texttt{ab}}\texttt{b}\pounds \vdash \$\underline{\texttt{ab}}\pounds \vdash \$\pounds$$

- $\mathbb{L}(\{\texttt{ab} \vdash \lambda;\ \texttt{ba} \vdash \lambda\}, \lambda)$ is the language $\{w \in \Sigma^\star : |w|_{\texttt{a}} = |w|_{\texttt{b}}\}$, because every rewriting step erases one a and one b.

- $\mathbb{L}(\{\texttt{aabb} \vdash \texttt{ab}; \texttt{\$ab£} \vdash \texttt{\$£}\}, \lambda) = \{\texttt{a}^n\texttt{b}^n : n \geq 0\}$. For instance,

$$\texttt{\$aa\underline{aabb}bb£} \vdash \texttt{\$a\underline{aabb}b£} \vdash \texttt{\$\underline{aabb}£} \vdash \texttt{\$\underline{ab}£} \vdash \texttt{\$£}$$

  Notice that the rule $\texttt{\$ab£} \vdash \texttt{\$£}$ is necessary for deriving λ (the last derivation step).
- $\mathcal{L}(\{\texttt{\$ab} \vdash \texttt{\$}\}, \lambda)$ is the regular language $(\texttt{ab})^*$. Indeed,

$$\underline{\texttt{\$ab}}\texttt{abab£} \vdash \underline{\texttt{\$ab}}\texttt{ab£} \vdash \underline{\texttt{\$ab}}\texttt{£} \vdash \texttt{\$£}$$

We claim that given any regular language L there is a system $\mathcal{R}$ such that $\mathbb{L}(\mathcal{R}) = L$. The fact that the rules can only be applied in a left-first fashion is a crucial reason for this. One can associate with every state in the DFA rules rewriting to the shortest string that reaches the state for the lex-length order.

15.4.2 *Algorithm* LARS

The algorithm we present (Learning Algorithm for Rewriting Systems) generates the possible rules among those that can be applied over the positive data, tries using them and keeps them if they do not create inconsistency (using the negative sample for that). Algorithm LARS (15.4) calls the function NEWRULE (15.3), which generates the next possible rule to be checked.

Algorithm 15.3: NEWRULE.

Input: S_+, rule ρ
Output: a new rule (l, r)
returns the first rule for $<_{\text{DSRS}}$ after ρ such that $\Sigma^\star\, l\, \Sigma^\star \cap L \neq \emptyset$

For this, one should choose *useful* rules, i.e. those that can be applied on at least one string from S_+. One might also consider useful a rule that allows us to diminish the size of the set S_+: a rule which, when added, has the property that two different strings rewrite into an identical string. The goal of usefulness is to avoid an exponential explosion in the number of rules to be checked. The function CONSISTENT (15.5) checks that by adding the new rule to the system, one does not rewrite a positive example and a negative example into a same string.

The goal is to be able to learn any DSRS with LARS. The simplified version proposed here can be used as basis for that, and does identify in the limit any DSRS. But, a formal study of the qualities of the algorithm (as far as mind changes and characteristic samples) is beyond the scope of this book.

Algorithm 15.4: LARS.

Input: S_+, S_-
Output: $\mathcal{R}$
$\mathcal{R} \leftarrow \emptyset$;
$\rho \leftarrow$ NEWRULE(S_+, (λ, λ));
while $|S_+| > 1$ **do**
  if LARS-CONSISTENT(S_+,S_-,$\mathcal{R} \cup \{\rho\}$) **then**
    $\mathcal{R} \leftarrow \mathcal{R} \cup \{\rho\}$;
    $S_+ \leftarrow$ APPLY$_{\mathcal{R}}$($\mathcal{R}$, S_+);
    $S_- \leftarrow$ APPLY$_{\mathcal{R}}$($\mathcal{R}$, S_-)
  end
  $\rho \leftarrow$ NEWRULE(S_+, r)
end
$w \leftarrow \min(S_+)$;
return $\langle \mathcal{R}, w \rangle$

Algorithm 15.5: LARS-CONSISTENT.

Input: S_+,S_-,$\mathcal{R}$
Output: a boolean indicating if the current system is consistent with (S_+,S_-)
if $\exists x \in S_+, y \in S_-$: APPLY$_{\mathcal{R}}$($\mathcal{R}$, x) = APPLY$_{\mathcal{R}}$($\mathcal{R}$, y) **then**
  return false
else
  return true
end

15.4.3 Running LARS

We give an example run of algorithm LARS on the following sample: $S_+ = \{$abb, b, aabbb, abababb$\}$, and $S_- = \{\lambda$, a, ab, ba, bab, abbabb$\}$. LARS tries the following rules:

- The smallest rule for the order proposed is a $\vdash \lambda$, which fails because ab and ba would both rewrite, using this rule, into b; but ab $\in S_-$ and b $\in S_+$.
- The next rule is \$a $\vdash$ \$, which fails because ab would again rewrite into b.
- No other rule based on the pair (a, λ) is tried, because the rule would apply to no string in S_+. So the next rule is b $\vdash \lambda$, and fails because b rewrites into λ.
- Again \$b $\vdash$ \$, b£ $\vdash$ £ and \$b£ $\vdash$ \$£ all fail because b would rewrite into λ.
- b $\vdash$ a, fails because b rewrites into a.

- $\mathtt{\$b} \vdash \mathtt{\$a}$, fails because b rewrites into a.
- $\mathtt{ab} \vdash \lambda$ is the next rule to be generated; it is rejected because bab would rewrite into b.
- $\mathtt{\$ab} \vdash \$$ is considered next and is this time accepted. The samples are parsed using this rule and are updated to $S_+ = \{\mathtt{b}, \mathtt{aabbb}\}$ and $S_- = \{\lambda, \mathtt{a}, \mathtt{ab}, \mathtt{ba}, \mathtt{bb}, \mathtt{bab}\}$.
- Rules with left-hand side ba, bb, aaa and variants are not analysed, because they cannot apply to S_+.
- The next rule to be checked (and rejected) is $\mathtt{aab} \vdash \lambda$ but then aabbb would rewrite into bb which (now) belongs to S_-.
- Finally, $\mathtt{aab} \vdash \mathtt{a}$ is checked, causes no problem, and is used to parse the samples obtaining $S_+ = \{\mathtt{b}\}$ and $S_- = \{\lambda, \mathtt{a}, \mathtt{ab}, \mathtt{ba}, \mathtt{bb}, \mathtt{bab}\}$. The algorithm halts with the system $(\{\mathtt{\$ab} \vdash \$;\ \mathtt{aab} \vdash \mathtt{a}\}, \mathtt{\$b£})$.

15.5 Some heuristics

The theoretical results about context-free grammar learning are essentially negative and state that no efficient algorithm can be found that can learn the entire class of the context-free languages in polynomial conditions, whether we want to learn with queries, from characteristic samples or in the PAC setting.

This has motivated the introduction of many specific heuristic techniques. It would be difficult to show them all, but some ideas are also given in Chapter 14. In order to present different approaches, we present here only two lines of work. The first corresponds to the use of minimum length encoding, the second to an algorithm that has not really been used in grammatical inference, but rather for compression tasks.

15.5.1 Minimum description length

We presented the minimum description length principle in a general way, but also for the particular problem of learning DFAs, in Section 14.4. The principle basically states that the size of an encoding should be computed as the sum of the size of the model (here the grammar) and the size of the object (here the strings) when encoded by the grammar.

We present the ideas behind an algorithm called GRIDS whose goal is to learn a context-free grammar from text.

The starting point is a grammar that generates the sample exactly, with exactly one rule $N_1 \to w$ per string w in the sample.

Then iteratively, the idea is to try to better the score of the current grammar by trying one of the two following operations:

- A ***creating*** operation takes a substring of terminals and non-terminals, invents a new non-terminal that derives into the string, and replaces the string by the non-terminal in all the rules. This operator does not modify the generated language.
- A ***merging*** operation takes two non-terminals and merges them into one; this operator can generalise the generated language.

Example 15.5.1 Suppose the current grammar contains:

- $N_1 \rightarrow \mathtt{a}N_2\mathtt{ba}$
- $N_2 \rightarrow N_3\mathtt{ba}$

Then the merging operation could replace these rules by:

- $N_1 \rightarrow \mathtt{a}N_2\mathtt{ba}$
- $N_2 \rightarrow N_2\mathtt{ba}$

whereas the creating operation might replace the rules by:

- $N_1 \rightarrow \mathtt{a}N_2N_4$
- $N_2 \rightarrow N_3N_4$
- $N_4 \rightarrow \mathtt{ba}$

Iteratively each possible merge/creation is tested and a score is computed: the score takes into account the size of the obtained grammar and the size of the data set, when generated by the grammar. The best score that betters the current score indicates which operation, if any, determines the new grammar.

The way the score is computed is important: it has to count the number of bits needed to optimally encode the candidate grammar, and the number of bits needed to encode (also optimally) a derivation of the text for that grammar. This should therefore take into account that a given non-terminal can have various derivations or not. Some initial ideas towards using the MDL can be found in Section 14.4, for the case of automata.

15.5.2 Grammar induction as compression

The idea of defining an operator over grammars that allows us to transform one grammar into another has also been used by an algorithm that, although not of grammar induction (there is no generalisation), is close to the ideas presented here. Moreover, the structure found by this algorithm can be used as a first step towards inferring a context-free grammar. The algorithm is called SEQUITUR and is used to compress (without loss) a text by finding the repetitions and structure of this text. The idea is to find a grammar that exactly generates one string, i.e. the text that should be compressed. If the grammar encodes the text in less symbols than the text length, then a compression is performed.

Two conditions have to be followed by the resulting grammar:

- each rule (but the 'initial' one) of the grammar has to be used at least twice,
- there can be no repeated substring of length more than one.

Example 15.5.2 The following grammar is accepted. Notice that each non-terminal is used just once in the left-hand side of a rule, and that all non-terminals, with the exception of N_1, are used at least twice.

- $N_1 \rightarrow N_2N_3N_4N_2$
- $N_2 \rightarrow \mathtt{a}N_3\mathtt{b}N_5$
- $N_3 \rightarrow N_5\mathtt{a}N_4$

- $N_4 \rightarrow$ baN_5
- $N_5 \rightarrow$ ab

The algorithm starts with just one rule, called the N_1 rule, which is $N_1 \rightarrow \lambda$. SEQUITUR works sequentially by reading one symbol of the text at a time, adds it to the N_1 rule and attempts to transform the running grammar by either:

- using an existing rule,
- creating a new rule,
- deleting some rule that is no longer needed.

Some of these actions may involve new actions taking place.

There is little point in giving the code of SEQUITUR because on one hand the algorithmic ideas are clear from the above explanation, but on the other hand, the implementation details are far beyond the scope of this section, although they are essential in order to keep the algorithm quasi-linear.

Let us instead run SEQUITUR on a small example.

Example 15.5.3 Suppose the entry string is aaabababaab.

- The initial grammar is $N_1 \rightarrow \lambda$, corresponding to having read the empty prefix of the entry string. SEQUITUR reads the first letter of the input (a). The current grammar therefore becomes $N_1 \rightarrow$ a. Nothing more happens (i.e. the grammar is accepted to far).
- SEQUITUR reads the next symbol (a). The total input is now a**a**. The current grammar becomes $N_1 \rightarrow$ aa.
- The next symbol is read (input is aa**a**). The grammar is updated to $N_1 \rightarrow$ aaa.
- Another symbol is read (input is aaa**b**). The grammar is therefore updated to $N_1 \rightarrow$ aaab. No transformation of the grammar is so far possible.
- The next symbol (a) is read for a total input of aaab**a**. The current grammar becomes $N_1 \rightarrow$ aaaba.
- The next symbol (b) is read. Input is now aaaba**b**. The current grammar becomes $N_1 \rightarrow$ aaabab, but substring ab appears twice. So a new non-terminal is introduced and the grammar is $N_1 \rightarrow$ aaN_2N_2, $N_2 \rightarrow$ ab.
- Another symbol is read. Input is now aaabab**a**. The current grammar becomes $N_1 \rightarrow$ aaN_2N_2a, $N_2 \rightarrow$ ab.
- Another symbol is read. Input is now aaababa**b**. The current grammar becomes $N_1 \rightarrow$ aaN_2N_2ab, $N_2 \rightarrow$ ab, but rule $N_2 \rightarrow$ ab can be used, so we have $N_1 \rightarrow$ aa$N_2N_2N_2$, $N_2 \rightarrow$ ab.
- Another symbol is read. Input is now aaabababa**a**. The current grammar becomes $N_1 \rightarrow$ aa$N_2N_2N_2$a, $N_2 \rightarrow$ ab.
- Another symbol is read. Input is now aaabababa**a**. The current grammar becomes $N_1 \rightarrow$ aa$N_2N_2N_2$aa, $N_2 \rightarrow$ ab.
- Another symbol is read. Input is now aaabababaa**b**. The current grammar becomes $N_1 \rightarrow$ aa$N_2N_2N_2$aab, $N_2 \rightarrow$ ab. We can now apply rule $N_2 \rightarrow$ ab and obtain $N_1 \rightarrow$ aa$N_2N_2N_2$aN_2, $N_1 \rightarrow$ ab but now aN_2 appears twice so we introduce $N_3 \rightarrow$ aN_2 and the grammar is $N_1 \rightarrow$ a$N_3N_2N_2N_3$, $N_2 \rightarrow$ ab, $N_3 \rightarrow$ aN_2.
- As the string is entirely parsed, the algorithm halts.

15.6 Exercises

15.1 Is the following grammar G very simple deterministic?
$G = \langle \Sigma, V, R, N_1 \rangle$, $\Sigma = \{\mathtt{a}, \mathtt{b}, \mathtt{c}, \mathtt{d}\}$, $V = \{N_1, N_2, N_3, N_4\}$, $R = \{N_1 \rightarrow \mathtt{b}N_2 + \mathtt{a}N_2N_4;\ N_2 \rightarrow \mathtt{c}N_2N_2 + \mathtt{d};\ N_3 \rightarrow \mathtt{a}N_3\mathtt{b};\ N_4 \rightarrow \mathtt{ab}\}$.

15.2 Complete Theorem 15.1.3: does the proposed algorithm for very simple deterministic grammars have polynomial characteristic samples? Does it make a polynomial number of mind changes?

15.3 Consider the learning sample containing trees:

- $N_1(N_2(\mathtt{b\,b}\,N_3(\mathtt{a\,b})\mathtt{a})N_3(\mathtt{b\,a}))$
- $N_1(N_3(\mathtt{a\,b})\mathtt{a})$
- $N_1(\mathtt{b\,a\,b}N_3(\mathtt{a\,b}))$
- $N_1(\mathtt{b\,b}\,N_2(\mathtt{a\,b})\mathtt{a})$
- $N_1(N_3(\mathtt{b\,a})N_2(\mathtt{b\,a}))$

Take k=1 and k=2. What are the k-ancestors of N_3 that we can deduce from the sample? Draw the corresponding k-contexts of N_3.

15.4 Why can we not deduce from Theorem 15.2.1 that context-free grammars are identifiable from text?

15.5 Is the following grammar $G = \langle \Sigma, V, R, N_1 \rangle$ k-reversible? For what values of k? $\Sigma = \{\mathtt{a}, \mathtt{b}\}$, $V = \{N_1, N_2, N_3, N_4\}$, $R = \{N_1 \rightarrow \mathtt{b}N_2 + \mathtt{a}N_4 + \mathtt{bab};\ N_2 \rightarrow \mathtt{a}N_2\mathtt{b}N_2 + \mathtt{a};\ N_3 \rightarrow \mathtt{a} + \mathtt{b};\ N_4 \rightarrow \mathtt{a}N_3\mathtt{b}N_2 + \mathtt{ab}\}$.

15.6 Find a context-free grammar for which the corresponding 0-reversible grammar is of exponential size.

15.7 Learn a pure grammar (see Section 15.4) from the following sample: $S_+ = \{\mathtt{aba}, \mathtt{a}^3\mathtt{ba}^3, \mathtt{a}^6\mathtt{a}^6\}$. Suppose the grammar is k-uniform.

15.8 Learn a pure grammar from the sample $S_+ = \{\mathtt{aba}, \mathtt{a}^3\mathtt{ba}^3, \mathtt{a}^6\mathtt{a}^6\}$. Suppose the grammar is deterministic.

15.9 Write a learning algorithm corresponding to the proof of Theorem 15.3.2. What is its complexity? Prove that it admits polynomial characteristic samples.

15.10 Run algorithm SEQUITUR over the following input texts. What sort of grammatical constructs does SEQUITUR find? Which are the ones it cannot find?

$$w_1 = \mathtt{aaaabbbb}$$
$$w_2 = \mathtt{a}^{256}\mathtt{b}^{256}$$
$$w_3 = (\mathtt{ab})^{256}$$

15.7 Conclusions of the chapter and further reading

15.7.1 Bibliographical background

The discussion about whether natural language is context-free has been going on since the class of context-free languages was invented (Chomsky, 1957). For a more grammatical inference flavour, see (Becerra-Bonache, 2006, Roark & Sproat, 2007). The discussion about the specific difficulties of learning context-free grammars relies mostly on work by Rémi Eyraud (Eyraud, 2006) and Colin de la Higuera (de la Higuera, 2006a). There are few negative results corresponding to learning context-free grammars. Dana Angluin conjectured that these were not learnable by using an MAT and the proof appears in (Angluin & Kharitonov, 1991). Colin de la Higuera (de la Higuera, 1997) proved that polynomial characteristic sets were of unboundable size.

A first line of research has consisted of limiting the class of context-free grammars to be learnt: even linear grammars (Takada, 1988), deterministic linear grammars (de la Higuera & Oncina, 2002) and very simple grammars (Yokomori, 2003) have been proved learnable in different settings.

Much work has been done on the problems relating to even linear grammars (Koshiba, Mäkinen & Takada, 1997, Mäkinen, 1996, Sempere & García, 1994, Sempere & Nagaraja, 1998, Takada, 1988, 1994). Positive results concerning subclasses of even linear languages have been obtained when learning from text (Koshiba, Mäkinen & Takada, 2000). Takashi Yokomori (Yokomori, 2003) introduced the class of simple deterministic grammars and obtained the different results reported here.

The special class of deterministic linear languages was introduced by Colin de la Higuera and Jose Oncina (de la Higuera & Oncina, 2002). The class was later adapted in order to take probabilities into account (de la Higuera & Oncina, 2003).

The algorithm for learning k-reversible grammars is initially due to Yasubumi Sakakibara (Sakakibara, 1992), based on Dana Angluin's algorithm for regular languages (Angluin, 1982). Further work by Jérôme Besombes and Jean-Yves Marion (Besombes & Marion, 2004a) and Tim Oates *et al.* (Oates, Desai & Bhat, 2002) is used in this presentation. The proof of Theorem 15.2.1 is constructive but beyond the scope of this book. It can be found in the above cited papers.

Pure grammars are learnt by Takeshi Koshiba *et al.* (Koshiba, Mäkinen & Takada, 2000) whereas the work we describe on the rewriting systems is by Rémi Eyraud *et al.* (Eyraud, de la Higuera & Janodet, 2006). In both cases the original ideas and algorithms have been (over) simplified in this chapter, and many alternative ideas and explanations can be found in the original papers.

Based on the MDL principle, Gerry Wolf (Wolf, 1978) introduced an algorithm called GRIDS whose ideas were further investigated by Pat Langley and Sean Stromsten (Langley & Stromsten, 2000), and then by George Petasis *et al.* (Petasis *et al.*, 2004a). Alternatively, the same operators can be used with a genetic algorithm (Petasis *et al.*, 2004b), but the results are no better.

Algorithm SEQUITUR is due to Craig Nevill-Manning and Ian Witten (Nevill-Manning & Witten, 1997a). Let us note that (like in many grammar manipulation programmes) it relies on hash tables and other programming devices to enable the algorithm to work in linear time. Experiments were made by SEQUITUR over a variety of sequential files, containing text or music: compression rates are good, but more importantly, the structure of the text is discovered. If one runs SEQUITUR on special context-free grammars (such as Dyck languages) results are poor: SEQUITUR is good at finding repetitions of patterns, but not necessarily at finding complex context-free structures. On the other hand, in our opinion, no algorithm today is good at this task.

The question of the relationship between learning (and specifically grammar learning) and compression can also be discussed. SEQUITUR performs a lossless compression: no generalisation, or loss, takes place. One can argue that other grammar induction techniques also perform a compression of the text (the learning data) into a grammar. But in that case it is expected that the resulting language is not equal to the initial text. Now the questions that arise from these remarks are: is there a continuum between the SEQUITUR type of lossless compression techniques and the GRIDS type of compression with loss techniques? In other words could we tune SEQUITUR to obtain a generalisation of the input text, or GRIDS to obtain a grammar equivalent to the initial one? How do we measure the loss due to the generalisation, in such a way as to incorporate it into the learning/compression algorithm?

15.7.2 Some alternative lines of research

If to the idea of simplifying the class of grammars we add that of using queries, there are positive results concerning the class of simple deterministic languages. A language is simple deterministic when it can be recognised by a deterministic push-down automaton by empty store, that only uses one state. All languages in this class are thus necessarily deterministic, λ-free and prefix.

Hiroki Ishizaka (Ishizaka, 1995) learns these languages using 2-standard forms: his algorithm makes use of membership queries and extended equivalence queries.

There have been no known positive results relating any form of learning with usual queries and the entire class of context-free languages. It is shown that context-free grammars have *approximate fingerprints* and therefore are not learnable from equivalence queries alone (Angluin, 1990), but also that membership queries alone (Angluin & Kharitonov, 1991) are insufficient even if learning in a PAC setting (under typical cryptographic assumptions), and it is conjectured (Angluin, 2001) that an MAT is not able to cope with context-free languages.

If one reads proceedings of genetic algorithms or evolutionary computing conferences dating from the seventies or eighties, one will find a number of references concerning grammatical inference. Genetic algorithms require a linear encoding of the solutions (hence here of the grammars) and a careful definition of the genetic operators one wants to use, usually a crossing-over operator and a mutation operator. Some (possibly numerical) measure of the quality of a solution Is also required.

The mutation operator takes a grammar, modifies a bit somewhere and returns a new grammar. The crossing-over operator would take two grammars, cut these into two halves and build two new grammars by mixing the halves (Wyard, 1994). One curious direction (Kammeyer & Belew, 1996) to deal with this issue is to admit that in that case, parts of the string will not be used to encode any more, and would correspond to what is known as *junk*

DNA. Other ideas correspond to very specific encodings of the partitions of non-terminals and offer resistant operators (Dupont, 1994, Sakakibara & Kondo, 1999). Yasubumi Sakakibara *et al.* (Sakakibara & Kondo, 1999, Sakakibara & Muramatsu, 2000) represented the grammars by parsing tables and attempted to learn these tabular representations by a genetic algorithm.

Among the several pragmatic approaches, let us mention the BOISDALE algorithm (Starkie & Fernau, 2004) which makes use of special forms of grammars, SYNAPSE (Nakamura & Matsumoto, 2005) which is based on parsing, LARS (Eyraud, de la Higuera & Janodet, 2006) which learns rewriting systems, and Alexander Clark's (Clark, 2004) algorithm (which won the 2004 OMPHALOS competition (Starkie, Coste & van Zaanen, 2004a)) concentrates on deterministic languages. An alternative is to learn a k-testable tree automaton, and estimate the probabilities of the rules (Rico-Juan, Calera-Rubio & Carrasco, 2002). An earlier line of research is based on exhaustive search, either by use of a *version space* approach (Vanlehn & Ball, 1987), or by using operators such as the *Reynolds cover* (Giordano, 1994).

There have also been some positive results concerning learning context-free languages from queries. Identification in the limit is of course trivially possible with the help of strong equivalence queries, through an enumeration process. A more interesting positive result concerns grammars in a special normal form, and when the queries will enable some knowledge about the structure of the grammar and not just the language. The algorithm is a natural extension of Dana Angluin's LSTAR algorithm (Angluin, 1987a) and was designed by Yasubumi Sakakibara (Sakakibara, 1990) who learns context-free grammars from structural queries (a query returns the context in which a substring is used).

In the field of computational linguistics, efforts have been made to learn context-free grammars from more informative data, such as trees (Charniak, 1996), following theoretical results by Yasubumi Sakakibara (Sakakibara, 1992). Learning from structured data has been a line followed by many: learning tree automata (Fernau, 2002, Habrard, Bernard & Jacquenet, 2002, Knuutila & Steinby, 1994), or context-free grammars from bracketed data (Sakakibara, 1990) allows to obtain better results, either with queries (Sakakibara, 1992), regular distributions (Carrasco, Oncina & Calera-Rubio, 2001, Kremer, 1997, Rico-Juan, Calera-Rubio & Carrasco, 2002), or negative information (García & Oncina, 1993). This has also led to different studies concerning the probability estimation of such grammars (Calera-Rubio & Carrasco, 1998, Lari & Young, 1990).

A totally different direction of research has been followed by authors working with categorial grammars. These are as powerful as context-free grammars. They are favoured by computational linguists who have long been interested in working on grammatical models that do not necessarily fit into Chomsky's hierarchy. Furthermore, their objective is to find suitable models for syntax and semantics to be interlinked, and provide a logic-based description language. Key ideas relating such models with the questions of language identification can be found in Makoto Kanazawa's book (Kanazawa, 1998), and discussion relating this to the way children learn language can be found in papers by a variety of authors, for instance Isabelle Tellier (Tellier, 1998). The situation is still unclear, as positive results can only be obtained for special classes of grammars (Forêt & Le Nir, 2002), whereas, here again, the corresponding combinatorial problems (for instance that of finding the smallest consistent grammar) appear to be intractable (Costa Florêncio, 2002).

The situation concerning learnability of context-free grammars has evolved since 2003 with renewed interest caused by workshops (de la Higuera *et al.*, 2003), and more importantly by the OMPHALOS context-free language learning competition (Starkie, Coste & van Zaanen, 2004b), where state of the art techniques were unable to solve even the easiest tasks. The method (Clark, 2007) that obtained best results used a variety of information about the distributions of the symbols, substitution graphs and context. The approach is mainly empirical and does not provide a convergence theorem.

About mildly context-sensitive languages, we might mention Leo Becerra *et al.* (Oates *et al.*, 2006): these systems can learn languages that are not context-free. But one should be careful: mildly context-sensitive languages usually do not contain all the context-free languages. A related approach is through kernels: semi-linear sets (Abe, 1995) and planar languages (Clark *et al.*, 2006) are alternative ways of tackling the problem.

15.7.3 Open problems and possible new lines of research

Work on learning context-free grammars is related to several difficult aspects: language representation issues, decidability questions, tree automata... Yet the subject is obviously very much open with several teams working on the question. Among the most notable open questions, here are some that are worth looking into:

(i) In the definition of identification in the limit from polynomial time and data, the size of a set of strings is taken as the number of bits needed to encode the set. If we take, as some authors propose (Carme *et al.*, 2005), the size of the set as the number of strings in the sample, then it is not known if context-free grammars admit characteristic samples of polynomial size.

(ii) The question of learning probabilistic context-free grammars is widely open with very few recent results, but no claim that these may be learnable in some way or another. Yet these grammars are of clear importance in fields such as computational biology (Jagota, Lyngsø & Pedersen, 2001) or linguistics (Charniak, 1996). In a way, these grammars would bridge the gap between learning languages and grammars, as the very definition of a probabilistic context-free language requires a grammar.

(iii) Another question that has been left largely untouched is that of learning pushdown automata. Yet these provide us with mechanisms that would allow us to add extra bias: the determinism of the automata, the number of turns, and the number of symbols in the stack can be controlled.

(iv) As may be seen in the section about pure grammars, there are several open questions and problems to be solved in that context: what is identifiable and what is not? What about polynomial bounds? The reference paper where these problems are best described is by Takeshi Koshiba *et al.* (Koshiba, Mäkinen & Takada, 2000).

(v) A curious problem is that of learning a context-free language L which is the intersection between another context-free language L_C and a regular language L_R, when the context-free language is given (de la Higuera, 2006b).

(vi) Finally, an interesting line of research was proposed in (Sakakibara & Muramatsu, 2000), where the learning data were partially structured, i.e. some brackets were given. Although the purpose of the paper was to prove that the number of generations of the genetic learning algorithm was lower when more structure was given, the more general question of how much structure is needed to learn is of great interest.

(vii) Most importantly, there is a real need for good context-free grammar-learning algorithms. There are three such algorithms used in computational linguistics (Adriaans & van Zaanen, 2004) we have not described here, in part because they rely on too many different complex algorithmic mechanisms:

- Algorithm EMILE, by Pieter Adriaans and Marco Vervoort (Adriaans & van Zaanen, 2002) relies on substitutability as its chief element: can a string be substituted by another? Is a context equivalent to another? Graph clustering techniques are used to solve these questions.
- Algorithm ABL (alignment-based learning) by Menno van Zaanen (van Zaanen, 2000) relies on aligning the different sentences in order to then associate non-terminals with clusters of substrings.
- Algorithm ADIOS, by Zach Solan *et al.*(Solan *et al.*, 2002) also represents the information by a graph, but aims to find the best paths in the graphs.
- The key idea of substitutability (which is essential in the above works) is studied in more detail by Alexander Clark and Rémi Eyraud (Clark & Eyraud, 2007). Two substrings are congruent if they can be replaced by each other in any context. This allows us, by discovering this congruence, to build a grammar. This powerful mechanism enables us to assemble context-free grammars, and corresponds to one of the most promising lines of research in the field.

16
Learning probabilistic finite automata

En efecto, las computadoras parten del sofisma, políticamente inaceptable, de que dos y dos son cuatro. Su conservadurismo es feroz en este respeto.

***Pablo de la Higuera**, In, Out, Off... ¡Uf!*

The promiscuous grammar has high a priori probability, but assigns low probability to the data. The ad-hoc grammar has very low a priori probability, but assigns probability one to the data. These are two extreme grammar types: the best choice is usually somewhere between them.

***Ray Solomonoff** (Solomonoff, 1997)*

A language is a set of strings, and a string either belongs to a language or does not. An alternative is to define a distribution over the set of all strings, and in some sense the probability of a string in this set is indicative of its importance in the language. This distribution, if learnt from data, can in turn be used to disambiguate, by finding the most probable string corresponding to a pattern, or to predict by proposing the next symbol for a given prefix.

We propose in this chapter to investigate ways of learning distributions representable by probabilistic automata from strings. No extra knowledge is usually provided. In particular the structure of the automaton is unknown. The case where the structure is known corresponds to the problem of *probability estimation* and is discussed in Chapter 17.

We only work in this chapter with deterministic probabilistic finite automata. Positive learning results for distributions defined by other classes of grammars are scarce; we nevertheless comment upon some in the last section of the chapter. Definitions of the probabilistic automata can be found in Chapter 5, whereas the convergence criteria were introduced and discussed in Chapter 10, where also a number of negative results were studied.

16.1 Issues

There are a number of reasons for which one may want to learn a probabilistic language and it should be emphasised that these reasons are independent and therefore do not necessarily add up.

16.1.1 Dealing with noise

If the task is about learning a language in a noisy setting, the algorithms presented in earlier chapters for learning from text, from an informant or with an Oracle have been shown not to be suited. Probabilities offer what looks like a nice alternative: why not learn a probabilistic language in which (hopefully) the weights of the strings in the language will be big and if any noisy string interferes in the learning process the probability of this string will be small and therefore not penalise the learnt hypothesis too much?

But then, and this is a common mistake, the probabilities we are thinking about in this noisy setting correspond to those indicating whether or not a string belongs to the language. The classes of probabilistic languages to be learnt (see Chapter 5) do not use this idea of stochasticity (one should look into fuzzy automata for this, a class for which there are very few learning results).

16.1.2 Prediction

There are situations when what is really important is the symbol and not the string itself. An important and typical question will then be:

Having seen so far abbaba, what is the next symbol?

If we are able to access a distribution of probabilities over $\Sigma^\star$ (and thus the notation $Pr_{\mathcal{D}}(w)$ makes sense), we are able to deduce a distribution over Σ in a given *context*. If we denote by $Pr_{\mathcal{D}}(a|w)$ the probability of generating an a after having generated w when following distribution $\mathcal{D}$ (and $Pr(a|w)$ when the distribution is clear),

$$Pr_{\mathcal{D}}(a|w) = \frac{Pr_{\mathcal{D}}(wa\,\Sigma^\star)}{Pr_{\mathcal{D}}(w\,\Sigma^\star)}$$

The converse is also true: if prediction is possible, then we can use the ***chain rule*** in order to define an associated distribution over Σ^n, for each value of n, and also $\Sigma^\star$ provided the probability of halting given a string is also defined. For this we usually introduce the symbol \$ used to terminate strings (strings will be in the set $\Sigma^\star\,\$$). Formally, given a string $a_1a_2\ldots a_n\$$,

$$Pr_{\mathcal{D}}(w\$) = Pr(a_1|\lambda)\cdot Pr(a_2|a_1)\cdots Pr(a_{i+1}|a_1\cdots a_i)\cdots Pr(a_n|a_1\cdots a_{n-1})\cdot Pr(\$|w)$$

16.2 Probabilities and frequencies

If the goal is to learn probabilistic finite automata, in practice we have no access to probabilities, but to frequencies. We will be told that in the sample, 150 out of the 500 strings start with the letter a, not that $\frac{3}{10}$ of the strings in the language start with a.

The first crucial issue is that even if mathematically $\frac{1}{2} = \frac{500}{1000}$, this does not fit the picture when we think of the fraction as a relative frequency. More precisely, if this fraction is supposed to indicate the empirical probability of an event, then the first fraction would mean 'we have made two tests and out of these two, one was heads', whereas the second fraction is supposed to indicate that 1000 tests were done and 500 resulted in heads.

The difference is that if wanting to say something like, 'I guess we get heads half of the time', there are probably more reasons to say this in the second case than the first.

The second point is that the automaton the learning algorithm will manipulate during the inference process should both keep track of the frequencies (because of the above reasons) and have an easy and immediate probabilistic interpretation.

For this reason, we introduce a new type of automaton, which we call a ***frequency finite automaton*** (FFA).

We will call *frequency* the number of times an event occurs, and *relative frequency* the number of times an event occurs over the number of times it could have occurred.

Definition 16.2.1 A **deterministic frequency finite automaton** (DFFA) is a tuple $\mathcal{A} = \langle \Sigma, Q, \mathbb{I}_{fr}, \mathbb{F}_{fr}, \delta_{fr} \rangle$ where

- Σ is the alphabet,
- Q is a finite set of *states*
- $\mathbb{I}_{fr} : Q \to \mathbb{N}$ (initial-state frequencies); since the automaton is deterministic there is exactly one state q_λ for which $\mathbb{I}_{fr}(q) \neq 0$,
- $\mathbb{F}_{fr} : Q \to \mathbb{N}$ (final-state frequencies),
- $\delta_{fr} : Q \times \Sigma \times Q \to \mathbb{N}$ is the transition frequency function,
- $\delta_{\mathcal{A}} : Q \times \Sigma \to Q$ is the associated transition function.

The definition is intended to be adjusted to that of a DPFA (Definition 5.2.3, page 89). The notation $\delta_{fr}(q, a, q') = n$ can be interpreted as 'there is a transition from q to q' labelled with a that is used n times'. Obviously, because of determinism, there is at most one state q' such that $\delta_{fr}(q, a, q') > 0$, and this state is such that $q' = \delta_{\mathcal{A}}(q, a)$.

There is a relationship, analogous to what can be found in flow diagrams, between the frequencies of the transitions leading to a state and those leaving a state:

Definition 16.2.2
A DFFA $\mathcal{A} = \langle \Sigma, Q, \mathbb{I}_{fr}, \mathbb{F}_{fr}, \delta_{fr}, \delta_{\mathcal{A}} \rangle$ is said to be **consistent** or **well defined** *if*$_{def}$ $\forall q \in Q,\ \mathbb{I}_{fr}(q) + \sum_{\substack{q' \in Q \\ a \in \Sigma}} \delta_{fr}(q', a, q) = \mathbb{F}_{fr}(q) + \sum_{\substack{q' \in Q \\ a \in \Sigma}} \delta_{fr}(q, a, q')$.

When the DFFA is well defined, the number of strings entering and leaving a given state is identical. We denote by FREQ(q) the total number both of entering strings and of leaving strings at state q.

Algorithm 16.1: CONSTRUCTING a DPFA from a DFFA.

Input: a DFFA $\mathcal{A} = \langle \Sigma, Q, \mathbb{I}_{fr}, \mathbb{F}_{fr}, \delta_{fr}, \delta_{\mathcal{A}} \rangle$
Output: a DPFA $\mathcal{B} = \langle \Sigma, Q, q_\lambda, \mathbb{F}_{\mathbb{P}}, \delta_{\mathbb{P}}, \delta_{\mathcal{B}} \rangle$
for $q \in Q$ **do** `/* Final probabilities */`
  FREQ[q] $\leftarrow \mathbb{F}_{fr}(q)$;
  for $a \in \Sigma$ **do** FREQ[q] $\leftarrow$ FREQ[q] $+ \delta_{fr}(q, a, \delta_{\mathcal{A}}(q, a))$ $\mathbb{F}_{\mathbb{P}}(q) \leftarrow \frac{\mathbb{F}_{fr}(q)}{\text{FREQ}[q]}$;
  for $a \in \Sigma$ **do**
    $\delta_{\mathbb{P}}(q, a, q') \leftarrow \frac{\delta_{fr}(q,a,q')}{\text{FREQ}[q]}$;
    $\delta_{\mathcal{B}}(q, a) \leftarrow \delta_{\mathcal{A}}(q, a)$
  end
end
return $\mathcal{B}$

Definition 16.2.3 $\forall q \in Q$,

$$\begin{aligned}\text{FREQ}(q) &= \mathbb{I}_{fr}(q) + \sum_{\substack{q' \in Q \\ a \in \Sigma}} \delta_{fr}(q', a, q) \\ &= \mathbb{F}_{fr}(q) + \sum_{\substack{q' \in Q \\ a \in \Sigma}} \delta_{fr}(q, a, q').\end{aligned}$$

In Definition 16.2.2, consistency is defined as *maintaining the flows*: any string that enters a state (or starts in a state) has to leave it (or end there). If this condition is fulfilled, the corresponding PFA can be constructed by Algorithm 16.1.

Example 16.2.1 We depict in Figure 16.1 an FFA (16.1(a)) and its associated PFA (16.1(b)).

On the other hand, we will sometimes need to compare frequencies over an FFA. But for this we would have to be able to associate with an FFA a sample in order to say something about the sample. But this is usually impossible, even if the automaton is deterministic (see Exercise 16.1). If the automaton is not deterministic, the situation corresponds to that of estimating the probabilities of a model given a sample. This crucial question will be discussed in more detail in Chapter 17.

When needed, since the transformation from a DFFA to a DPFA is straightforward with Algorithm 16.1, if we denote for a given FFA $\mathcal{A}$ the corresponding PFA by $\mathcal{B}$, $Pr_{\mathcal{A}}(w) = Pr_{\mathcal{B}}(w)$. Therefore we will be allowed to compute directly, for a DFFA $\mathcal{A}$, the probability of any string.

If concentrating on a specific state q of an FFA $\mathcal{A}$, we will denote by $Pr_{\mathcal{A},q}(w)$ the probability of string w obtained when parsed (on the associated PFA from state q with initial probability one). Formally: let $\mathcal{B}$ = FFA_TO_PFA($\mathcal{A}$) = $\langle \Sigma, Q, \mathbb{I}_{\mathbb{P}}, \mathbb{F}_{\mathbb{P}}, \delta_{\mathbb{P}}, \delta_{\mathcal{A}} \rangle$,

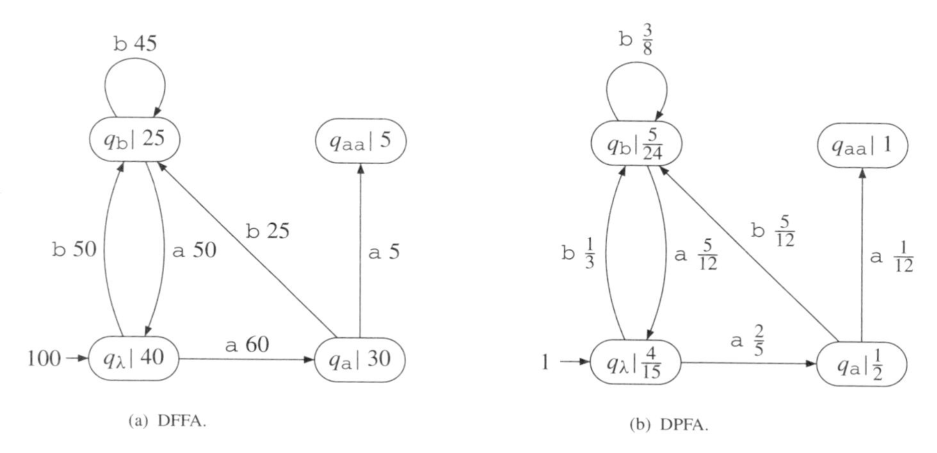

(a) DFFA. (b) DPFA.

Fig. 16.1. FFA to PFA.

$Pr_{\mathcal{A},q}(w) = Pr_{\mathcal{C},q}(w)$ where $\mathcal{C} = \langle \Sigma, Q, \mathbb{I}_{\mathbb{P}_q}, \mathbb{F}_{\mathbb{P}}, \delta_{\mathbb{P}}, \delta_{\mathcal{C}} \rangle$ and $\mathbb{I}_{\mathbb{P}_q}(q) = 1$, $\forall q' \neq q$, $\mathbb{I}_{\mathbb{P}_q}(q') = 0$.

Example 16.2.2 In the DPFA $\mathcal{A}$ represented in Figure 16.1(b), $Pr_{\mathcal{A},q_\lambda}(\text{ab}) = Pr_{\mathcal{A}}(\text{ab}) = 1 \cdot \frac{2}{5} \cdot \frac{5}{12} \cdot \frac{5}{24} = \frac{5}{144}$ whereas $Pr_{\mathcal{A},q_2}(\text{ab}) = 1 \cdot \frac{5}{12} \cdot \frac{1}{3} \cdot \frac{5}{24} = \frac{25}{864}$.

16.3 State merging algorithms

As in the case of learning finite state automata, the best studied algorithms to learn probabilistic automata rely on state merging.

16.3.1 From samples to FFAs

Definition 16.3.1 Let S be a multiset of strings from $\Sigma^\star$. The **frequency prefix tree acceptor** FPTA(S) is the DFFA : $\langle \Sigma, Q, \mathbb{I}_{fr}, \mathbb{F}_{fr}, \delta_{fr}, \delta_{\mathcal{A}} \rangle$ where

- $Q = \{q_u : u \in \text{PREF}(S)\}$,
- $\mathbb{I}_{fr}(q_\lambda) = |S|$,
- $\forall ua \in \text{PREF}(S),\ \delta_{fr}(q_u, a, q_{ua}) = |S|_{ua\,\Sigma^\star}$,
- $\forall ua \in \text{PREF}(S),\ \delta(q_u, a) = q_{ua}$,
- $\forall u \in \text{PREF}(S),\ \mathbb{F}_{fr}(q_u) = |S|_u$.

For example, in Figure 16.2, we have represented the FPTA built from the sample S, which is a multiset of size 100. $S = \{(\lambda, 53), (\text{a}, 12), (\text{b}, 15), (\text{aa}, 4), (\text{ab}, 5), (\text{ba}, 2), (\text{bb}, 2), (\text{aaa}, 2), (\text{aab}, 1), (\text{abb}, 1), (\text{bba}, 2), (\text{bbb}, 1)\}$.

16.3.2 Merging and folding

As in Chapter 12, we describe algorithms that start with a prefix tree acceptor and iteratively try to merge states, with a recursive folding process.

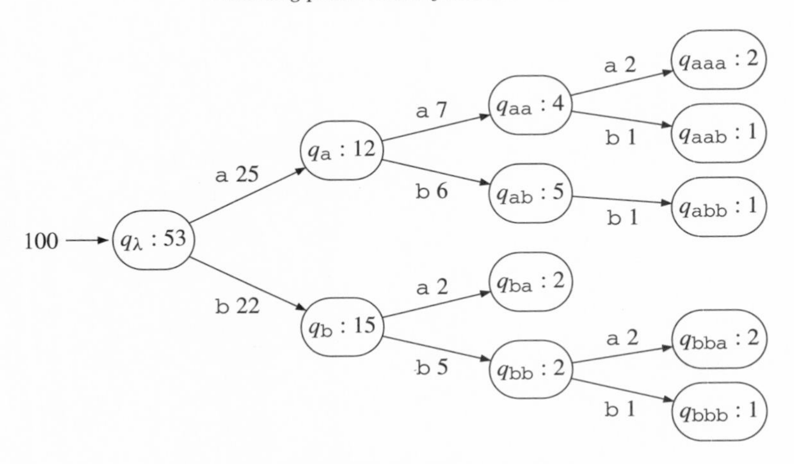

Fig. 16.2. An FPTA constructed from 100 strings.

Given a DFFA the merging operation takes two states q and q', where q is a RED state and $q' = \delta(q_f, a)$ is a BLUE state, the root of a tree. This also means that $\delta(q_f, a) \neq q_\lambda$. The operation is described by Algorithm STOCHASTIC-MERGE (16.2). First, transition $\delta_{\mathcal{A}}(q_f, a)$ is redirected to q. Then the folding of the tree rooted in q' into the automaton at state q is recursively called.

Algorithm 16.2: STOCHASTIC-MERGE.

Input: an FFA $\mathcal{A}$, 2 states $q \in$ RED, $q' \in$ BLUE
Output: $\mathcal{A}$ updated
Let (q_f, a) be such that $\delta_{\mathcal{A}}(q_f, a) = q'$;
$n \leftarrow \delta_{fr}(q_f, a, q')$;
$\delta_{\mathcal{A}}(q_f, a) \leftarrow q$;
$\delta_{fr}(q_f, a, q) \leftarrow n$;
$\delta_{fr}(q_f, a, q') \leftarrow 0$;
return STOCHASTIC-FOLD($\mathcal{A}, q, q'$)

For the resulting automaton to be deterministic, recursive folding is required. In this case Algorithm STOCHASTIC-FOLD (16.3) is used.

The situation is schematically represented in Figures 16.3(a) and 16.3(b): suppose we want to merge state q' (BLUE) with state q (RED). First q_f is the unique predecessor of q', and the transition from q_f labelled by a (and used n times) is redirected to q. The folding can then take place.

Example 16.3.1 Let us examine the merge-and-fold procedure with more care on an example. Suppose we are in the situation represented in Figure 16.4. We want to merge state $q_{\mathtt{aa}}$ with q_λ.

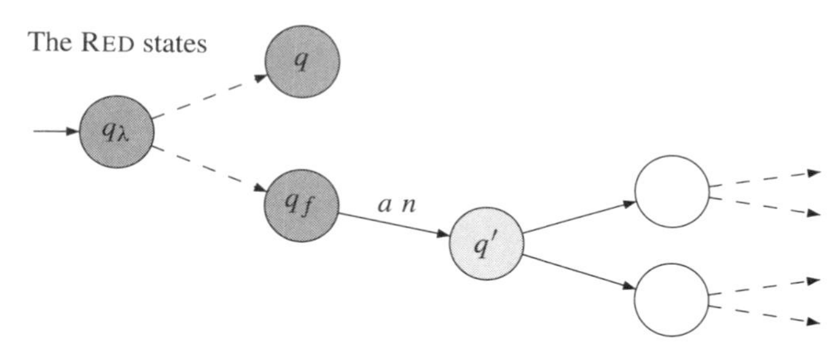

(a) The situation before merging.

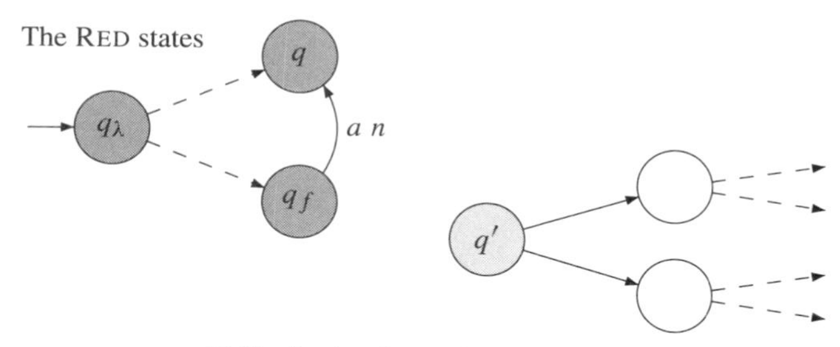

(b) The situation after merging and before folding.

Fig. 16.3. Merge-and-fold.

Algorithm 16.3: STOCHASTIC-FOLD.

Input: a DFFA $\mathcal{A}$, 2 states q and q'
Output: $\mathcal{A}$ updated, where subtree in q' is folded into q
$\mathbb{F}_{fr}(q) \leftarrow \mathbb{F}_{fr}(q) + \mathbb{F}_{fr}(q')$;
for $a \in \Sigma$ *such that* $\delta_{\mathcal{A}}(q', a)$ *is defined* **do**
  if $\delta_{\mathcal{A}}(q, a)$ *is defined* **then**
    $\delta_{fr}(q, a, \delta_{\mathcal{A}}(q, a)) \leftarrow \delta_{fr}(q, a, \delta_{\mathcal{A}}(q, a)) + \delta_{fr}(q', a, \delta_{\mathcal{A}}(q', a))$;
    $\mathcal{A} \leftarrow$STOCHASTIC-FOLD($\mathcal{A}, \delta_{\mathcal{A}}(q, a), \delta_{\mathcal{A}}(q', a)$)
  else
    $\delta_{\mathcal{A}}(q, a) \leftarrow \delta_{\mathcal{A}}(q', a)$;
    $\delta_{fr}(q, a, \delta_{\mathcal{A}}(q, a)) \leftarrow \delta_{fr}(q', a, \delta_{\mathcal{A}}(q', a))$
  end
end
return $\mathcal{A}$

Once the redirection has taken place (Figure 16.5), state $q_{\texttt{aa}}$ is effectively merged (Figure 16.6) with q_λ (leading to modification of the values of FREQ$_{\mathcal{A}}(q_\lambda)$, $\delta_{fr}(q_\lambda, \texttt{a})$ and $\delta_{fr}(q_\lambda, \texttt{b})$). Dashed lines mark the recursive merges that are still to be done.

Then (Figure 16.7) $q_{\texttt{aaa}}$ is folded into $q_{\texttt{a}}$, and finally $q_{\texttt{aab}}$ into $q_{\texttt{b}}$. The result is represented in Figure 16.8.

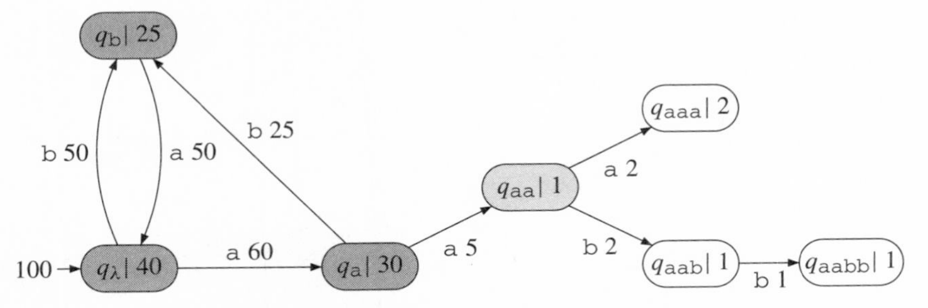

Fig. 16.4. Before merging q_{aa} with q_λ and redirecting transition $\delta_{\mathcal{A}}(q_{\mathrm{a}}, \mathrm{a})$ to q_λ.

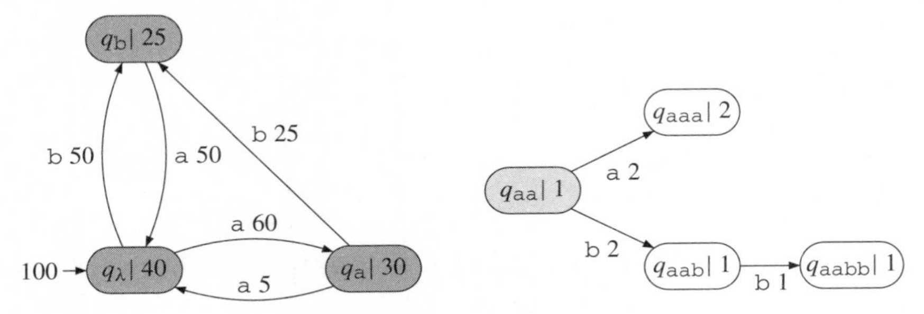

Fig. 16.5. Before folding q_{aa} into q_λ.

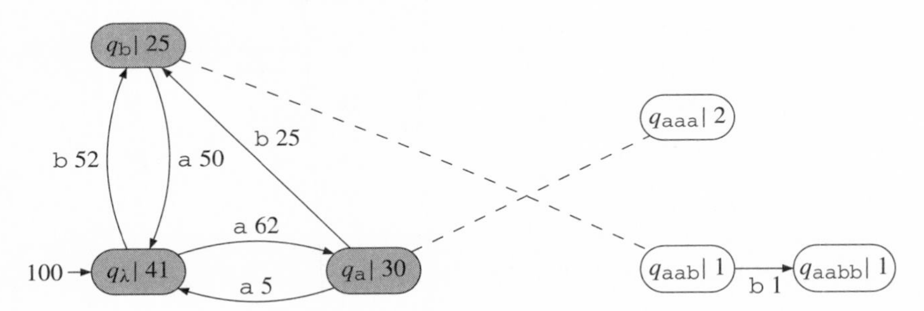

Fig. 16.6. Folding in q_{aa}.

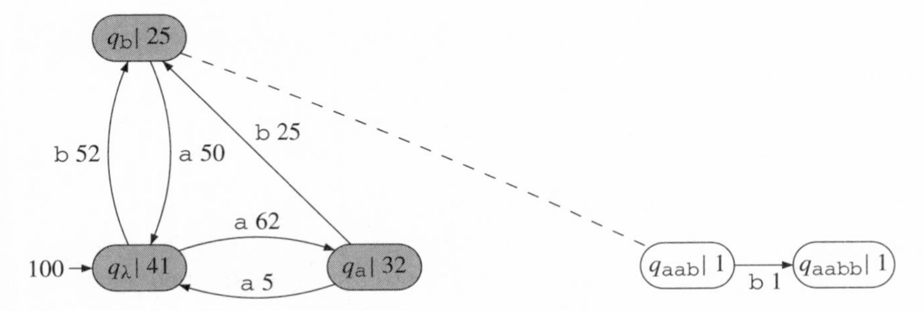

Fig. 16.7. Folding in q_{aaa}.

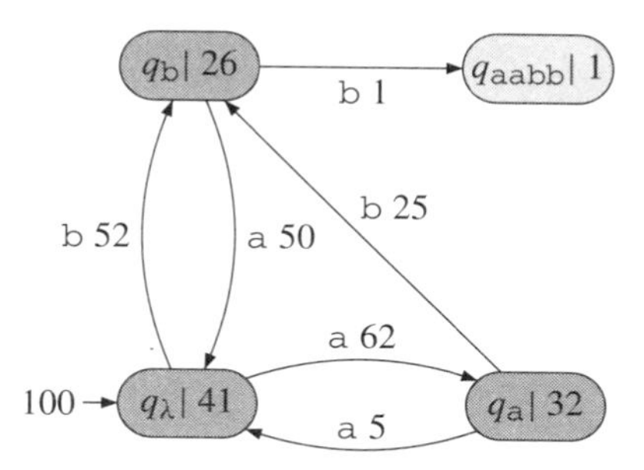

Fig. 16.8. Folding in $q_{\texttt{aab}}$.

16.3.3 Deciding equivalence

The next difficult problem is that of deciding if two states q_u and q_v should be merged. This usually corresponds to testing in an automaton if the two distributions obtained by using the same automaton but changing the initial frequency function are equivalent (better said, if they are equivalent when taking into account the fact that these two distributions are only approximately described by samples).

So the question is really about deciding if two states of a given DFFA $\mathcal{A} = \langle \Sigma, Q, \mathbb{I}_{fr}, \mathbb{F}_{fr}, \delta_{fr}, \delta_{\mathcal{A}} \rangle$ (one BLUE and one RED) should be merged, which really means discovering from both the sampling process and the merges that have been done up to that point if $\mathcal{D}_{\mathcal{A},q_u}$ and $\mathcal{D}_{\mathcal{A},q_v}$ are sufficiently close. One possibility is to use here the results and algorithms described in Section 5.4 where we studied distances between distributions, and in a way, this is what is done.

16.3.4 Equivalence on prefixes

Let us describe the key idea behind the test used by Algorithm ALERGIA, which will be described in detail in Section 16.4. Obviously, if the two distributions given by automata $\mathcal{A}_1$ and $\mathcal{A}_2$ (whether PFA or FFA) are equivalent, then we should have:

$$\forall w \in \Sigma^\star, \;\; Pr_{\mathcal{A}_1}(w) = Pr_{\mathcal{A}_2}(w)$$

or at least, when testing over random samples, these observed relative frequencies should be sufficiently close:

$$\forall w \in S, \;\; Pr_{\mathcal{A}_1}(w) \approx Pr_{\mathcal{A}_2}(w)$$

But this is not going to be possible to test, partly because the number of strings is infinite, but mainly because the values involved can be very small whenever the distribution is over many different strings.

A better idea is to compare the prefixes:

$$\forall w \in \Sigma^\star \;\; Pr_{\mathcal{A}_1}(w\,\Sigma^\star) \approx Pr_{\mathcal{A}_2}(w\,\Sigma^\star)$$

This test will be used (over a finite sample) but it should be noted that it can mean having to test on all possible strings w; so we want to do this in such a way as to keep the number of tests small (and somehow bounded by the size of the sample).

16.3.5 Added bias

A second altogether different way of testing if two distributions are similar corresponds to taking the bet (we call it bias) that there will be a distinguishing string. This idea will be exploited in Algorithm DSAI, described in Section 16.5. Note that the above quantities should diminish at exponential rate with the length of the prefixes. This means that if the difference between two states depends on a long prefix, we may not have enough information in the sample to be able to notice this difference. In order to deal with this problem, we can hope that two states are different because of some string which has different probabilities when reading from each of these states. We call this a *distinguishing string*. Obviously if it is normal that such distinguishing strings exist, the bias lies is in believing that the difference of probabilities is large.

Definition 16.3.2 Given $\mu > 0$, a DPFA (or DFFA) $\mathcal{A} = \langle \Sigma, Q, q_\lambda, \mathbb{F}_\mathbb{P}, \delta_\mathbb{P} \rangle$ is **μ-distinguishable** if_{def} for every pair of different states q_u and q_v in Q there exists a string w in $\Sigma^\star$ such that $|Pr_{\mathcal{A},q_u}(w) - Pr_{\mathcal{A},q_v}(w)| \geq \mu$.

This means that one can check, during the merge-and-fold process, if two states are μ-distinguished or not. It should be noted that if two states are μ-distinguishable, then every time we want to merge two states q_u and q_v, with high enough probability there is a string w such that $|Pr_{\mathcal{A}_{q_u}}(w) - Pr_{\mathcal{A}_{q_v}}(w)| > \mu$.

In order for this test to make sense, a threshold (denoted by t_0) will be used. The threshold provides a limit: how many strings entering the BLUE state are needed in order to have enough statistical evidence to consider making a merge? Only states having enough evidence will be checked.

An important question concerns the future of those states that do not meet this quantity. There are a number of possibilities, some more theoretically founded (like just merging all the remaining states at the end of the learning process in one unique state), and some using all types of heuristics. We shall not specifically recommend any here.

16.3.6 Promoting

Just as in the RPNI case, promotion takes place when the current (BLUE) state can be merged with no red state (RED). In that case, first the BLUE state is promoted and becomes RED. Then all the successors of the RED states that are not RED themselves become BLUE. Notice also that no frequencies are modified.

In practice it is possible to avoid exploring all the states in order to redefine the colourings, as, usually, only the successors of the promoted state can change to BLUE.

16.3.7 Withdrawing

There is, in the case of probabilistic automata learning, a last operation that didn't make much sense when working with non-probabilistic automata. At some moment the algorithm may have to take a decision about a state (or an edge) used by very few strings. This may seem like a trifling (there is not much data so even if we made a mistake it could seem it would not be a costly one), but this is not true: to take an extreme example suppose that just one string is generated from a state, perhaps $\mathtt{a}^n$ with n very large (much larger than the actual size of the sample). Then making a mistake and perhaps merging this state with some RED state can change the frequencies in a far too drastic way because the same string will be folded many times into the automaton. For details, see Exercise 16.2.

This phenomenon deserves to be analysed with much more care, but for the moment we will just solve the question by using a threshold t_0 with each algorithm: if the amount of data reaching a BLUE state is less than the threshold, then the state (and all its successors) is withdrawn. It remains in the FFA to maintain consistency, but will not be checked for merging.

16.4 ALERGIA

In Chapter 12 we described Algorithm RPNI. Algorithm ALERGIA (16.6) follows the same ideas: a predefined ordering of the states, a compatibility test and then a merging operation. The general algorithm visits the states through two loops and attempts (recursively) to merge states.

16.4.1 The algorithm

Algorithm 16.4: ALERGIA-TEST.

Input: an FFA $\mathcal{A}$, $f_1, n_1, f_2, n_2, \alpha > 0$
Output: a Boolean indicating if the frequencies $\frac{f_1}{n_1}$ and $\frac{f_2}{n_2}$ are sufficiently close
$\gamma \leftarrow \left|\frac{f_1}{n_1} - \frac{f_2}{n_2}\right|$;
return $\left(\gamma < (\sqrt{\frac{1}{n_1}} + \sqrt{\frac{1}{n_2}}) \cdot \sqrt{\frac{1}{2}\ln\frac{2}{\alpha}}\right)$

The compatibility test makes use of the Hoeffding bounds. The algorithm ALERGIA-COMPATIBLE (16.5) calls the test ALERGIA-TEST (16.4) as many times as needed, this number being finite due to the fact that the recursive calls visit a tree.

The basic function CHOOSE is as follows: take the smallest state in an ordering that has been done at the beginning (on the FPTA). But it has been noticed by different authors that using a data-driven approach offered better guarantees. The test that is used to decide if the states are to be merged or not (function COMPATIBLE) is based on the Hoeffding test made on the relative frequencies of the empty string and of each prefix.

Algorithm 16.5: ALERGIA-COMPATIBLE.

Input: an FFA $\mathcal{A}$, two states $q_u, q_v, \alpha > 0$
Output: q_u and q_v compatible?
Correct$\leftarrow$ **true**;
if ALERGIA-TEST($\mathbb{F}_{\mathbb{P}\mathcal{A}}(q_u)$, FREQ$_\mathcal{A}(q_u)$, $\mathbb{F}_{\mathbb{P}\mathcal{A}}(q_v)$, FREQ$_\mathcal{A}(q_v)$, α) **then**
Correct$\leftarrow$ **false**;
for $a \in \Sigma$ **do**
  if ALERGIA-TEST($\delta_{fr}(q_u, a)$, FREQ$_\mathcal{A}(q_u)$, $\delta_{fr}(q_v, a)$, FREQ$_\mathcal{A}(q_v)$, α) **then**
  Correct$\leftarrow$ **false**
end
return Correct

Algorithm 16.6: ALERGIA.

Input: a sample S, $\alpha > 0$
Output: an FFA $\mathcal{A}$
Compute t_0, threshold on the size of the multiset needed for the test to be be statistically significant;
$\mathcal{A} \leftarrow$ FPTA(S) ;
RED $\leftarrow \{q_\lambda\}$;
BLUE $\leftarrow \{q_a : a \in \Sigma \cap$ PREF$(S)\}$;
while CHOOSE q_b *from* BLUE *such that* FREQ$(q_b) \geq t_0$ **do**
  if $\exists q_r \in$ RED : ALERGIA-COMPATIBLE($\mathcal{A}, q_r, q_b, \alpha$) **then**
    $\mathcal{A} \leftarrow$ STOCHASTIC-MERGE($\mathcal{A}, q_r, q_b$)
  else
    RED $\leftarrow$ RED $\cup \{q_b\}$
  end
  BLUE $\leftarrow \{q_{ua} : ua \in$ PREF$(S) \wedge q_u \in$ RED$\} \setminus$ RED
end
return$\mathcal{A}$

The quantities that are compared are:

$$\frac{\mathbb{F}_{fr}(q)}{\text{FREQ}(q)} \text{ and } \frac{\mathbb{F}_{fr}(q')}{\text{FREQ}(q')}$$

and also for each symbol a in the alphabet Σ,

$$\frac{\delta_{fr}(q, a)}{\text{FREQ}(q)} \text{ and } \frac{\delta_{fr}(q', a)}{\text{FREQ}(q')}.$$

Table 16.1. *The data.*

λ	490	abb	4	abab	2	aaaaa	1
a	128	baa	9	abba	2	aaaab	1
b	170	bab	4	abbb	1	aaaba	1
aa	31	bba	3	baaa	2	aabaa	1
ab	42	bbb	6	baab	2	aabab	1
ba	38	aaaa	2	baba	1	aabba	1
bb	14	aaab	2	babb	1	abbaa	1
aaa	8	aaba	3	bbaa	1	abbab	1
aab	10	aabb	2	bbab	1		
aba	10	abaa	2	bbba	1		

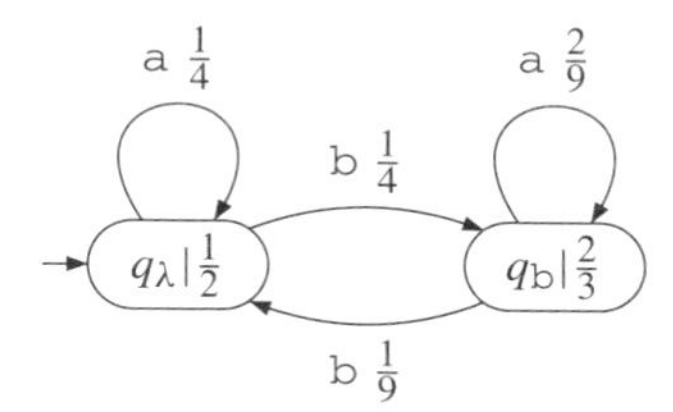

Fig. 16.9. The target.

16.4.2 Running the algorithm

Running probabilistic automata learning algorithms in a clear and understandable way is a difficult task: on one hand the targets should be meaningful and therefore contain cycles, and on the other hand the samples should be large enough to offer statistical evidence for the decisions that are taken by the algorithm. This results in a number of states in the FPTA much too large to be represented graphically.

In order to explain the way ALERGIA works, we shall use a sample of 1000 strings of (artificially) bounded length (just in order to make things simpler, but without loss of generality). The sample is represented in Table 16.1. The target, supposed to have generated the sample, is represented in Figure 16.9. The starting point is the DFFA given in Figure 16.10. State q_λ is RED and states q_a and q_b are BLUE.

Parameter α is arbitrarily set to 0.05. We choose 30 as a value for the threshold t_0. In theory, the value can be computed: it corresponds to a value for which both errors (to make a wrong merge or not to merge at all) can be bounded (see Section 16.9).

ALERGIA tries merging q_a and q_λ, which means comparing $\frac{490}{1000}$ with $\frac{128}{257}$, $\frac{257}{1000}$ with $\frac{64}{257}$, $\frac{253}{1000}$ with $\frac{65}{257}, \ldots$

For example, $\left|\frac{\delta_{fr}(q_\lambda,a)}{\text{FREQ}_{\mathcal{A}_q}(q_\lambda)} - \frac{\delta_{fr}(q_a,a)}{\text{FREQ}_{\mathcal{A}_{q_a}}(q_a)}\right| = \left|\frac{257}{1000} - \frac{64}{257}\right| = 0.008$ is less than $\left(\sqrt{\frac{1}{1000}} + \sqrt{\frac{1}{257}}\right)\sqrt{\frac{1}{2}\ln\frac{2}{\alpha}}$ which for $\alpha = 0.05$ is about 0.1277.

The merge is accepted and the new automaton is depicted in Figure 16.11.

The algorithm tries merging q_b (the only BLUE state at that stage) and q_λ. This fails after comparing $\frac{\mathbb{F}_{fr}(q_\lambda)}{\text{FREQ}(q_\lambda)} = \frac{661}{1340}$ and $\frac{\mathbb{F}_{fr}(q_b)}{\text{FREQ}(q_b)} = \frac{222}{339}$ (giving in the test 0.162 (for the

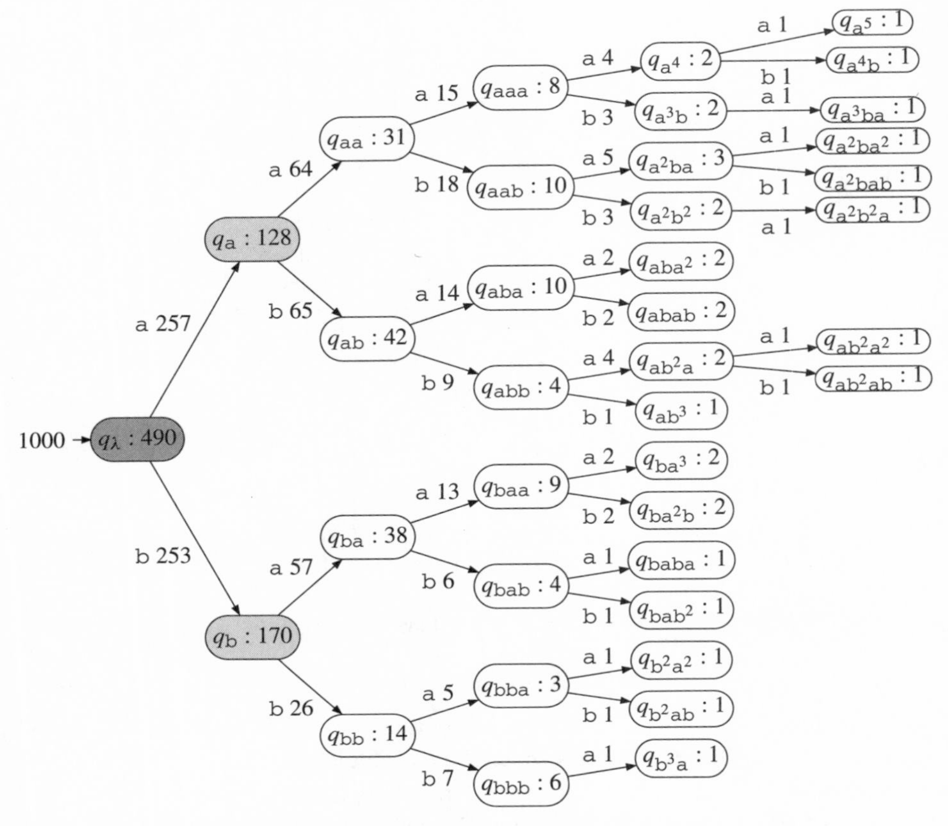

Fig. 16.10. The full FPTA, with coloured states.

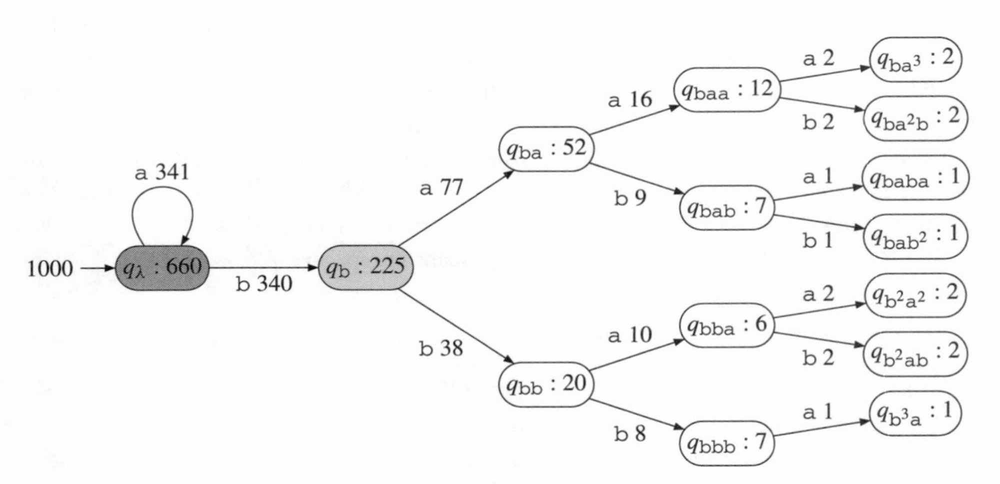

Fig. 16.11. After merging q_a and q_λ.

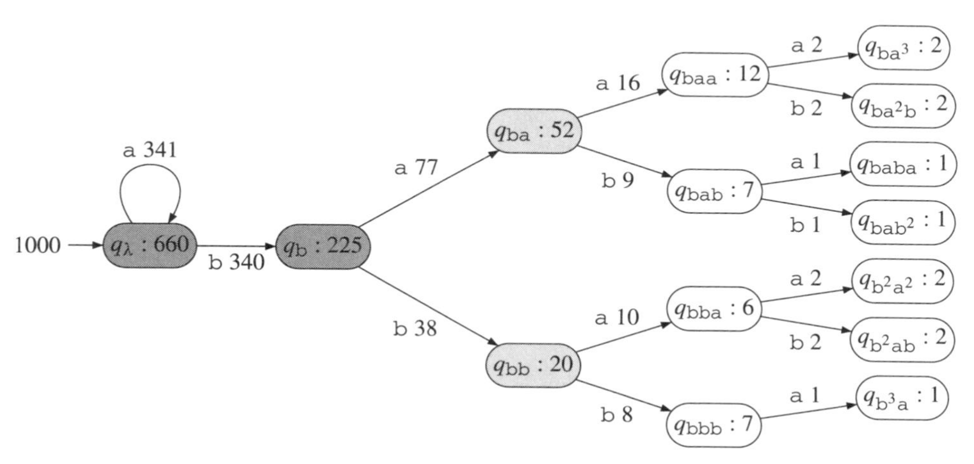

Fig. 16.12. Promoting q_{b}.

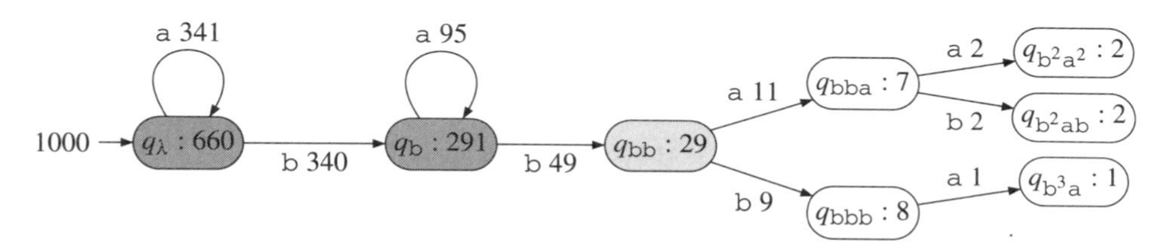

Fig. 16.13. After merging q_{ba} and q_{b}.

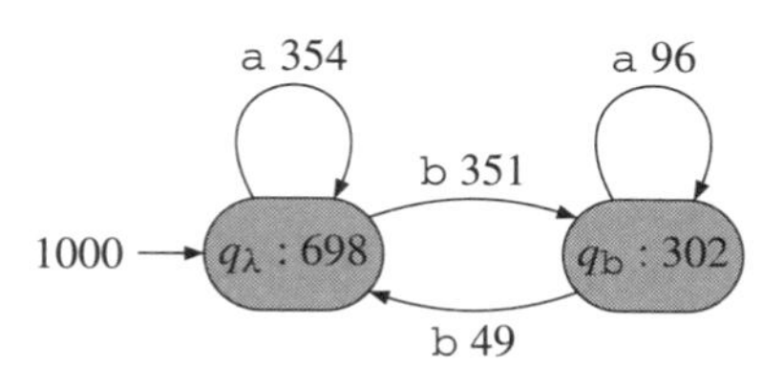

Fig. 16.14. Merging q_{bb} and q_λ.

γ) against 0.111). So state q_{b} gets promoted; the current automaton is represented in Figure 16.12.

The algorithm now tries merging q_{ba} and q_λ. This fails when comparing $\frac{660}{1341}$ with $\frac{52}{77}$. The difference between the two relative frequencies is too large, and when algorithm ALERGIA-TEST is called (with, as before, a value for α of 0.05), the result is negative.

The algorithm now tries merging q_{ba} and q_{b}. This merge is accepted, partly because there is little information to tell us otherwise. The new automaton is depicted in Figure 16.13.

We now try merging q_{bb} and q_λ. This is also accepted. The new automaton is depicted in Figure 16.14.

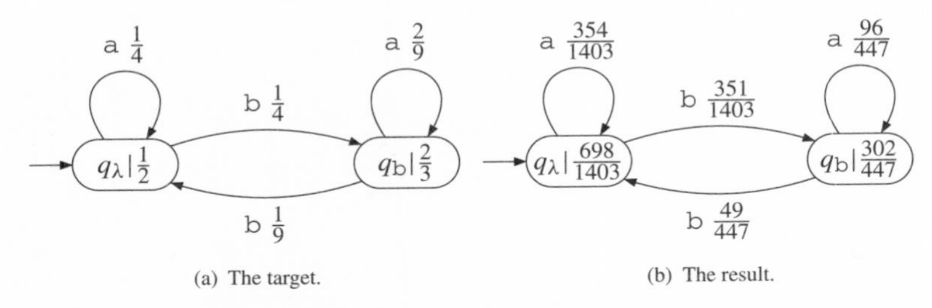

Fig. 16.15. Comparing the target and the result.

Only two states are left. By using Algorithm DFFA_TO_DPFA (16.1) we obtain the DPFA represented in Figure 16.15(b), which can be compared with the target (Figure 16.15(a)). One should notice how little the bad merges have been avoided, in a relatively simple case. This tends to show that the number of examples needed to run the algorithm in reasonable conditions has to be much larger.

16.4.3 Why ALERGIA works

We do not intend to prove the properties of algorithm ALERGIA here. These can be found (albeit for a more theoretical version of the algorithm) in different research papers. Nevertheless we can state the following facts:

Proposition 16.4.1 *Algorithm* ALERGIA *identifies any* DPFA *in the limit with probability one and runs in time polynomial in* $\|S\|$.

The compatibility test is in fact much cheaper than the one for RPNI. No merges need to be done before the test. Therefore this can be done through counting and in time linear with the size of the tree rooted in the BLUE node. This is the reason for which the data-driven approach described in Section 14.5.2 (page 293) has been used with good results.

Since this is done inside a double loop, the overall complexity is very excessively bounded by a cubic polynomial. Users report that the practical complexity is linear.

Convergence is much more of a tricky matter. There are two issues here:

- The first is to identify the structure.
- The second is to identify the probabilities. This can be done through different techniques, such as Algorithm 10.2 for Stern-Brocot trees (page 204).

There have been many variants of algorithm ALERGIA with modifications of:

- the order of the merges,
- the statistical test which is used,
- the way we deal with the uncertain states.

16.5 Using distinguishing strings

A different test can be used provided an extra bias is added: given a DPFA $\mathcal{A}$, two states q and q' are said to be *μ-distinguishable* if there exists a string x such that $\left|Pr_{\mathcal{A},q}(x) - Pr_{\mathcal{A},q'}(x)\right| \geq \mu$ where $Pr_{\mathcal{A},q}(x)$ is the probability of generating x by $\mathcal{A}$ when taking q as the initial state. This is equivalent to saying

$$\mathbf{L}_\infty(\mathcal{D}_{\mathcal{A},q}, \mathcal{D}_{\mathcal{A},q'}) \geq \mu$$

The chosen bias corresponds to believing that when things are different, this should be so for explainable or visible reasons: at least one string should testify to why there are two states instead of one. One may sometimes relax this condition in several ways, for instance by asking this to be true only for states that themselves have a fair chance of being reached.

When trying to detect this in an algorithm, it means that we decide upon a merge by testing:

Does there exist in $\Sigma^\star$ a string x such that $\left|Pr_{\mathcal{A},q}(x) - Pr_{\mathcal{A},q'}(x)\right| \geq \mu$?

Notice that again it is an 'optimistic' argument: a decision of merging is taken when there is nothing saying it is a bad idea to do so.

16.5.1 The algorithm

We describe here an algorithm we call DSAI (***Distinguishing Strings Automata Inference***), of which several variants exist. The one we present here is simple, but in order to obtain convergence results, one should be prepared to handle a number of extra parameters.

The idea is, given a RED state q_r and a BLUE state q_b (which is visited a minimum number of times), to compute $\mathbf{L}_\infty(\mathcal{D}_{\mathcal{A},q_r}, \mathcal{D}_{\mathcal{A},q_b})$. If this quantity is less than $\frac{\mu}{2}$ the two states are compatible and are merged. A state that is not reached enough times is kept until the end in order to avoid an accumulation of small errors.

In the version presented here, these less important states are just kept as a tree. For obvious reasons, merging together all the states reached a number of times inferior to the threshold may be a better strategy.

In general, computing $\mathbf{L}_\infty$ between two deterministic DPFAs or FFAs is not simple. But in this case the computation is between a DFFA and an FPTA. Let k be the number of strings in the FPTA with root q_b. Let us denote by S_b the set of all strings x such that $Pr_{\mathcal{A},q_b}(x) \neq 0$. Note that $k < |S_b| \leq |S|$. $\mathbf{L}_\infty(\mathcal{D}_{\mathcal{A},q_r}, \mathcal{D}_{\mathcal{A},q_b})$ is either reached by a string x in S_b, and there are only a finite number of such strings, or by one of the $k+1$ most probable strings for $\mathcal{D}_{\mathcal{A},q_r}$. We denote this set by $\text{mps}(\mathcal{A}_{q_r}, k+1)$.

Algorithm 16.7: DSAI-COMPATIBLE.

Input: an FFA $\mathcal{A}$, two states $q \in$ RED and $q' \in$ BLUE, $\mu > 0$
Output: a Boolean indicating if q and q' are compatible
for $x \in \Sigma^\star : Pr_{\mathcal{A},q'}(x) \neq 0$ **do**
  if $\left|Pr_{\mathcal{A},q}(x) - Pr_{\mathcal{A},q'}(x)\right| > \frac{\mu}{2}$ **then return false**
end
for *each* x *in* mps$(\mathcal{A}_{q_r}, k+1)$ **do**
  if $Pr_{\mathcal{A},q'}(x) > \frac{\mu}{2}$ **then return false**
end
return true

Algorithm 16.8: DSAI.

Input: a sample S, $\mu > 0$
Output: an FFA $\mathcal{A}$
Compute t_0, threshold on the size of the multiset needed for the test to be be statistically significant;
$\mathcal{A} \leftarrow$ FPTA(S) ;
RED $\leftarrow \{q_\lambda\}$;
BLUE $\leftarrow \{q_a : a \in$ PREF$(S)\}$;
while CHOOSE q_b *from* BLUE *such that* FREQ$(q_b) \geq t_0$ **do**
  if $\exists q_r \in$ RED : DSAI-COMPATIBLE$(\mathcal{A}, q_r, q_b, \mu)$ **then**
    $\mathcal{A} \leftarrow$ STOCHASTIC-MERGE$(\mathcal{A}, q_r, q_b)$
  else
    RED $\leftarrow$ RED $\cup \{q_b\}$
  end
  BLUE $\leftarrow \{q_{ua} : ua \in$ PREF$(S) \wedge q_u \in$ RED$\} \setminus$ RED
end
return $\mathcal{A}$

Given a DPFA $\mathcal{A}$ and an integer n, one can compute in polynomial time the set of the n most probable strings in $\mathcal{D}_\mathcal{A}$ (see Exercise 5.4).

Therefore, in this case, since q_b is a BLUE state, and therefore the root of a tree, DSAI-COMPATIBLE computes this distance $(\mathbf{L}_\infty(\mathcal{D}_{\mathcal{A},q_r}, \mathcal{D}_{\mathcal{A},q_b}))$.

Algorithm DSAI (16.8) is structured similarly to ALERGIA.

The key idea is that on one hand, if two states are not to be merged, the distinguishing string exists and will be seen, and, on the other hand, if the two states compared correspond in fact to the same state in the target, then the μ bound is sufficiently big to avoid reaching the wrong conclusion by 'bad luck'.

Table 16.2. *The data.*

λ	490	abb	4	abab	2	aaaaa	1
a	128	baa	9	abba	2	aaaab	1
b	170	bab	4	abbb	1	aaaba	1
aa	31	bba	3	baaa	2	aabaa	1
ab	42	bbb	6	baab	2	aabab	1
ba	38	aaaa	2	baba	1	aabba	1
bb	14	aaab	2	babb	1	abbaa	1
aaa	8	aaba	3	bbaa	1	abbab	1
aab	10	aabb	2	bbab	1		
aba	10	abaa	2	bbba	1		

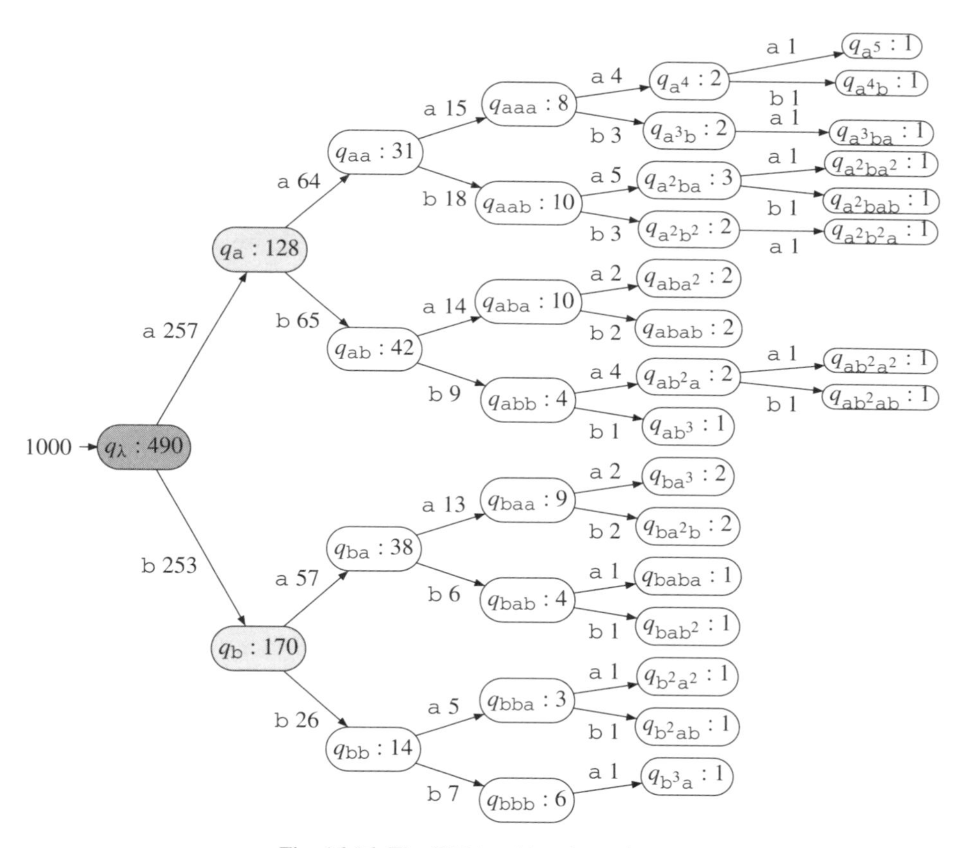

Fig. 16.16. The FPTA, with coloured states.

16.5.2 Running the algorithm

We run the algorithm DSAI on the same sample as in Section 16.4.2. The sample, of size 1000, is recalled in Table 16.2, whereas the FPTA is recalled in Figure 16.16.

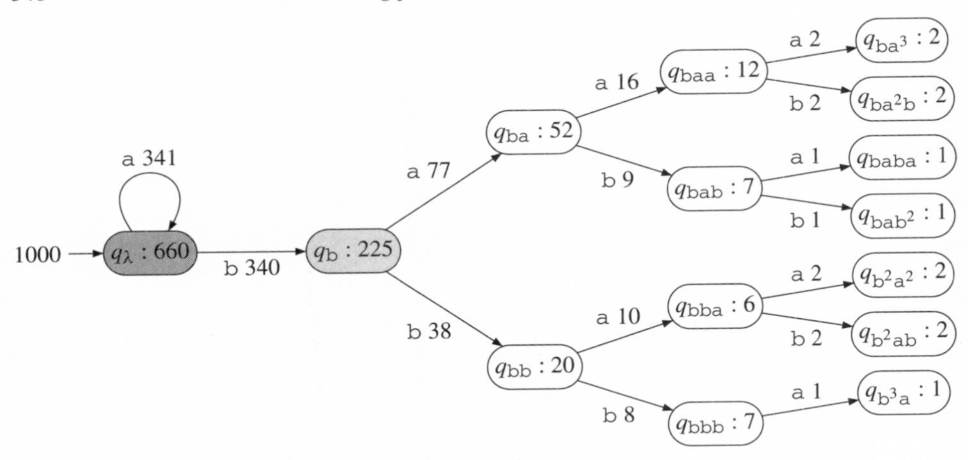

Fig. 16.17. After merging q_a and q_λ.

The starting point is represented in Figure 16.16.

We set μ to 0.2 and the threshold has value $t_0 = 30$. This means that each time the difference between the two relative frequencies is less than $\mu/2$, we are reasonably sure that this is because the two probabilities are not further apart than μ, and therefore they are identical.

Algorithm DSAI then tries merging q_a and q_λ. This corresponds to computing $\mathbf{L}_\infty(\mathcal{D}_{\mathcal{A},q_\lambda}, \mathcal{D}_{\mathcal{A},q_a})$. The maximum value of $|Pr_{\mathcal{A}_{q_\lambda}}(x) - Pr_{\mathcal{A}_{q_a}}(x)|$ is obtained for $x = \lambda$ and is 0.008. The merge is accepted.

The new automaton is depicted in Figure 16.17.

DSAI tries merging q_b and q_λ. But for string λ we have $Pr_{\mathcal{A}_{q_\lambda}}(\lambda) = \frac{660}{1341}$, $Pr_{\mathcal{A}_{q_b}}(\lambda) = \frac{225}{340}$ and the difference is 0.170 which is more than $\frac{\mu}{2}$. State q_b is therefore promoted.

The algorithm now tries merging q_{ba} and q_λ. This fails because $|Pr_{\mathcal{A}_{q_\lambda}}(\lambda) - Pr_{\mathcal{A}_{q_{ba}}}(\lambda)| = 0.675 - 0.492 > \frac{\mu}{2}$. But the possible merge between q_{ba} and q_b is accepted (the largest difference is again for string λ, with 0.014). The new DFFA is represented in Figure 16.18. We now try merging q_{bb} and q_λ. This is accepted, since $\mathbf{L}_\infty(\mathcal{D}_{\mathcal{A},q_\lambda}, \mathcal{D}_{\mathcal{A},q_{bb}}) = |Pr_{\mathcal{A}_{q_\lambda}}(\lambda) - Pr_{\mathcal{A}_{q_{bb}}}(\lambda)| = 0.044$.

The final automaton is shown in Figure 16.19 (DFFA on the left, DPFA on the right).

The result is exactly the same as that returned by ALERGIA. But, of course, this need not be so in general.

16.5.3 Why it works

In the literature, a version of Algorithm DSAI, in which the merges are only tested if they do not introduce cycles, is proved to have a number of features, of which some are positive PAC learning properties. Their results do not extend to the general case where there is no condition set on the topology of the automata.

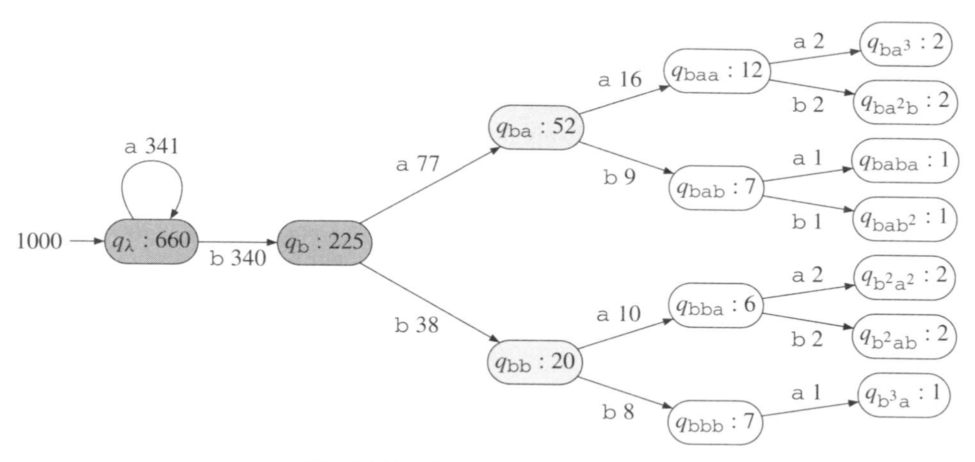

Fig. 16.18. After merging q_{ba} and q_b.

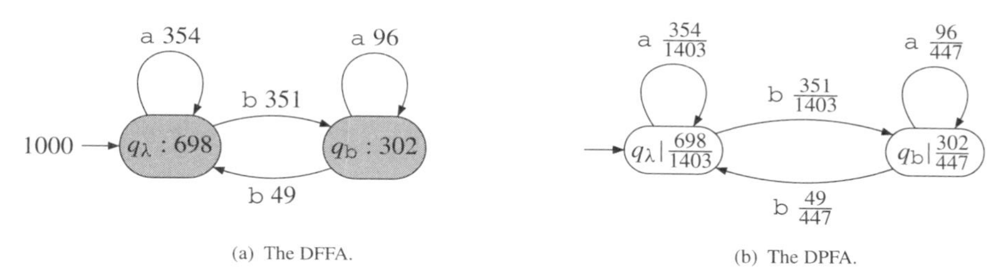

Fig. 16.19. After merging q_{ba} and q_λ.

Nevertheless, we do have the following result provided some small changes are made:

Proposition 16.5.1 *Algorithm* DSAI *identifies any μ-distinguishable* DPFA *in the limit with probability one and runs in polynomial time.*

Proof If we fix the threshold, with the number of strings increasing, the tests are going to return **true** for all but a finite number of values. Then, with techniques studied in earlier sections, the probabilities can be identified instead of being estimated.

The complexity of the algorithm is clearly polynomial with the size of the sample. □

16.6 Hardness results regarding ALERGIA and DSAI

16.6.1 A difficult target for ALERGIA and DSAI

Let us note two things concerning the approaches presented in Sections 16.4 and 16.5. Limiting oneself to acyclic automata offers an important theoretical advantage. This allows one to effectively consider that the strings have been randomly and independently sampled (which is a necessary condition for the statistical tests considered). On the other hand, in the case of ALERGIA, from the moment a merge has introduced a cycle, the same string is going to be used over and over. Intuitively this is probably not a bad thing, as the example is going to reinforce, if anything, the next set of tests.

In both cases convergence is achieved, but in the second case identification is met from a polynomial number of examples. Yet the μ parameter is important. It is then not too difficult to find a class of languages for which everything will fail. This family is indexed by an integer n, and consists of languages generated by the (indexed) probabilistic grammar G_n which generates:

- if $|u| < n$ $Pr_G(u) = 0$,
- if $|u| = n$ and $m \in \mathtt{a}\Sigma^*$ $Pr_G(u) = 2^{-n}$,
- if $|u| = n$ and $m \in \mathtt{b}\Sigma^*$ $Pr_G(u) = 0$,
- if $|u| = n + 1$ and $m \in \mathtt{a}\Sigma^*$ $Pr_G(u) = 0$,
- if $|u| = n + 1$ and $u \in \mathtt{b}\Sigma^*$ $Pr_G(u) = 2^{-n+1}$,
- if $|u| > n + 1$ and $u \in \Sigma^*$ $Pr_G(u) = 0$.

We represent in Figure 16.20 the automaton corresponding to the case where $n = 5$.

States 1 and 2 are not μ-distinguishable, or better said are only so for values of μ that increase exponentially with n, yet the different languages are very different. So an exponential number of examples will be needed for identification (of the structure) to take place.

16.6.2 The equivalence problem revisited

Finally let us attempt to generalise here both approaches in a common way, by using the intersection with regular sets. When studying algorithms ALERGIA and DSAI, the outside structure is similar, and only the compatibility test changes. Each time a state merging algorithm takes a decision concerning compatibility of two states q and q', we are attempting to compare the distributions when using q and q' as initial states. In one case (ALERGIA), we do this by testing prefixes, and in the case of Algorithm DSAI we are testing the significant strings. In other words, we are selecting a language L, and computing the value $\left|Pr_{\mathcal{A},q}(L) - Pr_{\mathcal{A},q'}(L)\right|$. This value is compared with some other value depending on the evidence, typically by using the Hoeffding bounds.

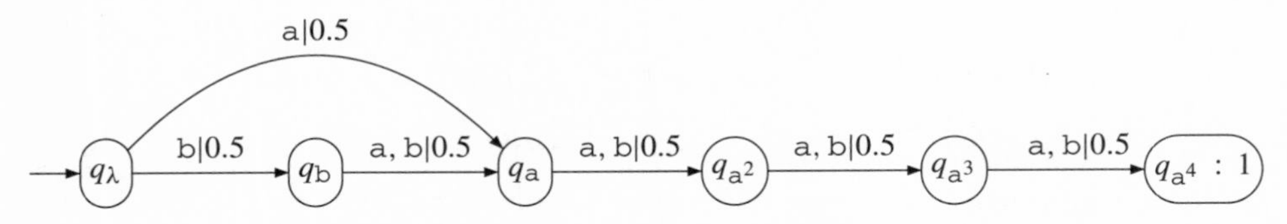

Fig. 16.20. A probabilistic automaton that is hard to learn, for $n = 5$.

In the case of ALERGIA, L is either just one string or a set $u\,\Sigma^\star$. In the case of DSAI, L is just one string.

A generalised test therefore consists of:

(i) Select some regular language L
(ii) Using the Hoeffding bounds, compare $|Pr_{\mathcal{A},q}(L) - Pr_{\mathcal{A},q'}(L)|$ with $\left(\sqrt{\frac{1}{\text{FREQ}(q)}} + \sqrt{\frac{1}{\text{FREQ}(q')}}\right) \cdot \sqrt{\frac{1}{2}\ln\frac{2}{\alpha}}$.

16.7 MDI and other heuristics

Algorithms ALERGIA and DSAI decided upon merging (and thus generalisation) through a local test: substring frequencies are compared and if it is not unreasonable to merge, then merging takes place. A more pragmatic point of view could be to merge whenever doing so is going to give us an advantage. The goal is of course to reduce the size of the hypothesis while keeping the predictive qualities of the hypothesis (at least with respect to the learning sample) as good as possible. For this we can use the likelihood (see Section 5.5.5). The goal is to obtain a good balance between the gain in size and the loss in perplexity.

Attempting to find a good compromise between these two values is the main idea of algorithm MDI (Minimum Divergence Inference).

16.7.1 The scoring

These ideas follow those presented in the MDL-inspired algorithms (see Section 14.4), so we need to define a score for an automaton that would somehow reflect the balance we have described. This can be done with respect to the learning sample by computing the perplexity of the sample (see Section 5.5.5, page 110) and dividing it by the size of the automaton, when counting the number of states.

$$\text{score}(S, \mathcal{A}) = \frac{\sum_{w \in S} \text{cnt}_S(w) \log Pr_{\mathcal{A}} w}{\|\mathcal{A}\|}$$

We defined and studied these measures in Section 5.5.5. Optimising the denominator will lead to merging, whereas the optimum perplexity is obtained by the FPTA(S).

16.7.2 The algorithm

The structure of the algorithm is similar to that used by the algorithms presented in the previous sections, but we modify the compatibility test. Algorithm MDI (16.10) uses the function MDI-COMPATIBLE (Algorithm 16.9).

Algorithm 16.9: MDI-COMPATIBLE.

Input: an FFA $\mathcal{A}$, two states q and q', $S, \alpha > 0$
Output: a Boolean indicating if q and q' are compatible
$\mathcal{B} \leftarrow$STOCHASTIC-MERGE($\mathcal{A}, q, q'$);
return $\big(\text{score}(S, \mathcal{B}) < \alpha\big)$

Algorithm 16.10: MDI.

Input: a sample $S, \alpha > 0$
Output: an FFA $\mathcal{A}$
Compute t_0, the threshold on the size of the multiset needed for the test to be be statistically significant;
$\mathcal{A} \leftarrow$ FPTA(S) ;
RED $\leftarrow \{q_\lambda\}$;
BLUE $\leftarrow \{q_a : a \in \text{PREF}(S)\}$;
current_score$\leftarrow$ score(S, FPTA(S));
while CHOOSE q_b *from* BLUE *such that* FREQ(q_b) $\geq t_0$ **do**
  if $\exists q_r \in$ RED : MDI-COMPATIBLE($q_r, q_b, \mathcal{A}, S, \alpha$) **then**
    $\mathcal{A} \leftarrow$ STOCHASTIC-MERGE($\mathcal{A}, q_r, q_b$)
  else
    RED $\leftarrow$ RED $\cup \{q_b\}$
  end
  BLUE $\leftarrow \{q_{ua} : ua \in \text{PREF}(S) \wedge q_u \in \text{RED}\} \setminus$ RED
end
return $\mathcal{A}$

The key difference is that the recursive merges are to be made inside Algorithm MDI-COMPATIBLE (16.9) and before the new score is computed instead of in the main algorithm.

Several variants of MDI can be tried: one may prefer trying several possible merges and keeping the best, or using a parameter α that can change in time and depend on the quantity of information available. Of course, the score function can also be modified. Different things that need finer tuning: instead of taking the sum of the logarithms of the probabilities, one may want to take the perplexity directly.

A difficult question is that of setting the tuning parameter (α): if set too high, merges will take place early, which will perhaps include a wrong merge, prohibiting later necessary merges, and the result can be bad. On the contrary, a small α will block all merges, including those that should take place, at least until there is little data left. This is the *safe* option, which leads in most cases to very little generalisation.

16.7.3 Why it works

MDI is known to work in practice, and it tends to work better than the algorithms we described earlier. On the other hand, no convergence property has been proved.

One of the key features of MDI is that the computation of the new scores (after a particular merge) can be done incrementally, which allows complexity to be kept low.

16.8 Exercises

16.1 Consider the FFA represented in Figure 16.21. Build two samples from which this FFA can be obtained.

16.2 Suppose we have a sample containing the string $\mathtt{a}^{1000}$ and the strings b, bb and bbb each appearing 100 times. Build the FPTA corresponding to this sample and merge state $q_{\mathtt{a}}$ with state q_λ. What has happened? How can we avoid this problem?

16.3 Compute the number of examples needed in a Hoeffding test to allow for an error ϵ of 0.05 and $\delta = 0.01$. What happens if you wish to decrease ϵ to 0.01?

16.4 Construct the FFA corresponding to a 3-state automaton ($\{q_1, q_2, q_3\}$) with sample $\{\mathtt{aab}, \mathtt{aba}, \mathtt{bbb}(3)\}$ and for the two following sets of constraints:
$C_1 = \{(q_1, \mathtt{a}, q_2), (q_2, \mathtt{a}, q_3), (q_3, \mathtt{a}, q_1), (q_1, \mathtt{b}, q_2), (q_2, \mathtt{b}, q_3), (q_3, \mathtt{a}, q_3)\}$
$C_2 = \{(q_1, \mathtt{a}, q_2), (q_1, \mathtt{a}, q_3), (q_2, \mathtt{a}, q_2), (q_3, \mathtt{b}, q_3), (q_2, \mathtt{a}, q_1), (q_3, \mathtt{b}, q_1)\}$

16.9 Conclusions of the chapter and further reading

16.9.1 Bibliographical background

The first work in learning probabilistic grammars is due to Jim Horning (Horning, 1969), who put up the setting and showed that with enumeration algorithms, identification with probability one was possible. Among the other early papers, probabilistic automata and their theory are described by Azaria Paz in his book (Paz, 1971). Dana Angluin made an important (even if unpublished) contribution in her study on identification with probability one (Angluin, 1988). In the mid-nineties a variety of results were shown: Andreas Stolcke (Stolcke, 1994) proposed a method not only of estimating the probabilities but also of learning the structure of hidden Markov models. Naoki Abe and Manfred Warmuth exposed the hardness of learning or approximating distributions (Abe & Warmuth, 1992); the results are combinatorial: given a set of constraints, estimating probabilities is a hard problem.

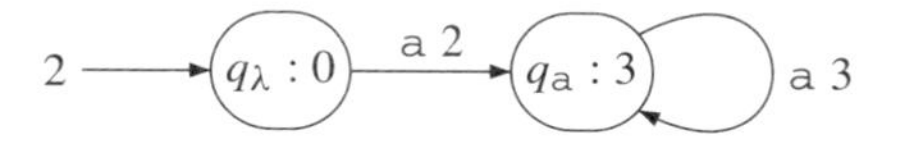

Fig. 16.21. What is the sample?

The DFFAs introduced in this chapter are inspired by the multiplicity automata that have been used in grammatical inference (Beimel *et al.*, 2000, Bergadano & Varricchio, 1996). Only the semantics are different.

The ideas from Section 16.4 come from (Carrasco & Oncina, 1994b): ALERGIA was invented by Rafael Carrasco and Jose Oncina. They proved the convergence of a simpler version of that algorithm, called RLIPS (Carrasco & Oncina, 1999). An extension of ALERGIA by Colin de la Higuera and Franck Thollard not only identifies the structure, but also the actual probabilities (de la Higuera & Thollard, 2000).

Extensions of ALERGIA to the tree case was done by the same authors in (Carrasco, Oncina & Calera-Rubio, 2001). Another extension to deal with the richer class of probabilistic deterministic linear languages can be found in (de la Higuera & Oncina, 2003). The same authors propose a study of the learnability of probabilistic languages for a variety of queries in (de la Higuera & Oncina, 2004).

Among the applications of algorithms that learn DPFAs, one can find text and document analysis (Young-Lai & Tompa, 2000), and web page classification (Goan, Benson & Etzioni, 1996).

The use of distinguishing strings (Section 16.5) was introduced by Dana Ron *et al.* (Ron, Singer & Tishby, 1995). Another specificity of this algorithm (and we have not followed this here) is to learn only acyclic automata. In fact the reason for this lies more in the necessity of a proof. In order to do this, two states q_u and q_v will be immediately declared incompatible if $|u| \neq |v|$. There have been several results following this paper. Improvements in (Guttman, 2006, Palmer & Goldberg, 2005, Thollard & Clark, 2004) concerned the different parameters that are needed in order to obtain PAC type results, but also on the bounds and the fact that acyclic conditions can be dropped. An incremental version was proposed in (Gavaldà *et al.*, 2006).

Algorithm MDI (Section 16.7) was invented by Franck Thollard *et al.* and used since then on a variety of tasks, with specific interest in language modelling (Thollard, 2001, Thollard & Dupont, 1999, Thollard, Dupont & de la Higuera, 2000).

16.9.2 Some alternative lines of research

A different approach is that of considering that the strings do not come from a set but from a sequence (each string can only be generated once). This has been analysed in (de la Higuera, 1998).

There are of course alternative ways to represent distributions over strings: recurrent neural networks (Carrasco, Forcada & Santamaria, 1996) have been tried, but comparisons with the direct grammatical-based approaches have not been made on large datasets.

We have concentrated here on deterministic automata, but there have been several authors attempting to learn non-deterministic probabilistic finite automata.

A first step has consisted of studying the class of the probabilistic residual finite state automata, introduced by Yann Esposito *et al.* (Esposito *et al.*, 2002), and finding the probabilistic counterparts to the residual finite state automata introduced by François Denis *et al.* (Denis, Lemay & Terlutte, 2000, 2001).

The richness of the class of the probabilistic finite automata has led to the introduction by François Denis *et al.* (Denis, Esposito & Habrard, 2006, Habrard, Denis & Esposito, 2006) of the innovative algorithm DEES that learns a multiplicity automaton (the weights can be negative) by iteratively solving equations on the residuals. Conversion to a PFA is possible.

Christopher Kermorvant (Kermorvant, de la Higuera & Dupont, 2004) learns DPFAs with additional knowledge. On trees (Rico-Juan, Calera-Rubio & Carrasco, 2000) Rico *et al.* learn k-testable trees and then compute probabilities.

Omri Guttman's thesis (Guttman, 2006, Guttman, Vishwanathan & Williamson, 2005) and published work add a geometric point of view to the hardness of learning PFAs.

We left the hard question of smoothing untouched in the discussion. The idea is that after using any of the algorithms presented within this chapter, there will usually be strings whose probability, when parsing with the resulting PFA, is going to be 0. This is source of all kinds of problems in practice. Pierre Dupont and Juan-Carlos Amengual (Dupont & Amengual, 2000), and also Franck Thollard (Thollard, 2001), studied this difficult question to which there still is a lot of room for answers.

16.9.3 Open problems and possible new lines of research

There are a number of directions one can follow and problems to be solved in this field:

(i) Fuzzy automata have sometimes also been called probabilistic automata (Rabin, 1966). In these, what is computed is not the weight of a string in the language but the probability that a string belongs to a language. The probability is thus that of being recognised, not of being generated. If there have been some (a few) attempts to learn such objects, there have not been (to our knowledge) any systematic results here.

(ii) A central question in this section has been that of taking the decision to merge, or not, two states. This has depended on the idea the learner has that the two states are equivalent, an idea based on the information it has at that moment, i.e. the fact that some earlier merges have been done and the samples. But this in turn leads to a natural question: given two samples S and S',

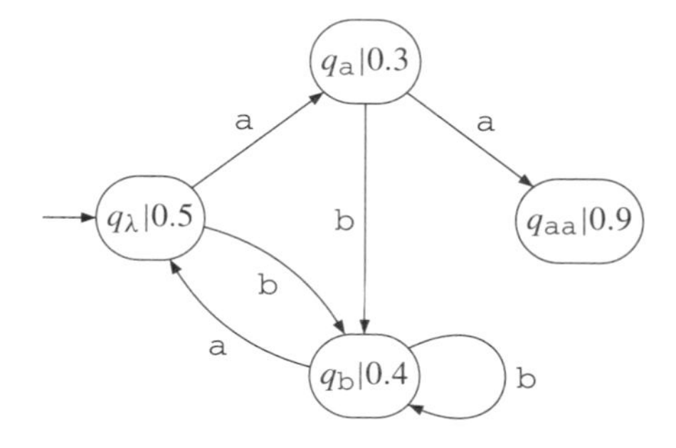

Fig. 16.22. Graphical representation of a deterministic MDP.

have they been produced by the same automaton? An answer to this question (obviously taking specific bias) is necessary in order to better understand what is feasible and what is not. Work in the statistical community deserves to be looked into for this reason.

(iii) Learning probabilistic context-free grammars is an open question. There were some early studies (Horning, 1969, Maryanski, 1974), some heuristics, often relying on constructing a grammar and then estimating the probabilities using the INSIDE-OUTSIDE algorithm (see Section 17.5, page 367) but no 'usable' algorithm exists.

(iv) Markov decision processes (MDPs) are used in reinforcement learning. They allow us to use finite state machines to manipulate probabilities in an active learning setting.

Definitions A **POMPD** (partially observable Markov decision process) is defined by a set of states Q, an input alphabet Σ, an initial state q_λ and two functions:

- A probabilistic transition function $\delta : Q \times \Sigma \times Q \rightarrow \mathbb{R}^+$ with $\forall q \in Q$, $\sum_{q \in Q, a \in \Sigma} \delta(q, a, q') = 1$.
- A probabilistic ***reward*** function $r : Q \rightarrow \mathbb{R}$.

A POMPD is deterministic if the function δ is deterministic, i.e. if $\forall q \in Q, \forall a \in \Sigma, \exists q' \in Q$ such that $\delta(q, a, q') = 1$. See Figure 16.22 for an example of a MDP. Notice that the rewards are also used as a goal function. POMPDs are used to model a situation: the 'learning' problem is then to construct the best strategy, given a POMPD. The other question, of finding the POMPD from some data (perhaps in a setting close to the one used in the introduction, page 6) has so far been left untouched.

17
Estimating the probabilities

We cannot seriously propose that a child learns the values of 10^9 parameters in a childhood lasting only 10^8 seconds.

George A. Miller and Noam Chomsky *(Miller & Chomsky, 1963).*

Par exemple, il arrive qu'après les douze chiffres du milieu sortent les douze derniers chiffres ; deux fois, mettons, le coup porte sur ces douze derniers chiffres et passe aux douze premiers. Une fois qu'il est tombé sur les douze premiers, il revient sur les douze du milieu ; trois, quatre fois de suite, les chiffres du milieu sortent, puis ce sont de nouveau les douze derniers ; après deux tours, on retombe sur les premiers, qui ne sortent qu'une fois, et les chiffres du milieu sortent trois fois de suite ; cela continue ainsi pendant une heure et demie ou deux heures. Un, trois et deux ; un, trois et deux. C'est très curieux.

Fedor Dostoïevski*, Le joueur.*

Let us suppose we are given a sample and an automaton. By automaton we mean the structure or at least some constraints on the number of states and some restrictive syntactical conditions on the transitions we are allowed to use. We are interested in finding a systematic way of converting the automaton into a probabilistic generator such as those we studied in Chapter 5. It would also be interesting to be able to do something similar for grammars instead of automata. Moreover we would like the probabilities of the automaton or of the grammar to be best suited to that sample, which, we hope, is supposed to be representative because it is generated randomly.

So the problem is not really any longer about finding the best automaton or grammar for a given sample, but about tuning the numerical parameters in the best way so that the chosen grammar or automaton works best.

Let us denote by $\mathcal{G}$ the set of all probabilistic machines that fit our set of syntactic constraints. Typically $\mathcal{G}$ contains automata that all share the same set of transitions but differ on the numerical parameters. We are searching, in $\mathcal{G}$, for the best suited grammar.

The first question is: 'What does *best suited* mean?' If we suppose that the sample (S) has been drawn following the distribution we wish to model, then we might like to have the automaton or grammar that maximises the probability of drawing this sample. This corresponds to finding the model with ***maximum likelihood***. The best parameters, for this criterion, are those which maximise the probability that the sample S is indeed generated by the probabilistic grammar.

We may therefore consider that the problem is about finding the grammar in $\mathcal{G}$ such that $Pr_G(S)$ is maximum. In other words:

$$\text{argmax}_{G\in\mathcal{G}}\{Pr_G(S)\} \tag{17.1}$$

But the problem with Expression 17.1 is that the probability of a sample is the sum of the probabilities of the strings that belong to it. Since each string in the sample is independently drawn, this would be a very incorrect way of measuring the probability of generating this sample exactly (see Exercises 17.4 and 17.5 for example): if we take the sum, we will be tempted to just provide a set of values such that the most frequent string receives the entire mass of probabilities.

Therefore a more correct expression is:

$$\text{argmax}_{G\in\mathcal{G}}\{\prod_{x\in S} Pr_G(x)^{\text{cnt}_S(x)}\} \tag{17.2}$$

where $\text{cnt}_S(x)$ denotes the number of occurrences of string x in multiset S.

There is an implicit bias in the above: we are supposing no *prior* regarding the probabilities. Of course, there can be applications where the probabilities are bounded or to be chosen inside a finite set, in which case the problem should be solved in a different way.

17.1 The deterministic case

A first case is easy to settle straight away: if we are given a DFA structure over which we are to estimate the probabilities and the language is therefore supposed to be generated by a DPFA, then each string in the sample has a unique parse that can be used to compute the probability of each transition or derivation rule by counting. For that, we just use the underlying DFA to parse the examples, count how many times each transition is used (by frequencies) and transform this deterministic FFA into a DPFA. This process was described in Section 16.2. The same arguments can be used if the automaton or the grammar is unambiguous, that is, if each string in the sample can be associated with a unique parse, but with more care.

Let us suppose that each string in the sample admits a unique parse. Precisely, if we are given a sample S and denote (as in Section 5.4.2, page 103) by $|S|_c$ the number of times condition c is met when parsing S, we will use the following notations:

- $|S|_{(\nrightarrow q)} = \sum_{x\in S} \text{cnt}_S(x) \cdot |\{u \in \Sigma^\star : \exists v \in \Sigma^\star\ uv = x \land \delta(q_\lambda, u) = q\}|$: the number of times state q is reached.
- $|S|_{(\downarrow q)} = \sum_{x\in S:\,\delta(q_\lambda,x)=q} \text{cnt}_S(x)$: the number of times state q is used as a final state.
- $|S|_{(\hookrightarrow q,a,q')} = \sum_{x\in S} \text{cnt}_S(x) \cdot |\{u \in \Sigma^\star : \exists v \in \Sigma^\star\, \exists a \in \Sigma\ uav = x \land \delta(q_\lambda, u) = q \land \delta(q, a) = q'\}|$: the number of times transition (q, a, q') is used by the sample.

Note that in each case, the number of times a transition or a state is used can actually be a larger value than the number of strings in the sample. Each of the above values can then be computed simply by counting. We can directly compute:

$$Pr_{\mathcal{A}}(q, a, q') = \frac{|S|_{(\hookrightarrow q, a, q')}}{|S|_{(\circlearrowleft q)}} \tag{17.3}$$

and also

$$\mathbb{F}_{\mathbb{P}\mathcal{A}}(q) = \frac{|S|_{(\downarrow q)}}{|S|_{(\circlearrowleft q)}} \tag{17.4}$$

So we build a DPFA from this by converting the relative frequencies. It can be shown (in the case of a DPFA, an unambiguous PFA or a PCFG) that the use of Equations 17.3 and 17.4 allow us to reach the DPFA with maximum likelihood.

Example 17.1.1 We depict in Figure 17.1 a sample (Figure 17.1(a), each string appears with its frequency in the sample) with the deterministic structure, or set of constraints the sample is to be parsed on (Figure 17.1(b)). Using Equations 17.3 and 17.4 the associated FFA is obtained (Figure 17.1(c)) and the resulting DPFA (Figure 17.1(d)) can easily be obtained.

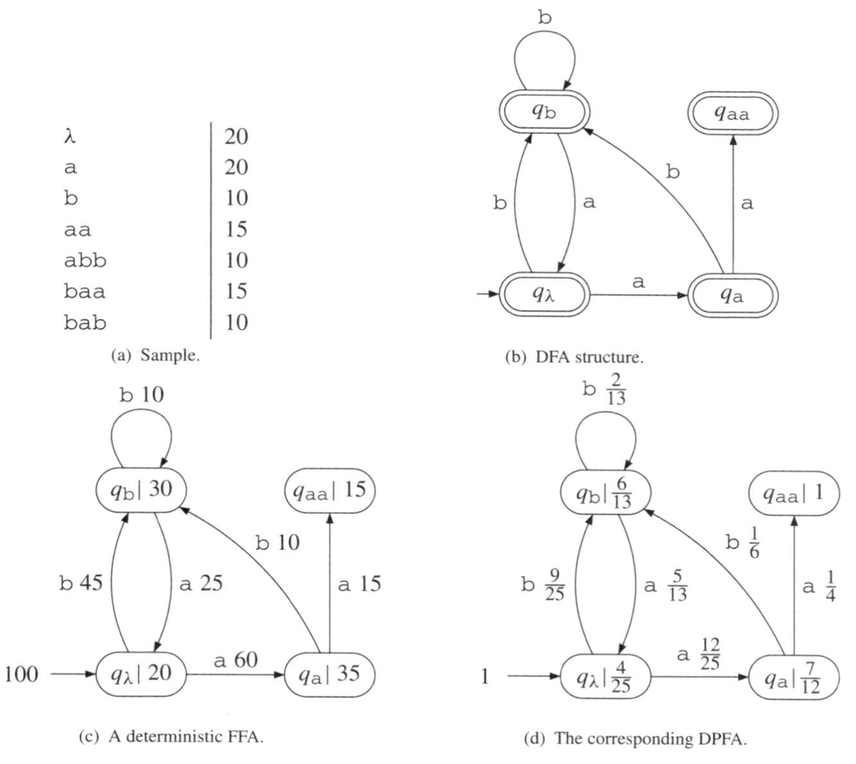

Fig. 17.1. Obtaining a DPFA from a sample and a DFA.

17.2 Towards non-determinism

Let us now consider a slightly more difficult case. Suppose for the same sample used in Example 17.1.1 we want to find the set of probabilities matching a non-deterministic automaton, or a context-free grammar. Clearly, the simple counting arguments don't work.

In theory, in an NFA we can put probabilities on all the transitions. But the knowledge we may have about the structure may again consist of imposing the constraint that some edges have a null value.

Definition 17.2.1 A (PFA) **constraint** is a NFA $\mathcal{A} = \langle \Sigma, Q, \mathbb{I}, \mathbb{F}_{\mathbb{A}}, \mathbb{F}_{\mathbb{R}}, \delta_N \rangle$. We say that the PFA $\mathcal{B} = \langle \Sigma, Q, \mathbb{I}_{\mathbb{P}}, \mathbb{F}_{\mathbb{P}}, \delta_{\mathbb{P}} \rangle$ respects the constraint $\mathcal{A}$ *if*$_{def}$ $\forall q, q' \in Q$, $\forall a \in \Sigma$,

- $\mathbb{I}_{\mathbb{P}}(q) > 0 \implies q \in \mathbb{I}$,
- $\mathbb{F}_{\mathbb{P}}(q) > 0 \implies q \in \mathbb{F}_{\mathbb{A}}$,
- $\delta_{\mathbb{P}}(q, a, q') > 0 \implies (q, a, q') \in \delta_N$.

One can notice that the set $\mathbb{F}_{\mathbb{R}}$ pays no part here as we are dealing with positive examples only.

Example 17.2.1 In Figure 17.2(b) a PFA constraint is represented. The PFA from Figure 17.3(a) respects the constraint (17.2(b)), whereas the right hand side PFA (17.3(b)) does not.

Suppose the sample is represented on the left (Figure 17.2(a)). Then clearly string baa admits three different parses.

The problem with non-determinism or ambiguity is that we may not be able to associate a path in the automaton to a string in the sample. Therefore, it is no longer just about counting. One should notice the difference with parsing: when we parse, all the paths are counted to sum up and give the probability of a string, whereas when the string is generated only one particular path has been used. So we cannot attribute the weight of the parse to any particular transition or rule since we only have access to the information that the string has been generated, and not how it has been generated. The problem is therefore much harder.

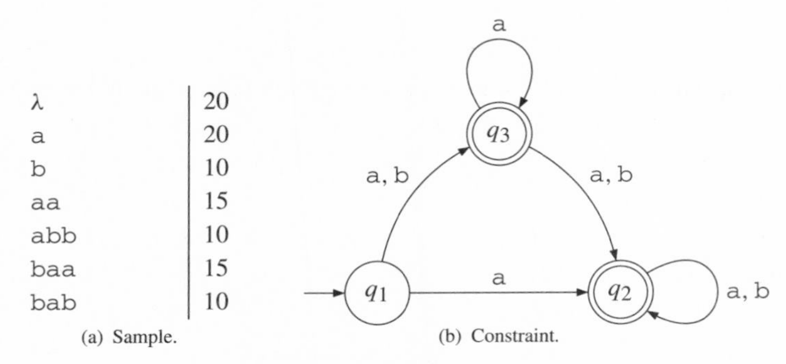

λ	20
a	20
b	10
aa	15
abb	10
baa	15
bab	10

(a) Sample. (b) Constraint.

Fig. 17.2. A context NFA for a given sample.

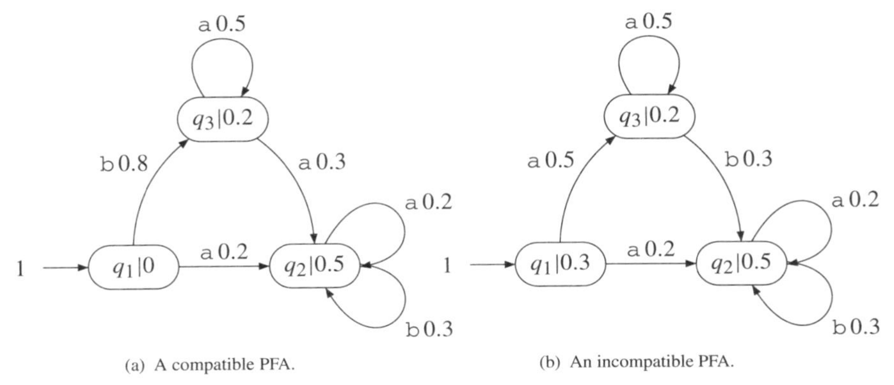

(a) A compatible PFA. (b) An incompatible PFA.

Fig. 17.3. Two PFAs for constraint in Figure 17.2(b).

17.3 The EM algorithm

So we now suppose that the underlying structure is ambiguous (some strings can have various possible generations). Given information about which derivation has been used, the computation would be easy, but we do not have access to that information. In fact, inferring this information would work just as well to solve our problem, since we could then return to the previous method, and convert the FFA into a PFA (or deal with probabilistic context-free grammars).

The general method used to infer the weights is called *expectation maximisation* (EM). The goal is to optimise the maximum likelihood (ML):

$$\text{argmax}_{G\in\mathcal{G}} \prod_{x\in S}\{Pr_G(x)^{\text{cnt}_S(x)}\}$$

where $\mathcal{G}$ is the set of all grammars consistent with the given constraints.

The idea is to start with an initial set of weights, and use these with the sample S to update the weights of the rules by using, instead of the strings, random variables associated with each string and with each possible parse.

To be as general as possible we will say that a string x has n different parses or explanations $e_{x,1},\ldots,e_{x,n}$ and call Explanations(x) the set $\{e_{x,1},\ldots,e_{x,n}\}$.

An explanation will be a path in a PFA or a derivation in a context-free grammar. We associate with each explanation $e_{x,i}$ the random variable $z_{x,i}$ indicating whether the explanation $e_{x,i}$ was followed or not. The expected value of $z_{x,i}$ given a current grammar G can be computed as follows:

$$\begin{aligned}E[z_{x,i}] &= Pr(z_{x,i}=1)\\ &= Pr(e_{x,i} \text{ is the real explanation})\\ &= \frac{Pr(e_{x,i}|G)}{\sum_{e_{x,j}\in\text{Explanations}(x)} Pr(e_{x,j}|G)}\end{aligned}$$

Therefore, if we can generate all the possible explanations, the equation above can be used to produce estimated counts. These can be used to update the different probability weights used in the grammar. This leads to iterating:

- the ***expectation step***, consisting of using the weights to estimate the counts,
- the ***maximisation step***, consisting of using the counts to produce new weights.

17.3.1 General remarks

Iterating the expectation step and the maximisation step is guaranteed to lead to a local maximum (or saddle point) in the likelihood surface.

But, both for non-deterministic automata and for context-free grammars, the number of possible explanations can be exponential with the length of the string; therefore, generating the different explanations is far too expensive. A dynamic programming technique allows us to avoid this generation and to compute the parameters, and therefore to implement the EM algorithm. In the first case this is known as the BAUM-WELCH algorithm; in the second the algorithm is called ***inside-outside***.

Notice that the expected counts have to be initialised to some arbitrary value. And in this sort of heuristic problem, the initialisation value is in fact an important question which we will have to return to later. Since the guarantee is only that a *local* maximum is reached, heuristics might allow us to be able to escape the local maximum, or hope to visit alternative ones.

There are a variety of heuristic approaches for escaping a local maximum such as using, for example, several different random initial estimates, or applying simulated annealing.

17.4 The Baum-Welch algorithm

The BAUM-WELCH algorithm uses the EM computations and deals with probabilistic finite state machines. Let $\mathcal{A}$ be the constraint. Let S be a finite sample of training strings assumed to be drawn from a regular distribution $\mathcal{D}$. The problem is to estimate the probabilistic parameters $\mathbb{I}_\mathbb{P}$, $\delta_\mathbb{P}$ and $\mathbb{F}_\mathbb{P}$ of $\mathcal{A}$ in such a way that $\mathcal{D}_\mathcal{A}$ approaches $\widehat{S}$.

The expectation step can now be done as follows in this context. Let $x = a_1 \cdots a_i \cdots a_{|x|}$. We estimate the count $\widehat{c_x}(\hookrightarrow q, a_i, q')$ as

$$\frac{Pr(q|a_1 \cdots a_{i-1}) \cdot \delta_\mathbb{P}(q, a_i, q') \cdot Pr_{\mathcal{A}_{q'}}(a_{i+1} \cdots a_{|x|})}{\sum_{\substack{s \in Q \\ s' \in Q}} Pr(s|a_1 \cdots a_{i-1}) \cdot \delta_\mathbb{P}(s, a_i, s') \cdot Pr_{\mathcal{A}_{s'}}(a_{i+1} \cdots a_{|x|})} \tag{17.5}$$

In the same way, we estimate $\widehat{c_x}(\downarrow q)$:

$$\frac{Pr(q|a_1 \cdots a_n) \cdot \delta_\mathbb{P}(q)}{\sum_{s \in Q} Pr(s|a_1 \cdots a_n) \cdot \delta_\mathbb{P}(s)} \tag{17.6}$$

Note:

- $Pr(q|a_1 \cdots a_{i-1})$ denotes the probability of being in state q after reading $a_1 \cdots a_{i-1}$. This is also called the forward probability and is computed by the FORWARD algorithm (Algorithm 5.2, page 91).
- $Pr_{\mathcal{A}_{q'}}(a_{i+1} \cdots a_{|x|})$ is the probability, of computing string $a_{i+1} \cdots a_{|x|}$ when starting from state q'. This is known as the backward probability, which is computed by the BACKWARD algorithm (Algorithm 5.4, page 92).

This has to be done for each symbol a_i in each string in S. The number of occurrences of each string in the sample (which is a multiset!) also has to be taken into account.

17.4.1 The algorithm

The computations can be combined into the BAUM-WELCH algorithm. This algorithm iteratively runs the expectation phase (17.1) and the maximisation phase (17.2). In the algorithms, we have chosen to mix the algorithmic tables with the notations we have been using. We have also supposed that the initial values correspond to the constraints: a value equal to 0 enforces the absence of the transition, or the fact that the state cannot be initial or final. This may not be what is wanted, in which case a slightly modified (but more complex) algorithm should be designed. Extra care should in any case be taken when attempting to implement it.

Initially, the PFA $\mathcal{A}$ has to respect the imposed constraints. Dynamic programming is used in order to compute the different possible paths.

The actual convergence of the algorithm can be to a local maximum, which has no reason to be close to the optimum. Only by trying out several initial values for the different parameters can we hope to better the results, but theory tells us that there exist cases where only luck with the initial choices will allow us to obtain a correct estimation.

The time and space complexities of the BAUM-WELCH algorithm are respectively $\mathcal{O}(\mid Q \mid \cdot B \cdot \sum_{x \in S} \mid x \mid)$ and $\mathcal{O}(\mid Q \mid \cdot \max_{x \in S} \mid x \mid)$, where B is the average number of transitions per state (*branching factor*) of $\mathcal{A}$.

There is a weak convergence result: it can be shown that, if the distribution is generated by some PFA $\mathcal{A}$ with the same *structural component* as the one we are using (meaning that if the constraints are correct, then the BAUM-WELCH algorithm is guaranteed to return a distribution $\mathcal{D}$ which approaches $\mathcal{D}_{\mathcal{A}}$ as the size of the sample grows to infinity).

On the other hand (see Exercise 17.6) if the constraints are bad, no parameter setting will allow us to obtain a good estimation.

17.4.2 An example run

We give a small run of the first two steps of the BAUM-WELCH algorithm in the case of a very simple automaton.

Algorithm 17.1: Algorithm BAUM-WELCH (expectation step).

```
Data: a PFA 𝒜, a sample S
Result: The different counts ĉ_S(↪ q, a, q'), ĉ_S(↛ q) and ĉ_S(↓ q)
for q ∈ Q do
    ĉ_S(↓ q) ← 0;
    ĉ_S(↛ q) ← 0;
    for q' ∈ Q do
        for a ∈ Σ do ĉ_S(↪ q, a, q') ← 0;
    end
end
for x = a_1 a_2 ··· a_n ∈ support(S) do
    Nbocc ← cnt_S(x);
    for q ∈ Q do                    /* Computation of BACKWARD and FORWARD */
        F[0][q] ← 𝕀_ℙ(q);
        B[n][q] ← 𝔽_ℙ(q);
        for i ∈ [n] do
            F[i][q] ← 0;
            B[n−i][q] ← 0;
            for q' ∈ Q do F[i][q] ← F[i][q]+F[i−1][q'] · δ_ℙ(q', a_i, q);
            B[n−i][q] ← B[n−i][q]+B[n−i+1][q'] · δ_ℙ(q, a_{n−i+1}, q')
        end
    end
    Total ← 0;                      /* Total corresponds to Pr_𝒜(x) */
    for q ∈ Q do Total ← Total+F[n][q] · 𝔽_ℙ(q);
    for i ∈ [n] do
        for q ∈ Q do
            for q' ∈ Q do
                Val ← Nbocc · (F[i−1][q]·δ_ℙ(q,a_i,q')·B[i][q']) / Total;
                ĉ_S(↪ q, a_i, q') ← ĉ_S(↪ q, a_i, q') + Val;
                ĉ_S(↛ q') ← ĉ_S(↛ q') + Val
            end
        end
    end
    for q ∈ Q do
        ĉ_x(↓ q) ← ĉ_x(↓ q) + Nbocc · (F[n][q]·𝔽_ℙ(q)) / Total;
        ĉ_S(↛ q) ← ĉ_S(↛ q) + Nbocc · (B[0][q]·𝕀_ℙ(q)) / Total
    end
end
return(ĉ_S(↪), ĉ_S(↓), ĉ_S(↛))
```

Algorithm 17.2: Algorithm BAUM-WELCH (maximisation step).

Data: a PFA $\mathcal{A}$, the estimated counts $\widehat{c_S}(\hookrightarrow q, a, q')$, $\widehat{c_S}(\looparrowright q)$ and $\widehat{c_S}(\downarrow q)$
Result: PFA $\mathcal{A}$ updated.

```
for q ∈ Q do    /* Compute in Val the number of times we enter a
state during parsing */
| Val[q] ← 0
end
for q ∈ Q do
    F_P(q) ← ĉ_S(↓q) / ĉ_S(↬q);
    for a ∈ Σ do
        for q' ∈ Q do
            δ_P(q, a, q') ← ĉ_S(↪q,a,q') / ĉ_S(↬q);
            Val[q'] ← Val[q'] + ĉ_S(↪ q, a, q')
        end
    end
end
for q ∈ Q do I_P(q) ← (ĉ_S(↬q) − Val[q]) / ĉ_S(↬q);
return A
```

In this example, to keep things visible, we have a one-letter alphabet, and just one initial state. Suppose we have a (very small) learning sample $S = \{(\lambda, 2), (\mathtt{a}, 3), (\mathtt{aa}, 1)\}$. Suppose now that the constraints are that the structure is described by the graph represented in Figure 17.4(a). We use as an initialisation the weights given in Figure 17.4(b).

We compute separately the BACKWARD and FORWARD tables (Table 17.1). Then the estimated counts are recomputed (Table 17.2). In each case, the number of occurrences of each string in the sample is used.

The values of the estimated counts for each transition and state are as follows:

$$\text{FREQ}(q_1) = \widehat{c_S}(\hookrightarrow q_1, \mathtt{a}, q_1) + \widehat{c_S}(\hookrightarrow q_1, \mathtt{a}, q_2) + \widehat{c_S}(\downarrow q_1) = \frac{924}{95}$$

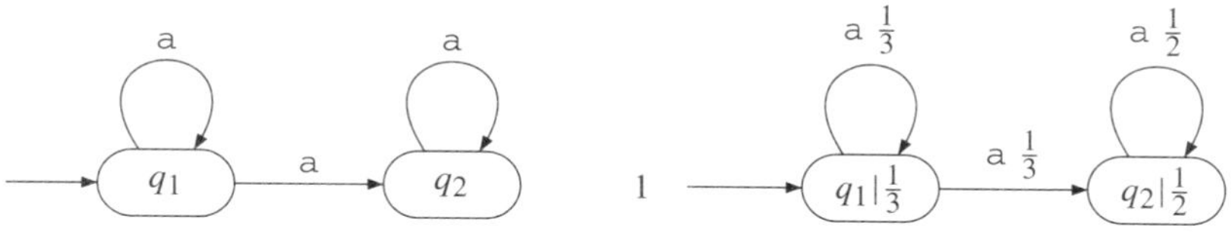

Fig. 17.4. An example structure and initialisation for BAUM-WELCH.

Table 17.1. FORWARD *and* BACKWARD *computations.*

x	$\text{cnt}_S(x)$	i	$\text{B}[\lvert x\rvert - i][q_1]$	$\text{B}[\lvert x\rvert - i][q_2]$	$\text{F}[i][q_1]$	$\text{F}[i][q_2]$	$Pr(x)$
λ	2	0	$\frac{1}{3}$	$\frac{1}{2}$	1	0	$\frac{1}{3}$
a	3	0	$\frac{1}{3}$	$\frac{1}{2}$	1	0	
		1	$\frac{5}{18}$	$\frac{1}{4}$	$\frac{1}{3}$	$\frac{1}{3}$	$\frac{5}{18}$
aa	1	0	$\frac{1}{3}$	$\frac{1}{2}$	1	0	
		1	$\frac{5}{18}$	$\frac{1}{4}$	$\frac{1}{3}$	$\frac{1}{3}$	
		2	$\frac{19}{108}$	$\frac{1}{8}$	$\frac{1}{9}$	$\frac{5}{18}$	$\frac{19}{108}$

Table 17.2. *The estimated counts.*

Rule		λ	+	a	+	aa		
$\widehat{c_x}(\hookrightarrow q_1, \text{a}, q_1)$	=			$\frac{2}{5}$	+	$\frac{14}{19}$	=	$\frac{354}{95}$
$\widehat{c_x}(\hookrightarrow q_1, \text{a}, q_2)$	=			$\frac{3}{5}$	+	$\frac{15}{19}$	=	$\frac{246}{95}$
$\widehat{c_x}(\hookrightarrow q_2, \text{a}, q_2)$	=					$\frac{9}{19}$	=	$\frac{9}{19}$
$\widehat{c_x}(\downarrow q_1)$	=	1	+	$\frac{2}{5}$	+	$\frac{4}{19}$	=	$\frac{324}{95}$
$\widehat{c_x}(\downarrow q_2)$	=			$\frac{3}{5}$	+	$\frac{15}{19}$	=	$\frac{246}{95}$

$$\text{FREQ}(q_2) = c_{\hookrightarrow q_2,\text{a},q_2} + c_{\downarrow q_2} = \frac{291}{95}$$

$$\mathbb{F}_{\mathbb{P}}(q_1) = \frac{324}{95} / \frac{924}{95} = 0.351$$

$$\delta_{\mathbb{P}}(q_1, \text{a}, q_1) = \frac{354}{95} / \frac{924}{95} = 0.383$$

$$\delta_{\mathbb{P}}(q_1, \text{a}, q_2) = \frac{246}{95} / \frac{924}{95} = 0.266$$

$$\mathbb{F}_{\mathbb{P}}(q_2) == \frac{246}{95} / \frac{291}{95} = 0.845$$

$$\delta_{\mathbb{P}}(q_2, \text{a}, q_2) = \frac{9}{19} / \frac{291}{95} = 0.155$$

A new PFA (Figure 17.5) can be built and compared with the original one. The algorithm has not converged yet. Several other iterations are needed.

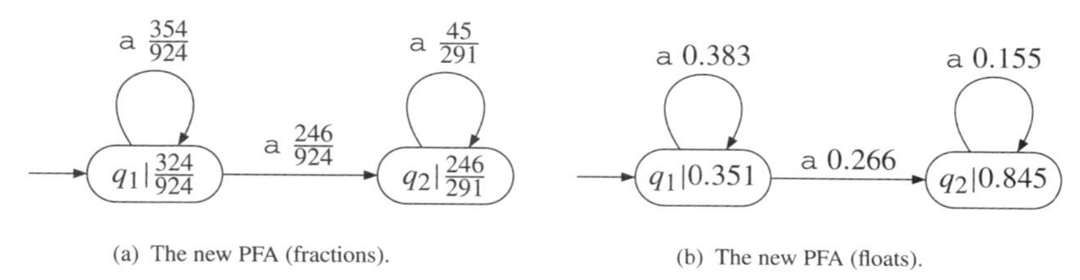

(a) The new PFA (fractions). (b) The new PFA (floats).

Fig. 17.5. An example structure and initialisation for BAUM-WELCH.

17.5 The INSIDE-OUTSIDE algorithm

The EM approach has also been adapted to estimate the probabilities of a context-free grammar. This is called the INSIDE-OUTSIDE algorithm.

For a string x and a non-terminal N, let $x_{i,j}$ be the substring between positions i and j. Let $Pr(x_{i,j}|N)$ indicate the probability that N derives into $x_{i,j}$. We define:

- the INSIDE probability $\beta_N(i, j) = Pr(x_{i,j} \mid N)$,
- the OUTSIDE probability $\alpha_N(i, j) = Pr(x_{1,i-1} N x_{j+1,|x|} \mid N_1)$.

The INSIDE probability $\beta_N(i, j)$ is the probability for non-terminal N to derive string $x_{i,j}$, whereas the OUTSIDE probability is the probability for string $x_{1,i-1} N x_{j+1,|x|}$ to be derived from the start symbol. We represent the situation by Figure 17.6.

The expectation and maximisation steps are implemented as follows (for a rule $N \rightarrow N_1 N_2$):

$$\widehat{c_x}(N \rightarrow N_1 N_2|x) = \sum_{j=2}^{j=|x|} \sum_{i=1}^{i=j-1} \sum_{k=i}^{k=j-1} \beta_{N_1}(i,k)\beta_{N_2}(k+1, j)\alpha_N(i, j) Pr(N \rightarrow N_1 N_2)$$

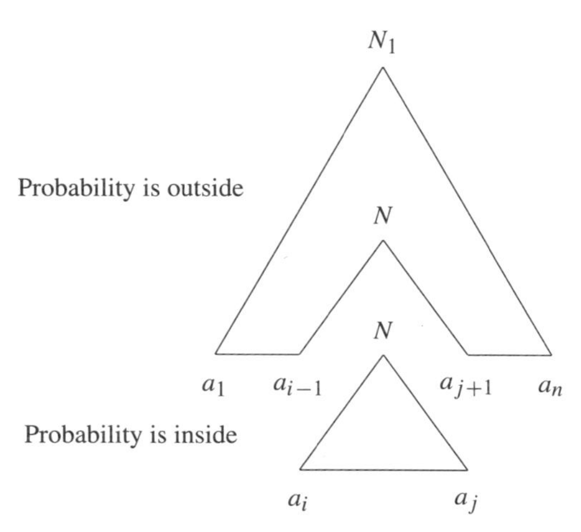

Fig. 17.6. INSIDE and OUTSIDE.

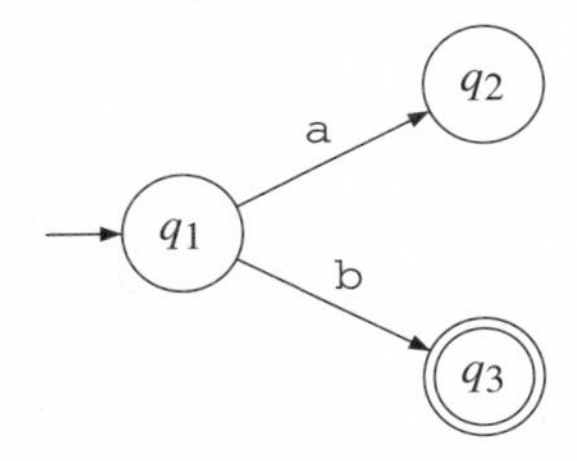

Fig. 17.7. A set of constraints represented as an automaton.

Now the estimation for each $x \in S$ of the new probability corresponding to the rule $N \rightarrow N_1 N_2$ is as follows:

$$\frac{\sum_{j=1}^{j=|x|} \sum_{i=1}^{i=j-1} \sum_{k=i}^{k=j} \beta_{N_1}(i,k)\beta_{N_2}(k+1,j)\alpha_N(i,j)}{\sum_{i<j} \beta_N(i,j)\alpha_N(i,j)}$$

Evaluating this equation requires time in $\mathcal{O}(n^3)$, to which must be added the time needed for computing values of α and β. Total runtime is again cubic.

17.6 Exercises

17.1 Write the algorithm which computes $|S|_{(\not\rightarrow q)}$, for a given DPFA.

17.2 Write the algorithm which computes $|S|_{(\downarrow q)}$, for a given DPFA.

17.3 Write the algorithm which computes $|S|_{(\hookrightarrow q,a,q')}$, for a given DPFA.

17.4 Let $S = \{(\mathtt{a}, 5), (\mathtt{b}, 1)\}$ and suppose the constraints are that the structure is the automaton represented in Figure 17.7. If we try to maximise $Pr_{\mathcal{A}}(S)$, what set of weights should you select?

17.5 Consider a sample X containing once a, three times b and six times c. Suppose the goal is to put a probability on the only possible rules $N_1 \rightarrow \mathtt{a}$, $N_1 \rightarrow \mathtt{b}$ and $N_1 \rightarrow \mathtt{c}$. What is the solution to $\text{argmax}_{G\in\mathcal{G}}\{Pr_G(S)\}$? What is the solution to $\text{argmax}_{G\in\mathcal{G}}\{\prod_{x\in X} Pr_G(x)\}$?

17.6 Suppose the sample is generated by the PFA represented in Figure 17.8(a). Suppose that the constraint is given by the NFA represented in Figure 17.8(b). What is the best PFA you can hope for when using Algorithm BAUM-WELCH?

17.7 Consider the structure depicted in Figure 17.9(a). Suppose the sample consists of $\{(\lambda, 5), (\mathtt{a}, 5), (\mathtt{a}^2, 5), (\mathtt{a}^3, 5), (\mathtt{a}^5, 5)\}$. Run Algorithm BAUM-WELCH on this sample, with initial probabilities as in Figure 17.9(b), then as in Figure 17.9(c).

17.8 Adjust the probabilities from automata 17.10(a) and 17.10(b) given the sample $S = \{(\mathtt{a}, 10), (\mathtt{b}^{10}, 1)\}$.

17.7 Conclusions of the chapter and further reading

A typical way of obtaining a probabilistic grammar or automaton is by providing a structure (the graph or the set of rules) and then attempting to find a set of parameters (the initial,

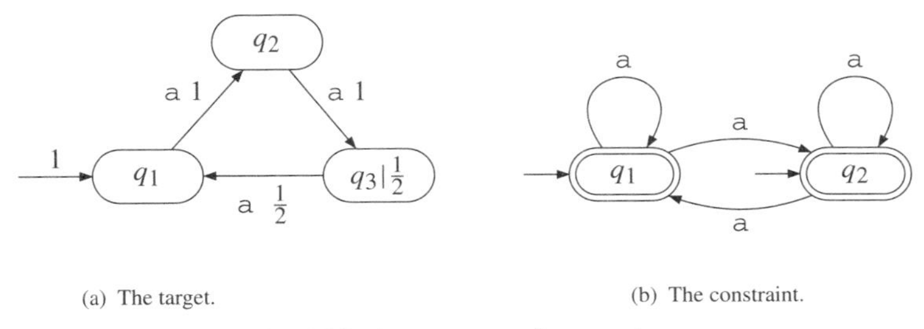

(a) The target. (b) The constraint.

Fig. 17.8. A wrong set of constraints.

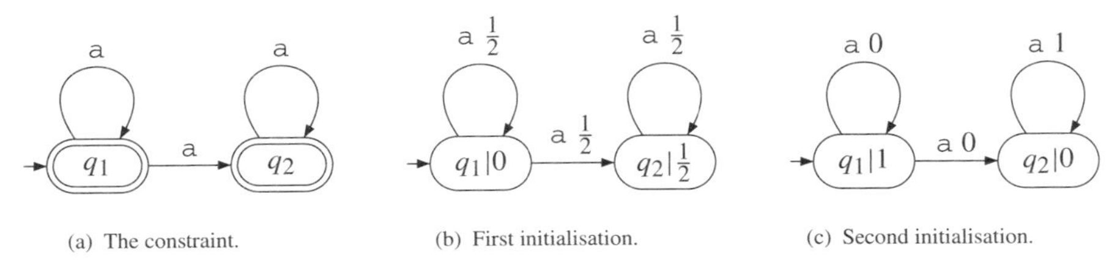

(a) The constraint. (b) First initialisation. (c) Second initialisation.

Fig. 17.9. Different initialisations for BAUM-WELCH.

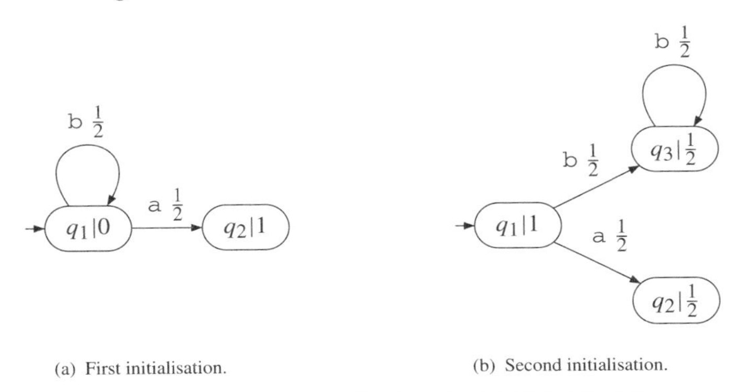

(a) First initialisation. (b) Second initialisation.

Fig. 17.10. Different initialisations for BAUM-WELCH.

final and transition probabilities in the case of an automaton; the probability of each rule in the case of a grammar). The best-known technique to achieve this is by the EM algorithm which starts with an initial set of parameters, then computes the probabilities of the strings using these initial parameters (called counts) and then uses these counts as new parameters. The actual implementation of this idea is called the BAUM-WELCH algorithm in the case of PFAs and INSIDE-OUTSIDE in the case of PCFGs.

17.7.1 Bibliographical background

The fact that the optimisation problem (Equation 17.3) is quite simple if the given automaton is deterministic was first noticed by Charles Wetherell (Wetherell, 1980). The expectation maximisation ideas were first presented for hidden Markov models and PFAs by (Baum, 1972, Baum *et al.*, 1970). Convergence issues of the algorithm are discussed in (Chaudhuri & Rao, 1986).

One alternative is to use the maximum path instead of the total probability in the function to be optimised: this allows us to use a simpler algorithm, called the ***Viterbi re-estimation algorithm***. This is discussed by Francisco Casacuberta in (Casacuberta, 1996a), while re-estimation algorithms for other criteria different from ML can be found in (Casacuberta, 1995a, 1995b, Picó & Casacuberta, 2000, 2001).

The problem itself of obtaining the optimum is probably $\mathcal{NP}$-hard, as partly shown in (Abe & Warmuth, 1992). That is why only *local-optimum* solutions to the optimisation problem (Equation 17.3) are possible.

The INSIDE-OUTSIDE algorithm, described in Section 17.5, is based on work by James Baker, and popularised by Karim Lari and Steve Young (Baker, 1979, Lari & Young, 1990, 1991). A careful analysis can be found in (Casacuberta, 1994, 1995b).

The relation between the probability of the optimal path of states and the probability of generating a string has been studied in (Sánchez, Benedí & Casacuberta, 1996).

The work presented in this chapter depends strongly on having the structure of the automaton or the grammar to work with. This assumption is sometimes taken as a *sine qua non* to learning probabilistic languages. Alternatively, many authors will argue that an alternative path consists of taking as a starting point a complete graph with as many states as needed (whatever that means), and using the EM from there. Obviously this approach only makes sense if we accept the bias that the structure is indeed small. Yet there are many applications where this should not be the case.

We have presented in Chapter 16 the alternative approach consisting of learning both the structure and the probabilities.

17.7.2 Some alternative lines of research

As the finite state models become more and more complex, the algorithms for estimating the probabilities have been adapted. There are EM algorithms for a variety of tasks, which can profit from better knowledge of the structure. Obviously, we should mention here that all the approaches in this chapter rely on the heavy assumption that the correct structure is provided. We have discussed in Chapter 16 the alternative approach of learning the probabilities and the structure at the same time.

17.7.3 Open problems and possible new lines of research

One important question that is not settled is to know how bad the estimation can be. Better said, given a target automaton, given a sample, and given either correct or incorrect knowledge of the structure, how good is the best possible estimation going to be?

Alternatively, a worst case analysis would be of help and one would be interested in designing a target for which EM is going to fail by quite a lot.

Among the open lines of research corresponding to estimating probabilities, one obviously concerns speeding up the INSIDE-OUTSIDE algorithm. This algorithm is widely used in practice, but there have been few systematic studies of its robustness.

18
Learning transducers

Pulpo a Feira, Octopus at a party

***Anonymous**, From a menu in O Grove, Galicia*

Die Mathematiker sind eine Art Franzosen: Redet man zu ihnen, so übersetzen sie es in ihre Sprache, und dann ist es alsbald etwas anderes.

***Johann Wolfgang von Goethe**, Maximen und Reflexionen*

18.1 Bilanguages

There are many cases where the function one wants to learn doesn't just associate a label or a probability with a given string, but should be able to return another string, perhaps even written using another alphabet. This is the case in translation, of course, between two 'natural' languages, but also of situations where the syntax of a text is used to extract some semantics. And it can be the situation in many other tasks where machine or human languages intervene.

There are a number of books and articles dealing with *machine translation*, but we will only deal here with a very simplified setting consistent with the types of finite state machines used in the previous chapters; more complex translation models based on context-free or lexicalised grammars are beyond the scope of this book.

The goal is therefore to infer special finite *transducers*, those representing *subsequential functions*.

18.1.1 Rational transducers

Even if in natural language translation tasks the alphabet is often the same for the two languages, this needs not be so. For the sake of generality, we will therefore manipulate two alphabets, typically denoted by Σ for the input alphabet and Γ for the output one.

Definition 18.1.1 (Transduction) A **transduction** from $\Sigma^\star$ to $\Gamma^\star$ is a relation $t \subseteq \Sigma^\star \times \Gamma^\star$.

Even if it is defined as a relation, we choose to give it a direction (from $\Sigma^\star$ to $\Gamma^\star$) to emphasise the asymmetric nature of the operation. Therefore, transductions are defined by pairs of strings, the first over the input alphabet, and the other over the output alphabet.

A first finite state machine used to recognise transductions is the rational transducer:

Definition 18.1.2 (Rational transducer) A **rational transducer** is a 5-tuple $\mathcal{T}=\langle Q, \Sigma, \Gamma, q_\lambda, E\rangle$:

- Q is a finite set of states,
- Σ, Γ are the input and output alphabets,
- $q_\lambda \in Q$ is the unique initial state,
- $E \subset (Q \times \Sigma^\star \times \Gamma^\star \times Q)$ is a finite set of transitions.

Rational transducers can be used like usual finite state machines; they *recognise* transductions. Given a transducer $\mathcal{T}=\langle Q, \Sigma, \Gamma, q_\lambda, E\rangle$, the transduction recognised by $\mathcal{T}$, denoted by $t_\mathcal{T}$ is the set of all pairs that one can read on a path starting in state q_λ. If we want to use a different initial state (say q), the transduction will be denoted by $t_{\mathcal{T}_q}$.

Note that rational transducers have no final states; the parse can halt in any state. Since these are finite state machines, they will be drawn as such. Each transition will be represented by an edge labelled by a pair $x :: y$.

Definition 18.1.3 (Translation) The string y is a **translation** of the string x if_{def} $x = x_1 \cdots x_n$, $y = y_1 \cdots y_n$ ($x_i \in \Sigma^\star$, $y_i \in \Gamma^\star, \forall i \in [n]$) and there is a sequence of states $q_{i_0}, \ldots, q_{i_n}$ such that $\forall j \in [n]$, $(q_{i_{j-1}}, x_j, y_j, q_{i_j}) \in E$, with $q_{i_0} = q_\lambda$.

In Figure 18.1 a very simplified rational transducer to translate from French to English is represented. Note that in the above definition the x_j and y_j are substrings, not individual symbols. Notation $x :: y$ gives an idea of determinism that is sometimes desirable and that we will build upon in the next section.

18.1.2 Sequential transducers

Following on from the earlier definition, the next step consists of reading the input symbols one by one and in a deterministic way.

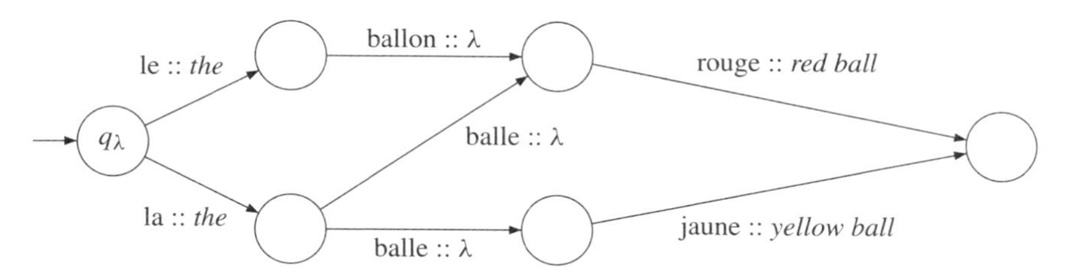

Fig. 18.1. Rational transducer.

Definition 18.1.4 (Sequential transducers) A **sequential transducer** $\mathcal{T}=\langle Q,\Sigma,\Gamma,q_\lambda,E\rangle$ is a rational transducer such that $E\subset Q\times\Sigma\times\Gamma^\star\times Q$ and $\forall(q,a,u,q'),(q,a,v,q'')\in E \Rightarrow u=v\wedge q'=q''$.

In the definition above, the transition system has become deterministic. This will allow us to do two things:

- Associate with E a new function τ_E such that $\tau_E(q,a)=(v,q')$ for $(q,a,u,q')\in E$. Let us also associate with E the two projections $\tau_1: Q\times\Sigma\to\Gamma^\star$ and $\tau_2: Q\times\Sigma\to Q$, with $(q,a,w,q')\in E \iff \tau_E(q,a)=(w,q'), \tau_1(q,a)=w$, and $\tau_2(q,a)=q'$.

  In the same way as with deterministic automata, we can extend τ_E to a function $Q\times\Sigma\to\Gamma^\star\times Q$ and write $\tau_E(q,\lambda)=(\lambda,q)$ and $\tau_E(q,a\cdot u)=\big(\tau_1(q,a)\cdot\tau_1(\tau_2(q,a),u),\ \tau_2(\tau_2(q,a),u)\big)$
- Name as usual each state by the shortest prefix over the input alphabet that reaches the state.

Example 18.1.1 The transducer represented in Figure 18.1 is not sequential because the inputs are strings and not symbols. In this case an alternative sequential transducer can be built, but this is not the general case because of the possible lack of determinism. The finite state machine from Figure 18.2 is a sequential transducer. One has, for instance, $\tau_E(q_1,0101)=(\mathtt{0111},q_\lambda)$.

Properties 18.1.1

- The transduction produced by a sequential transducer is a relation t over $\Sigma^\star\times\Gamma^\star$ that is functional and total, i.e. given any $x\in\Sigma^\star$ there is exactly one string $y\in\Gamma^\star$ such that $(x,y)\in t$.
  With the better adapted functional notation: a transduction t is a total function: $\Sigma^\star\to\Gamma^\star$.
- The sequential transductions preserve the prefixes, i.e. $t(\lambda)=\lambda$ and $\forall u,v\in\Sigma^\star$, $t(u)\in \textsc{Pref}(t(uv))$.

We can deduce from the above properties that not all finite transductions can be produced by a sequential transducer. But some computations are possible: for instance in Figure 18.2 we present a transducer that can 'divide' by 3 an integer written in base 2. For example, given the input string $u=\mathtt{100101}$ corresponding to 37, the corresponding output is $t(u)=\mathbf{001100}$, which is a binary encoding of 12. The operation is the integer division; one can notice that each state corresponds to the rest of the division by 3 of the input string.

In order to be a bit more general, we introduce subsequential transducers, where a string can be generated at the end of the parse.

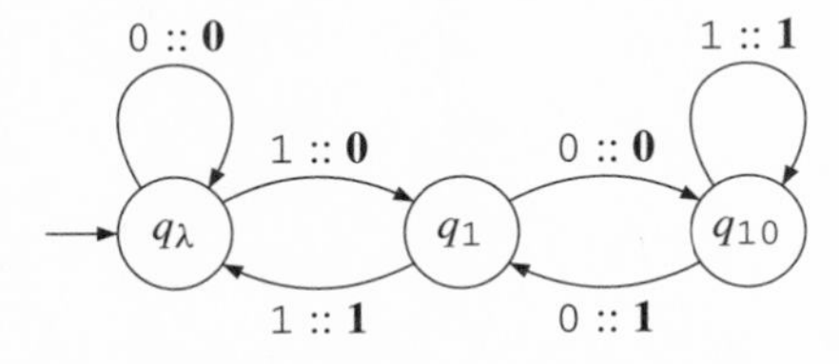

Fig. 18.2. Sequential transducer dividing by 3 in base 2.

18.1.3 Subsequential transducers

We only add to the previous definition a new function σ, called the ***state output***, which in any state can produce a string when halting in that state.

Definition 18.1.5 (Subsequential transducer) A **subsequential transducer** is a 6-tuple $\langle Q, \Sigma, \Gamma, q_\lambda, E, \sigma\rangle$ such that $\langle Q, \Sigma, \Gamma, q_\lambda, E\rangle$ is a sequential transducer and $\sigma : Q \to \Gamma^\star$ is a total function.

The transduction $t : \Sigma^\star \to \Gamma^\star$ is now defined as $t(x) = t'(x)\sigma(q)$ where $t'(x)$ is the transduction produced by the associated sequential transducer and q is the state reached with the input string x. We again denote by $t_\mathcal{T}(q)$ the transduction realised by the transducer $\mathcal{T}$ using q as the initial state: $t_\mathcal{T}(q) \subseteq \Sigma^\star \times \Gamma^\star$.

Definition 18.1.6 (Subsequential transducer) A transduction is **subsequential** if_{def} there exists a subsequential transducer that recognises it.

Intuitively, a subsequential transduction is one that can be produced from left to right using a bounded amount of memory, in the same way as a regular language is composed of strings recognised by a device reading from left to right and also using bounded memory. But it also corresponds to an optimistic (and thus naive) parse: the associated function has to be total, and thereby, translation on the fly is always possible (since we know it is not going to fail).

Example 18.1.2 A first example is that of multiplication by any number in any base. The case where we are in base 2 and we multiply by 3 is represented in Figure 18.3. In this example we have: $\tau(q_\lambda, 1) = (\mathbf{1}, q_1)$, $\tau_1(q_1, 0) = \mathbf{1}$, and $\tau_2(q_1, 1) = q_{11}$.

Example 18.1.3 Another example is that of the replacement of a pattern by a special string. Here we give the example of pattern abaa in Figure 18.4.

A third example concerns the translation from numbers written in English to Roman numerals. We do not explicit the rather large transducer here, but hope to show that indeed a bounded memory is sufficient to translate all the numbers less than 5000, like 'four hundred and seventy three' into 'CDLXXIII'.

On the other hand, it is easy to show that not all transductions can be recognised by subsequential transducers. A typical example of a transduction that is not subsequential

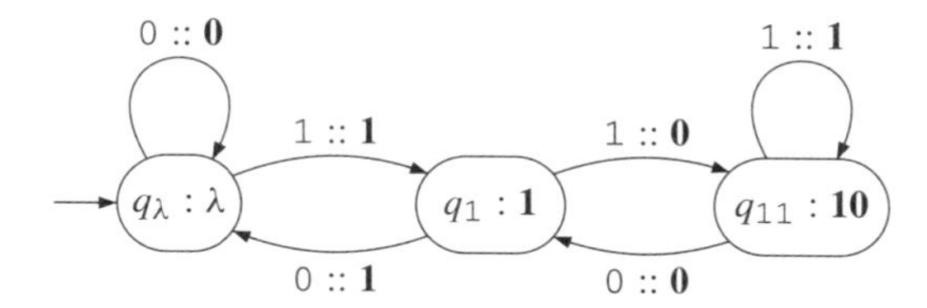

Fig. 18.3. Subsequential transducer multiplying by 3 in base 2. For input 101, output is **1111**.

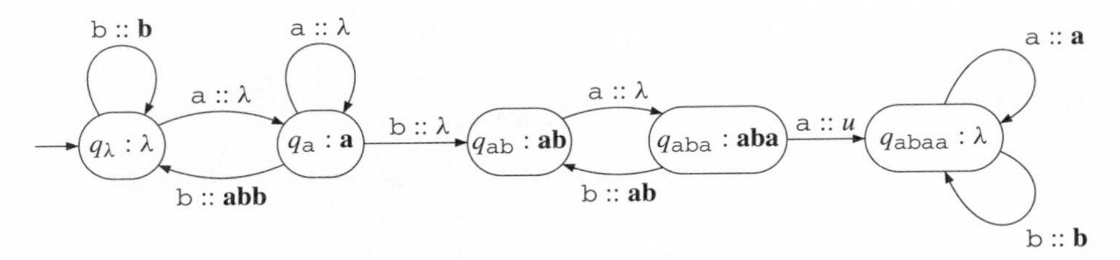

Fig. 18.4. Transducer replacing in each string the first occurrence of abaa by string u. All other symbols remain identical.

concerns reversing a string. Suppose we would like to consider the transduction containing all pairs of strings (w, w^R) where w^R is the string w written in reverse order, i.e. for $w = \texttt{abcd}$, $w^R = \texttt{dcba}$. We can use a traditional pumping lemma from formal language theory to prove that this transduction is not subsequential.

18.2 OSTIA, a first algorithm that learns transducers

The first algorithm to learn transducers, we introduce OSTIA (*Onward Subsequential Transducer Inference Algorithm*), a state-merging algorithm based on the same ideas as Algorithms RPNI (page 256) ALERGIA (page 339). The algorithm builds a special prefix-tree transducer, and from there, through both state-merging operations and advancing the translations as early as possible, a transducer is obtained (in polynomial time), which both is consistent with the learning data and is the correct transducer each time a characteristic sample is contained in the learning sample.

18.2.1 Incomplete transducers

Even if the goal is to learn total functions, we need to be able to denote the fact that in a given state, the information is still unknown. We therefore add to $\Gamma^\star$ a new symbol, $\perp$, to indicate that the information is still unknown. We denote by $\widehat{\Gamma^\star}$ the set $\Gamma^\star \cup \{\perp\}$.

The symbol $\perp$ should be interpreted as the empty set: when searching for a common prefix between $\perp$ and other strings, $\perp$ plays no part (it is neutral for the union). On the other hand, if we want to concatenate $\perp$ with another string, $\perp$ is absorbent. Summarising,

$$\forall u \in \widehat{\Gamma^\star} \perp \cdot u = u \cdot \perp = \perp$$
$$\forall A \in 2^{\widehat{\Gamma^\star}} \mathrm{lcp}(A \cup \{\perp\}) = \mathrm{lcp}(A).$$

18.2.2 The prefix-tree transducer

Like in the other state-merging techniques, the starting point for the algorithms is a tree-like finite state machine, called a ***prefix-tree transducer*** (PTT). There are two steps to building this machine. We use the new symbol, $\perp$, to indicate that the information is still unknown.

Algorithm 18.1: BUILD-PTT.

Input: a sample S, finite subset of $\Sigma^\star \times \Gamma^\star$
Output: a PTT : $\mathcal{T} = \langle Q, \Sigma, \Gamma, q_\lambda, E, \sigma\rangle$
$Q \leftarrow \{q_u : u \in \text{PREF}(\{x : (x, y) \in S\})\}$;
for $q_{u\cdot a} \in Q$ **do** $\tau(q_u, a) \leftarrow (\lambda, q_{u\cdot a})$;
for $q_u \in Q$ **do**
| **if** $\exists w \in \Gamma^\star : (u, w) \in S$ **then** $\sigma(q_u) \leftarrow w$ **else** $\sigma(q_u) \leftarrow \bot$
end
return $\mathcal{T}$

Proposition 18.2.1 *Given any finite set of input-output pairs $S \subset \Sigma^\star \times \Gamma^\star$, one can build a* prefix-tree transducer *$\langle Q, \Sigma, \Gamma, q_\lambda , E, \sigma\rangle$ where:*

- $Q = \{q_w : (ww', v) \in S\}$,
- $E = \{(q_w, a, \lambda, q_{wa}) : q_w, q_{wa} \in Q\}$,
- $\sigma(q_u) = \{v \in \widehat{\Gamma^\star} : (u, v) \in S\}$.

such that the transduction described by S is generated by it.

The trick is to only translate a string by using the function σ. This makes the proof of Proposition 18.2.1 straightforward by using Algorithm 18.1.

18.2.3 Advancing: onward transducers

The next important notion is that of *advancing*: the idea is to privilege translation as soon as possible. This gives us a normal form for transducers.

Definition 18.2.1 (Onward transducer) A transducer is **onward** *if*$_{def}$ $\forall q \in Q, \forall a \in \Sigma$, $\text{lcp}\big(\{u : (q, a, u, q') \in E\} \cup \{\sigma(q)\}\big) = \lambda$.

Remember that $\text{lcp}(W)$ designs the longest common prefix of set W. We only have to figure out how to deal with the value $\bot$. Since $\bot$ means that the information is unknown, we always have $\text{lcp}(\{\bot, u\}) = u$.

This means that the output is assigned to the transitions in such a way as to be produced as soon as we have enough information to do so. For example, the transducer from Figure 18.5(a) is not onward, since some prefixes can be advanced. The transducer from Figure 18.5(b) represents the same transduction, and is this time onward.

Unless $\text{lcp}(\{w \in \Gamma^\star : \exists x \in \Sigma^\star \text{ such that } (x, w) \in t_\mathcal{T}\}) = \lambda$, any transducer can be made onward. If this is not the case, this means that all output strings have a common prefix. This prefix can easily be removed before starting and we can therefore suppose that we are always in the situation where $\text{lcp}(\{w \in \Gamma^\star : \exists x \in \Sigma^\star \text{ such that} (x, w) \in t_\mathcal{T}\}) \neq \lambda$.

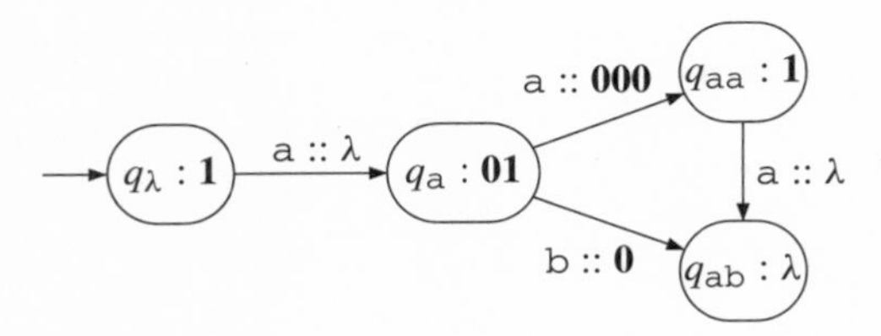

(a) Before advancing. $\text{lcp}(q_{\text{a}}) = \text{lcp}(\{\mathbf{01}, \mathbf{000}, \mathbf{0}\}) = \mathbf{0}$

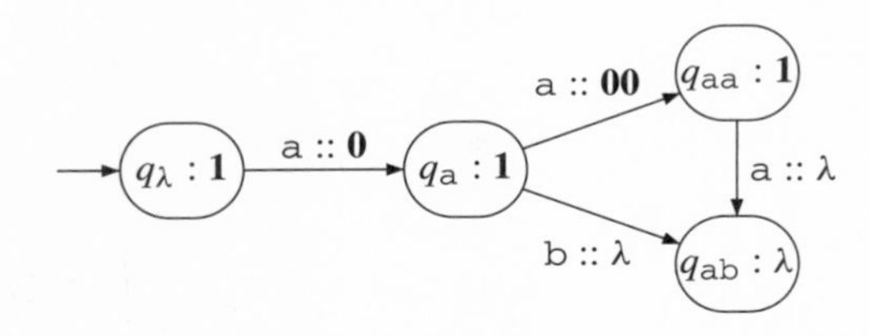

(b) After advancing. Now $\text{lcp}(q_{\text{a}}) = \lambda$.

Fig. 18.5. Making a transducer onward.

Algorithm 18.2: ONWARD-PTT.

Input: a PTT : $\mathcal{T} = \langle Q, \Sigma, \Gamma, q_\lambda, E, \sigma\rangle$, $q \in Q$, $u \in \Sigma^\star$
Output: an equivalent onward PTT : $\mathcal{T} = \langle Q, \Sigma, \Gamma, q_\lambda, E, \sigma\rangle$, a string $f = \text{lcs}\{q_u\}$
for $a \in \Sigma$ **do**
  if $\tau_2(q, a) \in Q$ **then** $(\mathcal{T}, q, w) \leftarrow$ONWARD-PTT($\mathcal{T}$, $\tau_2(q, a)$, a);
  $\tau_1(q, a) \leftarrow \tau_1(q, a) \cdot w$
end
$f \leftarrow \text{lcp}\big(\{\tau_1(q, a)\} \cup \{\sigma(q)\}\big)$;
if $f \neq \lambda$ **then**
  for $a \in \Sigma$ **do** $\tau_1(q, a) \leftarrow f^{-1}\tau_1(q, a)$;
  $\sigma(q) \leftarrow f^{-1}\sigma(q)$
end
return ($\mathcal{T}$,q,f)

18.2.4 The onward PTT

Building an onward prefix-tree transducer from a general prefix-tree transducer is easy.

Algorithm ONWARD-PTT takes three arguments: the first is a PTT $\mathcal{T}$, the second is a state q and the third a string f such that f is the longest common prefix of all outputs when starting in state q. When first called in order to make the PTT onward, f should of course be λ and q should be set to q_λ. The value returned is a pair $(\mathcal{T}, f)$.

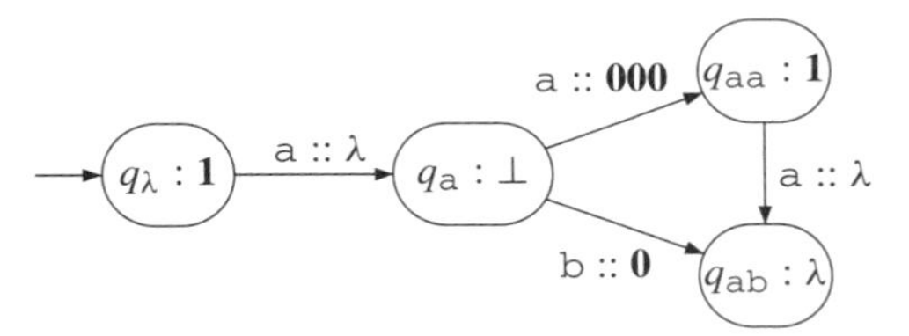

(a) Before advancing. $\mathrm{lcp}(q_{\mathrm{a}}) = \mathrm{lcp}(\{\perp, \mathbf{000}, \mathbf{0}\}) = \mathbf{0}$

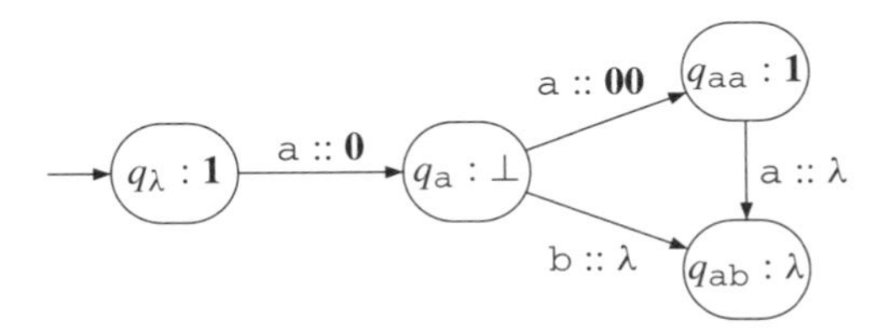

(b) After advancing. Now $\mathrm{lcp}(q_{\mathrm{a}}) = \lambda$.

Fig. 18.6. Making a transducer onward: the case with $\perp$.

$\mathcal{T}$ is the corresponding onward PTT rooted in u, and f is the prefix that has been forwarded.

18.2.5 A unique onward normal form

Theorem 18.2.2 *For any subsequential transduction there exists an onward subsequential transducer with a minimum number of states which is* unique up to isomorphisms.

Proof We only sketch the proof, which consists of studying the equivalence relation $\equiv_{\mathcal{T}}$ over $\Sigma^\star$ based on the transduction $t_{\mathcal{T}}$, where, if we write $\mathrm{lcps}(u) = \mathrm{lcp}\{x \in \Gamma^\star : \exists y \in \Sigma^\star \wedge (uy, x) \in t_{\mathcal{T}}\}$,

$$u \equiv_{\mathcal{T}} v \ \mathit{if}_{def}\ \forall z \in \Sigma^\star,\ (uz, \mathrm{lcps}(u)u') \in t_{\mathcal{T}} \wedge (vz, \mathrm{lcps}(v)v') \in t_{\mathcal{T}} \implies u' = v'.$$

In the above $\mathrm{lcps}(u)$ is a unique string in $\Gamma^\star$ associated with u, corresponding to: translations of any string starting with u all start with $\mathrm{lcps}(u)$. One can then prove that this relation has finite index. Furthermore, a unique (up to isomorphism) subsequential transducer can be built from $\equiv_{\mathcal{T}}$. □

18.3 OSTIA

The transducer learning algorithm OSTIA (18.7) makes use of Algorithm OSTIA-MERGE (18.5), which will merge the different states and, at the same time, ensure that the result is onward. The merging algorithm is a merge-and-fold variant: it first computes the longest common prefix of every two outputs it is going to have to merge, and then makes the necessary merges.

The first thing we need to be able to check is if two state outputs are identical (or one is $\perp$). Formally, we can use Algorithm OSTIA-OUTPUTS (18.3) for this.

Algorithm 18.3: OSTIA-OUTPUTS.

Input: $w, w' \in \widehat{\Gamma^\star}$
Output: a string
if $w = \perp$ **then return** w'
else if $w' = \perp$ **then return** w
else if $w = w'$ **then return** w
else return fail

18.3.1 Pushing back

Yet sometimes, we might miss some possible merges because of the onward process. The idea is then that if we can have another symmetrical process (called *pushing back*) that will somehow differ the outputs, a merge may still be feasible.

We first describe the idea using the simple example represented in Figure 18.7.

Typically $\perp$ just absorbs any pushed-back suffix. This corresponds to running Algorithm OSTIA-PUSHBACK (18.4).

Let us explain a little about Algorithm OSTIA-PUSHBACK. It takes as inputs two transitions both labelled by the same symbol a (one starting in state q_1 and the other in state q_2) and returns a transducer equivalent to the initial one in which the output has been unified. This is done by *pushing back* whatever uncommon suffixes the algorithm finds. There are cases where, due to loops, the result is that we don't have $\tau_1(q_1, a) = \tau_2(q_2, a)$.

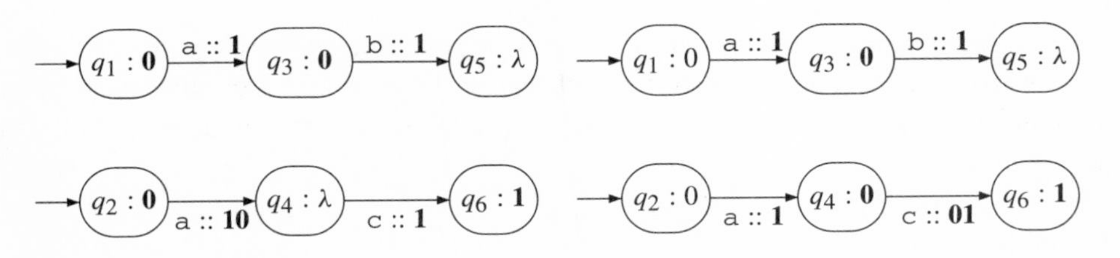

(a) Before pushing back. q_1 and q_2 can't be merged.

(b) After pushing back on $(q_2, \text{a}, \mathbf{10}, q_4)$. String $\mathbf{0}$ has been pushed back. q_1 and q_2 can now be merged.

Fig. 18.7. Pushing back.

Algorithm 18.4: OSTIA-PUSHBACK.

Input: a transducer $\mathcal{T}$, two states $q_1, q_2, a \in \Sigma$
Output: $\mathcal{T}$ updated
$u \leftarrow \text{lcs}\{\tau_1(q_1, a), \tau_1(q_2, a)\}$;
$u_1 \leftarrow u^{-1}\tau_1(q_1, a)$;
$u_2 \leftarrow u^{-1}\tau_1(q_2, a)$;
$\tau_1(q_1, a) \leftarrow u$;
$\tau_1(q_2, a) \leftarrow u$;
for $b \in \Sigma$ **do**
  $\tau_1(\tau_2(q_1, a), b) \leftarrow u_1 \cdot \tau_1(\tau_2(q_1, a), b)$;
  $\tau_1(\tau_2(q_2, a), b) \leftarrow u_2 \cdot \tau_1(\tau_2(q_1, a), b)$
end
$\sigma(\tau_2(q_1, a)) \leftarrow u_1 \cdot \sigma(\tau_2(q_1, a))$;
$\sigma(\tau_2(q_2, a)) \leftarrow u_2 \cdot \sigma(\tau_2(q_2, a))$;
return $\mathcal{T}$

18.3.2 Merging and folding in OSTIA

The idea is to adapt the merge-and-fold technique introduced in Chapter 12, which we already used in the stochastic setting (Chapter 16).

We can now write Algorithm OSTIA-MERGE:

Algorithm 18.5: OSTIA-MERGE.

Input: a transducer $\mathcal{T}$, two states $q \in$ RED, $q' \in$ BLUE
Output: $\mathcal{T}$ updated
Let q_f, a and w be such that $(q_f, a, w, q') \in E$;
$\tau(q_f, a) \leftarrow (w, q)$;
return OSTIA-FOLD($\mathcal{T}, q, q'$)

This algorithm calls OSTIA-FOLD:

Example 18.3.1 Let us run this merge-and-fold procedure on a simple exampe. Consider the transducer represented in Figure 18.8. Suppose we want to merge state q_{aa} with state q_λ. Notice that q_{aa} is the root of a tree.

We first redirect the edge $(q_{\text{a}}, \text{a}, \mathbf{1}, q_{\text{aa}})$, which becomes $(q_{\text{a}}, \text{a}, \mathbf{1}, q_\lambda)$ (Figure 18.9).

Algorithm 18.6: OSTIA-FOLD.

Input: a transducer $\mathcal{T}$, two states q and q'
Output: $\mathcal{T}$ updated, where subtree in q' is folded into q
$w \leftarrow$OSTIA-OUTPUTS($\sigma(q), \sigma(q')$);
if $w =$ **fail then**
  return fail
else
  $\sigma(q) \leftarrow w$;
  for $a \in \Sigma$ **do**
    if $\tau(q', a)$ *is defined* **then**
      if $\tau(q, a)$ *is defined* **then**
        if $\tau_1(q, a) \neq \tau_1(q', a)$ **then** `/* due to loops */`
          return fail
        else
          $\mathcal{T} \leftarrow$OSTIA-PUSHBACK($\mathcal{T}, q, q', a$);
          $\mathcal{T} \leftarrow$OSTIA-FOLD($\mathcal{T},\ \tau_2(q, a), \tau_2(q', a)$)
        end
      else
        $\tau(q, a) \leftarrow \tau(q', a)$
      end
    end
  end
  return $\mathcal{T}$
end

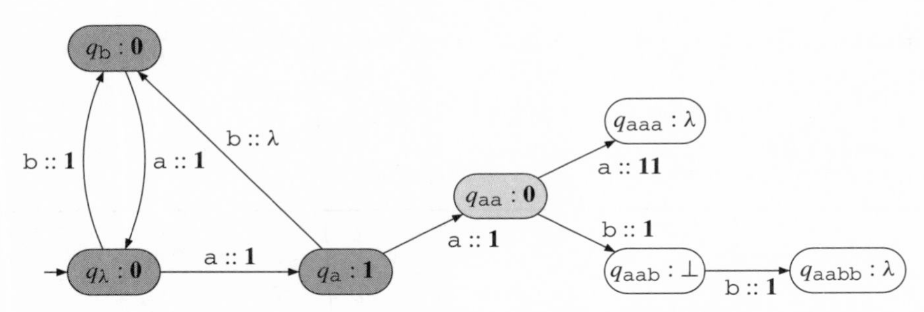

Fig. 18.8. Before merging q_{aa} with q_λ and redirecting transition $\tau_2(q_{\text{a}}, \text{a})$ to q_λ.

We now can fold q_{aa} into q_λ. This leads to pushing back the second **1** on the edge $(q_{\text{aa}}, \text{a}, \mathbf{1}, q_{\text{aaa}})$. The resulting situation is represented in Figure 18.10.

Then (Figure 18.11) q_{aaa} is folded into q_{a}, and finally q_{aab} into q_{b}. The result is represented in Figure 18.12.

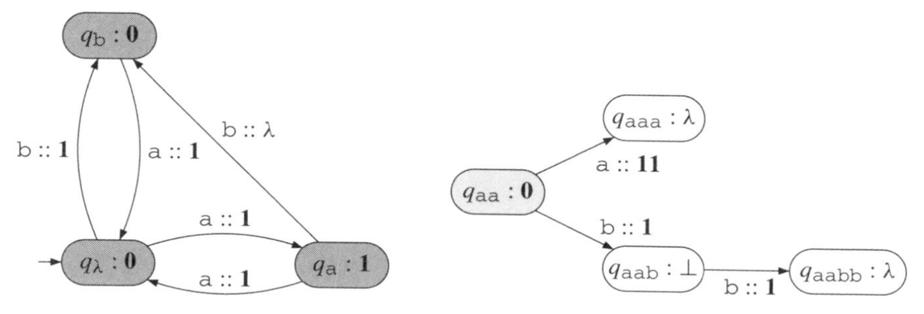

Fig. 18.9. Before folding q_{aa} into q_λ.

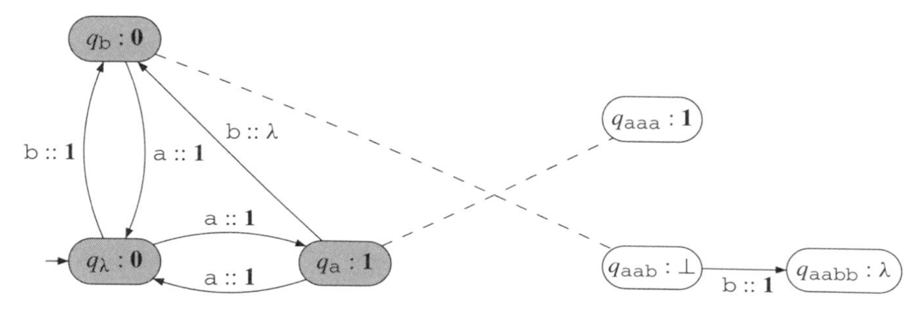

Fig. 18.10. Folding in q_{aa}.

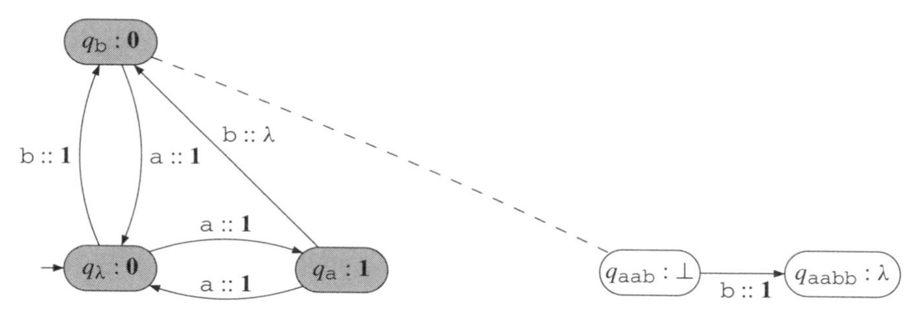

Fig. 18.11. Folding in q_{aaa}.

18.3.3 Properties of the algorithm

Algorithm OSTIA works as follows. First a PTT is built. Then it is made onward by running Algorithm ONWARD-PTT (18.2). There are two nested loops. The outer loop visits all BLUE states, whereas the inner loop visits the RED states to try to find a compatible state. The algorithm halts when there are only RED states left.

At each iteration, a BLUE state is chosen and compared with each of the RED states. If no merge is possible, the BLUE state is promoted to RED and all its successors become BLUE.

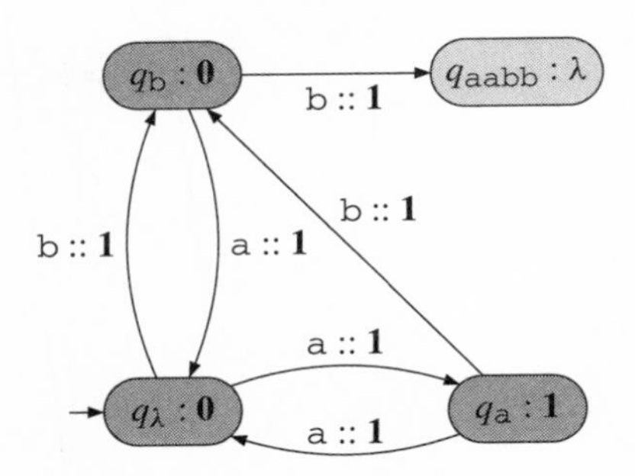

Fig. 18.12. Folding in q_{aab}.

Algorithm 18.7: OSTIA.

Input: a sample $S \in \Sigma^\star \times \Gamma^\star$
Output: $\mathcal{T}$
$\mathcal{T} \leftarrow$ ONWARD-PTT(BUILD-PTT(S));
RED $\leftarrow \{q_\lambda\}$;
BLUE $\leftarrow \{q_a : (au, v) \in S\}$;
while BLUE $\neq \emptyset$ **do**
  choose q in BLUE;
  if $\exists p \in$ RED : OSTIA-MERGE($\mathcal{T}, p, q$) $\neq$ **fail then**
    $\mathcal{T} \leftarrow$ OSTIA-MERGE($\mathcal{T}, p, q$)
  else
    RED $\leftarrow$ RED $\cup \{q\}$
  end
  BLUE $\leftarrow \{p : (q, a, v, p) \in E, q \in$ RED$\} \setminus$ RED
end
return $\mathcal{T}$

Properties 18.3.1

- *Algorithm* OSTIA *identifies in the limit any (total) subsequential transduction.*
- *The complexity is* $O(n^3(m + |\Sigma|) + nm|\Sigma|)$ *where*
  - *n is the sum of the input string lengths,*
  - *m is the length of the longest output string.*

Proof The proof closely follows the one for RPNI. We only sketch the main elements:

- Identification in the limit is ensured as soon as a characteristic sample is given, which avoids all the inconvenient merges.
- The complexity bound is estimated very broadly. In practice it is much less. □

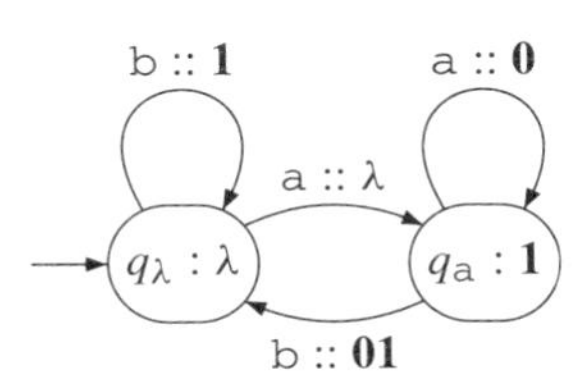

Fig. 18.13. The target.

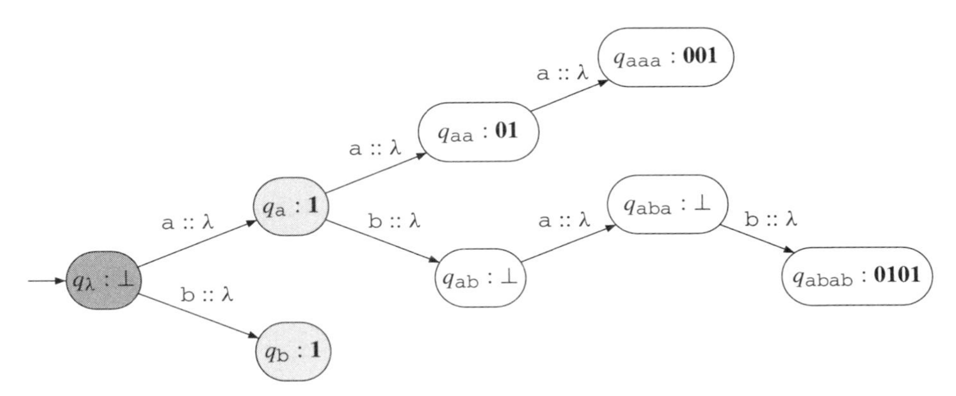

Fig. 18.14. The PTT constructed from $S = \{(\mathtt{a}, \mathbf{1}), (\mathtt{b}, \mathbf{1}), (\mathtt{aa}, \mathbf{01}), (\mathtt{aaa}, \mathbf{001}), (\mathtt{abab}, \mathbf{0101})\}$.

18.3.4 A run of algorithm OSTIA

Suppose we want to identify the transducer represented in Figure 18.13 using OSTIA. The target transducer takes as input any sequence of symbols from $\Sigma = \{\mathtt{a}, \mathtt{b}\}$, and replaces each a which is **not** followed by another a, by a **0**. If not a becomes **0** and b becomes **1**.

Typical examples (which make the learning sample), are $(\mathtt{a}, \mathbf{1})$, $(\mathtt{b}, \mathbf{1})$, $(\mathtt{aa}, \mathbf{01})$, $(\mathtt{aaa}, \mathbf{001})$, $(\mathtt{abab}, \mathbf{0101})$.

We first use Algorithm BUILD-PTT (18.1) and build the corresponding PTT represented in Figure 18.14. Algorithm ONWARD-PTT (18.2) is then called and we obtain the onward PTT (Figure 18.15).

The first merge that we test is between the state $q_{\mathtt{a}}$ (which is BLUE) and the unique RED state q_λ; the merge is rejected because the finals λ and 1 are different, so the states cannot be merged together. So $q_{\mathtt{a}}$ is promoted.

OSTIA then tries to merge $q_{\mathtt{b}}$ with q_λ; this merge is accepted, and the new transducer is depicted in Figure 18.16.

The next attempt is to merge $q_{\mathtt{aa}}$ with q_λ; this merge is rejected again because the finals λ and 1 are incompatible. Note that you cannot push back on the finals.

So in the transducer represented in Figure 18.16, OSTIA merges q_{aa} with q_a. It is accepted and the automaton represented in Figure 18.17 is built.

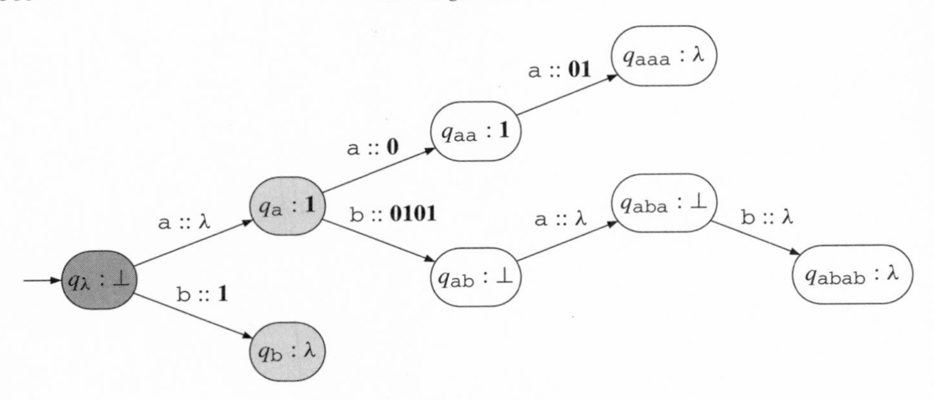

Fig. 18.15. Making the PTT onward.

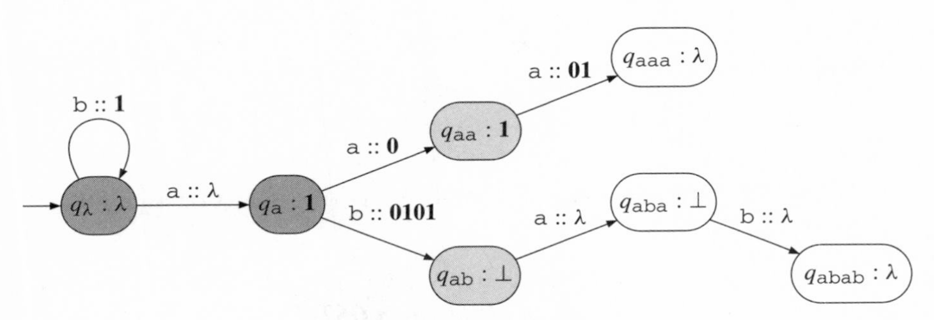

Fig. 18.16. After merging q_b and q_λ.

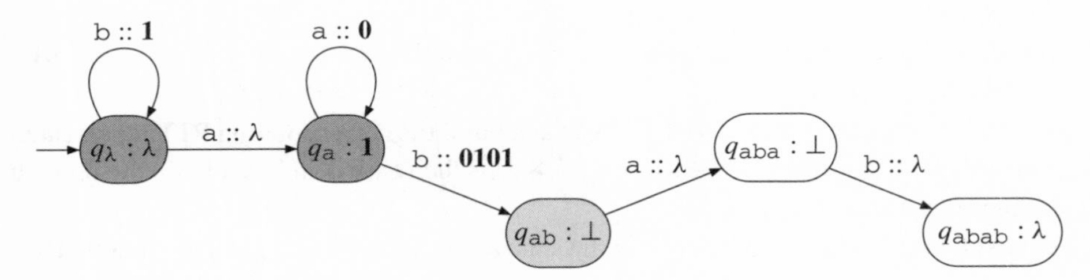

Fig. 18.17. After merging q_{aa} and q_a.

The next (and last) BLUE state is q_{ab}, which OSTIA tries to merge with q_λ. The merge is accepted. There are some pushing back operations that get done (and that we do not detail here).

The algorithm halts since there are no BLUE states left. The result is the target.

18.4 Identifying partial functions

It is easy to see that the OSTIA algorithm cannot identify partial functions, i.e. those that would not have translations for each input string. For that, just take a function that for a given regular language L translates w into **1** if $w \in L$, and has no translation for strings outside L. If we could identify these functions, the algorithm would also be capable of identifying regular languages from text, which is impossible (see Theorem 7.2.3, page 151).

Yet if one wants to use a transducer for a problem of morphology or of automatic translation, it is clear that total functions make little sense: not every random sequence of symbols should be translated. Furthermore, the absence of some input strings should be used to the advantage of the learning algorithm. As it is, this is not the case: the absence of an input string will mean the absence of an element forbidding a merge, in which case the (bad) merge will be made.

In order to hope to learn such partial functions, we therefore need some additional information. This can be of different sorts:

- using negative samples,
- using knowledge about the domain, or about the range of the function.

18.4.1 Using negative samples

We just explain the idea. The algorithm is given two samples: one contains transductions, i.e. pairs of strings in $\Sigma^\star \times \Gamma^\star$, and the second contains strings from $\Sigma^\star$ that do not admit translation. The algorithm is adapted so that, when checking if a merge is possible, we should check if one of the prohibited strings has obtained a translation. If not, the merge is rejected. Hence, more possibilities of rejecting strings exist.

The definition of *characteristic sample* can clearly be adapted to this setting.

18.4.2 Using domain knowledge

In Algorithm OSTIA (18.7) the function must be total. For straightforward reasons (see Exercise 18.2) it is impossible to identify partial functions in the limit. But if we are given the domain of the partial function then this can be used as background or expert knowledge in the learning function. Moreover, if the domain is a regular language, then the transducer has to respect the structure of the language in some way. This can be used during the learning phase: indeed, when seeking to merge two states, not only should the transductions correspond (or not be inconsistent), but also the types of the two states should coincide.

Example 18.4.1 Let us consider the subsequential transducer represented in Figure 18.18(a). Let us suppose that the learning sample contains the strings ($\mathtt{a}$, **1**), ($\mathtt{b}$, **1**), ($\mathtt{abab}$, **0101**) and the information that the domain of the function is $(\mathtt{ab}+\mathtt{b})^*(\mathtt{a}+\lambda)^*$.

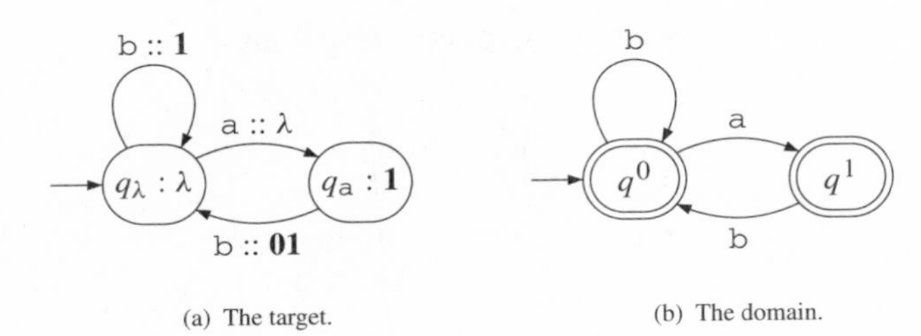

(a) The target. (b) The domain.

Fig. 18.18. Target and domain for OSTIA – D.

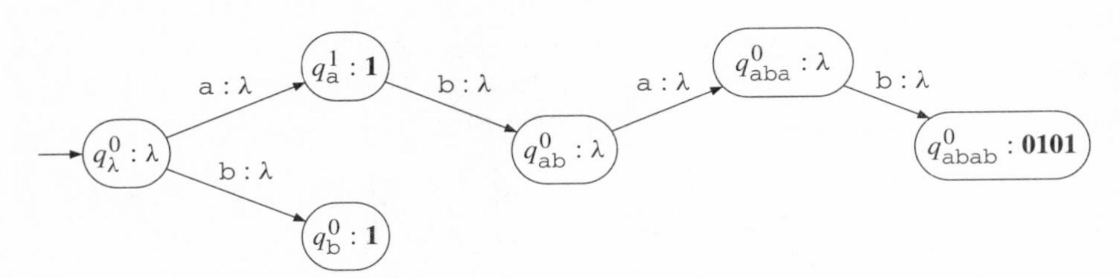

Fig. 18.19. The PTT labelled by the domain.

Then, the DFA recognising the domain of the language can be represented as in Figure 18.18(b). The initial point of the adapted algorithm consists of labelling the PTT with the indexes of the DFA corresponding to the domain. This is done in Figure 18.19. From there, a merge will only be tested between states that share the same superscript. For example the merge between q^0_λ and q^1_a won't even be tested.

18.5 Exercises

18.1 Build the PTT for {(abba, **abba**), (abaaa, **aa**), (bbaba, **bba**), (aa, **aa**)}.

18.2 Prove that learning partial subsequential transducers is hard: they cannot be identified in the limit. Hint: consider transducers that translate every string into λ.

18.3 Run OSTIA on the following sample {(aa, **a**), (aaa, **b**), (baa, **a**)}.

18.4 Suppose we know that the domain of the function contains only those strings of length at most 4. What is learnt?

18.5 Run OSTIA on the PTT from Figure 18.19, then run the adapted version using the domain information.

18.6 Conclusions of the chapter and further reading

18.6.1 Bibliographical background

Transducers were introduced by Marcel-Paul Schützenberger and studied by a number of authors since then (Reutenauer & Schützenberger, 1991, 1995), with Jean Berstel's book being the reference (Berstel, 1979). Mehryar Mohri has been advocating similarities between the different types of finite state machines, and has been defending the point of view that transducers represent the *initial object*, in the sense that a DFA (or NFA) can be seen as a machine for a transduction over $\Sigma^\star \times\{0, 1\}$, multiplicity automata as machines for transductions over $\Sigma^\star \times\mathbb{R}$ and PFAs the same over $\Sigma^\star \times(\mathbb{R} \cap [0; 1])$ (Mohri, 1997). Applications of transducers to natural language processing are still unclear, some specialists believing the mechanisms of a transducer to be too poor to express the subtleties of language. Conversely, Brian Roark and Richard Sproat (Roark & Sproat, 2007) argue that nearly all morphological rules can be described by finite state transducers. Applications to machine translation were done by Enrique Vidal, Francisco Casacuberta and their team (Amengual *et al.*, 2001, Casacuberta & Vidal, 2004).

The algorithm OSTIA that we describe in this chapter was designed by Jose Oncina, Pedro García and Enrique Vidal (Oncina, García & Vidal, 1993). The different extensions described in Section 18.4 are called OSTIA-N for the one that uses negative examples and OSTIA-D for the algorithm that makes use of domain knowledge. The domain version (Algorithm 18.4.2, page 387) was introduced by Jose Oncina and Miguel Angel Varó (Oncina & Varó, 1996). Similar ideas were explored later by Christopher Kermorvant *et al.* (Coste *et al.*, 2004, Kermorvant & de la Higuera, 2002, Kermorvant, de la Higuera & Dupont, 2004) in more general grammatical inference settings under the name of *'learning with help'*.

A version of OSTIA with queries was written by Juan Miguel Vilar (Vilar, 1996). Making OSTIA practical has been an important issue. Using dictionaries and word alignments has been tested (Vilar, 2000). This also allowed probabilities to be added to the transducers.

Theoretical results concerning stochastic transducers can be found in (Casacuberta & de la Higuera, 2000): some decoding problems are proved to be $\mathcal{NP}$-hard, but these hold specifically in the non-deterministic setting. Results follow typically from results concerning probabilistic automata.

Among the hardest attempts to learn difficult transducers, Alex Clark won the Tenjinno competition in 2006 by using OSTIA and several other ideas (Clark, 2006).

18.6.2 Some alternative lines of research

There are of course many alternative approaches than building finite state machines for translation tasks.

Learning probabilistic transducers is an important topic: these can be used to define stochastic edit distances (Bernard, Janodet & Sebban, 2006). Tree transducers are also important as they can be used with XML. Another crucial issue is that of smoothing, for which little is known. Researchers attempting to learn wrappers need to learn a function that transforms a tree into another, where the important information is made clear (Carme *et al.*, 2005).

18.6.3 *Open problems and possible new lines of research*

Machine translation tools are going to be increasingly important over the next years. But as shown during the Tenjinno competition, effort has to be made in many directions if transducer learning can be successfully used in applications.

An important area that deserves more attention is extending the definitions and algorithms to the probabilistic case.

19
A very small conclusion

> On two occasions I have been asked [by members of Parliament], 'Pray, Mr. Babbage, if you put into the machine wrong figures, will the right answers come out?' I am not able rightly to apprehend the kind of confusion of ideas that could provoke such a question.
>
> ***Charles Babbage***

> "Was it all inevitable, John?" Reeve was pushing his fingers across the floor of the cell, seated on his haunches. I was lying on the mattress. "Yes," I said. "I think it was. Certainly, it's written that way. The end of the book is there before the beginning's hardly started."
>
> ***Ian Rankin**, Knots and Crosses*

When ending this manuscript, the author decided that a certain number of things had been left implicit in the text, and could perhaps be written out clearly in some place where this would not affect the mathematical reading of the rest.

Let us discuss these points briefly here.

19.1 About convergence

Let us suppose, for the sake of argument, that the task we were developing algorithms for was the construction of random number generators. Suppose now that we had constructed such a generator, that given a seed s would return an endless series of numbers. Some of the questions we may be faced with might be:

- Is 25 more or less random than 17?
- Is 23 random?
- If in a practical situation my generator gives me the number 999999, should I decide this number is not random at all and therefore re-run the generator until a truly random number appears?

These questions are, of course, not the ones one should be considering when working with random numbers (remember they play an important part in encrypting those safe buys you make on the internet!). No one would dream of not establishing a mathematical proof of the generator.

Now let's return to grammatical inference and accept that our goal is to learn (or infer, or induce) a grammar from data about a language. Concretely, we are often given some text and have to work from it. Then, what is sometimes done is to extract some grammar from the text, say something about the fact that it is small, that we have followed some sound principle to extract this grammar, and that therefore we have learnt.

The point we make here is not that we haven't learnt. We possibly have, for that matter. The argument is that there is an alternative to looking at the data and the result: looking at the learning algorithm *per se*, showing some general property about it, the conditions needed for it to work, the biases we are using. This is all done *a priori*. Then one has to consider if what one knows about the data fits with this, and one can then hope to use the algorithm in these conditions.

One of the key ideas defended in this manuscript has been that this is precisely what grammatical inference is about: without being able to say something precise about the convergence of the algorithm, we are not learning.

19.2 About complexity

We have developed here many ideas concerning the complexity of language and grammar learning. As can be seen throughout the pages, there still is a lot of work to be done.

A partial conclusion we have reached during the writing of the manuscript, but also through discussion with many experts from very different fields, is that in this setting there is no unique definition of complexity. Several definitions exist and don't coincide.

What we still believe in is that we should know how to count, in different ways, in order to better match the problems that we can be facing.

There may be another issue with the questions regarding complexity for learning problems. Traditional complexity theory has been used for classical problems, where the data are given all at once and some function has to be optimised or some criterion has to be met. Yet there are many applications, today, where the data are obtained little by little and the algorithms are supposed to both optimise some immediate criterion and some long-term one. This is clearly the case in machine learning but also appears in a number of fields: in networks, in video streaming, in all adaptative interfaces...

The feeling we have is that the ideas studied in grammatical inference and summarised in this text can perhaps be used to participate in the building of a new complexity theory, able to take these issues into account.

19.3 About trees and graphs and more structure

We have left untouched (or nearly untouched) the question of learning from data that would be more structured than strings. There are many researchers working on learning tree grammars and tree automata. In some cases the work consists of adapting a string language inference algorithm to suit the tree case, but in many others the problems are new and novel algorithms are needed. Furthermore, in practice, in many cases the tree structures allow us to model the data in a much more accurate fashion.

The graph extension is more troublesome. If clearly graphs are used in modelling tasks, and it is therefore of interest to be able to have graph generators and recognisers, the amount of basic graph problems that are immediately intractable is fearsome: graph isomorphism, graph matching or alignment, finding the common subgraph, checking if one graph is a subgraph of another. But despite these difficulties, moving from strings to both trees and graphs is undoubtedly the path to be followed.

19.4 About applications

We have presented far too few applications in this manuscript. One reason for this is that the point of view one may have is going to become outdated most rapidly if using the applications too closely. The fact that new tools exist to manipulate automata with (today) a few million states means that problems that seemed unreachable yesterday have now become interesting.

A final word says that an algorithm that may be bad for application one may be well suited for application two. This of course is true not only for grammatical inference. One encouraging characteristic of grammatical inference is that these two applications can be set in such different fields, ranging from reverse engineering to linguistics, from automatic translation to computational biology, or from robotics to ethology.

19.5 About learning itself

A view defended by some is that learning is about compressing; a compression with loss, where the loss itself corresponds to the gain in learning.

Throughout the book we have viewed algorithms whose chief goal was to get hold of enormous amounts of data and somehow digest this into a simple set of rules which in turn allowed us to somehow replace the data by the grammar. In other words, the feeling we have reached is that learning is all about forgetting.

References

N. Abe. Characterizing PAC-learnability of semilinear sets. *Information and Computation*, 116:81–102, 1995.

N. Abe, R. Khardon, and T. Zeugmann, editors. *Proceedings of ALT 2001*, number 2225 in LNCS. Springer-Verlag, 2001.

N. Abe and H. Mamitsuka. Predicting protein secondary structure using stochastic tree grammars. *Machine Learning Journal*, 29:275–301, 1997.

N. Abe and M. Warmuth. On the computational complexity of approximating distributions by probabilistic automata. *Machine Learning Journal*, 9:205–260, 1992.

P. Adriaans. *Language Learning from a Categorical Perspective*. PhD thesis, Universiteit van Amsterdam, 1992.

P. Adriaans, H. Fernau, and M. van Zaanen, editors. *Grammatical Inference: Algorithms and Applications, Proceedings of ICGI '02*, volume 2484 of LNAI. Springer-Verlag, 2002.

P. Adriaans and C. Jacobs. Using MDL for grammar induction. In Sakakibara *et al.* (2006), pages 293–307.

P. Adriaans and M. Vervoort. The EMILE 4.1 grammar induction toolbox. In Adriaans, Fernau and van Zaanen (2002), pages 293–295.

P. Adriaans and M. van Zaanen. Computational grammar induction for linguists. *Grammars*, 7:57–68, 2004.

A .V. Aho. *Handbook of Theoretical Computer Science*, pages 290–300. Elsevier, Amsterdam, 1990.

H. Ahonen, H. Mannila, and E. Nikunen. Forming grammars for structured documents: an application of grammatical inference. In Carrasco and Oncina (1994a), pages 153–167.

B. Alpern, A. J. Demers, and F. B. Schneider. Defining liveness. *Information Processing Letters*, 21:181–185, 1985.

R. Alquézar and A. Sanfeliu. A hybrid connectionist-symbolic approach to regular grammatical inference based on neural learning and hierarchical clustering. In Carrasco and Oncina (1994a), pages 203–211.

H. Alshawi, S. Bangalore, and S. Douglas. Head transducer model for speech translation and their automatic acquisition from bilingual data. *Machine Translation*, 15(1-2):105–124, 2000a.

H. Alshawi, S. Bangalore, and S. Douglas. Learning dependency translation models as collections of finite state head transducers. *Computational Linguistics*, 26(1):45–60, 2000b.

J. C. Amengual, J. M. Benedí, F. Casacuberta, A. Castaño, A. Castellanos, V. M. Jiménez, D. Llorens, A. Marzal, M. Pastor, F. Prat, E. Vidal, and J. M. Vilar. The EuTrans-I speech translation system. *Machine Translation*, 15(1):75–103, 2001.

D. Angluin. On the complexity of minimum inference of regular sets. *Information and Control*, 39:337–350, 1978.

D. Angluin. Finding patterns common to a set of strings. In *Conference Record of the Eleventh Annual ACM Symposium on Theory of Computing*, pages 130–141. ACM Press, 1979.

D. Angluin. Inductive inference of formal languages from positive data. *Information and Control*, 45:117–135, 1980.

D. Angluin. A note on the number of queries needed to identify regular languages. *Information and Control*, 51:76–87, 1981.

D. Angluin. Inference of reversible languages. *Journal of the Association for Computing Machinery*, 29(3):741–765, 1982.

D. Angluin. Learning regular sets from queries and counterexamples. *Information and Control*, 39:337–350, 1987a.

D. Angluin. Queries and concept learning. *Machine Learning Journal*, 2:319–342, 1987b.

D. Angluin. Identifying languages from stochastic examples. Technical Report YALEU/DCS/RR-614, Yale University, March 1988.

D. Angluin. Negative results for equivalence queries. *Machine Learning Journal*, 5:121–150, 1990.

D. Angluin. Queries revisited. In Abe, Khardon and Zeugmann, pages 12–31.

D. Angluin. Queries revisited. *Theoretical Computer Science*, 313(2):175–194, 2004.

D. Angluin and M. Kharitonov. When won't membership queries help? In *Proceedings of 24th ACM Symposium on Theory of Computing*, pages 444–454. ACM Press, 1991.

D. Angluin and C. Smith. Inductive inference: theory and methods. *ACM Computing Surveys*, 15(3):237–269, 1983.

H. Arimura, H. Sakamoto, and S. Arikawa. Efficient learning of semi-structured data from queries. In Abe, Khardon and Zeugmann, pages 315–331.

J. Autebert, J. Berstel, and L. Boasson. Context-free languages and pushdown automata. In A. Salomaa and G. Rozenberg, editors, *Handbook of Formal Languages*, volume 1, Word Language Grammar, pages 111–174. Springer-Verlag, 1997.

J. K. Baker. Trainable grammars for speech recognition. In D. H. Klatt and J. J. Wolf, editors, *Speech Communication Papers for the 97th Meeting of the Acoustical Society of America*, pages 547–550, 1979.

V. Balasubramanian. Equivalence and reduction of hidden Markov models. Master's thesis, Department of Electrical Engineering and Computer Science, MIT, 1993. Issued as AI Technical Report 1370.

J. L. Balcázar, J. Diaz, R. Gavaldà, and O. Watanabe. An optimal parallel algorithm for learning DFA. In *Proceedings of the 7th COLT*, pages 208–217. ACM Press, 1994a.

J. L. Balcázar, J. Diaz, R. Gavaldà, and O. Watanabe. The query complexity of learning DFA. *New Generation Computing*, 12:337–358, 1994b.

L. E. Baum. An inequality and associated maximization technique in statistical estimation for probabilistic functions of Markov processes. *Inequalities*, 3:1–8, 1972.

L. E. Baum, T. Petrie, G. Soules, and N. Weiss. A maximization technique occurring in the statistical analysis of probabilistic functions of Markov chains. *Annals of Mathematical Statistics*, 41:164–171, 1970.

L. Becerra-Bonache. *On the Learnability of Mildly Context-sensitive Languages using Positive Data and Correction Queries*. PhD thesis, University of Tarragona, 2006.

L. Becerra-Bonache, C. Bibire, and A. Horia Dediu. Learning DFA from corrections. In H. Fernau, editor, *Proceedings of the Workshop on Theoretical Aspects of Grammar Induction (TAGI)*, WSI-2005-14, pages 1–11. Technical Report, University of Tübingen, 2005.

L. Becerra-Bonache, C. de la Higuera, J. C. Janodet, and F. Tantini. Learning balls of strings with correction queries. In *Proceedings of ECML '07*, LNAI, pages 18–29. Springer-Verlag, 2007.

L. Becerra-Bonache, C. de la Higuera, J. C. Janodet, and F. Tantini. Learning balls of strings from edit corrections. *Journal of Machine Learning Research*, 9:1841–1870, 2008.

L. Becerra-Bonache, A. Horia Dediu, and C. Tîrnauca. Learning DFA from correction and equivalence queries. In Sakakibara *et al.* (2006), pages 281–292.

L. Becerra-Bonache and T. Yokomori. Learning mild context-sensitiveness: toward understanding children's language learning. In Paliouras and Sakakibara (2004), pages 53–64.

A. Beimel, F. Bergadano, N. H. Bshouty, E. Kushilevitz, and S. Varricchio. Learning functions represented as multiplicity automata. *Journal of the ACM*, 47(3):506–530, 2000.

T. Berg, B. Jonsson, and H. Raffelt. Regular inference for state machines with parameters. In *Proceedings of FASE 2006*, volume 3922 of *LNCS*, pages 107–121. Springer-Verlag, 2006.

F. Bergadano and S. Varricchio. Learning behaviors of automata from multiplicity and equivalence queries. *SIAM Journal of Computing*, 25(6):1268–1280, 1996.

M. Bernard and C. de la Higuera. Apprentissage de programmes logiques par inférence grammaticale. *Revue d'Intelligence Artificielle*, 14(3):375–396, 2001.

M. Bernard and A. Habrard. Learning stochastic logic programs. International Conference on Inductive Logic Programming, Work in progress session, 2001.

M. Bernard, J.-C. Janodet, and M. Sebban. A discriminative model of stochastic edit distance in the form of a conditional transducer. In Sakakibara *et al.* (2006), pages 240–252.

J. Berstel. *Transductions and Context-free Languages*. Teubner, 1979.

J. Besombes and J.-Y. Marion. Learning reversible categorial grammars from structures. In *Proceedings of the International IIS: IIPWM '04 Conference*, Advances in Soft Computing, pages 181–190. Springer-Verlag, 2004a.

J. Besombes and J.-Y. Marion. Learning tree languages from positive examples and membership queries. In S. Ben-David, J. Case, and A. Maruoka, editors, *Proceedings of ALT 2004*, volume 3244 of *LNCS*, pages 440–453. Springer-Verlag, 2004b.

G. J. Bex, F. Neven, T. Schwentick, and K. Tuyls. Inference of concise DTDs from XML data. In *Proceedings of the 32nd International Conference on Very Large Data Bases*, pages 115–126, 2006.

A. Biermann. A grammatical inference program for linear languages. In *4th Hawaii International Conference on System Sciences*, pages 121–123, 1971.

A. Birkendorf, A. Boeker, and H. U. Simon. Learning deterministic finite automata from smallest counterexamples. *SIAM Journal on Discrete Mathematics*, 13(4):465–491, 2000.

V. D. Blondel and V. Canterini. Undecidable problems for probabilistic automata of fixed dimension. *Theory of Computer Systems*, 36(3):231–245, 2003.

L. E. Blum and M. Blum. Toward a mathematical theory of inductive inference. *Information and Control*, 28(2):125–155, 1975.

J. C. Bongard and H. Lipson. Active coevolutionary learning of deterministic finite automata. *Journal of Machine Learning Research*, 6:1651–1678, 2005.

J. Borges and M. Levene. Data mining of user navigation patterns. In B. Masand and M. Spiliopoulou, editors, *Web Usage Mining and User Profiling*, number 1836 in LNCS, pages 92–111. Springer-Verlag, 2000.

H. Boström. Theory-guided induction of logic programs by inference of regular languages. In *13th International Conference on Machine Learning*. Morgan Kaufmann, 1996.

H. Boström. Predicate invention and learning from positive examples only. In Nédellec and Rouveirol (1998), pages 226–237.

A. Brazma. *Computational Learning Theory and Natural Learning Systems*, volume 4, pages 351–366. MIT Press, 1997.

A. Brazma and K. Cerans. Efficient learning of regular expressions from good examples. In *AII '94: Proceedings of the 4th International Workshop on Analogical and Inductive Inference*, pages 76–90. Springer-Verlag, 1994.

A. Brazma, I. Jonassen, J. Vilo, and E. Ukkonen. Pattern discovery in biosequences. In Honavar and Slutski (1998), pages 257–270.

L. Bréhélin, O. Gascuel, and G. Caraux. Hidden Markov models with patterns to learn boolean vector sequences and application to the built-in self-test for integrated circuits. *Pattern Analysis and Machine Intelligence*, 23(9):997–1008, 2001.

A. Brocot. Calcul des rouages par approximation, nouvelle méthode. *Revue Chonométrique*, 3:186–194, 1861.

P. Brown, V. Della Pietra, P. de Souza, J. Lai, and R. Mercer. Class-based N-gram models of natural language. *Computational Linguistics*, 18(4):467–479, 1992.

J. R. Büchi. On a decision method in restricted second order arithmetic. In *Proceedings of the Congress in Logic Method and Philosophy of Science*, Stanford Univ. Press, 1960.

H. Bunke and A. Sanfeliu, editors. *Syntactic and Structural Pattern Recognition, Theory and Applications*, volume 7 of *Series in Computer Science*. World Scientific, 1990.

J. Calera-Rubio and R. C. Carrasco. Computing the relative entropy between regular tree languages. *Information Processing Letters*, 68(6):283–289, 1998.

J. Carme, R. Gilleron, A. Lemay, and J. Niehren. Interactive learning of node selecting tree transducer. In *IJCAI Workshop on Grammatical Inference*, 2005.

D. Carmel and S. Markovitch. Model-based learning of interaction strategies in multi-agent systems. *Journal of Experimental and Theoretical Artificial Intelligence*, 10(3):309–332, 1998.

D. Carmel and S. Markovitch. Exploration strategies for model-based learning in multi-agent systems. *Autonomous Agents and Multi-agent Systems*, 2(2):141–172, 1999.

R. C. Carrasco. Accurate computation of the relative entropy between stochastic regular grammars. *RAIRO (Theoretical Informatics and Applications)*, 31(5):437–444, 1997.

R. C. Carrasco, M. Forcada, and L. Santamaria. Inferring stochastic regular grammars with recurrent neural networks. In Miclet and de la Higuera (1996), pages 274–281.

R. C. Carrasco and J. Oncina, editors. *Grammatical Inference and Applications, Proceedings of ICGI '94*, number 862 in LNAI. Springer-Verlag, 1994a.

R. C. Carrasco and J. Oncina. Learning stochastic regular grammars by means of a state merging method. In Carrasco & Oncina (1994b), pages 139–150.

R. C. Carrasco and J. Oncina. Learning deterministic regular grammars from stochastic samples in polynomial time. *RAIRO (Theoretical Informatics and Applications)*, 33(1):1–20, 1999.

R. C. Carrasco, J. Oncina, and J. Calera-Rubio. Stochastic inference of regular tree languages. *Machine Learning Journal*, 44(1):185–197, 2001.

R. C. Carrasco and J. R. Rico-Juan. A similarity between probabilistic tree languages: application to XML document families. *Pattern Recognition*, 36(9):2197–2199, 2003.

F. Casacuberta. Statistical estimation of stochastic context-free grammars using the inside-outside algorithm and a transformation on grammars in grammatical inference and applications. In Carrasco and Oncina (1994a), pages 119–129.

F. Casacuberta. Probabilistic estimation of stochastic regular syntax-directed translation schemes. In R. Moreno, editor, *VI Spanish Symposium on Pattern Recognition and Image Analysis*, pages 201–297. AERFAI, 1995a.

F. Casacuberta. Statistical estimation of stochastic context-free grammars. *Pattern Recognition Letters*, 16:565–573, 1995b.

F. Casacuberta. Growth transformations for probabilistic functions of stochastic grammars. *International Journal on Pattern Recognition and Artificial Intelligence*, 10(3):183–201, 1996a.

F. Casacuberta. Maximum mutual information and conditional maximum likelihood estimation of stochastic regular syntax-directed translation schemes. In Miclet and de la Higuera (1996b), pages 282–291.

F. Casacuberta and C. de la Higuera. Optimal linguistic decoding is a difficult computational problem. *Pattern Recognition Letters*, 20(8):813–821, 1999.

F. Casacuberta and C. de la Higuera. Computational complexity of problems on probabilistic grammars and transducers. In de Oliveira (2000), pages 15–24.

F. Casacuberta and E. Vidal. Machine translation with inferred stochastic finite-state transducers. *Computational Linguistics*, 30(2):205–225, 2004.

A. Castellanos, I. Galiano, and E. Vidal. Application of OSTIA to machine translation tasks. In Carrasco and Oncina (1994a), pages 93–105.

A. Castellanos, E. Vidal, M. A. Varó, and J. Oncina. Language understanding and subsequential transducer learning. *Computer Speech and Language*, 12:193–228, 1998.

J. Castro. A note on bounded query learning. Universitat Politécnica de Catalunya, 2001.

M. J. Castro and F. Casacuberta. The morphic generator grammatical inference methodology and multilayer perceptrons: a hybrid approach to acoustic modeling. In SSPR, volume 1121 of LNCS, pages 21–29. Springer-Verlag, 1996.

J. Castro and R. Gavaldá. Towards feasible PAC-learning of probabilistic deterministic finite automata. In Clark, Coste and Miclet (2008), pages 163–174.

J. Castro and D. Guijarro. PACS, simple-PAC and query learning. *Information Processing Letters*, 73(1–2):11–16, 2000.

G. J. Chaitin. On the length of programs for computing finite binary sequences. *Journal of the ACM*, 13(4):547–569, 1966.

G. J. Chaitin. *Thinking about Gödel and Turing*. World Scientific, 2007.

E. Charniak. *Statistical Language Learning*. Cambridge: MIT Press, 1993.

E. Charniak. Tree-bank grammars. In *AAAI/IAAI*, volume 2, pages 1031–1036, 1996.

R. Chaudhuri and S. Rao. Approximating grammar probabilities: Solution to a conjecture. *Journal of the ACM*, 33(4):702–705, 1986.

E. Chávez, G. Navarro, R. Baeza-Yates, and J. L. Marroquin. Searching in metric spaces. *ACM Computing Surveys*, 33(3):273–321, 2001.

B. Chidlovskii. Wrapper generation by k-reversible grammar induction. In *Proceedings of the Workshop on Machine Learning and Information Extraction*, 2000.

B. Chidlovskii. Schema extraction from XML: a grammatical inference approach. In M. Lenzerini, D. Nardi, W. Nutt, and D. Suciu, editors, *Proceedings of KRDB 2001*, volume 45 of *CEUR Workshop Proceedings*, 2001.

B. Chidlovskii, J. Ragetli, and M. de Rijke. Wrapper generation via grammar induction. In *Proceedings of ECML 2000*, volume 1810, pages 96–108. Springer-Verlag, 2000.

J. Chodorowski and L. Miclet. Applying grammatical inference in learning a language model for oral dialogue. In Honavar and Slutski (1998), pages 102–113.

N. Chomsky. *The Logical Structure of Linguistic Theory*. PhD thesis, Massachusetts Institute of Technology, 1955.

N. Chomsky. *Syntactic Structure*. Mouton, 1957.

A. Clark. Learning deterministic context free grammars: the Omphalos competition. Technical report, 2004.

A. Clark. Large scale inference of deterministic transductions: Tenjinno problem 1. In Sakakibara *et al.* (2006), pages 227–239.

A. Clark. Learning deterministic context-free grammars: the Omphalos competition. *Machine Learning Journal*, 66(1):93–110, 2007.

A. Clark, C. Costa Florêncio, and C. Watkins. Languages as hyperplanes: grammatical inference with string kernels. In Fürnkranz, Scheffer and Spiliopoulou, pages 90–101.

A. Clark, C. Costa Florêncio, C. Watkins, and M. Serayet. Planar languages and learnability. In Sakakibara *et al.* (2006), pages 148–160.

A. Clark, F. Coste, and L. Miclet, editors. *Grammatical Inference: Algorithms and Applications, Proceedings of ICGI '08*, volume 5278 of LNCS. Springer-Verlag, 2008.

A. Clark and R. Eyraud. Polynomial identification in the limit of substitutable context-free languages. *Journal of Machine Learning Research*, 8:1725–1745, 2007.

R. Collobert and J. Weston. A unified architecture for natural language processing: deep neural networks with multitask learning. In W. W. Cohen, A. McCallum, and S. T. Roweis, editors, *Proceedings of ICML 2008*, volume 307 of *ACM International Conference Proceedings Series*, pages 160–167. ACM, 2008.

H. Comon, M. Dauchet, R. Gilleron, F. Jacquemard, D. Lugiez, S. Tison, and M. Tommasi. *Tree automata techniques and applications*, 1997.

C. Cortes, L. Kontorovich, and M. Mohri. Learning languages with rational kernels. In N. H. Bshouty and C. Gentile, editors, *Proceedings of COLT 2007*, volume 4539 of LNCS, pages 349–364. Springer-Verlag, 2007.

C. Cortes, M. Mohri, and A. Rastogi. On the computation of some standard distances between probabilistic automata. In *Proceedings of CIAA 2006*, volume 4094 of *LNCS*, pages 137–149. Springer-Verlag, 2006.

C. Costa Florêncio. Consistent identification in the limit of rigid grammars from strings is NP-hard. In Adriaans, Fernau and van Zaanen (2002), pages 49–62.

C. Costa Florêncio. *Learning Categorial Grammars*. PhD thesis, University of Utrecht, 2003.

F. Coste and D. Fredouille. Unambiguous automata inference by means of state-merging methods. In N. Lavrac, D. Gramberger, H. Blockeel, and L. Todorovski, editors, *Proceedings of ECML '03*, number 2837 in LNAI, pages 60–71. Springer-Verlag, 2003.

F. Coste, D. Fredouille, C. Kermorvant, and C. de la Higuera. Introducing domain and typing bias in automata inference. In Paliouras and Sakakibara (2004), pages 115–126.

F. Coste and J. Nicolas. How considering incompatible state mergings may reduce the DFA induction search tree. In Honavar and Slutski (1998a), pages 199–210.

F. Coste and J. Nicolas. Inference of finite automata: reducing the search space with an ordering of pairs of states. In Nédellec and Rouveirol (1998b), pages 37–42.

B. Courcelle. Recursive queries and context-free graph grammars. *Theoretical Computer Science*, 78(1):217–244, 1991.

T. Cover and J. Thomas. *Elements of Information Theory*. John Wiley and Sons, 1991.

V. Crescenzi and G. Mecca. Automatic information extraction from large websites. *Journal of the ACM*, 51(5):731–779, 2004.

V. Crescenzi and P. Merialdo. Wrapper inference for ambiguous web pages. *Applied Artificial Intelligence*, 22(1–2):21–52, 2008.

M. Crochemore, C. Hancart, and T. Lecroq. *Algorithmique du texte*. Vuibert, 2001.

M. Crochemore, C. Hancart, and T. Lecroq. *Algorithms on Strings*. Cambridge University Press, 2007.

P. Cruz and E. Vidal. Learning regular grammars to model musical style: comparing different coding schemes. In Honavar and Slutski (1998), pages 211–222.

P. Cruz-Alcázar and E. Vidal. Two grammatical inference applications in music processing. *Applied Artificial Intelligence*, 22(1–2):53–76, 2008.

T. Dean, K. Basye, L. Kaelbling, E. Kokkevis, O. Maron, D. Angluin, and S. Engelson. Inferring finite automata with stochastic output functions and an application to map learning. In W. Swartout, editor, *Proceedings of the 10th National Conference on Artificial Intelligence*, pages 208–214. MIT Press, 1992.

C. de la Higuera. Characteristic sets for polynomial grammatical inference. *Machine Learning Journal*, 27:125–138, 1997.

C. de la Higuera. Learning stochastic finite automata from experts. In Honavar and Slutski (1998), pages 79–89.

C. de la Higuera. A bibliographical study of grammatical inference. *Pattern Recognition*, 38:1332–1348, 2005.

C. de la Higuera. Data complexity issues in grammatical inference. In M. Basu and T. Kam Ho, editors, *Data Complexity in Pattern Recognition*, pages 153–172. Springer-Verlag, 2006a.

C. de la Higuera. Ten open problems in grammatical inference. In Sakakibara *et al.* (2006b), pages 32–44.

C. de la Higuera, P. Adriaans, M. van Zaanen, and J. Oncina, editors. *Proceedings of the Workshop and Tutorial on Learning Context-free grammars, at ECML '03*. 2003.

C. de la Higuera and M. Bernard. Apprentissage de programmes logiques par inférence grammaticale. *Revue d'Intelligence Artificielle*, 14(3):375–396, 2001.

C. de la Higuera and F. Casacuberta. Topology of strings: median string is NP-complete. *Theoretical Computer Science*, 230:39–48, 2000.

C. de la Higuera and J-C. Janodet. Inference of ω-languages from prefixes. *Theoretical Computer Science*, 313(2):295–312, 2004.

C. de la Higuera, J.-C. Janodet, and F. Tantini. Learning languages from bounded resources: the case of the DFA and the balls of strings. In Clark, Coste and Miclet (2008), pages 43–56.

C. de la Higuera and L. Micó. A contextual normalised edit distance. In E. Chávez and G. Navarro, editors, *Proceedings of the First International Workshop on Similarity Search and Applications*, pages 61–68. IEEE Computer Society, 2008.

C. de la Higuera and J. Oncina. Learning deterministic linear languages. In Kivinen and Sloan (2002), pages 185–200.

C. de la Higuera and J. Oncina. Identification with probability one of stochastic deterministic linear languages. In Gavaldà *et al.* (2003), pages 134–148.

C. de la Higuera and J. Oncina. Learning probabilistic finite automata. In Paliouras and Sakakibara (2004), pages 175–186.

C. de la Higuera, J. Oncina, and E. Vidal. Identification of DFA: data-dependent versus data-independent algorithm. In Miclet and de la Higuera (1996), pages 313–325.

C. de la Higuera and F. Thollard. Identication in the limit with probability one of stochastic deterministic finite automata. In de Oliveira (2000), pages 15–24.

F. Denis. Learning regular languages from simple positive examples. *Machine Learning Journal*, 44(1):37–66, 2001.

F. Denis, C. d'Halluin, and R. Gilleron. PAC learning with simple examples. In *13th Symposium on Theoretical Aspects of Computer Science '96*, LNCS, pages 231–242, 1996.

F. Denis, Y. Esposito, and A. Habrard. Learning rational stochastic languages. In *Proceedings of COLT 2006*, volume 4005 of *LNCS*, pages 274–288. Springer-Verlag, 2006.

F. Denis and R. Gilleron. PAC learning under helpful distributions. In Li and Maruoka (1997), pages 132–145.

F. Denis, A. Lemay, and A. Terlutte. Learning regular languages using non deterministic finite automata. In de Oliveira (2000), pages 39–50.

F. Denis, A. Lemay, and A. Terlutte. Learning regular languages using RFSA. In Abe, Khardon and Zeugmann (2001), pages 348–363.

A. L. de Oliveira, editor. *Grammatical Inference: Algorithms and Applications, Proceedings of ICGI '00*, volume 1891 of LNAI. Springer-Verlag, 2000.

A. L. de Oliveira and J. P. Marques Silva. Efficient search techniques for the inference of minimum size finite automata. In *Proceedings of the 1998 South American Symposium on String Processing and Information Retrieval*, pages 81–89. IEEE Computer Society Press, 1998.

A. L. de Oliveira and J. P. M. Silva. Efficient algorithms for the inference of minimum size DFAs. *Machine Learning Journal*, 44(1):93–119, 2001.

A. Dubey, P. Jalote, and S. Kumar Aggarwal. Inferring grammar rules of programming language dialects. In Sakakibara *et al.* (2006), pages 201–213.

P. Dupont. Regular grammatical inference from positive and negative samples by genetic search: the GIG method. In Carrasco and Oncina (1994a), pages 236–245.

P. Dupont. Incremental regular inference. In Miclet and de la Higuera (1996), pages 222–237.

P. Dupont and J.-C. Amengual. Smoothing probabilistic automata: an error-correcting approach. In de Oliveira (2000), pages 51–62.

P. Dupont, F. Denis, and Y. Esposito. Links between probabilistic automata and hidden markov models: probability distributions, learning models and induction algorithms. *Pattern Recognition*, 38(9):1349–1371, 2005.

P. Dupont, B. Lambeau, C. Damas, and A. van Lamsweerde. The QSM algorithm and its application to software behavior model induction. *Applied Artificial Intelligence*, 22(1–2):77–115, 2008.

P. Dupont, L. Miclet, and E. Vidal. What is the search space of the regular inference? In Carrasco and Oncina (1994a), pages 25–37.

T. Erlebach, P. Rossmanith, H. Stadtherr, A. Steger, and T. Zeugmann. Learning one-variable pattern languages very efficiently on average, in parallel, and by asking queries. In Li and Maruoka (1997), pages 260–276.

Y. Esposito, A. Lemay, F. Denis, and P. Dupont. Learning probabilistic residual finite state automata. In Adriaans, Fernau and van Zaanen (2002), pages 77–91.

R. Eyraud. *Context-free Grammar Learning*. PhD thesis, Université de Saint-Etienne, 2006.

R. Eyraud, C. de la Higuera, and J.-C. Janodet. LARS: a learning algorithm for rewriting systems. *Machine Learning Journal*, 66(1):7–31, 2006.

J. Feldman. Some decidability results on grammatical inference and complexity. *Information and Control*, 20:244–262, 1972.

H. Fernau. Identification of function distinguishable languages. In H. Arimura, S. Jain, and A. Sharma, editors, *Proceedings of ALT 2000*, volume 1968 of LNCS, pages 116–130. Springer-Verlag, 2000.

H. Fernau. Learning XML grammars. In P. Perner, editor, *Proceedings of MLDM '01*, number 2123 in LNCS, pages 73–87. Springer-Verlag, 2001.

H. Fernau. Learning tree languages from text. In Kivinen and Sloan (2002), pages 153–168.

H. Fernau. Identification of function distinguishable languages. *Theoretical Computer Science*, 290(3):1679–1711, 2003.

H. Fernau. Algorithms for learning regular expressions. In Jain, Simon and Tomita (2005), pages 297–311.

H. Fernau and C. de la Higuera. Grammar induction: an invitation to formal language theorists. *Grammars*, 7:45–55, 2004.

J. A. Ferrer, F. Casacuberta, and A. Juan-Císcar. On the statistical estimation of stochastic finite-state transducers in machine translation. *Applied Artificial Intelligence*, 22(1–2):4–20, 2008.

A. Forêt and Y. Le Nir. On limit points for some variants of rigid Lambek grammars. In Adriaans, Fernau and van Zaanen (2002), pages 106–119.

A. Fred. Computation of substring probabilities in stochastic grammars. In de Oliveira (2000), pages 103–114.

K. S. Fu. *Syntactic Methods in Pattern Recognition*. Academic Press, 1974.

K. S. Fu. *Syntactic pattern recognition and applications*. Prentice Hall, 1982.

K. S. Fu and T. L. Booth. Grammatical inference: introduction and survey. Part I and II. *IEEE Transactions on Systems, Man and Cybernetics*, 5:59–72 and 409–423, 1975.

J. Fürnkranz, T. Scheffer, and M. Spiliopoulou, editors. *Proceedings of ECML '06*, volume 4212 of *LNCS*. Springer-Verlag, 2006.

P. García and J. Oncina. Inference of recognizable tree sets. Technical Report DSIC-II/47/93, Departamento de Lenguajes y Sistemas Informáticos, Universidad Politécnica de Valencia, Spain, 1993.

P. García, E. Segarra, E. Vidal, and I. Galiano. On the use of the morphic generator grammatical inference (MGGI) methodology in automatic speech recognition. *International Journal of Pattern Recognition and Artificial Intelligence*, 4:667–685, 1994.

P. García and E. Vidal. Inference of K-testable languages in the strict sense and applications to syntactic pattern recognition. *Pattern Analysis and Machine Intelligence*, 12(9):920–925, 1990.

P. García, E. Vidal, and J. Oncina. Learning locally testable languages in the strict sense. In *Workshop on Algorithmic Learning Theory (ALT 90)*, pages 325–338, 1990.

R. Gavaldà. On the power of equivalence queries. In *Proceedings of the 1st European Conference on Computational Learning Theory*, volume 53 of *The Institute of Mathematics and its Applications Conference Series*, pages 193–203. Oxford University Press, 1993.

R. Gavaldà, K. Jantke, and E. Takimoto, editors. *Proceedings of ALT 2003*, number 2842 in LNCS. Springer-Verlag, 2003.

R. Gavaldà, P. W. Keller, J. Pineau, and D. Precup. PAC-learning of markov models with hidden state. In Fürnkranz, Scheffer and Spiliopoulou (2006), pages 150–161.

C. L. Giles, S. Lawrence, and A.C. Tsoi. Noisy time series prediction using recurrent neural networks and grammatical inference. *Machine Learning Journal*, 44(1):161–183, 2001.

J. Y. Giordano. Inference of context-free grammars by enumeration: structural containment as an ordering bias. In Carrasco and Oncina (1994a), pages 212–221.

J. Y. Giordano. Grammatical inference using tabu search. In Miclet and de la Higuera (1996), pages 292–300.

F. Glover and M. Laguna. *Tabu search*. Springer-Verlag, 1997.

T. Goan, N. Benson, and O. Etzioni. A grammar inference algorithm for the world wide web. In *Proceedings of AAAI Spring Symposium on Machine Learning in Information Access*. AAAI Press, 1996.

E. M. Gold. Language identification in the limit. *Information and Control*, 10(5):447–474, 1967.

E. M. Gold. Complexity of automaton identification from given data. *Information and Control*, 37:302–320, 1978.

S. A. Goldman and M. Kearns. On the complexity of teaching. *Journal of Computer and System Sciences*, 50(1):20–31, 1995.

S. A. Goldman and H. Mathias. Teaching a smarter learner. *Journal of Computer and System Sciences*, 52(2):255–267, 1996.

R. Gonzalez and M. Thomason. *Syntactic Pattern Recognition: an Introduction*. Addison-Wesley, 1978.

J. Goodman. A bit of progress in language modeling. Technical report, Microsoft Research, 2001.

R. L. Graham, D. E. Knuth, and O. Patashnik. *Concrete Mathematics: A Foundation for Computer Science*. Addison-Wesley, 1994.

D. Gusfield. *Algorithms on Strings, Trees, and Sequences*. Cambridge University Press, 1997.

O. Guttman. *Probabilistic Automata and Distributions over Sequences*. PhD thesis, The Australian National University, 2006.

O. Guttman, S. V. N. Vishwanathan, and R. C. Williamson. Learnability of probabilistic automata via oracles. In Jain, Simon and Tomita (2005), pages 171–182.

A. Habrard, M. Bernard, and F. Jacquenet. Generalized stochastic tree automata for multi-relational data mining. In Adriaans, Fernau and van Zaanen (2002), pages 120–133.

A. Habrard, F. Denis, and Y. Esposito. Using pseudo-stochastic rational languages in probabilistic grammatical inference. In Sakakibara *et al.* (2006), pages 112–124.

A. Hagerer, H. Hungar, O. Niese, and B. Steffen. Model generation by moderated regular extrapolation. In R. Kutsche and H. Weber, editors, *Proceedings of the 5th International Conference on Fundamental Approaches to Software Engineering (FASE '02)*, volume 2306 of LNCS, pages 80–95 Springer-Verlag, 2002.

T. Hanneforth. A memory-efficient epsilon-removal algorithm for weighted acyclic finite-state automata. In *Proceedings of FSMNLP 2008*, 2008.

T. Hanneforth and C. de la Higuera. Epsilon-removal by loop reduction for finite state automata over complete semirings. In *Pre-proceedings of FSMNLP 2009*, 2009.

M. H. Harrison. *Introduction to Formal Language Theory*. Addison-Wesley, 1978.

L. Hellerstein, K. Pillaipakkamnatt, V. Raghavan, and D. Wilkins. How many queries are needed to learn? *Journal of the ACM*, 43(5):840–862, 1996.

V. Honavar and G. Slutski, editors. *Grammatical Inference, Proceedings of ICGI '98*, number 1433 in LNAI. Springer-Verlag, 1998.

T. W. Hong and K. L. Clark. Using grammatical inference to automate information extraction from the Web. In *Principles of Data Mining and Knowledge Discovery*, volume 2168 of LNCS, pages 216–227. Springer-Verlag, 2001.

J. E. Hopcroft and J. D. Ullman. *Formal languages and their relation to automata*. Addison-Wesley, 1977.

J. E. Hopcroft and J. D. Ullman. *Introduction to Automata Theory, Languages, and Computation*. Addison-Wesley, 1979.

J. J. Horning. *A study of Grammatical Inference*. PhD thesis, Stanford University, 1969.

IAPR. *Proceedings of ICPR 2006*. IEEE Computer Society, 2006.

O. H. Ibarra, T. Jiang, and B. Ravikumar. Some subclasses of context-free languages in NC1. *Information Processing Letters*, 29(3):111–117, 1988.

Y. Ishigami and S. Tani. VC-dimensions of finite automata and commutative finite automata with k letters and n states. *Discrete Applied Mathematics*, 74:123–134, 1997.

H. Ishizaka. Polynomial time learnability of simple deterministic languages. *Machine Learning Journal*, 5:151–164, 1995.

A. Jagota, R. B. Lyngsø, and C. N. S. Pedersen. Comparing a hidden Markov model and a stochastic context-free grammar. In *Proceedings of WABI '01*, number 2149 in LNCS, pages 69–74. Springer-Verlag, 2001.

S. Jain, D. Osherson, J. S. Royer, and A. Sharma. *Systems That Learn*. MIT Press, 1999.

S. Jain, H.-U. Simon, and E. Tomita, editors. *Proceedings of ALT 2005*, volume 3734 of *LNCS*. Springer-Verlag, 2005.

C. A. James. Opensmiles specification. www.opensmiles.org/spec/open-smiles.html, 2007.

F. Jelinek. *Statistical Methods for Speech Recognition*. MIT Press, 1998.

T. Kammeyer and R. K. Belew. Stochastic context-free grammar induction with a genetic algorithm using local search. In R. K. Belew and M. Vose, editors, *Foundations of Genetic Algorithms IV*, Morgan Kaufmann, 1996.

M. Kanazawa. *Learnable Classes of Categorial Grammars*. CSLI Publications, 1998.

S. Kapur. *Computational Learning of Languages*. PhD thesis, Department of Computer Science, Cornell University, 1991.

M. J. Kearns, Y. Mansour, D. Ron, R. Rubinfeld, R. E. Schapire, and L. Sellie. On the learnability of discrete distributions. In *Proceedings of the 25th Annual ACM Symposium on Theory of Computing*, pages 273–282, 1994.

M. Kearns and L. Valiant. Cryptographic limitations on learning boolean formulae and finite automata. In *21st ACM Symposium on Theory of Computing*, pages 433–444, 1989.

M. J. Kearns and U. Vazirani. *An Introduction to Computational Learning Theory*. MIT Press, 1994.

C. Kermorvant and C. de la Higuera. Learning languages with help. In Adriaans, Fernau and van Zaanen, pages 161–173.

C. Kermorvant, C. de la Higuera, and P. Dupont. Improving probabilistic automata learning with additional knowledge. In A. Fred, T. Caelli, R. Duin, A. Campilho, and D. de Ridder, editors, *Structural, Syntactic and Statistical Pattern Recognition, Proceedings of SSPR and SPR 2004*, volume 3138 of *LNCS*, pages 260–268. Springer-Verlag, 2004.

E. B. Kinber. On learning regular expressions and patterns via membership and correction queries. In Clark, Coste and Miclet (2008), pages 125–138.

J. Kivinen and R. H. Sloan, editors. *Proceedings of COLT 2002*, number 2375 in LNAI. Springer-Verlag, 2002.

R. Kneser and H. Ney. Improved clustering techniques for class-based language modelling. In *European Conference on Speech Communication and Technology*, pages 973–976, 1993.

T. Knuutila and M. Steinby. Inference of tree languages from a finite sample: an algebraic approach. *Theoretical Computer Science*, 129:337–367, 1994.

S. Kobayashi. In Z. Esik, C. Martin-Vide and V. Mitrana, editors, *Recent Advances in Formal Languages and Applications*, pages 209–228. Springer-Verlag, 2003.

A. N. Kolmogorov. Three approaches to the quantitative definition of information. *Problems of Information Transmission*, 1(1):1–7, 1967.

G. Korfiatis and G. Paliouras. Modeling web navigation using grammatical inference. *Applied Artificial Intelligence*, 22(1–2):116–138, 2008.

R. Kosala, M. Bruynooghe, J. van den Bussche, and H. Blockeel. Information extraction from web documents based on local unranked tree automaton inference. In *Proceedings of IJCAI-03*, pages 403–408. Morgan Kaufmann, 2003.

T. Koshiba. Typed pattern languages and their learnability. In *Proceedings of Euro COLT '95*, number 904 in LNAI, pages 367–379. Springer-Verlag, 1995.

T. Koshiba, E. Mäkinen, and Y. Takada. Learning deterministic even linear languages from positive examples. *Theoretical Computer Science*, 185(1):63–79, 1997.

T. Koshiba, E. Mäkinen, and Y. Takada. Inferring pure context-free languages from positive data. *Acta Cybernetica*, 14(3):469–477, 2000.

S. C. Kremer. Parallel stochastic grammar induction. In *Proceedings of the 1997 International Conference on Neural Networks (ICNN '97)*, volume I, pages 612–616, 1997.

B. Lambeau, C. Damas, and P. Dupont. State merging DFA induction with mandatory merge constraints. In Clark, Coste and Miclet (2008), pages 139–153.

K. Lang. Random DFA's can be approximately learned from sparse uniform examples. In *Proceedings of COLT 1992*, pages 45–52, 1992.

K. Lang. Faster algorithms for finding minimal consistent DFAs. Technical report, NEC Research Institute, 1999.

K. Lang and B. A. Pearlmutter. *The Abbadingo One DFA Learning Competition*, 1997.

K. Lang, B. A. Pearlmutter, and F. Coste. *The Gowachin Automata Learning Competition*, 1998.

K. Lang, B. A. Pearlmutter, and R. A. Price. Results of the Abbadingo one DFA learning competition and a new evidence-driven state merging algorithm. In Honavar and Slutski (1998), pages 1–12.

S. Lange and S. Zilles. On the learnability of erasing pattern languages in the query model. In Gavaldà, Jantke and Takimoto (2003), pages 129–143.

P. Langley. Simplicity and representation change in grammar induction. Technical report, Stanford University, 1995.

P. Langley and S. Stromsten. Learning context-free grammars with a simplicity bias. In *Proceedings of ECML 2000, 11th European Conference on Machine Learning*, volume 1810 of LNCS, pages 220–228. Springer-Verlag, 2000.

K. Lari and S. J. Young. The estimation of stochastic context free grammars using the inside-outside algorithm. *Computer Speech and Language*, 4:35–56, 1990.

K. Lari and S. J. Young. Applications of stochastic context-free grammars using the inside-outside algorithm. *Computer Speech and Language*, 5:237–257, 1991.

F. Lerdahl and R. Jackendoff. An overview of hierarchical structure in music. *Music Perception*, 1(2):229–252, 1983.

V. I. Levenshtein. Binary codes capable of correcting deletions, insertions, and reversals. *Doklady Akademii Nauk SSSR*, 163(4):845–848, 1965.

M. Li and A. Maruoka, editors. *Proceedings of ALT '97*, volume 1316 of *LNCS*. Springer-Verlag, 1997.

M. Li and P. Vitanyi. Learning simple concepts under simple distributions. *Siam Journal of Computing*, 20:911–935, 1991.

M. Li and P. Vitanyi. *An Introduction to Kolmogorov Complexity and its Applications*. Springer-Verlag, 1993.

N. Littlestone. Learning quickly when irrelevant attributes abound: a new linear threshold. *Machine Learning Journal*, 2:285–318, 1987.

D. López and J. M. Sempere. Handwritten digit recognition through inferring graph grammars. A first approach. In *Proceedings of SSPR '98 and SPR '98*, volume 1451 of LNCS, pages 483–491. Springer-Verlag, 1998.

S. Lucas, E. Vidal, A. Amari, S. Hanlon, and J. C. Amengual. A comparison of syntactic and statistical techniques for off-line OCR. In Carrasco and Oncina (1994a), pages 168–179.

D. Luzeaux. String distances. In *Distancia 92*, 1992.

S. Lucas. Learning DFA from noisy samples. http://cswww.essex.ac.uk/staff/sml/gecco/NoisyDFA.html, 2004.

R. B. Lyngsø and C. N. S. Pedersen. Complexity of comparing hidden Markov models. In *Proceedings of ISAAC '01*, number 2223 in LNCS, pages 416–428. Springer-Verlag, 2001.

R. B. Lyngsø, C. N. S. Pedersen, and H. Nielsen. Metrics and similarity measures for hidden Markov models. In *Proceedings of ISMB '99*, pages 178–186, 1999.

E. Mäkinen. A note on the grammatical inference problem for even linear languages. *Fundamenta Informaticae*, 25(2):175–182, 1996.

O. Maler and A. Pnueli. On the learnability of infinitary regular sets. In *Proceedings of COLT*, pages 128–136. Morgan–Kauffman, 1991.

F. J. Maryanski. *Inference of Probabilistic Grammars*. PhD thesis, University of Connecticut, 1974.

A. Marzal and E. Vidal. Computation of normalized edit distance and applications. *Pattern Analysis and Machine Intelligence*, 15(9):926–932, 1993.

D. McAllester and R. Schapire. *Exploring Artificial Intelligence in the New Millenium*. Morgan Kaufmann, 2002.

L. Miclet. *Structural Methods in Pattern Recognition*. Chapman and Hall, New York, 1986.

L. Miclet. *Syntactic and Structural Pattern Recognition, Theory and Applications*, pages 237–290. World Scientific, 1990.

L. Miclet and C. de la Higuera, editors. *Proceedings of ICGI '96*, number 1147 in LNAI. Springer-Verlag, 1996.

L. Micó, J. Oncina, and E. Vidal. A new version of the nearest-neighbour approximating and eliminating search algorithm (AESA) with linear preprocessing time and memory requirements. *Pattern Recognition Letters*, 15:9–17, 1994.

G. A. Miller and N. Chomsky. *Handbook of Mathematical Psychology*, volume 2, pages 419–491. Wiley, 1963.

A. Mitchell, T. Scheffer, A. Sharma, and F. Stephan. The VC-dimension of subclasses of pattern languages. In O. Watanabe and T. Yokomori, editors, *Proceedings of ALT '99*, volume 1720 of LNCS, pages 93–105. Springer-Verlag, 1999.

M. Mohri. Finite-state transducers in language and speech processing. *Computational Linguistics*, 23(3):269–311, 1997.

M. Mohri, F. C. N. Pereira, and M. Riley. The design principles of a weighted finite-state transducer library. *Theoretical Computer Science*, 231(1):17–32, 2000.

M. Mosbah. Probabilistic graph grammars. *Fundamenta Informaticae*, 26(3/4):341–362, 1996.

T. Motoki, T. Shinohara, and K. Wright. The correct definition of finite elasticity: corrigendum to identification of unions. In *Proceedings of the fourth annual workshop on Computational Learning Theory*, page 375, 1991.

T. Murgue. Log pre-processing and grammatical inference for web usage mining. In *UM 2005 – Workshop on Machine Learning for User Modeling: Challenges*, 2005.

T. Murgue and C. de la Higuera. Distances between distributions: comparing language models. In A. Fred, T. Caelli, R. Duin, A. Campilho, and D. de Ridder, editors, *Structural, Syntactic and Statistical Pattern Recognition, Proceedings of SSPR and SPR 2004*, volume 3138 of LNCS, pages 269–277. Springer-Verlag, 2004.

K. Nakamura and M. Matsumoto. Incremental learning of context free grammars based on bottom-up parsing and search. *Pattern Recognition*, 38(9):1384–1392, 2005.

B. L. Natarajan. *Machine Learning: a Theoretical Approach*. Morgan Kauffman, 1991.

M. Nasu and N. Honda. Mappings induced by pgsm-mappings and some recursively unsolvable problems of finite probabilistic automata. *Information and Control*, 15:250–273, 1969.

G. Navarro. Searching in metric spaces by spatial approximation. *VLDB Journal*, 11(1):28–46, 2002.

C. Nédellec and C. Rouveirol, editors. *Proceedings of ECML '98*, number 1398 in LNAI. Springer-Verlag, 1998.

C. Nevill-Manning and I. Witten. Identifying hierarchical structure in sequences: a linear-time algorithm. *Journal of Artificial Intelligence Research*, 7:67–82, 1997.

H. Ney. Stochastic grammars and pattern recognition. In P. Laface and R. De Mori, editors, *Proceedings of the NATO Advanced Study Institute*, pages 313–344. Springer-Verlag, 1992.

H. Ney, S. Martin, and F. Wessel. *Corpus-Based Statistical Methods in Speech and Language Processing*, pages 174–207. Kluwer Academic Publishers, 1997.

A. Nowak, N. L. Komarova, and P. Niyogi. Computational and evolutionary aspects of language. *Nature*, 417:611–617, 2002.

T. Oates, T. Armstrong, L. Becerra-Bonache, and M. Atamas. Inferring grammars for mildly context sensitive languages in polynomial-time. In Sakakibara *et al.* (2006), pages 137–147.

T. Oates, D. Desai, and V. Bhat. Learning k-reversible context-free grammars from positive structural examples. In C. Sammut and A. G. Hoffmann, editors, *Proceedings of ICML 2002*, pages 459–465. Morgan Kaufmann, 2002.

T. Oates, S. Doshi, and F. Huang. Estimating maximum likelihood parameters for stochastic context-free graph grammars. In *Proceedings of ILP 2003*, volume 2835 of LNCS, pages 281–298. Springer-Verlag, 2003.

J. Oncina. The data driven approach applied to the OSTIA algorithm. In Honavar and Slutski (1998), pages 50–56.

J. Oncina and P. García. Identifying regular languages in polynomial time. In H. Bunke, editor, *Advances in Structural and Syntactic Pattern Recognition*, volume 5 of *Series in Machine Perception and Artificial Intelligence*, pages 99–108. World Scientific, 1992.

J. Oncina, P. García, and E. Vidal. Learning subsequential transducers for pattern recognition interpretation tasks. *Pattern Analysis and Machine Intelligence*, 15(5):448–458, 1993.

J. Oncina and M. Sebban. Learning stochastic edit distance: application in handwritten character recognition. *Pattern Recognition*, 39(9):1575–1587, 2006.

J. Oncina and M. A. Varó. Using domain information during the learning of a subsequential transducer. In Miclet and de la Higuera (1996), pages 301–312.

D. Osherson, D. de Jongh, E. Martin, and S. Weinstein. *Handbook of Logic and Language*, pages 737–775. MIT Press, 1997.

G. Paliouras and Y. Sakakibara, editors. *Grammatical Inference: Algorithms and Applications, Proceedings of ICGI '04*, volume 3264 of LNAI. Springer-Verlag, 2004.

N. Palmer and P. W. Goldberg. PAC-learnability of probabilistic deterministic finite state automata in terms of variation distance. In Jain, Simon and Tomita (2005), pages 157–170.

R. Parikh. On context-free languages. *Journal of the ACM*, 13(4):570–581, 1966.

A. Paz. *Introduction to probabilistic automata*. Academic Press, 1971.

G. Petasis, G. Paliouras, V. Karkaletsis, C. Halatsis, and C. Spyropoulos. E-grids: computationally efficient grammatical inference from positive examples. *Grammars*, 7:69–110, 2004a.

G. Petasis, G. Paliouras, C. D. Spyropoulos, and C. Halatsis. Eg-grids: context-free grammatical inference from positive examples using genetic search. In Paliouras and Sakakibara (2004b), pages 223–234.

D. Picó and F. Casacuberta. A statistical-estimation method for stochastic finite-state transducers based on entropy measures. In *Advances in Pattern Recognition*, volume 1876 of LNCS, pages 417–426. Springer-Verlag, 2000.

D. Picó and F. Casacuberta. Some statistical-estimation methods for stochastic finite-state transducers. *Machine Learning Journal*, 44(1):121–141, 2001.

L. Pitt. Inductive inference, DFA's, and computational complexity. In *Analogical and Inductive Inference*, number 397 in LNAI, pages 18–44. Springer-Verlag, 1989.

L. Pitt and M. Warmuth. Reductions among prediction problems: on the difficulty of predicting automata. In *3rd Conference on Structure in Complexity Theory*, pages 60–69, 1988.

L. Pitt and M. Warmuth. The minimum consistent DFA problem cannot be approximated within any polynomial. *Journal of the Association for Computing Machinery*, 40(1):95–142, 1993.

N. Poggi, T. Moreno, J.-L.Berral, R. Gavaldà, and J. Torres. Web customer modeling for automated session prioritization on high traffic sites. In *Proceedings of User Modeling 2007*, volume 4511 of LNCS, pages 450–454. Springer-Verlag, 2007.

B. Pouliquen. Similarity of names across scripts: edit distance using learned costs of N-grams. In *Proceedings of GOTAL 2008, Advances in Natural Language Processing*, LNCS, pages 405–416. Springer-Verlag, 2008.

H. Qiu and E. R. Hancock. Graph matching using commute time spanning trees. In *ICPR* (2006), pages 1224–1227.

M. O. Rabin. Probabilistic automata. *Information and Control*, 6:230–245, 1966.

M. Rabin and D. Scott. Finite automata and their decision problems. *IBM Journal of Research and Development*, 3:114–125, 1959.

L. Rabiner. A tutorial on hidden Markov models and selected applications in speech recoginition. *Proceedings of the IEEE*, 77:257–286, 1989.

H. Raffelt and B. Steffen. Learnlib: A library for automata learning and experimentation. In *Proceedings of FASE 2006*, volume 3922 of LNCS, pages 377–380. Springer-Verlag, 2006.

C. Reutenauer and M.-P. Schützenberger. Minimization of rational word functions. *SIAM Journal of Computing*, 20(4):669–685, 1991.

C. Reutenauer and M.-P. Schützenberger. Variétés et fonctions rationnelles. *Theoretical Computer Science*, 145(1&2):229–240, 1995.

J. R. Rico-Juan, J. Calera-Rubio, and R. C. Carrasco. Probabilistic k-testable tree-languages. In de Oliveira (2000), pages 221–228.

J. R. Rico-Juan, J. Calera-Rubio, and R. C. Carrasco. Stochastic k-testable tree languages and applications. In Adriaans, Fernau and van Zaanen, pages 199–212.

J. R. Rico-Juan and L. Micó. Comparison of AESA and LAESA search algorithms using string and tree-edit-distances. *Pattern Recognition Letters*, 24(9-10):1417–1426, 2003.

A. Rieger. Inferring probabilistic automata from sensor data for robot navigation. In M. Kaiser, editor, *Proceedings of the MLnet Familiarization Workshop and Third European Workshop on Learning Robots*, pages 65–74, 1995.

J. Rissanen. Modeling for shortest data description. *Automatica*, 14:465–471, 1978.

R. L. Rivest and R. E. Schapire. Inference of finite automata using homing sequences. *Information and Computation*, 103:299–347, 1993.

B. Roark and R. Sproat. *Computational Approaches to Syntax and Morphology*. Oxford University Press, 2007.

D. Ron and R. Rubinfeld. Learning fallible deterministic finite automata. *Machine Learning Journal*, 18:149–185, 1995.

D. Ron, Y. Singer, and N. Tishby. Learning probabilistic automata with variable memory length. In *Proceedings of COLT 1994*, pages 35–46. ACM Press, 1994.

D. Ron, Y. Singer, and N. Tishby. On the learnability and usage of acyclic probabilistic finite automata. In *Proceedings of COLT 1995*, pages 31–40. ACM Press, 1995.

P. Rossmanith and T. Zeugmann. Learning k-variable pattern languages efficiently stochastically finite on average from positive data. In Honavar and Slutski (1998), pages 13–24.

H. Rulot, N. Prieto, and E. Vidal. Learning accurate finite-state structural models of words through the ECGI algorithm. In *ICASSP-89*, volume 1, pages 643–646, 1989.

Y. Sakakibara. Inferring parsers of context-free languages from structural examples. Technical Report 81, Fujitsu Limited, International Institute for Advanced Study of Social Information Science, Numazu, Japan, 1987.

Y. Sakakibara. Learning context-free grammars from structural data in polynomial time. *Theoretical Computer Science*, 76:223–242, 1990.

Y. Sakakibara. Efficient learning of context-free grammars from positive structural examples. *Information and Computation*, 97:23–60, 1992.

Y. Sakakibara. Recent advances of grammatical inference. *Theoretical Computer Science*, 185:15–45, 1997.

Y. Sakakibara, M. Brown, R. Hughley, I. Mian, K. Sjolander, R. Underwood, and D. Haussler. Stochastic context-free grammars for tRNA modeling. *Nuclear Acids Research*, 22:5112–5120, 1994.

Y. Sakakibara, S. Kobayashi, K. Sato, T. Nishino, and E. Tomita, editors. *Grammatical Inference: Algorithms and Applications, Proceedings of ICGI '06*, volume 4201 of LNAI. Springer-Verlag, 2006.

Y. Sakakibara and M. Kondo. Ga-based learning of context-free grammars using tabular representations. In *Proceedings of 16th International Conference on Machine Learning (ICML-99)*, pages 354–360, 1999.

Y. Sakakibara and H. Muramatsu. Learning context-free grammars from partially structured examples. In de Oliveira (2000), pages 229–240.

J. Sakarovich. *Eléments de théorie des automates*. Vuibert, 2004.

A. Salomaa. On languages defined by numerical parameters. Technical Report 663, Turku Centre for Computer Science, 2005.

I. Salvador and J-M. Benedí. RNA modeling by combining stochastic context-free grammars and n-gram models. *International Journal of Pattern Recognition and Artificial Intelligence*, 16(3):309–316, 2002.

J. A. Sánchez, J. M. Benedí, and F. Casacuberta. Comparison between the inside-outside algorithm and the Viterbi algorithm for stochastic context-free grammars. In P. Perner, P. Wang, and A. Rosenfeld, editors, *Advances in Structural and Syntactical Pattern Recognition*, volume 1121 of LNCS, pages 50–59. Springer-Verlag, 1996.

A. Saoudi and T. Yokomori. Learning local and recognizable ω-languages and monadic logic programs. In *Proceedings of EUROCOLT, LNCS*, pages 157–169. Springer-Verlag, 1993.

L. Saul and F. Pereira. Aggregate and mixed-order Markov models for statistical language processing. In C. Cardie and R. Weischedel, editors, *Proceedings of the Second Conference on Empirical Methods in Natural Language Processing*, pages 81–89. Association for Computational Linguistics, Somerset, New Jersey, 1997.

K. U. Schulz and S. Mihov. Fast string correction with levenshtein automata. IJDAR, 5(1):67–85, 2002.

M. Sebban and J-C. Janodet. On state merging in grammatical inference: a statistical approach for dealing with noisy data. In *Proceedings of ICML*, 2003.

J. M. Sempere and P. García. A characterisation of even linear languages and its application to the learning problem. In Carrasco and Oncina (1994a), pages 38–44.

J. M. Sempere and G. Nagaraja. Learning a subclass of linear languages from positive structural information. In Honavar and Slutski (1998), pages 162–174.

J. Shawe-Taylor and N. Christianini. *Kernel Methods for Pattern Analysis*. Cambridge University Press, 2004.

M. Simon. *Automata Theory*. World Scientific, 1999.

Z. Solan, E. Ruppin, D. Horn, and S. Edelman. Automatic acquisition and efficient representation of syntactic structures. In *Proceedings of NIPS*, 2002.

R. Solomonoff. A preliminary report on a general theory of inductive inference. Technical Report ZTB-138, Zator Company, Cambridge, Mass., 1960.

R. Solomonoff. A formal theory of inductive inference. *Information and Control*, 7(1):1–22 and 224–254, 1964.

R. Solomonoff. The discovery of algorithmic probability. *JCSS*, 55(1):73–88, 1997.

B. Starkie, F. Coste, and M. van Zaanen. The Omphalos context-free grammar learning competition. In Paliouras and Sakakibara (2004a), pages 16–27.

B. Starkie, F. Coste, and M. van Zaanen. Omphalos context-free language learning competition. http://www.irisa.fr/Omphalos, 2004b.

B. Starkie and H. Fernau. The Boisdale algorithm – an induction method for a subclass of unification grammar from positive data. In Paliouras and Sakakibara (2004), pages 235–247.

B. Starkie, M. van Zaanen, and D. Estival. The Tenjinno machine translation competition. In Sakakibara *et al.* (2006), pages 214–226.

S. E. Stein, S. R. Heller, and D. V. Tchekhovskoi. The IUPAC chemical identifier technical manual. Technical report, Gaithersburg, Maryland, 2006.

M. A. Stern. Über eine zahlentheoretische funktion. *Crelle's Journal*, 55:193–220, 1858.

A. Stolcke. *Bayesian Learning of Probabilistic Language Models*. PhD dissertation, University of California, 1994.

A. Stolcke and S. Omohundro. Inducing probabilistic grammars by bayesian model merging. In Carrasco and Oncina (1994a), pages 106–118.

Y. Takada. Grammatical inference for even linear languages based on control sets. *Information Processing Letters*, 28(4):193–199, 1988.

Y. Takada. A hierarchy of language families learnable by regular language learners. In Carrasco and Oncina (1994a), pages 16–24.

F. Tantini, C. de la Higuera, and J.-C. Janodet. Identification in the limit of systematic-noisy languages. In Sakakibara *et al.* (2006), pages 19–31.

I. Tellier. Meaning helps learning syntax. In Honavar and Slutski (1998), pages 25–36.

I. Tellier. When categorial grammars meet regular grammatical inference. In *Proceedings of LACL 2005 (Logical Aspects of Computational Linguistics, Bordeaux, France)*, volume 3492 of LNCS, pages 301–316. Springer-Verlag, 2005.

F. Thollard. Improving probabilistic grammatical inference core algorithms with post-processing techniques. In *Proceedings 8th International Conference on Machine Learning*, pages 561–568. Morgan Kauffman, 2001.

F. Thollard and A. Clark. PAC-learnability of probabilistic deterministic finite state automata. *Journal of Machine Learning Research*, 5:473–497, 2004.

F. Thollard and P. Dupont. Entropie relative et algorithmes d'inférence grammaticale probabiliste. In M. Sebag, editor, *Actes de la conférence CAP '99*, pages 115–122, 1999.

F. Thollard, P. Dupont, and C. de la Higuera. Probabilistic DFA inference using Kullback-Leibler divergence and minimality. In *Proceedings of the 17th International Conference on Machine Learning*, pages 975–982. Morgan Kaufmann, 2000.

K. Thompson. Regular expression search algorithm. *Journal of the ACM*, 11(6):419–422, 1968.

C. Tirnauca. A note on the relationship between different types of correction queries. In Clark, Coste and Miclet (2008), pages 213–223.

B. Trakhtenbrot and Y. Bardzin. *Finite Automata: Behavior and Synthesis*. North Holland Publishing Company, 1973.

L. G. Valiant. A theory of the learnable. *Communications of the Association for Computing Machinery*, 27(11):1134–1142, 1984.

K. Vanlehn and W. Ball. A version space approach to learning context-free grammars. *Machine Learning Journal*, 2:39–74, 1987.

M. van Zaanen. ABL: alignment-based learning. In *Proceedings of COLING 2000*, pages 961–967. Morgan Kaufmann, 2000.

M. van Zaanen. The grammatical inference homepage. http://labh-curien.univ-st-etienne.fr/informatique/gi, 2003.

E. Vidal, E. Segarra, P. García, and I. Galiano. Multi-speaker experiments with the morphic generator grammatical inference methodology. In H. Niemann, M. Lang, and

G. Sagerer, editors, *Recent Advances in Speech Understanding and Dialog Systems*, pages 323–327. Springer-Verlag, 1988.

E. Vidal, F. Thollard, C. de la Higuera, F. Casacuberta, and R. C. Carrasco. Probabilistic finite state automata – part I. *Pattern Analysis and Machine Intelligence*, 27(7):1013–1025, 2005a.

E. Vidal, F. Thollard, C. de la Higuera, F. Casacuberta, and R. C. Carrasco. Probabilistic finite state automata – part II. *Pattern Analysis and Machine Intelligence*, 27(7):1026–1039, 2005b.

J. M. Vilar. Query learning of subsequential transducers. In Miclet and de la Higuera (1996), pages 72–83.

J. M. Vilar. Improve the learning of subsequential transducers by using alignments and dictionaries. In de Oliveira (2000), pages 298–312.

A. J. Viterbi. Error bounds for convolutional codes and an asymptotically optimum decoding algorithm. *IEEE transactions of the empirical distribution*, 13:260–269, 1967.

R. Wagner and M. Fisher. The string-to-string correction problem. *Journal of the ACM*, 21:168–178, 1974.

L. G. Wallace and D. M. Ball. An information measure for classification. *Computer Journal*, 11:185–194, 1968.

Y. Wang and A. Acero. Evaluation of spoken language grammar learning in the ATIS domain. In *Proceedings of the International Conference on Acoustics, Speech, and Signal Processing*, 2002.

J. T. Wang, S. Rozen, B. A. Shapiro, D. Shasha, Z. Wang, and M. Yin. New techniques for DNA sequence classification. *Journal of Computational Biology*, 6(2):209–218, 1999.

O. Watanabe. A framework for polynomial time query learnability. *Mathematical Systems Theory*, 27(3):231–256, 1994.

C. S. Wetherell. Probabilistic languages: a review and some open questions. *Computing Surveys*, 12(4):361–379, 1980.

G. Wolf. Grammar discovery as data compression. In *Proceedings of AISBI/GI Conference on Artificial Intelligence*, pages 375–379, 1978.

G.-Wolf. *Unifying computing and cognition*. Cognition research, 2006.

K. Wright. Identification of unions of languages drawn from an identifiable class. In *Proceedings of the Workshop on Computational Learning Theory*, pages 328–333. Morgan Kaufmann, 1989.

P. Wyard. Representational issues for context free grammar induction using genetic algorithms. In Carrasco and Oncina (1994a), pages 222–235.

T. Yokomori. Learning non-deterministic finite automata from queries and counterexamples. *Machine Intelligence*, 13:169–189, 1994.

T. Yokomori. On polynomial-time learnability in the limit of strictly deterministic automata. *Machine Learning Journal*, 19:153–179, 1995.

T. Yokomori. Learning two-tape automata from queries and counterexamples. *Mathematical Systems Theory*, pages 259–270, 1996.

T. Yokomori. Polynomial-time identification of very simple grammars from positive data. *Theoretical Computer Science*, 1(298):179–206, 2003.

T. Yokomori. Grammatical inference and learning. In C. Martín-Vide, V. Mitrana, and Gh. Păun, editors, *Formal Languages and Applications*, pages 507–528. Springer-Verlag, 2004.

T. Yokomori. *Formal languages*. Springer-Verlag, 2005.

T. Yokomori and S. Kobayashi. Inductive learning of regular sets from examples: a rough set approach. In *Proceedings of International Workshop on Rough Sets and Soft Computing*, 1994.

M. Young-Lai and F. W. Tompa. Stochastic grammatical inference of text database structure. *Machine Learning Journal*, 40(2):111–137, 2000.

H. Yu and E. R. Hancock. String kernels for matching seriated graphs. In *ICPR* (2006), pages 224–228.

L. Yujian and L. Bo. A normalized Levenshtein distance metric. *Pattern Analysis and Machine Intelligence*, 29(6):1091–1095, 2007.

T. Zeugmann. Can learning in the limit be done efficiently? In Gavaldà, Jantke and Takimoto (2003), pages 17–38.

T. Zeugmann. From learning in the limit to stochastic finite learning. *Theoretical Computer Science*, 364(1):77–97, 2006.

Index